ELSEVIER SCIENTIFIC PUBLISHING COMPANY
MOLENWERF 1
P.O. Box 211, 1000 AE AMSTERDAM, THE NETHERLANDS

Distributors for the United States and Canada:

ELSEVIER SCIENCE PUBLISHING COMPANY INC.
52, VANDERBILT AVENUE
NEW YORK, N.Y. 10017

Chemical Microscopy

Translated from the German by Dietrich Hucke

PRINTED IN THE GERMAN DEMOCRATIC REPUBLIC

Thermomicroscopy of Organic Compounds

PRINTED IN THE NETHERLANDS

LIBRARY OF CONGRESS CARD NUMBER: 58-10158

ISBN 0-444-41950-0 (Vol. 16)
ISBN 0-444-41735-4 (Series)

Contributors to Volume XVI

CHEMICAL MICROSCOPY

by

H.-H. Emons
Bergakademie Freiberg
9200 Freiberg, D.D.R.

H. Keune
Pädagogische Hochschule "Dr Theodor Neubauer"
Erfurt/Mühlhausen
5700 Mühlhausen, D.D.R.

H.-H. Seyfarth
Pädagogische Hochschule "Dr Theodor Neubauer"
Erfurt/Mühlhausen
5700 Mühlhausen, D.D.R.

THERMOMICROSCOPY OF ORGANIC COMPOUNDS

by

M. Kuhnert-Brandstätter
Institut für Pharmakognosie der Universität
Peter-Mayr-Strasse 1
A-6020 Innsbruck, Austria

Wilson and Wilson's

COMPREHENSIVE ANALYTICAL CHEMISTRY

Edited by

G. SVEHLA, PH.D., D.SC., F.R.S.C.

Reader in Analytical Chemistry
The Queen's University of Belfast

VOLUME XVI

Chemical Microscopy
Thermomicroscopy of Organic Compounds

ELSEVIER SCIENTIFIC PUBLISHING COMPANY
AMSTERDAM OXFORD NEW YORK
1982

WILSON AND WILSON'S

COMPREHENSIVE ANALYTICAL CHEMISTRY

VOLUMES IN THE SERIES

Vol. IA	Analytical Processes Gas Analysis Inorganic Qualitative Analysis Organic Qualitative Analysis Inorganic Gravimetric Analysis
Vol. IB	Inorganic Titrimetric Analysis Organic Quantitative Analysis
Vol. IC	Analytical Chemistry of the Elements
Vol. IIA	Electrochemical Analysis Electrodeposition Potentiometric Titrations Conductometric Titrations High-frequency Titrations
Vol. IIB	Liquid Chromatography in Columns Gas Chromatography Ion Exchangers Distillation
Vol. IIC	Paper and Thin-Layer Chromatography Radiochemical Methods Nuclear Magnetic Resonance and Electron Spin Resonance Methods X-Ray Spectrometry
Vol. IID	Coulometric Analysis
Vol. III	Elemental Analysis with Minute Samples Standards and Standardization Separations by Liquid Amalgams Vacuum Fusion Analysis of Gases in Metals Electroanalysis in Molten Salts
Vol. IV	Instrumentation for Spectroscopy Atomic Absorption and Fluorescence Spectroscopy Diffuse Reflectance Spectroscopy
Vol. V	Emission Spectroscopy Analytical Microwave Spectroscopy Analytical Applications of Electron Microscopy

Vol. VI	Analytical Infrared Spectroscopy
Vol. VII	Thermal Methods in Analytical Chemistry Substoichiometric Analytical Methods
Vol. VIII	Enzyme Electrodes in Analytical Chemistry Molecular Fluorescence Spectroscopy Photometric Titrations Analytical Applications of Interferometry
Vol. IX	Ultraviolet Photoelectron and Photoion Spectroscopy Auger Electron Spectroscopy Plasma Excitation in Spectrochemical Analysis
Vol. X	Organic Spot Test Analysis The History of Analytical Chemistry
Vol. XI	The Application of Mathematical Statistics in Analytical Chemistry Mass Spectrometry Ion Selective Electrodes
Vol. XII	Thermal Analysis Part A. Simultaneous Thermoanalytical Examinations by Means of the Derivatograph Part B. Biochemical and Clinical Applications of Thermometric and Thermal Analysis
Vol. XIII	Analysis of Complex Hydrocarbon Mixtures Part A. Separation Methods Part B. Group Analysis and Detailed Analysis
Vol. XIV	Ion Exchangers in Analytical Chemistry
Vol. XV	Methods of Organic Analysis
Vol. XVI	Chemical Microscopy Thermomicroscopy of Organic Compounds

Preface

In *Comprehensive Analytical Chemistry*, the aim is to provide a work which, in many instances, should be a self-sufficient reference work; but where this is not possible, it should at least be a starting point for any analytical investigation.

It is hoped to include the widest selection of analytical topics that is possible within the compass of the work, and to give material in sufficient detail to allow it to be used directly, not only by professional analytical chemists, but also by those workers whose use of analytical methods is incidental to their work rather than continual. Where it is not possible to give details of methods, full reference to the pertinent original literature is made.

Volume XVI is devoted entirely to microscopy. The first chapter on *Chemical Microscopy* is in fact a general introduction into the theory and practice of microscopic techniques in chemistry. The second, on *Thermomicroscopy of Organic Compounds*, is more specialised, aimed mainly at the organic chemist, for whom this technique is really essential. I would like to remind the reader that a longer chapter on the analytical applications of electron microscopy has already been published in Volume VI, and that Volume XV describes the traditional methods of organic analysis.

All the authors of the present volume are from German-speaking European institutions.

Dr. C.L. Graham of the University of Birmingham, England, assisted in the production of the present volume: his contribution is acknowledged with many thanks.

December 1981 G. Svehla

CONTENTS

Chemical Microscopy

by H.-H. Emons, H. Keune and H.-H. Seyfarth

Foreword

Microscopic techniques of measurement have a very great variety of uses in chemistry; in solving its multivarious problems, they offer considerable advantages over other physical measuring techniques.

The present work relies on many years of experience gathered in applying microscopy to chemical research and practice. In the endeavour especially to give the chemist an initial survey of the present state of microscopical applications, we have derived from our work a complex of instructions for sample preparation, measurement and examination methods, with as high a degree of generalization as to enable the reader to apply the methods described to other cases and problems. A great number of illustrations, tables and instructional charts and a comprehensive bibliography for further reading have been provided to support that objective.

For better understanding, the text is preceded by explanations of selected technical terms and a brief introduction into important theoretical fundamentals of crystal optics. We have largely refrained, however, from theoretically discussing the various crystal-optical problems, as the work is primarily intended to be a manual of instruction, a work of reference and a source of inspiration for the practical worker.

Fully aware of the difficulties involved in such an enterprise, we shall appreciate critical hints from the readership. We should be happy if the book appealed not only to chemists but also to materials scientists, physicists, crystallographers, geologists, physicians and pharmaceutists working in the field of microscopic substance examination.

We are indebted to the Light Microscopy Research Department of JENOPTIK JENA GmbH and especially to P. Döpel, J. Bergner and M. Neupert, for their close and fruitful cooperation and assistance.

The stimulating discussions and advice received from many colleagues at institutes and universities and in the chemical and potash industries are gratefully acknowledged, as is the scientific and technical help extended by our coworkers in the Chemical Processing Section, Department of Inorganic Technical Chemistry of our Institute, in preparing the manuscript.

The Authors

Contents

Glossary of selected microscopical terms

Analyzer. Device for analyzing the vibrational state of plane-polarized light that has been transmitted through the object.

Aperture diaphragm. Aperture-controlling iris diaphragm at the condenser. Also called condenser iris.

Bright-field illumination. Both the principal (zero-order) maximum and the diffracted light share contribute to image formation.

Brightness control. Brightness in the image plane is controlled by neutral density filters of graded transmission or by two circular neutral density wedges or polarizing filters revolving in opposite directions.

Collector. Optical system that gathers source light over a larger solid angle. By varying the source-to-collector distance, the light can be made convergent, parallel or divergent.

Condenser. Optical system that generates an evenly illuminated object field (image field) of high luminosity. Dark-field condensers have a central diaphragm to block out the principal maximum.

Crystal systems. Coordinate systems to which the lattices of all crystalline substances can be referred, viz.

cubic	$a_1 = a_2 = a_3$	$\alpha = \beta = \gamma = 90°$
tetragonal	$a_1 = a_2 \neq c$	$\alpha = \beta = \gamma = 90°$
hexagonal/trigonal	$a_1 = a_2 = a_3 \neq c$	$aa = 120°$, $ac = 90°$
rhombic	$a \neq b \neq c$	$\alpha = \beta = \gamma = 90°$
monoclinic	$a \neq b \neq c$	$\alpha = \gamma = 90°$, $\beta \neq 90°$
triclinic	$a \neq b \neq c$	$\alpha \neq \beta \neq \gamma \neq 90°$

α, β and γ are the angles between the coordinate axes.

Dark-field illumination. Illumination under such an angle that the principal (zero-order) maximum does not enter the objective. The image is formed by at least two maxima of higher orders by way of diffraction by the specimen.

Depth of field. Object depth range of satisfactory definition in front of and behind the plane at which the objective is focussed.

Depth of focus. Distance through which the image plane may be moved away (back or forth) from the film plane without causing blur that is noticeable by the sensor (light-sensitive emulsion, human retina).

Dispersion. Dependence of optical quantities (refractive index, optic axial angles, birefringence, extinction) on the wavelength of the light used or on the temperature prevailing during the measurement.

Dry objective. Objective corrected for air as a medium between objective, specimen and condenser, rather than an immersion liquid.

Eyepiece. Optical system for visual observation and secondary magnification of the intermediate image formed by the microscope objective.

Field stop. Diaphragm in front of the light source, for controlling the size of the luminous field.

Filters. Filters have a wide variety of uses in microscopy, e.g. daylight filters (correct colour balance of specimens), ground glass filters (effect homogeneous illumination of the visual field), UV barrier filters (eliminate unwanted UV light in fluorescence microscopy), heat-absorbing filters (reduce proportion of heat rays in illumination), contrast filters (complementary colour filters to improve contrast), colour filters (select a broad-band wavelength range from the spectrum, but do not provide strictly monochromatic light of one wavelength), graded interference filters (produce monochromatic light of continuously variable wavelength), dispersion filters (strictly monochromatic light in the visible, IR and UV ranges).

Imbibition medium. Liquid impregnant for the imbibition method.

Immersing medium. Embedding (mounting) substance, either liquid or melt, used in preparing microscopic specimens to reduce the difference in refractive index between the object and its surroundings in transmitted light and thus to eliminate disturbing total reflection fringes.

Immersion objective. Objective corrected for use with a liquid medium between the front lens and the specimen (or also between specimen and condenser) in order to reduce differences in refractive index. The method increases N.A. and, thus, resolving power and contrast.

Indicatrix. Ellipsoid constructed as an auxiliary figure to represent the difference in optical properties found if light passes the crystal lattice of an optically anisotropic substance in different spatial directions.

Kinematic microscopy. Investigation into changes of microscopical objects as a function of time.

Koehler illumination. Arrangement providing even lighting of the visual field. The *field stop,* projected into the specimen plane by the condenser, controls the area of the specimen to be illuminated. The *condenser iris* controls the N.A. of the illuminating beam and, thus, image contrast. Brightness control is by *attenuation filters.* Rather than by the light source directly, the specimen is illuminated by the source image projected into the condenser iris plane.

Luminance. Measure of the brightness per unit area of a light-emitting surface. As luminance may differ for different portions of a light source, one usually specifies mean luminance.

Magnification. *Visual magnification:* Ratio of apparent image size and object size in visual observation from a reference distance of 250 mm (*conventional visual distance*). Magnification is specified as a factor (e.g., × 320); it is found by multiplying the objective and eyepiece magnifications (and the tube factor). A more precise magnification calibration for any particular eyepiece-objective combination can be performed with eyepiece and stage micrometers. *Useful magnification* is a term for the limit after which further magnification ceases to provide more information. This limit is between 500 and 1000 times the objective N.A., with a maximum of ×1 500 for light microscopes. The magnification of the eyepiece used should not exceed a value that would make the product of eyepiece and objective magnifications exceed the useful magnification calculated for the objective. Special purposes may sometimes necessitate a higher, so-called »empty« magnification. *Photographic magnification:* In photomicrography, the ratio of linear image size and linear object size, usually specified as a factor, but sometimes as a ratio (e.g., 250:1). Also called scale (or ratio) of reproduction.

Monochromator. Optical system for obtaining light of a defined wavelength and highest possible intensity. A lens system and a variable slit direct the light to a rotary glass prism, by which it is spectrally dispersed. The desired part of the spectrum leaves through another slit and another arrangement of lenses, then hitting the microscope mirror.

Numerical aperture (N.A.). The ratio of half the aperture angle of an optical system (e.g., objective) and the refractive index of the medium in front of the front lens. The N.A. characterizes the widest cone of rays the optical system can accept. It is thus decisive for resolving power, contrast and image brightness.

Objective. The objective gathers the various diffraction maxima that form the image in the intermediate image plane, and thus is decisive for the quality of the microscopic image. Engraved on an objective are its magnification, N.A., tube length and the cover glass thickness for which the objective is optimally corrected. The cover glass thickness specification may also indicate that the objective is designed for uncovered specimens or is insensitive to cover glass thickness variation. Objectives for polarizing microscopy are very delicate, responding even to slight mechanical or thermal stress.

Polarizer. Device for producing plane-polarized light.

Projection eyepiece. Used instead of a visual-observation eyepiece, it projects the intermediate image of a microscope on to camera film plane or projection screen.

Resolving power. The smallest lateral distance that can be resolved in an image is determined by the ratio of the wavelength used and the N.A. of the objective. The higher the N.A. and the smaller the light wavelength, the finer are the structural details that can be discerned.

Scale. Indicates the ratio of a dimension in the image space to the correspondi
sion in the object space.

Section. (a) Imagined plane through the lattice of a microscopical specimen normal to the direction of light transmission. (b) Solid specimen cut, ground and polished for microscopical examination.
Polished sections are opaque, one-side polished specimens for reflected-light microscopy; *thin sections* are transparent and polished on both sides for transmitted-light examination. Very thin sections can be obtained with a *microtome.*

Stage micrometer. Glass plate graduated into 100 parts per millimetre.

Static microscopy. Examination of specimen properties that do not change during observation.

Zoom system. Optical system permitting continuous variations of magnification without changing the distance between specimen and image. In microscopes sometimes called pancratic system.

1 Introduction

In geology, biology and medicine, microscopical techniques are well established, with a multitude of standard problem solutions and prescribed procedures to ensure microscopy a rank equal to other physical investigation methods.

The use of microscopy in chemistry can be traced back in history just as far, though, but has so far been sporadic rather than generally accepted practice.

In most of the reports published, microscopy is applied to solve partial problems only; possibilities of developments towards automated measurement yielding highly informative results are widely unknown. Publications, mostly on isolated problems, are scattered and sometimes not readily available. The present volume quotes a number of publications on microscopical studies selected from more than 5 000.

There are, to be certain, several factors that impede the general spreading of microscopy in chemistry. With many microscopical instruments, taking a measurement still requires profound theoretical knowledge and long practical experience. Standard method sheets are widely missing. Studies on methodical generalization and development have not proceeded beyond an initial stage. A few classical cases in analytics excepted, the methods reported have not yet entered the curricula of college and university students. Nevertheless, development studies should ensure that, in the near future, microscopical techniques will be applicable to the same extent as roentgenographic, spectroscopic and thermoanalytical methods.

The specific differences from other fields of knowledge engaged in studying the solid state result from the properties of the objects investigated. These properties call for special sampling and preparation techniques as well as for modified methods and equipment for measurement and evaluation. The term »chemical microscopy« has been chosen for this complex to mark it off from the specific requirements of examining biological or geological specimens.

»Chemical microscopy« is predominantly concerned with examining solids, although it may provide useful information on crystalline order states in liquids, too.

Accordingly, examinations under normal conditions must be complemented by measurement and observation under high and low temperatures and pressures and in various gas atmospheres.

Analyses of phases, structures, grains and surfaces as well as kinetic studies of physicochemical processes can be employed to advantage in agrochemistry, silicate chemistry, polymer chemistry, pharmaceutical chemistry, and chemical engineering, especially in catalysis, provision of raw materials, and corrosion.

Unlike many other physical methods of investigation, microscopy offers two kinds of results at a time –

(1) knowledge from image observation, and
(2) measured data.

For example, a local degradation of the mechanical properties of a high polymer can show up in the microscope image by a change of structure; simultaneously, measurements will reveal a decrease in refractive index towards the defect.

This combination with visual image interpretation of a measured spot and its surroundings is highly beneficial for the interpretation of the measurement result, especially if important specimen features stand out in rich contrast. As an additional advantage, the size of measured spots and their distances can be varied between 0.2 μm and 1 μm by selecting suitable magnifications. The sample may be scanned successively by a scan line or by a two-dimensional grid of points, to suit the specimen. Microscopic measurement is a differential method that registers grain by grain, grain part by grain part, or point by point, contrary to the integral results obtained with x-ray, thermoanalytical and spectroscopic powder methods. Microscopic examination will yield

(1) results on the spot measured (spot area depending on magnification), and
(2) statistical results on the specimen as a whole by a summation of individual measurements.

Depending on the behaviour of the specimen with regard to time, microscopy may be

(1) **static,** for examining non-changing specimens (e.g., measurement of the optic axial angle of a crystal), or
(2) **kinematic,** for examining the change of specimens with time (e.g., the kinetics of a crystal melting process).

The two branches differ by their measuring techniques and methodology. Static specimens can, in a number of applications, be observed and measured by standard method sheets, whereas similar transferrable problem solutions for kinetic investigations are widely lacking.

Disadvantages of microscopical methods

Grain sizes below 1 μm can hardly be examined in practice. In special cases, specimens far below 1 μm yield usable results, such as in refractive index and birefringence measurements on submicroscopical particles, and path difference measurements down to 10 Å employing interference microscopy.

The instrumental outlay required for examination will increase enormously if specimens are highly absorbing (coloured or opaque). In such cases, other physical methods are superior.

An unfavourable effect on the use of microscopical methods is exerted by the lack of standardized method sheets in many fields.

Advantages of microscopical methods

By the possibility of viewing and measuring selected features or spots, microscopy has a specific field of applications, in which it cannot be replaced fully by other methods.

Since statements on a specimen as a whole are derived from a summation of individual measurements, such statements, in order to be accurate, require that a truly representative number of individual measurements be made. This number depends on specimen quantity, homogeneity and other factors.

In other fields, microscopical methods share equal rank with other methods; examples are the phase analysis of main constituents, and grain size analysis.

In most of the examinations, very little time is required. A decision whether a sample is an aggregate of two phases or consists of mixed crystals can frequently be made within a few minutes. Specimen preparation and the measurement of many parameters will take between 5 and 30 minutes as a rule.

The cost of instrumentation is between 10 and 30 % of the costs of other large-size physical instruments providing a comparable amount of information. In many cases, the accuracy of results found by microscopy is very high. Under favourable conditions, phase analysis of solid aggregates, e.g., can identify fractions at concentrations of some parts in thousand; surface differences on crystal faces down to 10 Å can be measured, and substance differences in mixed crystals be determined that defy chemical separation.

The smallest substance quantities that can be measured on are between 10^{-10} and 10^{-3} g.

Examinations are non-destructive (except hot-stage work). Thus, the specimen is available for other investigation techniques afterwards.

There is a wide variety of microscopical techniques; methods and equipment are highly flexible in adaptation to the problem in hand.

The terms »grain«, »solid«, »compound« and others rather than »crystal« are used in this book in order to give due consideration to the big group of non-crystalline or partially crystalline substances, especially to organic high polymers.

2 Crystal optics of visible light

Microscopy as a physical method of observation and measurement depends on the interactions between visible light and sample matter. Irrespective of the true, complicated nature of light, Maxwell's concept of an electromagnetic transverse wave provides the most vivid explanation of the optical phenomena involved. The following explications are therefore based on that theory.

Special cases excepted, microscopical measurements of optical phenomena on objects in chemistry require transmitted light methods. A first introduction into basic theory will facilitate their application. For comprehensive treatments, see the literature [2.1 to 2.37].

For chemical specimens, incident-light microscopy is predominantly a method of observation, especially with dark-field illumination. Measurement by incident light would require specimens to be mirror-finished, which is rarely possible (metals, ores). The theoretical foundations of such measurements can be found in the literature [2.38 to 2.45].

All microscopical examinations in chemistry should be made with plane-polarized light, as this provides a defined initial vibration direction for testing the interaction between a crystal lattice and irradiated light. The optical phenomena involved in light passage through a crystal vary greatly with the direction of passage relative to the lattice; this relationship must be carefully considered. Optical phenomena are related to lattice structure through morphological elements (crystal faces, cleavage faces, symmetry elements in twins). Amorphous or partially crystalline solids and liquids also permit microscopical examination.

2.1 Passage of light through amorphous solids

As soon as light passes from a vacuum (or air) into a solid, its velocity decreases. The ratio of light velocity in the vacuum to that in the solid is the refractive index of the solid substance. The more a substance retards light, the higher is its refractive index.

In an amorphous solid, light propagates at equal velocity in any direction. The refractive index of the substance is not direction-dependent.

2.2 Passage of light through crystalline solids

The phenomenon of optical anisotropy

A crystalline solid also retards light passing through it, the refractive index (>1) varying with different transmission directions relative to the crystal lattice.

The values n of the refractive index lie between two extrema, n_γ (maximum) and n_α (minimum). The two extrema are substance-specific. Thus, crystals are characterized by a light refraction interval. This behaviour is not shared by crystals having cubic lattices. Owing to their high lattice symmetry, the differences encountered by light passing in various directions are negligible. Those crystals behave like amorphous solids, exhibiting no direction dependence of their refractive indices.

By their optical behaviour, solids can be divided into two big groups:

(1) *Optically isotropic substances* (amorphous solids and cubic crystals): The refractive index is not direction-dependent, but substance-specific.

(2) *Optically anisotropic substances* (crystalline solids except those having cubic lattices): The refractive index is direction-dependent; the magnitudes of, and difference between, the extrema n_γ and n_α are substance-specific.

The phenomenon of birefringence

Generally, any light ray passing an optically anisotropic substance is resolved into two partial rays (birefringence), which have the following properties:
They are *plane-polarized.*
They *vibrate* in planes *perpendicular to each other.*
They propagate through the lattice at *different speeds* (though in the same direction).
Hence, they have *different refractive indices.*

The greater refractive index is denoted by n_γ', the smaller by n_α' (Fig. 2.1).

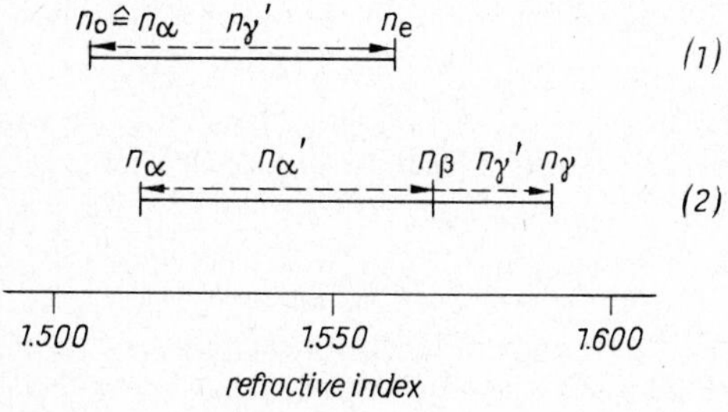

Figure 2.1. Notation of refractive indices of optically uniaxial (1) and biaxial (2) crystals
$n_e = n_\gamma$

Generally, n_γ' and n_α' are not equal to the substance-specific extrema n_γ and n_α, respectively, but lie somewhere within the interval between n_γ and n_α ($n_\gamma' < n_\gamma$, $n_\alpha' > n_\alpha$). They would coincide only in case of one particular transmission direction.

The difference of $n_\gamma' - n_\alpha'$ (in the general case) is the *birefringence* in the respective transmission direction. There is only one such direction in which birefringence attains the substance-specific extremum $n_\gamma - n_\alpha$.

Besides this transmission direction there are others that do not exhibit birefringence. These »directions of optical isotropy« are called optic axes. Depending on whether there are one or two such directions, the substance is optically *»uniaxial«* or optically *»biaxial«*.

Table 2.1 shows the relationship between optical behaviour and lattice structure.

Table 2.1. Relation between optical behaviour and lattice structure

Crystal symmetry	Optical behaviour isotropic	anisotropic	Uniaxial	Biaxial
Amorphous	×	–	–	–
Cubic	×	–	–	–
Tetragonal	–	×	×	–
Hexagonal, trigonal	–	×	×	–
Rhombic	–	×	–	×
Monoclinic	–	×	–	×
Triclinic	–	×	–	×

The two directions of isotropy (lack of birefringence) in optically biaxial substances include certain angles known as optic axial angles $2V$. The $2V$ values are important substance characteristics. The magnitude of the angle $2V$ is a function of temperature and of the light wavelength used (dispersion of the optic axial angle). Depending on the relationship between the optic axial direction and the direction of extreme birefringence, the substance is either *optically positive* or *optically negative*.

These direction-dependent optical relations are elucidated by the *indicatrix*, a three-dimensional optical surface of reference. Its coordinate vectors (semiaxes) correspond to the principal refractive indices n_γ, n_β, n_α (Fig. 2.2).

Construction of the indicatrix

Imagine a point source of light located anywhere inside a crystal lattice and emitting rays into all spatial directions. The magnitude of the refractive index of a wave vibrating in a certain lattice direction (perpendicular to its propagation direction) is represented by the length of the ray along that lattice direction. The end points of all rays can be united into a centrally symmetrical surface termed indicatrix or index ellipsoid, which has a defined position relative to the crystal lattice. With optically uniaxial substances it takes the form of an ellipsoid of revolution having two principal axes, n_γ and n_α;

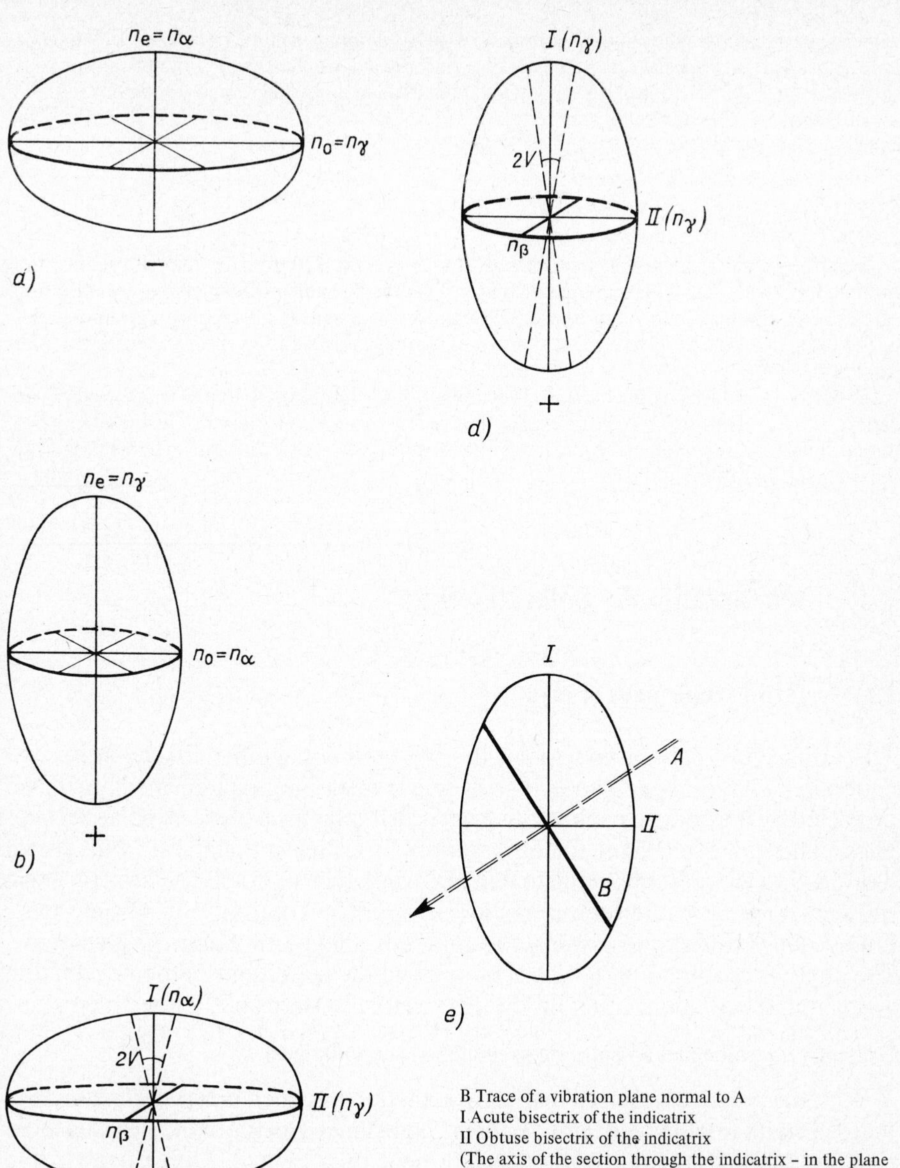

B Trace of a vibration plane normal to A
I Acute bisectrix of the indicatrix
II Obtuse bisectrix of the indicatrix
(The axis of the section through the indicatrix – in the plane of vibration – indicates the positions and magnitudes of n_α' and n_γ' for this direction of light transmission through the lattice.)

Figure 2.2. Direction dependence of optical properties in a crystal lattice, represented by means of the optical indicatrix a) b) Indicatrix of optically uniaxial crystals c) d) Indicatrix of optically biaxial crystals e) Relations of crystals a) through d) to the transmission direction in a two-dimensional section

with biaxial substances the ellipsoid is triaxial, with n_γ, n_β and n_α as the principal axes. A uniaxial indicatrix results as a special case of the biaxial one if the axial angle $2V = 0°$. For optically uniaxial substances, n_β is equal to one of the other two semiaxes, n_γ or n_α. In that case, the notations n_o and n_e are frequently used instead of n_γ and n_α.
Optically positive: $n_o = n_\alpha \quad n_e = n_\gamma$
Optically negative: $n_o = n_\gamma \quad n_e = n_\alpha$

Use of the indicatrix

The optical relationships for a given transmission direction result from the analysis of the section through the indicatrix in a plane perpendicular to that direction and including the centre of the indicatrix. Generally, the section is an ellipse. The lengths of its two axes correspond to the magnitudes of the refractive indices n_γ' and n_α' of the two partial rays formed from the incident ray by birefringence.

In specific transmission directions (= directions of the optic axes), the two axes of the ellipse are equal ($n_\gamma' = n_\alpha'$). The section then is a circle (birefringence $n_\gamma' - n_\alpha' = 0$). Fig. 2.2 illustrates, by a two-dimensional representation of the indicatrix, the relation between the direction of transmission and the sectional plane perpendicular to it.

2.3 Observation by polarized light

2.3.1 Orthoscopic path of rays

The orthoscopic path of rays shows the observer a magnified image of the specimen in its typical appearance. The eyepiece magnifies the intermediate image projected by the objective into a plane beyond its rear focal plane. A polarizer supplies plane-polarized light for observation. Inserting an analyzer, with its vibration direction perpendicular to that of the polarizer (»crossed polarizers«), will cause characteristic interference colours to appear on optically anisotropic specimens. They superimpose upon the inherent colours of coloured specimens. The analysis of these interference colours and the conditions of their extinction yields important information on the optical relationships of the specimen.

Optically isotropic and anisotropic specimens (one polarizer)

With the polarizer inserted in the ray path, the specimen image corresponds to that seen with unpolarized light. Where no substance is in the path, the visual field remains bright (transmitted-light bright-field microscopy). Differences in the specimen appear as differences of absorption.

Optically isotropic specimens (crossed polarizers)

Plane-polarized light passes through a specimen feature (grain) without changing its direction of vibration and is then extinguished by the analyzer. The visual field appears dark, whether with a specimen on the stage or without. Changing the direction of transmission (e.g., by turning the stage) will not effect any change.

Optically anisotropic specimens (crossed polarizers), Fig. 2.3

Birefringence in the anisotropic specimen resolves every ray into two partial rays that travel through the grain at different speeds (see 2.2). The difference in speed results in a certain optical path difference (retardation) between the partial rays, which varies in proportion with specimen thickness (path length for which the speed difference becomes effective) and with birefringence (magnitude of speed difference).

$R = t \cdot B$

R retardation
t specimen thickness
B birefringence

After being made unidirectional by the analyzer, the two partial beams interfere with each other. A certain retardation is represented by a certain interference colour of the specimen (illuminated by white light).

If, in a special case, the vibration directions of the partial rays in the specimen are parallel or normal to that of the polarizer (0° and 90° positions), the rays will be extinguished. That special case corresponds, in appearance, to an optically isotropic object. Rotating the specimen stage changes the relative positions of the vibration directions of the specimen and the polarizers. Four extinction positions will occur during a full 360° stage rotation. In all other stage positions the specimen is bright, showing the interference colours that correspond to the respective retardation. Halfways between any two successive extinction positions there is a position of maximum brightness (45° position).

To obtain distinct interference colours in the 45° position and maximum blackout in the 0° (90°) position, the effective aperture of the optical system has to be slightly reduced, e.g. by using low-power objectives of low N.A. Exessive reduc-

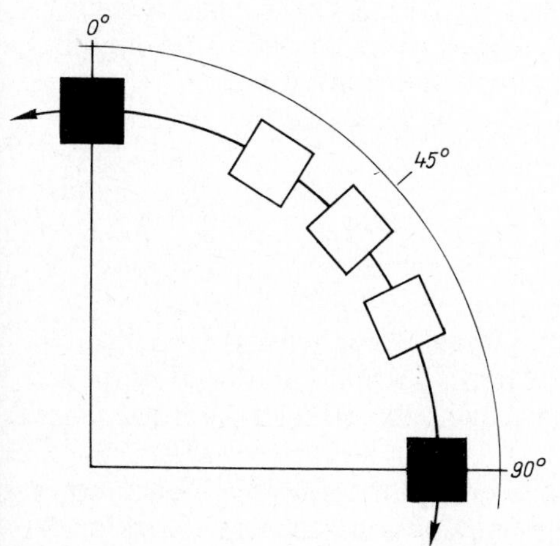

Figure 2.3. Bright and dark positions (45° and 90°/0° positions) of an optically anisotropic grain rotated with the stage between crossed nicols in orthoscopy

tion of aperture, however, will cause an unwanted deterioration of resolving power and of colour saturation.

If one compares grains of a phase of equal specimen thickness (approximately equal grain size) but different birefringence, the resulting retardation produces different interference colours. (With constant thickness t, R is a function of B only.) Table 2.2 correlates these colours with degrees of birefringence at constant t.

Table 2.2. Interference colours depending on birefringence, at constant specimen thickness

Interference colours	Birefringence	Colour of grain
1st order colours	Low	Greyish-yellow interference colours, no colour fringes at grain edge, Vivid colours after full-wave plate is inserted.
Medium-order colours	Medium	Vivid interference colours, one or more colour fringes about parallel to grain edge
Higher-order colours	High	Pale, whitish colours, wedge-shaped edges may have narrow fringes of vivid colours; no distinct coloration after full-wave plate is inserted

With constant birefringence, interference colours (path differences) vary in proportion to grain thickness, which may vary, e.g., from the middle towards a wedge-shaped edge.

An optically anisotropic grain subjected to light transmission along one of its optic axes behaves like an isotropic grain, remaining dark as the stage is revolved. Usually these grains are slightly brightened but do not change this low brightness during stage revolution. After the transmission direction has been changed, e. g. by inclining the grain or shifting the cover glass, the grain displays the normal behaviour of optically anisotropic grains (chart 2.1).

2.3.2 Conoscopic path of rays

The conoscopic path of rays does not show real-world images of the specimen but specimen-dependent interference patterns. They are produced in the rear focal plane of the objective, simultaneously as the orthoscopic intermediate image is formed beyond that plane.

The patterns cannot be viewed directly through the eyepiece, as it is not focussed in that plane. With the eyepiece removed, one can see them by looking into

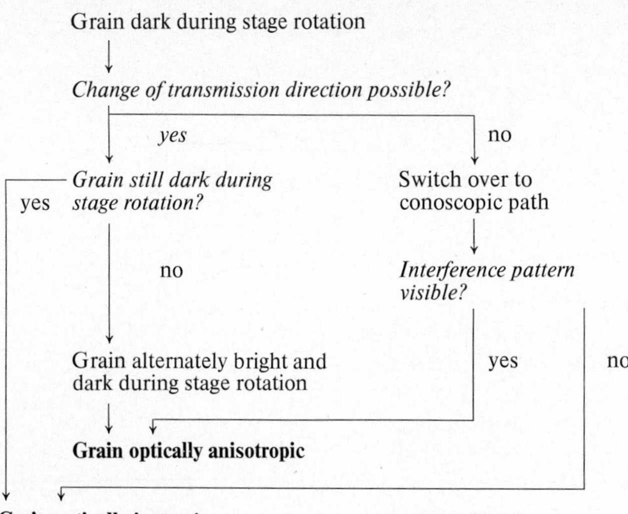

Chart 2.1. Analysis of grains remaining dark during stage rotation between crossed polarizers (orthoscopic procedure)

the tube. Combination of the eyepiece with a Bertrand lens swung into the ray path creates an auxiliary microscope, which is focussed on the rear focal plane and permits observation of the patterns at a magnified scale. This optical arrangement is known as »conoscopic« (or »indirect«) path of rays [2.7, 2.27, 2.30 to 2.37]. The configuration of an interference pattern bears a direct relation to the crystal lattice crossed by the ray path, while being independent from the outer, visible shape of the grain. Thus, polygonal and round grains show identical patterns in the conoscopic path if their lattices have the same direction relative to the ray path.

The analysis of an interference pattern allows information to be derived on optical properties in various transmission directions, whereas an image observed in the orthoscopic path is informative only with regard to the respective transmission direction. For obtaining clearly defined interference patterns, the procedure given in Chart 2.2 should be followed. Focussing on interference patterns must in no case be effected with the microscope's coarse and fine focussing controls. A limited amount of focussing is possible by means of a collar that varies the distance between eyepiece and Bertrand lens. The specimen must be capable of strictly homogeneous extinction in the orthoscopic ray path; specimens that are decomposed or extinguish undulatingly provide indistinct interference patterns.

The most informative patterns are those that correspond to a transmission direction near any of the optic axis directions. To obtain such patterns it is necessary for grains to be selected or arranged so as to have a certain section plane normal to the ray path. These grains are distinguished by low birefringence, i.e. of all

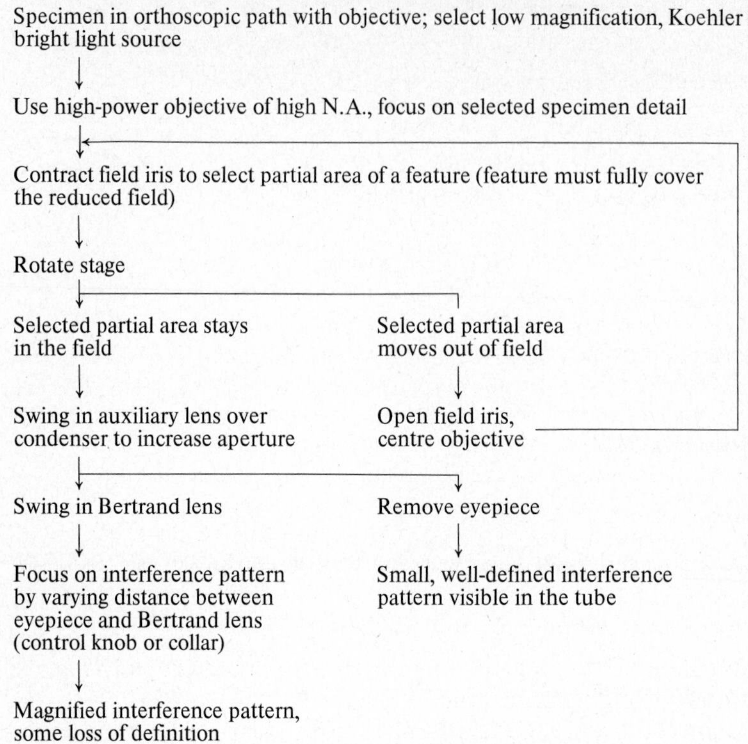

Chart 2.2. Setting for interference figures (conoscopic procedure)

equally sized grains of a phase they show the lowest-order interference colours (cf. 2.3.1). The most significant elements of an interference pattern are dark bands (isogyres), which may take the shape of straight bars or hyperbolic curves. They exhibit characteristic changes as the stage is rotated. If the isogyres remain parallel to the reticle crosshairs, the pattern is that of an optically uniaxial crystal.

Table 2.3. Optically uniaxial interference patterns

Transmission direction	Pattern configuration
Parallel to optic axis	Dark cross, parallel to crosshairs (Fig. 2.4)
Oblique at varying angles with the optic axis	Depending on the angle, the centre of the dark cross moves out of the centre of the field. Only the outer cross bars move in parallel with the crosshairs as the stage is rotated (Fig. 2.4).
Normal to optic axis	A blurred, broad dark cross fills nearly all of the visual field and opens quickly upon slight stage rotation (Fig. 2.4).

If the initially straight bars turn into a figure resembling a hyperbola, which does not remain parallel to the crosshairs, the grain is optically biaxial (Fig. 2.4).

Tables 2.3 and 2.4 relate interference pattern configurations to transmission directions.

In the patterns of optically biaxial substances, the vertices of the hyperbolas in the 45° position correspond to the exit points of the optic axes. The farther apart the vertices are in that position and the less curved the hyperbolas are, the greater is the optic axial angle. Quantitating this relation permits optic axial angles to be measured precisely. Pattern variations after the insertion of compensator plates indicate whether the indicatrix is configured optically positive or negative.

An interesting variety of a ray path that also permits the sending of light through a specimen in several directions at a time and measuring the respective optical properties was reported by Mozzherin [2.46] and termed *»variascopic microscopy«*.

Table 2.4. Optically biaxial interference patterns

Transmission direction	Pattern configuration	
	Small axial angle	Great axial angle
Parallel to an optic axis		
0° position	Vaguely discernible dark cross at margin of field	Dark horizontal or vertical bar in centre of field
45° position	Two dark hyperbolas, one vertex in centre of field (Fig. 2.4)	Slightly curved, oblique dark bar through centre of field (Fig. 2.4)
Oblique to optic axes		
At directions near the acute bisectrix		
0° position	Dark, indistinct cross formed by a broad and a narrow bar	Part of a dark, indistinct cross formed by a broad and a narrow bar
45° position	Two dark hyperbolas	Slightly curved, oblique dark bar through centre of field
Normal to the plane of the optic axes, corresponds to n_β axis of indicatrix	A blurred, broad dark cross fills nearly all of the visual field and opens quickly upon slight stage rotation (Fig. 2.4).	

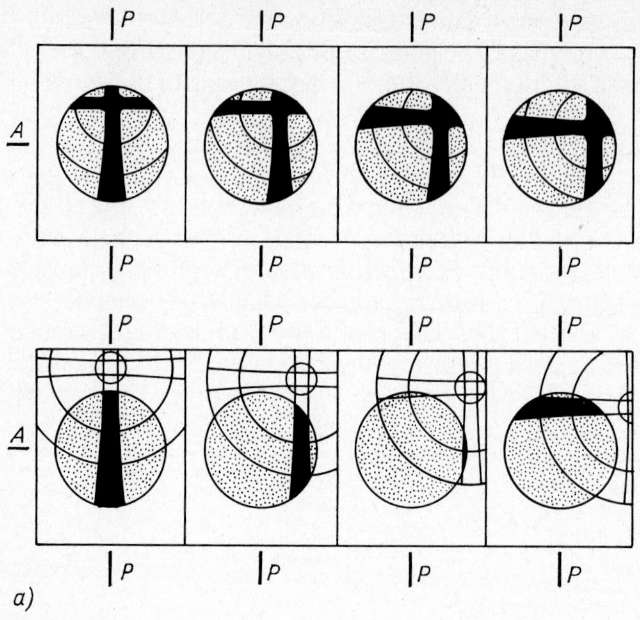

Figure 2.4

Table 2.5. Orthoscopic and conoscopic paths in the polarizing microscope

	Orthoscopic path		Conoscopic path	
	One polarizer	Crossed polarizers	Without eyepiece	With Bertrand lens
Optical outfit	Eyepiece – – Objective Object Condenser Polarizer Illuminator	Eyepiece – Analyzer Objective Object Condenser Polarizer Illuminator	– – Analyzer Objective Object Condenser Polarizer Illuminator	Eyepiece Bertrand lens Analyzer Objective Object Condenser Polarizer Illuminator
Important quantities measured	Refraction Colour Pleochroism Crystal morphology	Birefringence Extinction angle	Optically unaxial/biaxial? Optically positive/negative? Optic axial angle Dispersion of optic axes	
Image observed	Real-world specimen image, natural colour (if any)	Specimen image in interference colour	Small, highly defined interference patterns	Larger, less defined interference patterns

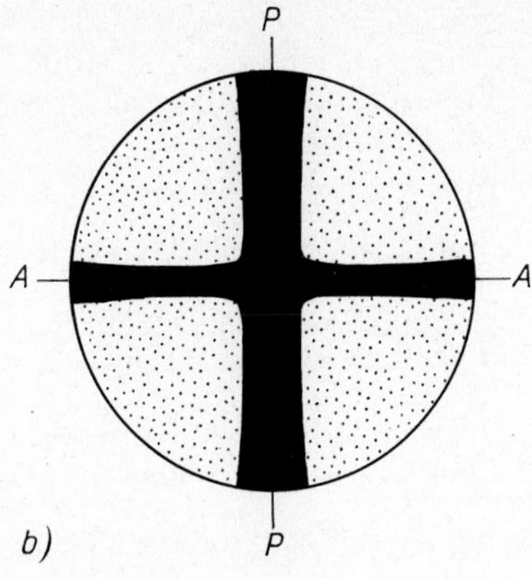

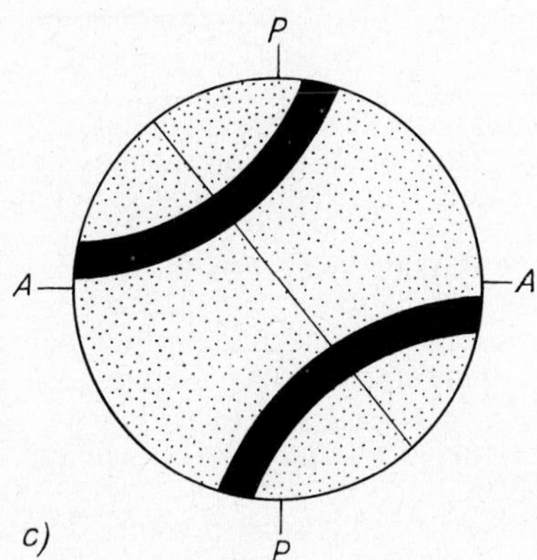

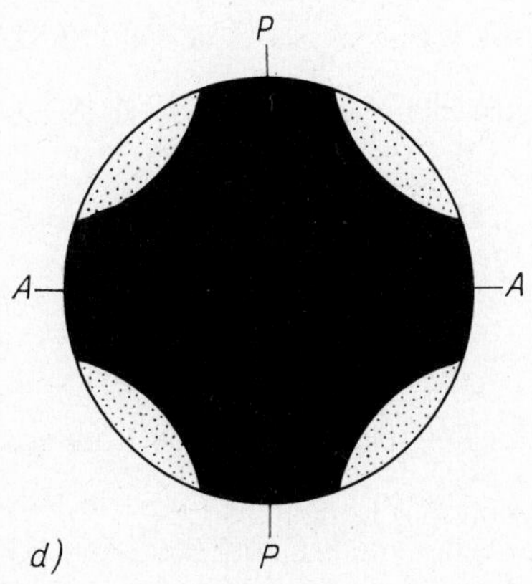

Figure 2.4. Interference figures seen in conoscopy

a) Uniaxial crystal, transmission oblique to optic axis, at various stage positions
b) Biaxial crystal, transmission parallel to acute bisectrix, in 0° (90°) position
c) As above, but 45° position
d) Transmission normal to the optic axis of an uniaxial crystal or parallel to n_β of a biaxial crystal
A analyzer, P polarizer

References

[2.1] *Appelt, H.:* Einführung in die mikroskopischen Untersuchungsmethoden. 3rd ed. Leipzig: Akademische Verlagsgesellschaft Geest und Portig KG 1955

[2.2] *Beljankin, D. S., and W. P. Petrow:* Kristalloptik. Berlin: VEB Verlag Technik 1954

[2.3] *Buchwald, E.:* Einführung in die Kristalloptik. 5th ed. Berlin: Sammlung Göschen, Bd. 619/619a, 1963

[2.4] *Burri, C.:* Das Polarisationsmikroskop. Lehrbücher und Monographien aus dem Gebiet der exakten Wissenschaften 25, Chemische Reihe Bd. V. Basel: Verlag Birkhäuser 1950

[2.5] *Freund, H.:* Handbuch der Mikroskopie in der Technik, Bd. I. Die optischen Grundlagen, die Instrumente und Nebenapparate für die Mikroskopie in der Technik. Frankfurt/Main: Umschau-Verlag 1957

[2.6] *Kleber, W.:* Einführung in die Kristallographie. 6th ed. Berlin: Akademie-Verlag 1962

[2.7] *Mosebach, R.:* Zur Kennzeichnung der grundlegenden Beobachtungsmethoden in der Polarisationsmikroskopie. Z. wiss. Mikroskopie 63 (1956) 2, 72–85

[2.8] *Mosebach, R.:* Mikroskopie und kristallchemische Untersuchungsmethoden. Houben-Weyl, Methoden der organischen Chemie, Bd. III, Teil 1. Stuttgart: Thieme-Verlag 1955

[2.9] *Rittmann, A., and B. Flaschenträger:* Untersuchung von Kristallen mit dem Polarisationsmikroskop in *Hoppe-Seyler/Thierfelder:* Handbuch der physiologischen und pathologischen chemischen Analyse I. 10th ed. Heidelberg: Springer-Verlag 1953

[2.10] *Rosenbusch, H., and O. Mügge:* Mikroskopische Physiographie. 5th ed. Stuttgart: Schweizerbartsche Verlagsbuchhandlung 1927

[2.11] *Rinne, F., and M. Berek:* Anleitung zu optischen Untersuchungen mit dem Polarisationsmikroskop. 2nd ed. Stuttgart: Schweizerbartsche Verlagsbuchhandlung 1953

[2.12] *Raaz, F., and H. Tertsch:* Einführung in die geometrische und physikalische Kristallographie. 3rd ed. Wien: Springer-Verlag 1958

[2.13] *Rösler, H.-J., and L. Pfeiffer:* Einführung in die Polarisationsmikroskopie. Bergakademie Freiberg, internes Weiterbildungsmaterial, 1971

[2.14] *Fedorov, F. J.:* Optika anizotropnych sred. Minsk: Izd. Akad. Nauk SSSR 1958

[2.15] *Bulok, L. N., and L. N. Nikulina:* Kristalloptičeskij analiz. Leningrad: Technolog. Institut Lensowjet 1967

[2.16] *Šubnikov, A. V.:* Osnovy optičeskoj kristallografii. Moskva: Izd. Akad. Nauk SSSR 1958

[2.17] *Šubnikov, A. V.:* Optičeskaja kristallografija. Moskva, Leningrad: Izd. Akad. Nauk SSSR 1950

[2.18] *Kerr, P. F.:* Optical Mineralogy. New York: McGraw-Hill 1959

[2.19] *Winchell, A. N., and H. Winchell:* Elements of Optical Mineralogy. New York: John Wiley & Sons 1927

[2.20] *Wahlstrom, E. E.:* Optical Crystallography. 4th ed. New York: John Wiley & Sons 1969

[2.21] *Hartshorne, N. H., and A. Stuart:* Practical Optical Crystallography. London: E. Arnold & Co. 1964

[2.22] *Hartshorne, N. H., and A. Stuart:* Crystals and the Polarizing Microscope. London: E. Arnolf & Co. 1960

[2.23] *Gay, P.:* An Introduction in Crystals Optics. London: Longmans 1967

[2.24] *Heinrich, E. W.:* Microscopic Petrography. New York and London: McGraw-Hill 1956

[2.25] *Wood, E. A.:* Crystals and Light. An Introduction to Optical Crystallography. New Jersey: van Nostrand 1964
[2.26] *Wang, S.:* The Optical Properties of Solids. New York: Academic Press 1966
[2.27] *Rittmann, A.:* Quantitative Konoskopie. Schweiz. mineralog. petrogr. Mitt. 43 (1963) 1, 11-36
[2.28] *Rösch, S.:* Über Interferenzfarben im polarisierten Licht an Substanzen mit Dispersion der Doppelbrechung (Brewster-Herschel-Farben). Tschermaks mineralog. petrog. Mitt. 8 (1962) 36
[2.29] *Gahm, J.:* Die Interferenzfarbtafel nach Michel-Levy. JENOPTIK JENA GmbH, Werkzeitschrift 46 (1962) 118
[2.30] *Schumann, H.:* Konoskopische Interferenzfiguren bei Beobachtung im Auflicht an nichtopaken Kristallen. Heidelberger Beitr. Mineralog., Petrogr. 4 (1954) 198-206
[2.31] *Rath, R., and D. Pohl:* Intensitätsverteilung in Interferenzbildern. Neues Jb. Mineralog. Mh. (1969) 73
[2.32] *Schumann, H.:* Über den Anwendungsbereich der konoskopischen Methodik. Fortschr. Mineralog. 25 (1941) 217
[2.33] *Becke, F.:* Die Skiodromen. Ein Hilfsmittel bei der Ableitung der Interferenzbilder. Tschermaks mineralog. petrogr. Mitt. 24 (1905) 1
[2.34] *Tertsch, H.:* Zur Frage der »Skiodromen«. Tschermaks mineralog. petrogr. Mitt. 3 F., Bd. IX (1964) 1/2, 1-6
[2.35] *Eales, H. V.:* Prüfung der von Polarisationsfiguren abgeleiteten optischen Daten. Econ Geol. 62 (1967) 5, 737-738
[2.36] *Kamb, W. B.:* Isogyres in Interference Figures. Amer. Mineralogist 43 (1958) 1029
[2.37] *Sáenz de, J. M., and J. C. Tessore:* Aus der Indikatrix abgeleitete Richtungskonstante und ihre Anwendung in der quantitativen Konoskopie. Schweiz. minera. log. petrogr. Mitt. 48 (1968) 2, 471-484
[2.38] *Hauser, F., Oettel, W. O., and H. Gause:* Das Arbeiten mit auffallendem Licht in der Mikroskopie. Leipzig: Akademische Verlagsgesellschaft Geest und Portig KG 1960
[2.39] *Schneiderhöhn, H., and P. Ramdohr:* Lehrbuch der Erzmikroskopie. Berlin: Verlag Bornträger 1934
[2.40] *Schneiderhöhn, H.:* Erzmikroskopisches Praktikum. Stuttgart: Schweizerbartsche Verlagsbuchhandlung 1952
[2.41] *Oettel, W. O.:* Grundlagen der Metallmikroskopie. Leipzig: Akademische Verlagsgesellschaft Geest und Portig KG 1959
[2.42] *Wachromejew, S. A.:* Erzmikroskopie. Berlin: VEB Verlag Technik 1954
[2.43] *Cameron, E. N.:* Rudnaja Mikroskopija. Moskva: Mir 1966
[2.44] *Vlasov, G. A., and E. V. Smirnova:* Bestimmung der optischen Konstanten stark absorbierender Stoffe. Optika i Spektroskopija 26 (1969) 1, 124-126 (Russian)
[2.45] *Rath, R., and D. Pohl:* Die Indexfläche stark absorbierender niedrig symmetrischer Kristalle. Optik 32 (1970) N 3, 266-270
[2.46] *Možžerin, J. V.:* Novyj metod variaskopičeskoj mikroskopii v kristalloptike. Zap. vses. mineralog. Obćś. XCVI (1967) 4, 369-376

3 Microscopical equipment

3.1 Development and characteristics

During the past 50 years, microscopes have been undergoing fundamental changes. Earlier (*first generation*) instruments served for the mere observation of magnified specimen images. The microscope of today is a measuring instrument that is expandable and flexible in its adaptation to the measuring problem, and permits most measurements to be automated (Fig. 3.1, see Annex). A basic microscope design today may be available in more than 30 configurations. Table 3.1 gives a survey of the development. For detailed accounts, see the literature [3.1 to 3.21].

Characteristically, the microscope more and more assumes the function of a support structure carrying the manifold accessories that are inserted into the ray path in combinations that provide an ever greater variety of measuring facilities. That way, the microscope gradually loses its familiar appearance. For examining

Table 3.1. Stages in microscope development

Generation	Abridged characteristics	Examples
1	**Main function: magnification** Simple microscopes, almost no accessories, little facilities for measurement; subjective data collection and image analysis	Telaval, Boetius hot stage microscope, Dialux, Neophan, Poladun II
2	**Main function: measurement** Versatile measuring facilities by many accessories; high convenience of measurement; subjective data collection and image analysis	Amplival pol, Orthoplan, Zetopan pol, MPJ 5, MIN-8
3	**Main function: measurement** Versatile measuring facilities; accessories are dominating, microscope serves as basic unit only to guide the ray path; automatic image analysis and data collection	Quantimet, Classimat, Digiscan, Morphoquant, Parmoquant

the spinning process of polyamide fibres, e.g., the individual components have to be installed in the spinning shaft normal to the running fibre, without that arrangement being readily recognizable as a microscope.

Instrumental possibilities today permit examinations at temperatures ranging from −272 °C to 2 900 °C, but more than 90% of the microscopical measurements reported until 1971 were made at normal room temperature.

General path of rays in a microscope

Light source →	Specimen →	Optical system →	Reproduction
(transmitter)	(modulator)	(carrier)	(receiver)

This ray path may be arranged vertically, horizontally or inclinedly (Table 3.2).

Principally, the light used should be linearly polarized. That is why a polarizer is invariably required in front of the specimen.

Fig. 3.2 is a schematic view of the design and ray path of a modern polarizing microscope using transmitted and reflected (incident) light, respectively. In the reflected-light microscope, light reaches the specimen via a prism (at low magnifications) or an inclined plane glass plate (at high magnifications).

The two main illumination techniques common to both microscope types are *»bright field«* and *»dark field«*, differing by whether or not part of the beam is cut out.

Chart 3.1 shows the possibilities of automatic image analysis and data acquisition offered by modern microscopy. Comprehensive treatments of equipment and techniques [3.1 to 3.21] and reports on special instrumental problems [3.22 to 3.37] can be found in the literature.

Image formation in a microscope

The more or less structured microscopic specimen corresponds to a grating that splits transmitted light into maxima and minima by diffraction at a slit. The diffraction pattern is received by the objective, but the angular aperture of the latter puts a restriction on the number of maxima captured. The bundle of light of

Table 3.2. Various ray path directions in a microscope

Ray path	Transmitted light	Reflected light
Vertical, downwards	Used in most transmitted-light microscopes (Fig. 3.1)	In many reflected-light microscopes
Vertical, upward	Rarely used, »chemist's microscope«	In many reflected-light microscopes (Fig. 3.1)
Horizontal	Optical bench for transmitted light (Fig. 3.1)	Optical bench for reflected light
Oblique	Special arrangement for special purposes	

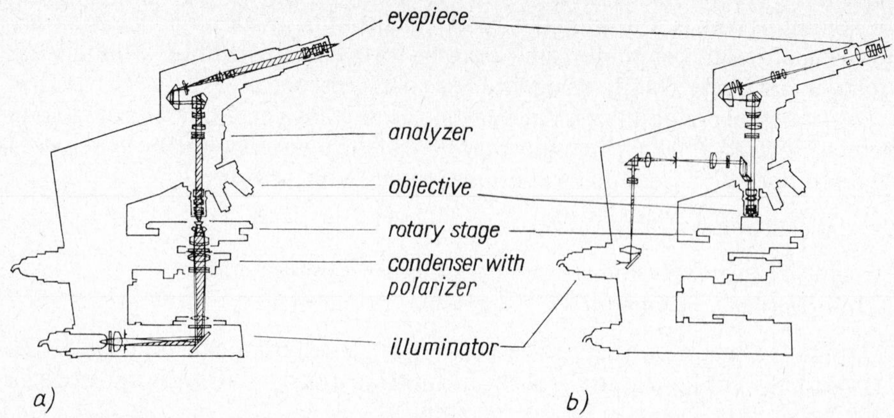

Figure 3.2. Light path in a microscope (schematic)
a) transmitted-light beam b) reflected-light beam

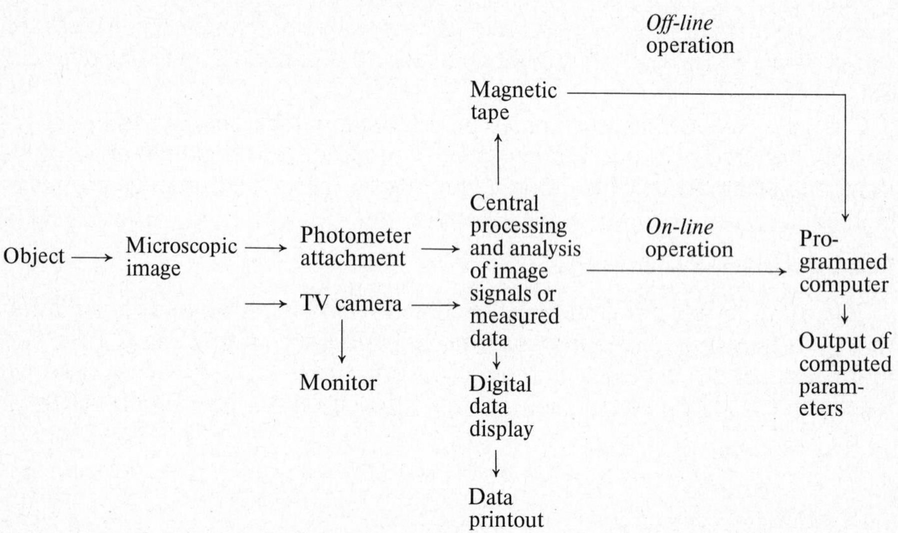

Chart 3.1. Modern microscopical facilities for automatic image analysis and data acquisition

each maximum captured converges in a point in the rear focal plane of the objective, thus forming diffraction patterns of the light source, to which the direct (undiffracted) light fraction and the maxima of the 1st, 2nd, 3rd ... orders contribute.

The waves emerging from the diffraction patterns interfere with each other in the intermediate image plane, forming an intermediate image similar to the specimen (»real-world« image). That image is magnified by the eyepiece.

The more diffraction maxima interfere with each other, the closer is the resemblance between image and specimen. One maximum singled out will not produce an image, but merely a field of uniform brightness.

3.2 Equipment for phase and interference microscopy

Both methods serve for examining phase objects that differ slightly from their surrounding medium in refractive index and thickness rather than in absorption. Combined with polarization, the two methods substantially expand the measuring and observing potential of transmitted and reflected light microscopy. Both methods rely on interference processes, but differ in essential parts of their optical paths. The phase contrast equipment imparts to the diffracted portion of the light a phase shift relative to the direct-pass fraction, whereas in the interference microscope a coherent comparison wavefront is superimposed on the imaging wavefront. Unlike normal amplitude objects (differing in absorption), phase objects cannot be examined satisfactorily by other microscopical methods. For important measuring problems involved in the two methods, see the literature [3.38 to 3.50].

3.2.1 Equipment for phase microscopy (transmitted light)

A phase outfit consists of a special condenser with annular diaphragm, and phase objectives with phase-retarding annuli.

Light passing a phase object suffers a slight change of its propagation velocity and thus of its phase position relative to the light passing the surrounding medium. To become visible, the phase difference has to be converted into an amplitude (brightness) difference.

In the microscope, the light is split up by the specimen into a direct and a diffracted portion, which reach the objective's rear focal plane in separate places. The direct fraction converges into a light ring, whereas the diffracted one is distributed over the entire plane.

If a phase annulus is located in that plane, the phase of the direct light changes by a defined amount ($\frac{1}{4}\lambda$), irrespective of the diffracted portion. The two portions then interfere with each other in the intermediate image plane, where the phase difference is visualized as an amplitude difference.

The intensity of phase contrast depends on the phase shift imparted to the light by the phase object.

$$\varphi = t\,(n_i - n_g)\,\frac{360}{\lambda}$$

t specimen thickness
λ wavelength
φ phase angle
n_i refractive index of immersing liquid
n_g refractive index of grain

In phase microscopy, the most favourable grain size (specimen thickness) is within 0.5 and 50 μm; the difference $n_i - n_g$ should be small.

Chart 3.2 instructs on how to prepare a phase outfit for measurement.

In phase contrast work, centration has to be checked whenever the objective or the specimen has been changed (Chart 3.2). The literature provides detailed instructions on phase microscopy [3.39 to 3.44].

3.2.2 Equipment for interference microscopy (transmitted light)

To visualize phase objects, an interferometer splits the beam of light into two coherent partial beams. Well-approved techniques are splitting behind the objective (Amplival-Interphako, JENOPTIK JENA GmbH) and splitting in front of

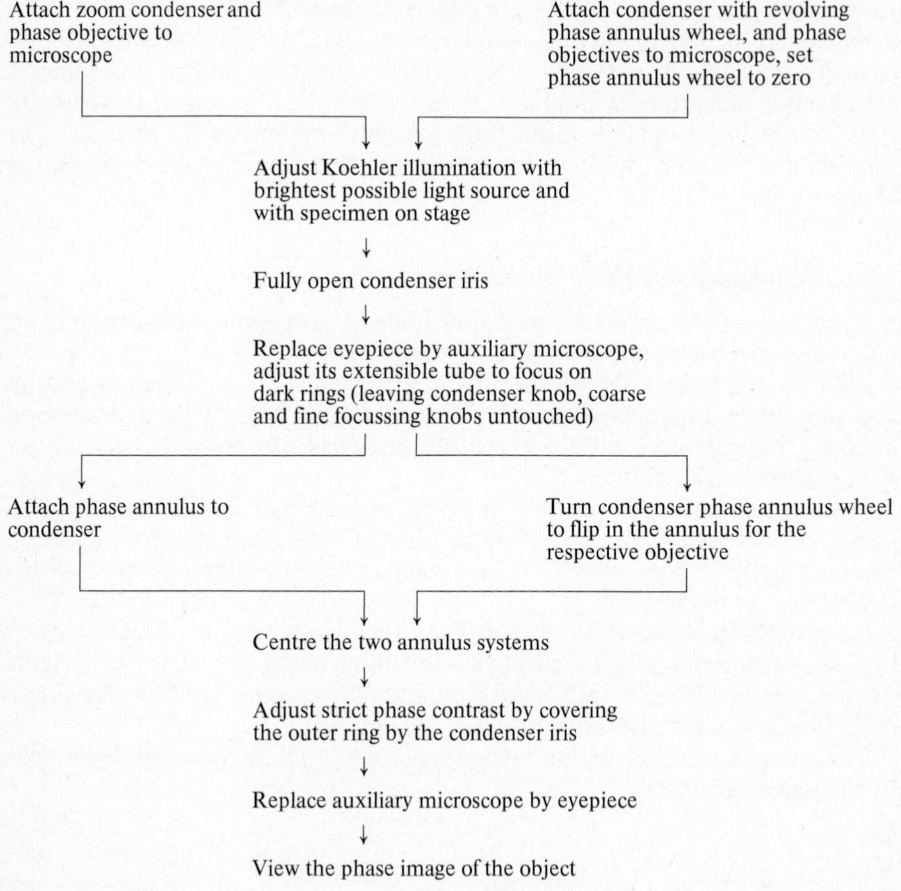

Chart 3.2. Adjustment of phase contrast outfit (JENOPTIK JENA GmbH)

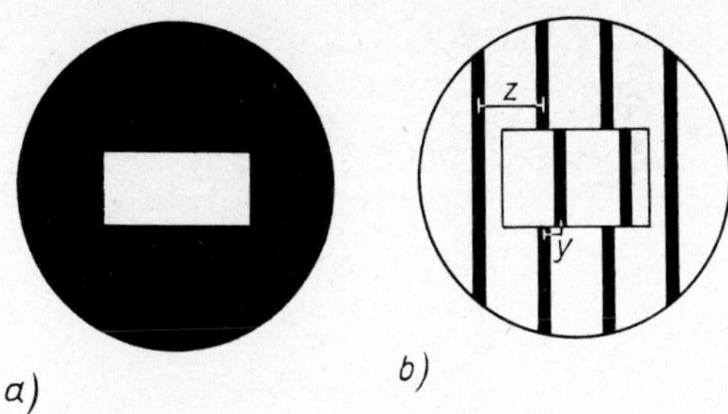

Figure 3.3. Object seen through a transmitted-light interference microscope

a) phase object in a homogeneous field (infinite fringe width)
b) fringe displacement by a homogeneous object (fringe method)

the object (interference microscope). In evaluating interference-microscopical images, one should bear in mind that they visualize optical path differences of the specimen. They correlate with the geometrical profile of the specimen only if this is uniform in refractive index. Special measuring problems are treated elsewhere [3.45 to 3.50].

Amplival-Interphako

Both partial beams interfere with each other in the intermediate image plane. The instrument offers a choice between lateral image splitting (*shearing*) or examination without splitting (*interference phase contrast*, Fig. 3.3).

(1) Shearing

Shearing may be differential or total. Differential shearing is by an amount close to the limit of resolution so that no double image is seen, whereas total shearing produces a distinct double image. The two partial wavefronts can be set to have different inclinations; that way, families of interference fringes of varying separation can be produced (fringe method). Fringe separation increases as beam inclination is reduced, with a fringeless, homogeneous field as a limit case. Tables 3.3 and 3.4 show the various settings possible.

Table 3.3. Combinations of interference-microscopical techniques

Angle between the two wavefronts	Zero shearing	Shearing techniques	
		Narrow (differential)	Wide (total)
Zero	Homogeneous field, no image splitting (Interphako method)	Homogeneous field, differential splitting (interference contrast in strict sense)	Homogeneous field, total splitting
Small (fringe method)	Wide-spaced fringes, no image splitting	Wide-spaced fringes, differential splitting	Wide-spaced fringes, total splitting
Great (fringe method)	Narrow-spaced fringes, no image splitting	Narrow-spaced fringes, differential splitting	Narrow-space fringes, total splitting

Table 3.4. Applicability of interference-microscopical techniques (cf. Table 7.13)

Technique	Characteristics	Applications
Differential shearing	No double image of object	Contrasting by relief effect, measurement of refractive index gradients, measurement of surface inclinations
Total shearing	Visible double image	Measurement of retardation differences
Shearing, fringe method	Object shifts fringes (Fig. 3.3), retardation measurable via fringe shifts	Applicable for large-surface objects
Interference contrast (homogeneous field)	Object changes colour or brightness compared to surrounding medium, retardation measurable via compensation	Applicable for small objects and small retardations

(2) Interference phase contrast (Interphako)

This technique can be regarded as a limit case of the *shearing technique,* with shearing = 0. By intervention into the diffraction pattern, a near-plane comparison wavefront is produced. In the intermediate image plane, it interferes with the wavefront that is influenced by the specimen.

Ray path 1. The diffracted light fraction is cut out. The remaining direct fraction provides a homogeneous background that is coherent with the specimen structures and corresponds, in effect, to a plane comparison wave. The direct fraction alone cannot image specimen structures, as the microscopic image is formed by interference between the diffracted and direct light fractions.

Ray path 2. The diffracted and direct light fractions form an image of the specimen structures.

The microscopic image (ray path 2) corresponds to a plane wave with a phase shift impressed on it by the object. If it is now superimposed on another wave constituting a coherent, homogeneous background (ray path 1), phase (or retardation) differences will occur between an object and its surrounding medium. These will be visualized as differences of intensity so that the phase object comes out in rich contrast. The retardation differences between object and surrounding medium can be measured by compensation and evaluated.

Interference microscope

Arranged as a double microscope with object and reference beams widely separated, this instrument permits, especially, the study of specimens in microcells, without size and distribution of objects in the specimen having any influence of the ray path. Fringe separation in the visual field can be varied continuously. In addition to the *fringe method*, the microscope may also be set for homogeneous field (*interference contrast method*).

3.3 Equipment for varying the direction of transmission through the specimen

3.3.1 The universal stage [3.51 to 3.79]

The universal stage is an accessory that attaches to the stage of a polarizing microscope, with which it must be exactly concentric (see Chart 3.3). A glass sphere can be rotated about several axes perpendicular to each other. Arranged in the centre between the two hemispherical glass segments forming the sphere is the specimen, sandwiched between slide and cover glass (Fig. 3.1, see Annex). The average light refraction of the glass segments, same as that of the immersion medium between the segments and the specimen, should match the average refraction of the crystal under examination. For that purpose, manufacturers supply hemispheres of different refractive indices. Suitable immersing liquids for joining the segments with the specimen are glycerol, cedar wood oil, or mixtures of cinnamic aldehyde and oxalic acid ester. If glycerol is used, the segments and the specimen may be rinsed with distilled water. The segments must not be treated with alcohol.

As the system of segments-immersion liquid-specimen is optically inhomogeneous, the angles of tilt have to be corrected. For appropriate corrections, see the literature [3.51 and 3.52].

A crystal mounted on the universal stage can be positioned for light transmission in various directions within the free range of rotation. This may vary between 80° and 120°, depending on stage design and specimen shape. A universal stage may rotate about three, four or five axes.

The device permits to measure the indicatrices of about 80% of all crystals of any orientation, and thus to register the spatial relationships between optically and crystal-morphologically privileged directions. These include, e.g., the directions of the principal refractive indices n_α, n_β, n_γ (semiaxes of the indicatrix), the directions of the optic axes and the angle between them, the directions of normals to crystal faces, cleavage faces, twinning planes, composition planes, etc.

Measurement can be performed in the orthoscopic path by analyzing the extinction behaviour of the crystals or in the conoscopic path by evaluating the interference patterns appearing while the stage is rotated about its various axes. Small

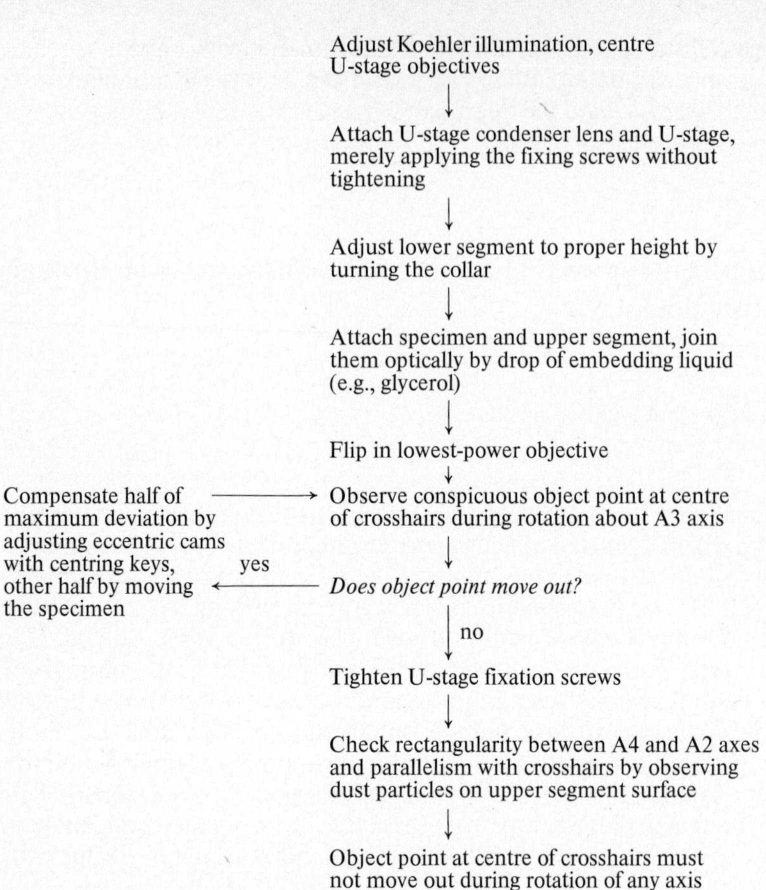

Chart 3.3. Adjustment of the four-axis universal stage (U-stage)

grains that would not yield interference patterns in the conoscopic path lend themselves to examination in the orthoscopic path.

For evaluation, the measured data are entered in suitable coordinate grids. The equal-area Schmidt grid is employed in texture analysis (analysis of direction distribution), whereas the indicatrix is constructed by means of the isogonal stereographic grid according to Wulff.

Methods of universal stage measurement and evaluation are described in the literature [3.54 to 3.65].

Special modifications of the universal stage have been developed especially by Soviet designers [3.66, 3.68 to 3.75]. In 1968, a patent was granted on a semi-automatic universal stage for serial measurements in petrofabric analysis [3.76]. Heatable universal stages permit, within limits, to measure the temperature dependence of optical properties. A flow-through segment can be used for

Emmons' temperature variation methods of refractive index determination within a range of +150 °C to −60 °C. Ehrenberg [3.79] modified the universal stage for observation by reflected light.

3.3.2 The spindle stage

The spindle stage is a horizontal-axis rotation device attached concentrically to the stage of a polarizing microscope. The rotary spindle ends in a needle point of metal or glass protruding into a cell filled with an immersion medium. A specimen grain is fixed to the point by means of a suitable adhesive (e.g., grease). The grain can be rotated about the needle axis and with the microscope stage.

The techniques and apparatuses employed are based, above all, on studies by Soviet and American scientists [3.80 to 3.102].

Soviet workers tend to refer to the method as »universal-theodolite immersion method« in order to point out the favourable possibilities of direct refractive index measurement in various rotary positions. The method is employed in two main versions.

(1) Evaluation of extinction measurements in the orthoscopic path, with crossed polarizers [3.80, 6.91 to 6.94, 6.103]

The extinction angles of n_α' and n_γ' measured at various angles of rotation result in curved lines (stereo patterns) in a stereographic projection. They constitute the geometrical locus of the stereographic projection of the extinction direction. By evaluating the graphs, the indicatrix can be plotted.

(2) Determination of the main refractive indices by combination of grain rotation and immersing method (cf. 6.1.1)

Phase and interference microscopy lend themselves well to this version. Table 3.5 offers a survey of major design modifications of spindle stages.

3.3.3 The Waldmann sphere

The Waldmann sphere [3.103 to 3.105] is an accessory to the universal stage permitting examination of isolated crystals within a wide range of transmission directions. It consists of a hollow glass sphere with a lateral aperture, which is closed by a metal segment. The grain is fastened to a needle attached to the segment so that it can be adjusted to be in the centre of the sphere (Fig. 3.4). The cavity contains an appropriate immersion medium. With a lateral holder, the sphere is mounted on the universal stage in place of the two glass hemispheres, or it is placed into a fitting ring. The device expands the 100° to 200° transmission range of the universal stage to about 270°; thus, a crystal can be set to the most favourable position before measurement with the universal stage.

Table 3.5. Spindle stage design versions

Design version	Characteristics	References and remarks
Double spindle stage	Etalon crystal on rotating axis protrudes into cell opposite the needle carrying the specimen grain. Amount of etalon rotation – until refraction is equal – is a measure (to be calibrated) of refractive index of cell liquid (corresponds to a microrefractometer).	[3.85]
Heatable cell	Serves for temperature variation method of determining refractive index (cf. 6.1.1.).	Experimental model
Rotated glass fibre	Grain cemented to a glass fibre and rotated with it.	[3.100]
Combination with micromanipulator	The needle holder of a micromanipulator standing next to the microscope also serves as axis of the spindle stage.	[3.102, 7.68]
Two-circle goniometer stage	The rotation device of the spindle stage moves a goniometer head that permits tilting the grain in addition to rotating. During rotating and tilting, the grain remains in a cell whose top and bottom are plane-parallel, the beam entering normal to them.	[3.86, 3.106]
Revolving cell turret	Several cells arranged on turret and filled with immersion liquids of graded refractive indices. One after the other cell swings in front of the needle point.	Self-made by Inst. of Mineralogy, Moscow University

A special preparation technique [3.105] enables very small or easily decomposing crystals to be examined. The grain is embedded amidst a polystryrene-xylene mixture filled in by two steps (Fig. 3.4).

3.3.4 The goniomicroscope

The path of rays in a goniomicroscope fundamentally differs from that of any other microscope. It corresponds to the goniometer known in crystallography, the

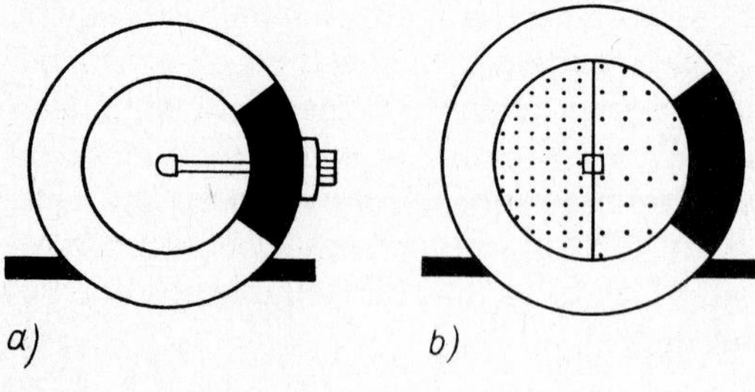

Figure 3.4. Mounting of a small grain inside the Waldmann sphere
a) by fixing to a needle point
b) by special filling technique

observation telescope used for measuring large crystals being replaced by a microscope that is rotatable about the specimen.

The instrument, preferably used with reflected light, serves to measure crystal faces, etching pit faces, twins inclined towards each other, and other morphological elements. The literature reports about various kinds of measurement [3.107 to 3.109].

3.4 Cold and hot stages

3.4.1 The cold stage

The cold stage is a transmitted and reflected light microscope stage modified for examining specimens at temperatures below 20 °C. Various models are known that employ different cooling methods [3.110 to 3.128].

Table 3.6 shows the most common coolants and the average specimen temperatures obtained with them. Higher temperatures than the minimum values specified require electrical heating in addition. Within limits, still lower temperatures can be attained by evacuation; the practically attainable cooling capacity, however, depends on stage design and on the objective used. For investigations by reflected light, direct thermal contact between specimen and coolant involves less problems, whereas in transmitted-light microscopy there will be higher temperature gradients between specimen and coolant. Where indirect or gas-flow cooling is employed, the temperature differences can be reduced by the use of special specimen slides made of metal with small glass insets. Fig. 3.5 (Annex) shows a modern device using Peltier cooling.

Requirements on stage design for low-temperature microscopy

(1) The stage surface should be as low as possible and the working distance of the objective as long as possible, so that high magnifications can be achieved.

Table 3.6. Cooling agents in low-temperature microscopy

Cooling agent	Lowest stage temperature (°C) (on average)
Peltier elements	−40
Refrigerant brines of various compositions	−50
CO_2	−60
CO_2/ether mixture	−80
N_2 at normal pressure	−185
Ne at normal pressure	−240
He at normal pressure	−265

(2) Specimen temperature must be measurable with high precision. Accordingly, the thermocouple should be as close to the specimen as possible. Favourable solutions are a specimen slide with the thermocouple embedded [7.32] or direct contact of the latter with the specimen.

(3) Specimen temperature must be controllable precisely and continuously by the smallest possible intervals.

The thermometric system of a cold stage can be calibrated by the melting points of various compounds (Table 3.7).

Frosting can be prevented by a coat of glycerol or isopropanol on the optical parts and scavenging the cooled parts with dried nitrogen.

3.4.2 The hot stage

Hot stages permit examination of specimens at temperatures above 20 °C. Transmitted-light models are known for temperatures up to 1300 °C, reflected-light ones for up to 1800 °C in continuous operation and up to 2800 °C in special cases. The designs reported in the literature [3.129 to 3.144] vary greatly, there being no trend towards standardization so far.

Table 3.7. Melting or freezing points of substances used for calibrating cold stages (according to the literature)

Melting or freezing point (°C)	Compound or system	Melting or freezing point (°C)	Compound or system
17	n-Decylamine	−23.0*	KI-H_2O
11.8	Dioxan	−26.0	Benzaldehyde
5.7	CH_2I_2(I)	−28.0*	NaBr-H_2O
5.2	CH_2I_2 (II)	−28.5	Nitromethane
2.5	N,N-Dimethylaniline	−33.6*	$MgCl_2$-H_2O
0.0	Water	−34.7	Ethyl benzoate
−1.1*	Na_2SO_4-H_2O	−36.5*	K_2CO_3-H_2O
−3.0*	KNO_3-H_2O	−39.9	Diethyl ketone
−3.9*	$MgSO_4$-H_2O	−40.0*	$CuCl_2$-H_2O
−6.1	Aniline	−43.0*	HNO_3-H_2O
−7.8*	$BaCl_2$-H_2O	−45.0	Acetic acid ester
−9.0	Acetonyl acetone	−54.0*	$CaCl_2$-H_2O
−10.7*	KCl-H_2O	−63.4	Chloroform
−12.4	Methyl benzoate	−65.0*	KOH-H_2O
−12.6*	KBr-H_2O	−76.3	n-Butyl acetate
−15.3	Benzyl alcohol	−82.4	Ethyl acetate
−15.4*	NH_4Cl-H_2O	−85.9	Methyl ethyl ketone
−18.3*	$(NH_4)_2SO_4$-H_2O	−94.6	Acetone
−21.1*	NaCl-H_2O	−112.0	Ethyl alcohol
−22.65	Carbon tetrachloride		

* Cryohydratic point of the respective salt-water system

Design versions differing in the method of holding the specimen

(1) Specimen rests on specimen slide. The slide material is fused quartz or sapphire for transmitted light; metal or special materials are used for reflected-light work. Technically mature models at present are
(a) a transmitted-light hot stage for 1250 °C,
(b) a reflected-light hot stage for 1750 °C, and
(c) a »Vacutherm« reflected-light hot stage for 1800 °C.

The illuminating rays are incident on the specimen from above (b and c) or below (a). In high-temperature microscopy, lateral illumination is restricted to special cases (heatable microcells). Fig. 3.5 (Annex) shows a hot stage for special examinations within a temperature range of 20 to 80 °C.

(2) Specimen held laterally. This method requires compact, electrically conductive specimens, such as metal strips. Heating is directly effected by current flowing through the specimen.

Table 3.8.
Melting points of some compounds used for calibrating hot stages (according to the literature)

Melting point (°C)	Compound	Melting point (°C)	Compound
308	$NaNO_3$	800.4	NaCl
333	KNO_3	814	As
385	CdI_2	844	$PbCrO_4$
392	Na_2CrO_4	851	Na_2CO_3
398	$K_2Cr_2O_7$	870	LiF
402	PbI_2	880	KF
419.4	Zn	884	Na_2SO_4
434	AgBr	891	K_2CO_3
455	AgCl	960.5	Ag
498	$CuCl_2$	962	$BaCl_2$
501	$PbCl_2$	975	K_2CrO_4
520	CdF_2	992	NaF
529.5	KBF_4	1063	Au
563.7	NaCN	1083	Cu
577	B_2O_3	1120	PbS
592	$Ba(NO_3)_2$	1127	SnO_2
614	LiCl	1185	$MgSO_4$
618	Li_2CO_3	1330	CaF_2
621	CsI	1396	MgF_2
634.5	KCN	1420	Si
636	CsBr	1650	$NaAlO_2$
646	CsCl	1700	Zr
651	Mg, NaI	1720	SiO_2
660	Al	1755	Pt
712	$MgCl_2$	1850	ZnS
715	PbCl	1900	Cr_2O_3
723	KI	1950	CeO_2
730	KBr	2000	La_2O_3
755	NaBr	2176	UO_2
772	$CaCl_2$	2430	SrO
790	KCl		
797	NiS		

(3) Specimen supported by the tip of a hot wire. A thermocouple simultaneously acts as heater, temperature measuring device and specimen holder. (The heating current is 50 Hz a.c.) Heating and thermovoltage measurement alternate.

Thermocouples (e.g., a Pt, Rh/Pt combination) can be used for temperatures up to 1800 °C. Calibrating substances for that range are listed in Table 3.8. Above 1800 °C, optical pyrometers have to be employed at present. Temperature measurement in that range is difficult and not very accurate.

The mode of heating the stage depends very much on the specimen properties, as outlined by Table 3.9.

Requirements on stage design for high-temperature microscopy

(1) The stage surface should be as low as possible and the working distance of the objectives as long as possible, so that high magnifications can be achieved.
(2) Specimen temperature must be controllable precisely and continuously, either by the smallest possible intervals or at a high speed within a wide range.
(3) The specimen chamber must allow a choice of evacuation down to below 10^{-2} Pa or normal-pressure gas atmosphere. The gas may be inert, reductive, oxidizing or caustic. Ultra-pure argon is a frequently used inert gas.
(4) Observation window and sample must not get fogged during examination.
(5) In maximum temperature work, the intensive radiation emitted by the specimen itself must be overcompensated by high-intensity lamps and blue filters in order to obtain images of rich contrast. Examinations by polarized light should be possible up to 1300 °C.

Reflected-light examinations involve additional problems caused by different processes on the visible specimen surface and in the invisible interior. Only those processes can be visualized that effect a change of the geometrical relief or the reflectivity of the specimen.

Table 3.9. Heating methods in high-temperature microscopy

	Indirect heating	Direct heating
Technical configuration	Heating strips of tantalum, tungsten, molybdenum or platinum sheet (contact heating), combined with radiation guide plates (radiation heating)	Heating by electrical resistance of specimen
Specimen type	Any	Electrically conductive
Examination	By transmitted and reflected light	By reflected light
Cooling of the heated space	Necessary	Unnecessary
Quantity of heat required	Relatively large	Relatively small
Thermal inertia	Differing	Very low
Stage build-up	Relatively complicated	Simple

3.5 Other special-purpose microscopes and accessories

A great number of accessories and special microscopes are available for special purposes. The most important ones and their applications are listed in Table 3.10.

Table 3.10. Functions and applications of special microscopes

Instrument	Operating principle	Applications	References
Dissecting microscope, Lanometer	Continuous change of magnification, long working distance of objective, wide field	Selection of grain specimens, specimen cell microscopy, survey examination of coarse-grained large-surface objects, chemical ultramicroanalysis, examination of textile fibres, industrial dusts etc.	[3.145]
Chemist's microscope	»Inverted« microscope with objective below object; illumination from above; quantitative measurements impossible	Examination of sediments in glass vessels	[2.5]
Double microscope acc. to Lau	Objective image on rotating ground glass observed with auxiliary microscope; low depth range	Improved image quality of fine specimen detail at limit resolution	[3.146]
Ultra-microscope	Imaging of light scattered by submicroscopical particles	Detection of sub-microscopic particles, such as in colloidal solutions, fine inclusions in glass etc.	[3.147]
Micro-beam microscope	Irradiation of selected object spots of 2 to 300 μm edge size with light of selected wavelength (250 to 1000 nm)	Examination of irradiation behaviour of light-sensitive objects	
Dark-field microscope	Image formed by no other light than that diffracted by the specimen	*Reflected-light dark field,* most important method to view unpolished specimens (grains)	[3.148]

Table 3.10, continued

Instrument	Operating principle	Applications	References
	Specimen texture appears bright on dark ground; direct beam cut out by dark-field condenser	*Bordering dark field* for determining the refractive index of powders. If this is equal to refractivity of embedding medium, grain margin shows characteristic colour effects.	
Schlieren microscope	Makes low refractive index gradients visible by special optical arrangement	Observation of mixing processes in liquids, crystal growth and dissolution processes etc.	[3.149]
Holographic microscope	Reveals spatial structure of micro-objects by wavefront reconstruction with hologram as intermediate stage. *Light source:* pulsed laser	Spatial representation of solids distribution in gases and liquids; ascertainment of spatial structure of aggregates	[6.154]
Fluorescence microscopy	*Light source:* UV lamp (for excitation). UV barrier filters pass on visible fluorescent light only; fluorescent object appears bright on dark ground. Optical system (with specimen slide, cover glass and immersion medium) must absorb but little UV light	Differentiation of objects by making fluorescent parts visible	[3.150]
Remote-control box microscope	Complex unit comprising preparation and examination parts; operation with rubber gloves or mechanical manipulators. Preparation and specimen stage compartments may be separated, and remote-controlled, from observation and data output parts (TV microscope or periscope). Little screening is required for radiation up to 0.005 curie. Above 10 curie: concrete cells.	Examination of decomposing, hygroscopic, aggressive, toxic or radioactive substances at high or low temperatures	[2.5]

Table 3.10, continued

Instrument	Operating principle	Applications	References
Microscope with pressure/ vacuum devices	1. High-pressure cell on stage 2. Vacuum chamber	1. Specimen behaviour under high pressure 2. Combined with hot stage, e.g., to study decomposition at underpressure	
Microscope with mechanical stress devices	1. Fibre twisting and stretching devices 2. Microhardness testers	1. Tensile behaviour of fibres 2. Testing the micro-hardness of substances	[3.151]
Integrating stage	*Point count:* object displaced by constant steps; crosshair position recorded by counters (Fig. 7.1) *Intercept count:* object displaced across cross-hairs by the screw micrometer correlated to object kind; intercept lengths of objects of a kind are automatically added up by the respective micro-meter.	Measuring the area of objects by statistical counting of points or intercepts	[3.152]
Micro-photometer eyepiece	*Microscope photometry:* measurement of the intensity of radiation coming from the speci-men *Microscope spectro-photometry:* measurement of spectral absorption of illumina-ting light of varied wavelength	Automatic image analysis (specimen features of differing absorption) Characterization of coloured substances by analyzing spectral absorbance behaviour; measurement of pleo-chroism	[3.153] to 3.155]
Micro-spectral eyepiece	*Microspectrography:* radiation coming from the object is spectrally dispersed and analyzed	Characterization of coloured substances by analyzing spectral absorbance behaviour; measurement of pleo-chroism	
Compensator	Accessory to polarizing microscope, for measuring retardation	Measurement of bire-fringence or specimen thickness; ascertain-ment of optical character and orien-tation of n_α' and n_γ'	[3.156, 3.157]
Test strip attachment	Visual comparison of specimen image with reference photographs	Investigation of structure and grain size	

Table 3.10, continued

Instrument	Operating principle	Applications	References
Microprojection attachment	Projection of microscope image by powerful lamp	High-quality projection image without loss of detail	
Micrometer eyepiece	Eyepieces with various micrometer discs	Visual measurement of objects viewed through the microscope	
Comparison eyepiece	Device for simultaneous observation of two specimens through two microscopes	Visual comparative examinations	

3.6 Light microscopy combined with other methods of measurement

The information extractable from microscopic examinations can be enhanced considerably by combination with the results of other investigations. The following instrumental versions are possible for combining two methods:

(1) Simultaneous registration of the measuring effects of both methods in a single measuring process on the same specimen.

(2) Successive examinations with the two methods by switching over the measuring beams, the specimen retaining its position.

(3) Successive examinations with the two methods, the specimen being moved from one instrument to the other.

The first version yields the most informative results.The instrument combinations that have become known so far are outlined in Table 3.11.

Table 3.11. Combination of light microscopy with other methods

Combination with	Operating principle	Applications	References
Laser micro-emission spectroanalysis	Finely focussed laser beam evaporates selected specimen spot a few μm wide; vapour is excited by spark discharge between electrodes to emit light, which is analyzed by spectrograph (single tests or specimen scanning)	Analysis of chemical composition of specimen details, detection of local chemical variations; qualitative and semi-quantitative identification of constituent elements	[3.158]

Table 3.11, continued

Combination with	Operating principle	Applications	References
Electron microscopy, electron diffraction, microprobe	Simultaneous inspection of microscopically selected object by electron microscopy, selected-area electron diffraction and x-ray fluorescence analysis; spot size 1 μm; single test or scanning; preferably for electrically conductive specimens	Chemical spot analysis of microscopic objects, detection of local chemical variations; qualitative and quantitative identification of constituent elements	[3.159]
Single crystal x-ray diffraction	1. Specimen simultaneously in microscopic and y-ray beam paths 2. Grain attached to or held inside glass capillary; microscopy and x-ray diffraction successively in separate places	Identification and structure analysis of microscopically selected grains; combined evaluation of microscopical and roentgenographic data	[3.160 to 3.162]
Differential thermo-analysis (DTA)	As specimens are small, a highly sensitive DTA scanner is required. *Arrangement of thermocouples:* 1. in the inert sample and in the specimen 2. inside the specimen and on the specimen surface; conclusion on different thermal processes inside and on the surface	Study of thermal behaviour of substances at high and low temperatures, correlation of *DTA* effects and visual specimen changes	[3.163, 3.164]
Magnet probe	Probe consisting of two Si ferrite cores in bobbins with equal primary and secondary windings; the intrinsic field of the coil interacts with ferromagnetic domains of the specimen; the harmonic of the voltage induced is amplified and measured	Study of metals at high temperatures, observation of spontaneous magnetization in ferromagnetic specimens, identification of the Curie point	[3.165]
UV microscope	Visual observation through image converters or by UV photomicrography; optical parts including slide, cover glass and immersion media must absorb but little UV (as do Canada balsam or glycerol)	Resolving power increased to 0.08 μm; high-contrast presentation of unstained objects differing by their absorption of UV light	[3.166]

Table 3.11, continued

Combination with	Operating principle	Applications	References
IR microscope	Projection of specimen image through catoptric objective on to IR photoplate or image converter photocathode	Visualization of sulfides, semiconductor materials etc. that stop visible light but transmit IR, such as Cu diffused into Si monocrystals; constitution analysis of organic crystals using polarized IR light	[3.167]

3.7 Refractometers for microscopy [3.169 to 3.172]

Many microscopical examinations require that the refractive index of the embedding liquid be known. Refractive index can be measured directly in the microscopical specimen, or after applying a drop of the liquid on to a refractometer.

Microscope refractometry

Direct measurement of refractive index on smallest amounts of liquid excludes errors due to refractive index changes during the applications of a drop of a liquid that is a mixture of components having different vapour pressures.

The literature reports on different versions of microscope refractometry:

(1) Measuring the critical angle of total internal reflection.
(2) Measuring optical path difference using a transmitted-light interference microscope with special specimen slides.
(3) Measuring the rotation angles of an etalon crystal in a measuring cell on the spindle stage.
(4) Temperature and wavelength variation methods using calibrated etalon crystals.

Microrefractometers are independent from the microscope. They require somewhat greater sample volumes for refractive index measurement. Suitable instruments are the Abbe Refractometer and the Jelley Microrefractometer.

References

[3.1] *Gause, H.:* Entwicklungstendenzen in der Polarisationsmikroskopie. Chem. d. Erde 22 (1962) 111–166

[3.2] *Reumuth, H.:* Auflicht – verglichen mit Durchlicht-Beobachtungen, ein Mikromethodenvergleich im Spiegel praktischer Ergebnisse. Z. Reyon, Zellwolle u. a. Chemiefasern (1965) 106–113

[3.3] *Reumuth, H.:* Leistungs- und Prinzipvergleiche zwischen Lichtmikroskopen, Durchstrahlungs-Elektronenmikroskopen und dem neuen Raster-Elektronenmikroskop. Melliand Textilber. (1967) 489–501

[3.4] *Moenke, H.:* Optische Bestimmungsverfahren und Geräte für Mineralogen und Chemiker. Leipzig: Akademische Verlagsgesellschaft Geest und Portig KG 1965

[3.5] *Francon, M.:* Einführung in die neuen Methoden der Lichtmikroskopie. Karlsruhe: Verlag G. Braun 1967

[3.6] *Romeis, B.:* Mikroskopische Technik. München, Wien: Verlag Oldenbourg 1968

[3.7] *Schild, E.:* Praktische Mikroskopie. 3rd ed. Wien: Verlag Maudrich 1955

[3.8] *Bloss, F,.-D.:* An Introduction to the Methods of Optical Crystallography. New York: Reinhart & Winston 1961

[3.9] *De Hoff, R. T., and F. N. Rhines:* Quantitative Microscopy. New York: McGraw Book Comp. 1968

[3.10] *Bartels, P. H.:* Polarized light analysis. Microchem. J. 7 (1963) 98–119

[3.11] *Chambers, R., and C. E. McClug:* Handbook of Microscopical Technique. New York: Harper 1950

[3.12] *Foster, L. V.:* Microscope Optics. Analytic. Chem. 21 (1949) 432

[3.13] *Gander, R.:* Hilfs- und Testpräparate in der Mikroskopie. Mikroskopie 20 (1965) 117–123

[3.14] *Lautenschläger, E.:* Praktische Mikroskopie. Basel: Verlag Wiss., Techn. u. Ind. 1970

[3.15] *Lehr, A.*, et al.: Über einige Fragen der Polarisationsmikroskopie. Bergakademie [Freiberg] 15 (1963) 394–399

[3.16] *Melancholin, N. M., and S. V. Grum-Grzimajlö:* Metody issledovanija optičeskich svojst Kristallov. Moskva: Izd. Akad. Nauk SSSR 1954

[3.17] *Ostrowskij, I. A.:* Ob otečestvennoj mikroskopičeskoj apparature. Trudy 4. sovesc. eksper. i techn. mineral. vyp II. Moskva: Izd. Akad. Nauk SSSR 1953

[3.18] *Ramdohr, P:* Erzmikroskopie, was sie zeigt und was wir mit ihren Methoden erschließen können. Aufschluß 20 (1969) 178–183

[3.19] *Swann, M., and J. M. Mitchison:* Refinements in Polarized Light Microscopy. J. exp. Biology 27 (1950) 226–273

[3.20] *Weber, K.:* Lichtmikroskope. Z. Ver. dtsch. Ing. 108 (1966) 805–807

[3.21] *West, P. W.:* Polarized Light Microscopy. Chemist-Analyst 34 (1945) 4–8, 28–35

[3.22] *Bonnke, H.:* Erzeugung und Betrachtung des Interferenzbildes kleiner Kristallkörper mit dem Poladun II und IV. Feingerätetechnik 8 (1959) 365

[3.23] *Freere, R. H.:* Filter und ihre Anwendung in der Mikroskopie. Das medizinische Laboratorium XXI (1968) 11, 261–294

[3.24] *Gersing, R.:* Eine selbstzündende Kohle-Lichtbogenlampe konstanter Lichtleistung für mikroskopische Zwecke. Z. wiss. Mikroskopie 63 (1957) 257–260

[3.25] *Günzel, H.:* Metallinterferenzfilter aus Jena. Wiss. u. Fortschr. 12 (1962) 8, 366

[3.26] *Harrison, R. K., and G. Day:* Continuous Monochromatic Interference Filter. Mineralog. Mag. 33 (1963) 517

[3.27] *Helbig, E.:* Über Lichtquellen hoher Leuchtdichte und die Bedeutung der Leuchtdichte bei optischen Geräten. Monatsschrift Feinmechanik-Optik 71 (1954) 5–13

[3.28] *Holgate, N.:* Monochromator or Graded Spectrum Filter? Mineralog. Mag. 33 (1963) 512

[3.29] *Jeglitsch, F., and R. Mitsche:* Die Anwendung optischer Kontrastmethoden in der Metallographie. Radex-Rdsch. 3/4 (1967) 587–596

[3.30] *Klier, R.:* Erzeugung monochromatischer Strahlung mit Spektrallampen. GIT-Fachz. Lab. 12 (1968) 1, 10–11

[3.31] *Liverseege, J. F.,* et al.: Quantitative Microscopy Chiet. Analyst 47 (1922) 430

[3.32] *Missmahl, H. P.:* Einfache Prüfung der Optik des Polarisationsmikroskops auf Spannungsdoppelbrechung. Z. wiss. Mikroskopie 66 (1964) 81–84

[3.33] *Mosebach, R.:* Die Bestimmung des optischen Schwerpunkts von Glühlicht-Filter-Kombinationen. Heidelberger Beitr. Mineralog. Petrogr. 2 (1951) 437–442

[3.34] *Randall, R. P., and R. E. H. Trevor:* Die Anwendung von Bildverstärkern mit hoher Verstärkung. GIT-Fachz. Lab. 12 (1968) 5, 494–504

[3.35] *Rösch, S.:* Der Variocolor, ein Hilfsmittel zur optischen Färbung im mikroskopischen Gesichtsfeld. Leitz-Mitt. Wiss. u. Techn. I (1958) 2

[3.36] *Schumann, H., and H. Piller:* Über die Verwendungmöglichkeit moderner Polarisationsfilter in mineralogischen Mikroskopen. Neues Jb. Mineralog. Mh. (1950) 1, 1–16

[3.37] *Stach, E.:* Auflicht-Mikroskopie mit Immersionskappen. Mikroskopie 12 (1957) 7/8, 232–242

[3.38] *Rienitz, J.:* Kritik des Positiv-Negativ-Verfahrens und der mikroskopischen Reliefverfahren (schiefe Beleuchtung, Schlierenmikroskopie, Shearing-Interferenzmikroskopie). Mikroskopie 22 (1967) 169–193

[3.39] *Allen, R. D., and J. W. Brault:* Image Contrast and Phasemodulated Light Methods in Polarization and Interference Microscopy. *Barer, R., and V. E. Cosslett:* Advances in Optical and Electron Microscopy, vol. 1. London and New York: Academic Press 1966, 77–114

[3.40] *Beyer, H.:* Theorie und Praxis des Phasenkontrastverfahrens. Leipzig: Akademische Verlagsgesellschaft Geest und Portig KG 1965

[3.41] *Bernhardt, W.:* Erfahrungen mit der Phasenkontrasteinrichtung am Neophot II. Jenaer Rdsch. 14 (1969) 4, 234–237

[3.42] *Frey-Wyssling, R.:* Kontrasteffekte dicker Objekte im Phasenmikroskop. Naturwissenschaften 39 (1952) 145–146

[3.43] *Heidermanns, G.:* Zur Technik der Pulvermikroskopie mit dem Phasenkontrastmikroskop. Staub – Reinhalt. Luft (1959) 104–105

[3.44] *Otto, L.:* Eine Phasenkontrasteinrichtung mit langer Schnittweite. Jenaer Rdsch. 1 (1965) 105–106

[3.45] *Schmidt, H.:* Interferenzmikroskopische Hilfsmittel in Forschung/Technik. Naturwiss. Rdsch. 18 (1965) 11, 445–455, Beil. Mitt. Verb. dtsch. Biologen Nr. 114

[3.46] *Uhlig, M.:* Untersuchungen von Durchlichtobjekten mit dem Auflicht-Interferenzmikroskop. Mikroskopie 21 (1967) 9/10, 269–270

[3.47] *Beyer, H., and G. Schöppe:* Interferenzeinrichtung für Durchlichtmikroskopie. Jenaer Rdsch. 10 (1965) 1, 99–105

[3.48] *Krug, W., Rienitz, J., and G. Sculz:* Beiträge zur Interferenzmikroskopie. Berlin: Akademie-Verlag 1961

[3.49] *Polze, S.:* Ein Laser-Interferenzmikroskop. Mber. dtsch. Akad. wiss. Berlin 7 (1965) 631–632

[3.50] *Metz, A.:* Ein neues Auflicht-Interferenzmikroskop. Silikat-J. Selb. 5 (1966) 12, 163–165

[3.51] *Kleemann, A. W.:* Nomograms for Correcting Angle of Tilt of the Universal Stage. Amer. Mineralogist 37 (1952) 115

[3.52] *Tröger, E.:* Nomogramme zur Reduktion von Kippwinkeln am U-Tisch. Zbl. Mineral. Petrogr. Abt. A (1939) 177

[3.53] *Berger, P. J.:* U-Tisch-Methoden, Erfahrungen mit dem Leitzschen U-Tisch-Refraktometer. Z. angew. Mineral. Bd. IV (1942/43) 2/3, 240
[3.54] *Berek, M.:* Mikroskopische Mineralbestimmung mit Hilfe der U-Tisch-Methoden. Berlin: Verlag Gebr. Bornträger 1924
[3.55] *Berek, M.:* Grundsätzliches zur Bestimmung der optischen Indikatrix mit Hilfe des U-Tisches. Schweiz. mineralog. petrogr. Mitt. 29 (1949) 1, 1
[3.56] *Berek, M.:* Neue Wege zur Universalmethode. Neues Jb. Mineralog. Petrogr., Beil. Bd. 48 (1923) 34
[3.57] *Dimler, R. J., and M. A. Stahrmann:* A Mount for the Universal Stage Study of Fragile Materials. Amer. Mineralogist 25 (1940) 502
[3.58] *Emmons, R. C.:* The Universal Stage Geol. Soc. America Mem. 8 (1943)
[3.59] *Fischbach, E.:* Some Technique of Universal Stage Conoscopy. Neues Jb. Mineralog. Mh. (1970) 3, 343–348
[3.60] *Hallimond, A. F.:* Universal Stage Methods I/II Mining Mag. 83, Juli/Dez. 1951
[3.61] *Leutwein, F.:* Der Universaldrehtisch nach Fedorow. Bergakademie [Freiberg] 7 (1955) 5, 236–241
[3.62] *Mellis, O.:* Ein Nomogramm zur Lokalisierung von Mineralkörnern auf dem U-Tisch. Mikroskopie 6 (1951) 157
[3.63] *Nickel, E., and G. Frenzel:* Apparative Bemerkungen zur Messung nach dem Drehtischprinzip. Heidelberger Beitr. Mineralog. Petrogr. 2 (1951) 552–555
[3.64] *Philipsborn v., H., and R. F. v. Hodenberg:* Der Universaldrehtisch als Instrument zum Vermessen sehr kleiner Kristalle. Z. Kristallogr. 111 (1959) 81–93
[3.65] *Reinhard, M.:* Universaldrehtischmethoden. Basel: Verlag Wepf u. Cie 1931
[3.66] *Sarantschina, G. M.:* Die Fedorow-Methode. Berlin: VEB Deutscher Verlag der Wissenschaften 1963
[3.67] *Schumann, H.:* Orthoskopische und konoskopische Beobachtungsweise im Universaldrehtisch. Mikroskopie 6 (1951) 104
[3.68] *Zavarickij, A. N.:* Dalnejšij šag vo primenenii Universalnogo stolika. Zap. vseross. mineralog. Obsc. 72 (1943) 2, 93–107
[3.69] *Vardanjanc, L. A.:* Teorija fedorowskogo metoda. Erevan: Izd. Akad. Nauk Arm. SSR 1959
[3.70] *Solotuchin, V. V.:* K voprosu ustanovlenija svjazi meždu immersionnyu i fedorovskom metodami. Zap. vses. mineralog. Obsc. vyp. 6 (1957) 86
[3.71] *Soboleva, V. S.:* Fedorovskij metod. Moskva: Izd. Nedra 1954
[3.72] Universalnyj stolik E. S. Fedorova. Team of authors. Moskva: Izd. Akad. Nauk SSSR 1953
[3.73] *Aršinov, V. V.:* K uproščeniju fedorovskogo metoda. Moskva: Trudy inst. kristallogr. Akad. Nauk SSSR, Vyp. 9. 353–366
[3.74] *Kucharenko, A. A.:* Opyt primenenija universalnogo stolika Federova k immersii. Zap. vseross. mineralog. Obsc. c. 68 (1939) 2, 224–226
[3.75] *Kajzer, S. A.:* Novyj variant pribora Kolotuškina dlja opredelnija trech glavnych pokazatelej prelomlenija mineralov v immersionnych židkostjach. Trudy kazach. nauk.-issl. institut. mineral. syr., vyp. 3, 1960
[3.76] DDR-Patentschrift 62467 of 20. 11. 1968
Watznauer, A., Behr, H. J., et al.: Vorrichtung für Mikroskope zum Einmessen der optischen Achsen von Quarzkristallen
[3.77] *Bautsch, H. J., Engel, A., and H. Voigtländer:* Ein neuer heizbarer Einsatz für den U-Tisch. Z. angew. Geol. 11 (1965) 7, 381–382
[3.78] *Vigfusson, V. A.:* An Improved Stage Cell. Amer. Mineralogist 25 (1940) 763
[3.79] *Ehrenberg, H.:* Ein neuer Universaldrehtisch für Anschliffe und seine Anwendungsmöglichkeiten. Z. Erzbergbau Metallhüttenwes. Bd. III (1950) 3, 65–69
[3.80] *Vardanjanc, L. A.:* Teorija i praktika metoda napravlenii pogasanija. Moskva: Izd. Nedra 1966

[3.81] *Kockin, J. N.:* Metod crasčajusčejsja igly na stolike Fedorova. Zap.vses. mineralog. Obsc., c. 89 (1960) 1

[3.82] *Fekliček, V. G., and I. V. Florinskij:* Universalnyji pribor dlja opredelenija pokazatelej prelomlenija teodolitno-immersionnym metodom. Izv. Akad. Nauk SSR, ser. geol. 30 (1965) 12, 106–111

[3.83] *Vedeneeva, N. E.:* Universalnaja apparatura dlja izmerenija pokazatelej prelomlenija pod mikroskopom. Trudy 2. Sovesc. po eksper. i techn. mineral. Moskva: Izd. Akad. Nauk SSSR 1937, 45–49

[3.84] *Fekličev, V. G.:* Novyj variant teodolitno-immersionnogo metoda. Trudy IMGR, vyp. 6, Moskva: 1961

[3.85] *Fekličev, V. G.:* Universalnyj teodolitno-immersionnyj metod. Moskva: Izd. Nauka 1967

[3.86] *Perdok, W. G., and P. Terpstra:* A Two-circle Optical Goniometer, specially adapted to Measuring Small Crystals of Organic Compounds. Recueil Trav. chim. Pays-Bas 73 (1954) 385–392

[3.87] *Saylor, C. P., and H. B. Lowey:* Zentrierbarer Rotor zur Messung von Kristalleigenschaften. J. Res. nat. Bur. Standards, Sect. C 69 (1965) 191–193

[3.88] *Dubeski, B. M.:* A Device for Holding Objects under the Microscope. Microscope 15 (1966) 4, 146

[3.89] *Fisher, D. J.:* Temperature Control Spindle Stage. Amer. Mineralogist 47 (1962) 649–664

[3.90] *Hartshorne, N. H.:* Single-axis Rotation Apparatus and Accesoory Devices for Studying the Optical Properties of Crystals. Microscope and Crystal Front 14 (1963) 81

[3.91] *Oppenheim, M. J.:* The spindle Stage – a Modification Using a Hypodermic Syringe. Amer. Mineralogist 47 (1962) 903–906

[3.92] *Hartshorne, N. H.:* A New Device for Mounting Crystals on the Single Axis Rotation Apparatus. Microscope and Crystal Front 14 (1964) 7, 282

[3.93] *Vincent, H. C. G.:* An Accessory to the Polarizing Microscope for the Optical Examination of Crystals. Mineralog. Mag. 30 (1955) 227, 513–517

[3.94] *Vincent, H. C. G.:* An Improved Orientation Stage for Microscopic Optical Crystallography. Nature 181 (1958) 693

[3.95] *Steinbach, B. J., and T. R. P. Gibb:* Device for Orienting Small Crystals under the Microscope. Analytic. Chem. 29 (1957) 860

[3.96] *Jones. F. T.:* Spindle Stage with Easily Changed Liquid and Improved Crystal Holder. Amer. Mineralogist 53 (1968) 7/8 1399–1404

[3.97] *Hartshorne, N. H., and P. M. Swift:* An Improved Single Axis Rotation Apparatus for the Study of Crystals under the Polarizing Microscope. J. Roy. Microscop. Soc. 75 (1954) 129

[3.98] *Hartshorne, N. H.:* A Simple Single Axis Rotation Apparatus. Mineralog. Mag. 33 (1963) 693

[3.99] *Rawlins, F. J. G., and C. W. Hawksley:* A Cell for Refractivity Measurements on Minute Crystals. J. Sci. Instruments 11 (1934) 282

[3.100] *Chromy, S.:* Tisch mit drehbarer Glasfaser zum Messen von Brechzahlen von Mineralien durch Variationsmethoden. Z. wiss. Mikroskopie 67 (1966) 71–77

[3.101] *Fisher D. J.:* A New Universal-type Microscope. Z. Kristallogr. 113 (1960) 77–93

[3.102] *Emons, H.-H., Seyfarth, H.-H., and B. Hahne:* Zur mikroskopischen Untersuchung zersetzlicher oder hygroskopischer Stoffe. Jenaer Rdsch. (1974) 4, 223–229

[3.103] *Waldmann, H.:* Glashohlkugeln für Kristall- und Edelsteinuntersuchungen. Schweiz. mineralog. petrogr. Mitt. 27 (1947) 2, 472

[3.104] *Fekličev, V. G.:* Prisposoblenie dlja izyčenija na stolike Fedorova mikroseren mineralov bez izgotovle nijaslifov. Trudy IMGRE vyp. 18 (1963) 139–141

[3.105] *Seyfarth, H.-H.:* Die Verwendung der »Waldmann-Kugel« in der chemischen Mikroskopie. Kristall und Technik 2 (1968) 4, K 81–K 83
[3.106] *Steck, A., and E. Glauser:* Universaldrehtisch für optische Untersuchungen von Mineralkörnern. Schweiz. mineralog. petrogr. Mitt. 48 (1968) 3, 815–820
[3.107] *Donnay, J. D. H., and W. A. Obrien:* Microscope goniometry. Chemist-Analyst 17 (1945) 593
[3.108] *Tadamitaoka* et al.: Ein Goniomikroskop und einige Anwendungen. Prakt. Metallogr. 5 (1968) 1, 22–33
[3.109] *Schumann, H.:* Ein Goniometer mit Mikroskopoptik. Leitz-Mitt. Wiss. u. Tech. 3 (1966) 161–192
[3.110] *Clothier, W. C.:* Microscope for Use of Temperatures Near −190 °C. J. Sci. Instruments 44 (1967) 7, 535
[3.111] *Bazarov, L. S.:* Ustanovka dlja zamorazivanija vključenij v mineralach. Materialy po genet. i eksper. mineral. T. IV. Novosibirsk: Akad. Nauk SSSR, sibir. otdel., inst. geolog., 1966
[3.112] *Gobrecht, H., Nelkowski, H., and R. Schlegelmilch:* Mikroskopie bei sehr tiefen Temperaturen. Z. angew. Physik 20 (1966) 194–195
[3.113] DDR-Patentschrift 37404 of 15. 3. 1965, Kl.: 42h, 15/03 IPK: GO2d
Heymer, A.: Kühlvorrichtung für Auflicht- und Durchlichtmikroskopie
[3.114] DDR-Patentschrift WP Nr. 113688, Kl.: 42h, Patentschrift 51714 of 5. 12. 1966
Heymer, A., et al.: Vorrichtung zur mikroskopischen Untersuchung unter hohem Druck bei tiefen Temperaturen
[3.115] *Hooss, E.:* Ein thermoelektrisches Gerät zur Kühlung und Heizung im Bereich von −3 bis +50 °C für Durchströmungstische. Mikroskopie 18 (1963) 10, 487–492
[3.116] *Hull, D., and R. D. Garwood:* Ein Tieftemperatur-Mikroskoptisch für Metallproben. J. sci. Instruments 32 (1955) 232–233
[3.117] *Hunt, H. D., Jackson, K. A., and H. Brown:* Mikroskopobjekttisch mit Temperaturgradienten für erstarrende Materialien mit Schmelzpunkten zwischen −100 und 200 °C. Rev. sci. Instruments 37 (1966) 805
[3.118] *Kolenko, E. A.:* Termoelektričeskie ochlaždajussie pribory. Moskva and Leningrad: Izd. Akad. Nauk SSSR 1963
[3.119] *Kofler, W., Kofler, A., and L. Kofler:* Über einen Mikrokühltisch und seine Anwendung. Mikrochim. Acta 38 (1951) 218–231
[3.120] DDR-Patentschrift Nr. 977 of 21. 3. 1952, AW 1029 K, 171520 IX a/42h, Kl.: 42h, Gr. 15/03
Kröger, C.: Kältekammer für mikroskopische Untersuchungen
[3.121] *Loasby, R. G.:* A Low Inertia Hot-cold Microscope Stage. J. sci. Instruments. ser. 2, vol. 1 (1968) 148–150
[3.122] *Markussen, J., and W. C. McCrone:* Thermoelectric Cold Stage for the Microscope. Microscope and Crystal Front 14 (1965) 395–402
[3.123] US-Patentschrift 3, 230, 773 of 25. 1. 1966
Matthews, C. E., et al.: Microscope Cold Stage.
[3.124] *McCrone, W. C., and S. Massenberg O Brandovic:* Mikroskopkühltisch für kontrollierte Untersuchungen in dem Bereich von −100 bis +100 °C. Analytic. Chem. 28 (1956) 1038–1040
[3.125] *Monier, J. C., and R. J. Hocart:* Eine Einrichtung für die mikroskopische Untersuchung von Metallen und Kristallen im polarisierten Licht bei Temperaturen von −130 bis +35 °C. J. Sci. Instruments 27 (1950) 302
[3.126] *Rey, L. R.:* Anordnung zur mikroskopischen Untersuchung bei tiefen Temperaturen. Experientia 13 (1957) 201–202
[3.127] *Virgin, W. W. and C. J. Massoni:* Ein neuer TT-Flüssigkeitsheiztisch. Amer. Mineralogist 43 (1958) 606–609

[3.128] *Curinov, G. G., and V. Z. Kolodjaznyj:* Stolik dlja kristallo-optičeskich izmerenij v temperaturnom intervale ot −160 do +250 °C. Peredovoj nauk.-techn. opyt, tema 37, Moskva: Izd. Filiala VJNJTJ 1957

[3.129] *Baumann, H. N.:* Microscopy of High-Temperature Phenomena. Amer. Ceramic Society, Bull. 27 (1948) 7, 267–271

[3.130] *Brenden, B. B., Newkirk, H. W., and J. L. Bates:* Principles of High-Temperature Microscopy. J. Amer. Ceram. Soc. 43 (1960) 246–251

[3.131] *Couroy, A. R., and J. A. Robertson:* Heizvorrichtung mit kontrollierter Atmosphäre für die mikroskopische Beobachtung in der Glasschmelze. Glas Ind. 44 (1963) 76–79 and 44 (1963) 139–143 and 175–177

[3.132] *Dimitriev, J., and M. Marinov:* Apparatur für mikroskopische Untersuchungen bei hohen Temperaturen. Godis. chim.-technol. Inst. (Sofia) 12 (1965), published 1966) 2, 219–230 (Bulgarian)

[3.133] *Dodd, J. G.:* A Simple High-Temperature Hot Stage for Controlled Temperature to 2000 °C. Microscope 14 (1965) 8, 302–305

[3.134] *Engel, H. J., and E. Zerbst:* Untersuchungen zur Präparattemperatur auf Mikroskopheiztischen. Z. wiss. Mikroskopie 64 (159/60) 384

[3.135] *Gutt, W.:* Mikroofen für Hochtemperaturmikroskopie und Röntgenanalyse bis 2150 °C. J. sci. Instruments 41 (1964) 393–394

[3.136] *Hamer, D. H.:* A Constant Temperature Device for the Kofler Hot Stage. Microscope 17 (1969) 2, 137

[3.137] *Knorr, W., Dudek, M., and P. Maschmeier:* Arbeiten mit dem Hochtemperaturmikroskop. Arch. Eisenhüttenwes. 38 (1967) 8, 635–638

[3.138] *Kulmburg, A.*, et al.: Zusatzeinrichtungen für die Hochtemperaturmikroskopie und praktische Anwendung. Radex-Rdsch. 3/4 (1967) 723–726

[3.139] *Lemmlein, G. G.:* Obraščennyj nagrevatelnyj mikroskop dlja nabljudenija i mikrokinematografirovanija pri vysokich temperaturach. Trudy 4 sovesc. po eksperim. i techn. mineral. vyp. II. Moskva: Izd. Akad. Nauk SSSR 1953

[3.140] *Lozinsky, M. G.:* High-Temperature Metallography. J. Royal Microsc. Soc. 86 (1967) Pt. 3, 211–234

[3.141] *Mathews, F. W.:* A Microscope Hot Stage. Analytic. Chem. 20 (1948) 1112

[3.142] *Mitsche, R., Gabler, F., and F. Jeglitsch:* Die Direktbeobachtung von Umwandlungsvorgängen in Festkörpern. Mikroskopie in der Technik, Bd. III, Teil 2, 269–311. Frankfurt/Main: Umschau-Verlag

[3.143] *Schmidt, M. B.:* On the History, Development and Application of the Warm Stage. J. Royal Microsc. Soc. II 79 (1960) 119–124

[3.144] *Stransky, K., and M. Kralova:* Surface Phenomena in High-Temperature Microscopy. Technical Digest 5 (1965) 323–331

[3.145] *Schultz, J., and H. Gause:* Stereomikroskope mit Polarisationseinrichtung für Präparation und Aufbereitung. Abh. Deutsch. Akad. Wiss. Berlin, Bergbau, Hüttenw., Montangeol. (1964) 2, 471–475

[3.146] *Lau, E., Schüller, A., and G. Rose:* Mineralogisch-petrographische Forschungen mit dem Mikroskop nach Lau. Geologie 9 (1960) 4, 426–439

[3.147] *Derjagin, B. V.*, et al.: Strömungsultramikroskopische Methode zur Untersuchung der Keimbildung einer neuen Phase im Kohlenwasserstoffmedium bei niedrigen Temperaturen. Pribory Techn. Eksperimenta (1967) 3, 108–110 (Russian)

[3.148] *Menzel, E.:* Dunkelfeldmikroskopie als Vorläufer moderner optischer Konzepte. Nachr. Akad. Wiss. Göttingen II. math. phys. Kl. (1966) 6, 115–121

[3.149] *Dodd, J. G.:* Observations with a »Schlieren« microscope. Microscope 17 (1969) 1, 1

[3.150] *Parker, C. A.:* Spectrophosphorimeter Microscopy: An Extension of Fluorescence Microscopy. Analyst 94 (1969) 1116, 161–176

[3.151] *Kaufmann, S., and R. Landgraf:* Untersuchungen mit einer Faserdreheinrichtung am Mikroskop. Faserforsch. u. Textiltechn. 20 (1969) 9, 442–449, and 20 (1969) 12, 593–597

[3.152] *Bonnke, H., and H. G.: Scheplitz:* Die Integriervorrichtung »Eltinor« des VEB Rathenower Optische Werke. Z. Feingerätetechnik 7 (1958) 1–6

[3.153] *Demirsoy, S.:* Entwicklung des Mikroskop-Photometers mit besonderer Berücksichtigung der Reflexionsmessung. Zeiss-Mitt. 4 (1967) 6, 254–279

[3.154] *Beyermann, K.:* Grundlagen und Arbeitstechnik der Mikrophotometrie. Fortschr. chem. Forsch. 11 (1969) 3, 473–506

[3.155] *Agroskin, L. S., and N. V. Korolev:* Microscope-Spectrophotometers. Optika i Spektroskopija 6 (1959) 544–545 (Russian)

[3.156] *Scheuner, G., and J. Hutschenreiter:* Anwendung der vierteiligen Quarzplatte nach Bertrand bei polarisationsoptischen Messungen von Gangunterschieden an biologischen Objekten.Z. wiss. Mikroskopie 68 (1967) 4, 224–227

[3.157] *Smithson, F.:* High-Order Plates for the Microscopic Examination of Mineral Grains. Mineralog. Mag. 30 (1955) 221, 145

[3.158] *Moenke, H., and L. Moenke:* Einführung in die Laser-Mikroemissionsspektralanalyse. Techn. phys. Monographie Bd. 21. Leipzig: Akademische Verlagsgesellschaft Geest und Portig KG 1966

[3.159] *Suwalski, G., and H. Vollstädt:* Anwendung der Röntgenmikroanalyse zur Lösung spezieller mineralogischer Probleme. Wiss. Ztschr. Humboldt-Universität Berlin, Math.-Naturwiss. Reihe XVI (1967) 5, 801–807

[3.160] *Perinet, G.:* Neue Apparatur zur Kombination von Heiztischmikroskop und Debye-Scherrer-Röntgenkamera. Bull. Soc. franç. Mineral. Cristallogr. 89 (1966) 325–328

[3.161] *Pikulin, S. A., and S. M. Olesnevic:* Gezielte Röntgenstrukturanalyse von Abschnitten eines Schliffes in einer Standardkamera. Zav. Labor. 34 (1968) 554 (Russian)

[3.162] *Haussühl, S.:* Quantitative Phasenanalyse und Korngrößenbestimmung von Dünnschliffen aus der Absorption von Röntgenstrahlen. Neues Jb. Mineralog. Mh. (1960) 9, 204–215

[3.163] *Dichtl, H. J., and F. Jeglitsch:* Kombination der Hochtemperaturmikroskopie mit der DTA. Radex-Rdsch. (1967) 3/4, 671–722

[3.164] *Miller, R. P., and G. Sommer:* A Hot Microscope Incorporating a Differential Thermal Analysis Unit. J. sci. Instruments 43 (1966) 293–297

[3.165] *Brandstaetter, F., Mitsche, R., and F. Gabler:* Kombination des Hochtemperaturmikroscops mit einer Magnetsonde. Radex-Rdsch. (1967) 2/4, 710–715

[3.166] *Hovnanian, A.*, et al.: UV-Television Microscopy and Microspectrophotometry. Microscope 14 (1964) 4, 141

[3.167] *Heerd, E.:* Technik der IR-Photographie mit dem Interferenzmikroskop. Z. wiss. Mikroskopie (1967) 4, 208–218

[3.168] *Turner, V. M.:* Quantitative Microscopic Ionophoresis and Chromatography. Nature 179 (1957) 964

[3.169] *Wilke, A.:* Ein einfaches Mikroskoprefraktometer für Flüssigkeiten besonders für das Einbettungsverfahren. Fortsch. Mineralog. 28 (1951, Jg. 1949) 84–86

[3.170] *Piller, H.:* Ein neues Mikroskoprefraktometer. Mikroskopie 16 (1961) 3/4, 88–92

[3.171] *Rath, R.:* Dispersionsbestimmungen mit Zeißschen Abbe-Refraktometern. Neues Jb. Mineralog. Abh. 87 (1954) 163 and 90 (1957) 1

[3.172] *Meixner, H.:* Erweiterte Arbeitsmöglichkeiten mit dem Leitz-Jelleyschen Mikrorefraktometer, insbesondere zur Einbettungsmethode für Mineralogen und Chemiker. Mikroskopie (1951) 1/2, 41

4 Sampling, pre-preparation and preparation of specimens

4.1 Sampling

Sampling means to separate from the total volume of a substance of interest a representative portion whose analysis may be taken as characteristic of the total volume.

Errors in that operation leading to a non-representative portion frequently exceed analysis errors by an order of magnitude. Thus, even before measurement, such errors largely determine the quality and informative value of measuring results.

Representative samples are difficult to obtain from

- bulk goods (fertilizers, building materials etc.),
- mixtures with heterogeneous grain composition (ores, coal, sintered products: the individual grains are aggregates formed by intergrowth of several phases), and
- mixtures of substances differing widely in density, grain shape, grain size or volume fraction of the constituent phases.

In microscopic examinations, the minuteness of the sample volume that can be measured increases the difficulty of obtaining information representative of the total substance volume.

To avoid sampling errors, as many individual samples as possible have to be taken from different places in a reactor, crystallizer, storage room and the like. Deviations between the analyses of these samples characterize the inhomogeneity of the total substance. The average of the analysis results comes the closer to the actual integral composition of the total substance the more individual samples have been examined.

Sampling for microscopic examinations varies depending on how the solid sample of interest is dispersed in various media (Table 4.1).

The individual grain of a solid substance consists of a monocrystal or of an aggregate formed of several monocrystals by intergrowth. Inhomogeneous monocrystals occur in non-stoichiometric compounds, zonal structure, inclusions, or similar phenomena.

A survey of sampling terms can be seen from the literature [4.1 to 4.7]. In [4.3 to 4.6], mathematical statements for computing the minimum sample quantity are derived. Sampling for phase analyses (Gy's model [4.3]) must be made in such a

Table 4.1. Sampling of solids dispersed in solids, liquids and gases

Solid, dispersed in	Sampling method
Solids	Direct sampling from loose grains (e.g., with sampling tube)
	matrix elimination, e.g. by dissolution, etching, cutting, grinding, drilling out
	crushing and disintegration of aggregates, followed by phase separation
Liquids	Sampling a liquid volume and direct examination of dispersed solids
	separation of solids from a liquid sample
	direct separation of solids from liquid
	sampling after precipitation of dispersed solids
Gases	Direct separation of specimen volume at the gas outlet (sedimentation or suction collector)
	sampling from precipitated solids

way that all grains have the same probability of getting into the sample. The grain fractions in the sample will then have equal proportions to those of the initial volume. The dominating influence factor in that statement is the size of the coarse fraction or the degree of disintegration, followed by other important factors, such as density, grain shape, homogeneity of material, and volume fraction.

Conditions are different in grain size analyses. According to Hutschenreiter's model [4.4], the probability for a grain to enter the sample must be inversely proportional to its mass. The masses of individual fractions in the sample will then correspond to those contained in the total substance.

Fig. 4.1 shows schematically the relation between the minimum sample quantity required and the spread of results for various degrees of homogeneity. Sampling equipment must be largely adapted to sampling conditions (Table 4.1). In serial production control, above all, there is a trend towards automating this operation. [4.8].

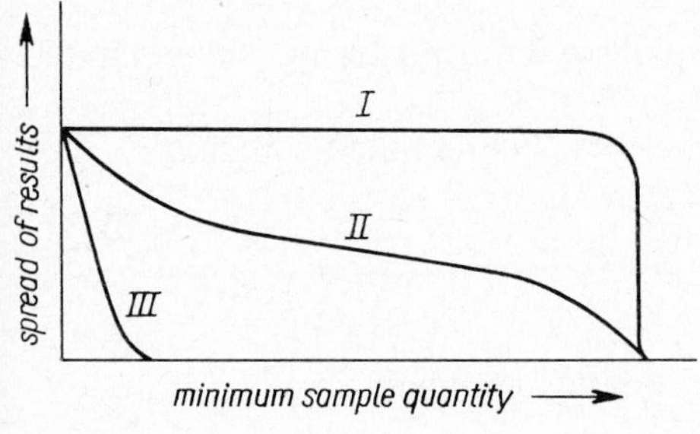

Figure 4.1. Relation (schematic) between minimum sample quantity and spread of results for several mixing types

I ideally unmixed sample
II real mixed sample
III ideally mixed sample

4.2 Pre-preparation of specimens

Pre-preparation comprises a variety of methods to adapt the samples taken to the examination conditions.

The pre-preparation method to be chosen depends on

(1) purpose of examination (phase, structural, petrofabric, grain or surface texture analysis),
(2) constitution of sample (grains cemented or not),
(3) method of final preparation,
(4) examination method chosen, and
(5) substance properties.

The *substance properties* determine special requirements for treating the specimen; they may be divided into four groups:

Group I – durable in dry state
Group II – conditionally durable in dry state
Group III – durable in parent solution (in the formation system) only
Group IV – durable only at higher or lower temperatures

The general pre-preparation procedure (Chart 4.1) is modified considerably depending on which group a substance belongs to. Individual operations in that procedure may be skipped to suit the special conditions prevailing. For a comprehensive treatment of pre-preparation methods, see the literature [4.9 to 4.11].

4.2.1 Substances durable in dry state

Separation of the liquid phase (drying)

A liquid phase can be separated by a centrifuge with mesh bottom strainer, by washing liquids, and by sucking dry gas through the specimen. Oven drying will frequently cause secondary reaction products and thus may destroy thermally delicate specimens.

Distinction between reactants and secondary products

Changed system conditions after sampling and during drying will precipitate substances dissolved in the liquid phase.

Existing deposits may decompose. The dry specimen then contains

Reactants: Residues of initial substances
Intermediate products formed
Reaction products

Secondary formations: Solid phases formed during or after sampling

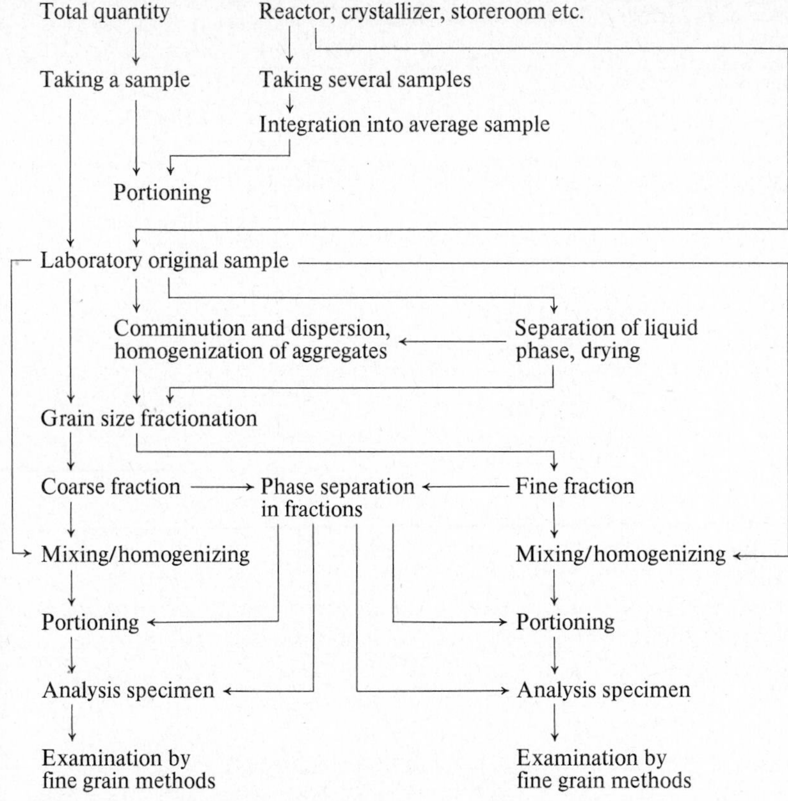

Chart 4.1. Sampling and pre-preparation of specimens

The analysis of grain size proportions, grain shapes, replacement patterns and intergrowth types yields information on how the phases of a sample were formed.

Comminution, dispersion [4.12 to 4.14]

The nature of grain cohesion in aggregates determines the methods of comminution and dispersion (Tables 4.2 and 4.3). Attention must be paid to changes under mechanical stress. Some hydrated salts, e.g., dehydrate even under slight pressure. If the mortar is used, grinding is less favourable than short, strong blows. These quickly disintegrate the aggregate into individual grains without producing noteworthy fine grain fractions. Finest grain formation is also prevented by frequently screening out the sample parts already comminuted.

Materials that are elastic at room temperature should be cooled with liquid nitrogen before comminution (paraffins, high polymers).

Table 4.2. Methods of aggregate dispersion depending on aggregation type

Type of grain aggregation	Separation method
Firm intergrowth, firm epitaxial growth	Comminution to the smallest grain size by rock saw and mortars made of steel, agate, hard rubber, lime wood
	microtoming, trimming stand, refrigerated microtome
	microdrill, acid saw, vacuum granulation, ultrasonic hammer
Entangled acicular grains, electrostatic adhesion of fine particles	Suspension in suitable liquids (e.g., glycerol, silicone oil, kerosene)
	addition of dispersants
	ultrasonic dispersion, specimen slide vibrator, micromanipulator

Table 4.3. Types of cut-off wheels for solid aggregates

Nature of aggregate	Cut-off wheel
Hard, coarsely grained fabrics	Bronze-sintered diamond wheel
Soft, coarsely grained fabrics	Steel-sintered diamond wheel
Homogeneous aggregates and monocrystals	Thin sheet metal wheels (0.5 mm) peripherally studded with diamonds, or thread cut-off machines
Radioactive material	Plastic-bonded corundum or silicon carbide wheels, ceramic wheels

Grain size fractionation

It is often sufficient to separate a sample into a coarse and a fine fraction, which require different examination methods. Errors may occur by incomplete comminution or new formation of aggregates, displacement of screen apertures, disposition of the fine fraction to adhere to larger grains, and phase separation due to fast comminution of soft fractions during screening.

Phase separation

In complex mixtures, individual fractions are further disintegrated by phase separation. Possible techniques [4.15 to 4.23] are picking with the micromanipulator, semi-automatic grain selectors, ultrasonic hammer, selective drilling under the microscope, selective dissolution, selective thermal disintegration, gravity separation (Table 4.4, Fig. 4.2), dense-medium centrifugation, suspension, sifting, electromagnetic separation, screening and flotation.

The use of the micromanipulator depends on the nature of the sample (see p. 63).

Table 4.4. Liquids for gravity separation of solid mixtures (acc. to the literature)

Liquid	Density (g/cm^3)	Dilutant
Bromoform	2.88	Methanol, benzene, xylene, toluene
Tetrabromoethane	2.95	Benzene, xylene, toluene
Thoulet's solution (aqueous solution of KI and HgI_2)	3.19	Water
Diiodomethane	3.33	Benzene, xylene, tetra-bromoethane
Duboin's solution (aqueous solution of NaI and HgI_2)	3.46	Water
Rohrbach's solution (aqueous solution of BaI_2 and HgI_2)	3.58	Water
Clerici's solution (aqueous solution of thallium malonate and thallium formiate)	4.2	Water
Thallium malonate	4.5	-

Dry picking of dispersed grains (Fig. 4.3 a and b, see Annex)

Disperse the grain mixture on the specimen slide with a highly volatile liquid that does not affect the grains (e.g., methanol). Small grains can be collected without an adhesive. They firmly adhere to the fine glass or steel needle point by electrostatics. X-ray examinations require to fix the grain with a droplet of adhesive; otherwise they will drop off during exposure. Suitable adhesives are epoxy resin, zapon lacquer, polystyrene-xylene and other slow-curing glues, as well as greases. If the isolated grain is required after roentgenography, use an easily soluble adhesive. To isolate smaller grains grown on bigger ones, fix the specimen to the slide with a thin strip of Scotch tape.

Picking of grains from a liquid or paste

If a grain of interest is not conspicuous by its morphological properties, picking without an immersion medium is difficult. An embedding liquid is required in order to optically enhance the grains of interest. In such work using no cover glass, the embedding medium should not have excessive vapour pressure. Highly volatile embedding media require a humidity chamber of the type used in chemical ultramicroanalysis.

Picking is made by shifting the grain to the periphery of the embedding liquid drop using hooked glass tools, or by micropipette sucking on the micromanipulator (Table 4.5). Ramsay grease is a suitable adhesive to stick a grain to a glass point in liquids.

The picking of grains from pasty specimens constitutes a border case. Apply a

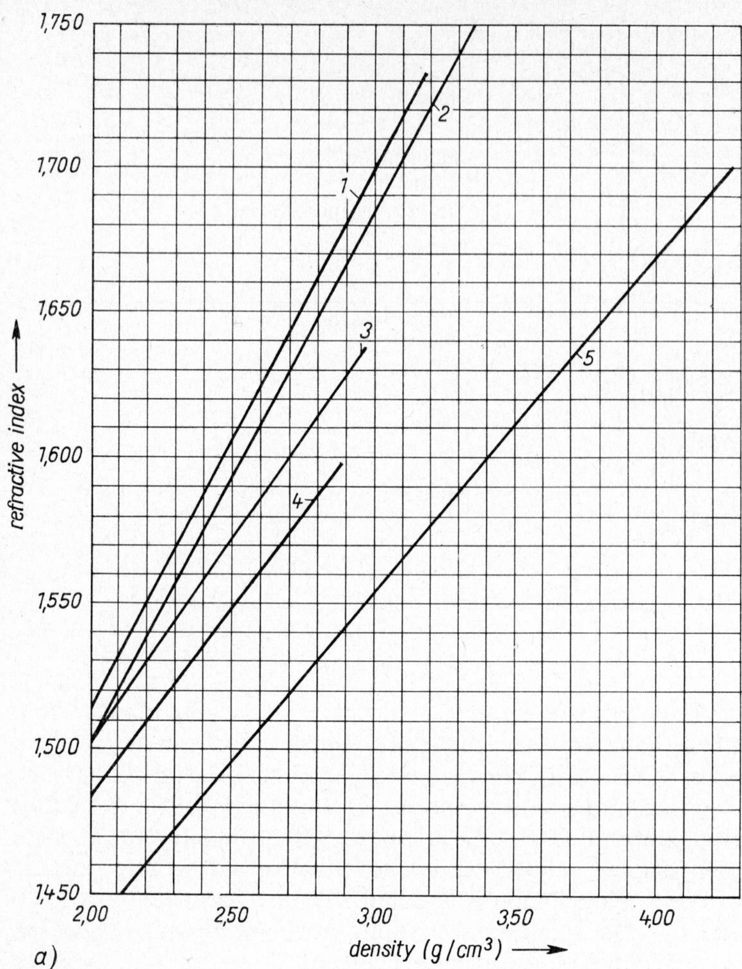

Figure 4.2

Table 4.5. Techniques of collecting selected grains of various sizes isolated from liquids with the micromanipulator

Collection technique	Size of grain (μm)
Grain pushed to the edge with a hook and picked up with micropincers	Coarse grains (> 100)
Suction by capillary action	Medium grains (50 to 100)
Sticking to steel or glass needle points using Ramsay grease	Small grains (< 50)

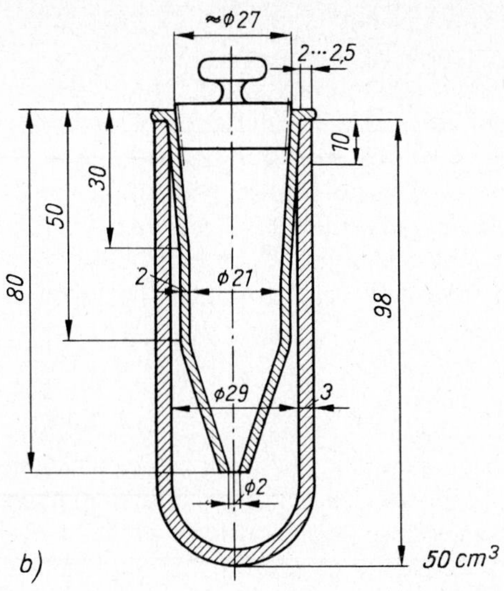

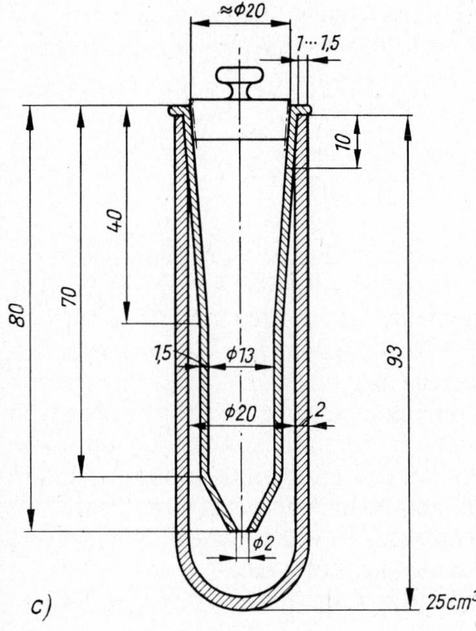

Figure 4.2. Phase separation by densities

a) Density vs. refractive index of some separation liquids

1 Thoulet's solution 2 Rohrbach's solution 3 Tetrabromoethane 4 Bromoform 5 Clerici's solution

b) Centrifugation tube for 50 cm^3 (dimensions in mm)

c) Centrifugation tube for 25 cm^3 (dimensions in mm)

thin smear of paste on to the specimen slide; isolate grains in a moist chamber using hooked tools, to avoid drying of the paste during work.

Separate phases not interesting by selective dissolution and etching, carefully observing the quantitative behaviour (cf. Table 6.33). Precipitation of secondary formations and the change of the phases of interest may lead to errors.

Separation and enrichment in phase mixtures can also be performed by differences in thermal stability, in reaction behaviour under gas flow, and in volatilization by sublimation. Employment of physical methods reduces the risk of interesting phases being changed during treatment, but also reduces the degree of selectivity.

Mixing, homogenization

Homogenization of the sample is a prerequisite for obtaining a representative specimen. Difficulties are involved in homogenizing mixtures of acicular and isometric grains and moist or hygroscopic substances; mixing is difficult if grain sizes are very small (due to tendency to agglomeration) or if the phases to be mixed differ greatly in density and grain size. Mixing of a deposit in a liquid phase is frequently more successful than dry mixing. Of the many types of equipment developed, the inclined rotary mixer and the column homogenizer [4.10] are particularly appropriate.

Sample division

The representative portion required can be divided most expediently with a fluted divider (Fig. 4.4, see Annex). For fine-grained and low-volume specimens, the dropping height can be reduced and the fluted part lined up. Quartering and other provisional fast methods result in distributions that are statistical to some approximation only.

4.2.2 Substances of limited stability

For such substances, the general sequence of sample pre-preparation varies depending on the nature of the stability restrictions [4.10]. Table 4.6 gives practical hints on how to treat those specimens. In many cases, a preparation box will be a useful auxiliary. The various models [4.24 to 4.26] correspond to two basic versions. Round chambers are easier to evacuate than rectangular cross sections and permit more exact vacuum setting. The manipulations required can be done with rubber gloves or mechanical manipulators. Important features of a preparation box include heat insulation, facility for precise temperature control, built-in observation microscope with micromanipulator, airlocks, and space for miniature specimen pre-preparation (mortars, sieves, pipettes). When using aggressive gas, make sure that materials are corrosion-proof.

Transferring the sample to the box and from there to the microscope can be

Table 4.6. Pre-preparation of specimens of limited stability

Operation	Limited stability in dry state	Stable in the formation system only	Stable at high or low temperatures only	
			Formed on hot or cool stage	Transferred at system temperature
General handling	*Hygroscopic substances:* in the box with dried inert gas *Decomposing substances:* in the cooled box with suitable inert gas *Very hygroscopic or easily decomposing substances:* handling under mother solution	In the box with suitable atmosphere, under mother solution	Handling with micromanipulator or micropipette possible to a limited extent Formation temperature precisely adjustable	Transfer to the pre-heated or pre-cooled specimen stage possible unless substance is affected by operational fluctuation of system conditions
Secondary formations	Grain decomposition, stating at surface	By *system change* during *pre-preparation:* new formation, dissolution or decomposition of grains may occur		By *system change* during *transfer of specimens:* new formation, dissolution or decomposition may occur
Comminution and dispersion of aggregates	Frequently not possible completely	Best to be done under mother solution	Possible to a limited extent only	
Grain size fractionation	Errors by aggregates not comminuted, especially in the fine fraction	Best to be done by wet screening		Not possible
Phase separation	Difficulties due to cohesion	Limited, by collection	Limited possibility by temperature change or selective dissolution	
Mixing, homogenization	Impeded by cohesion	Well homogenizable, mind gravity separation		Not possible
Specimen portioning	Not statistical due to insufficient homogenization	By quartering or with special specimen portioner		Not possible

done in a temperature-controlled transport container. A complex preparation box-cum-microscope unit is even more favourable (cf 7.1.4).

Toxic or radioactive substances are special cases for pre-preparation and preparation in the box.

4.3 Preparation of specimens

4.3.1 Static microscopy

Static microscopy is concerned with objects that do not change during examination. To allow registration, these objects have to be exposed to the incidence or transmission of an aimed beam of light by means of suitable preparation techniques. To avoid disturbing optical effects, transmitted light is made to pass vertically through specimen slide, cover glass and embedding medium.

The specimens occurring most frequently in chemistry are loose grains and microtome sections. In addition, a number of special techniques are employed whose development and standardization have not yet been completed (Tables 4.7 to 4.9 [4.27 to 4.44]).

Loose grains

Individual grains isolated from each other are embedded in a liquid or solidified embedding (immersing) medium between specimen slide and cover glass. Simple glass slides of 1 mm thickness will do for most examinations. Since glass absorbs much of the light between 200 and 360 nm, examinations in that wavelength range require quartz slides. High-temperature transmitted-light microscopy employs fused silica or corundum discs.

Specimen slides with a ground-in cavity are of advantage where the constituents of a phase mixture vary greatly in density. In a suitable dense liquid (mixtures see Table 4.4), the heavier phases sink and the less heavier ones rise; they can be examined subsequently by focussing separately on to either plane.

Optimum cover glass thickness is, as a rule, between 0.15 and 0.17 mm. Objectives are corrected to match these thicknesses [4.29]. Deviations will substantially deteriorate image quality, especially in case of high N.A.

The presence of a mounting medium prevents the appearance of disturbing dark margins of total reflection at the grains. These margins may make examination entirely impossible, especially with small grain sizes. The difference of refractive indices between grain and embedding medium should be 0.1 to 0.2 to ensure contrasty images. For optically attenuating a phase one would require an embedding medium of a refractive index equal to that of the object.

The mounting medium must neither react with the substance (dissolution, decomposition, incorporation into the lattice) nor crystallize. Incorporation into the

Table 4.7. Preparation methods in static microscopy

Type of specimen	Transmitted light	Reflected light	Single grain	Aggregate	Object durable in dry state	Object hygroscopic or easily decomposed	Object temperature	
							High	Low
Loose grains	×	×	×	(×)	×	×	×	×
Microtomed section	×	×	(×)	×	×	–	–	×
Etched microtomed section	×	×	–	×	×	–	–	×
Thin section	×	–	×	×	×	(×)	(×)	×
Polished section	–	×	×	×	×	(×)	(×)	×
Etched polished section	–	×	(×)	×	×	–	–	×
Polished thin section	×	×	×	×	×	(×)	–	–
Membrane-filtered specimen	×	–	×	–	×	–	–	(×)
Adhesive plate specimen	(×)	×	×	–	×	–	–	×
Paste smear, film specimen	×	×	–	×	×	×	–	×
Replica	–	×	×	–	×	(×)	–	×
Semi-embedded fibre specimen	×	×	×	–	×	–	–	×
Evaporated specimen	×	×	×	–	×	–	–	–

Table 4.8. Special preparations and their applicabilities

Type of specimen	Applicability	Preparation	Reference
Membrane-filter specimen	Examination of solid particles dispersed in gases or liquids	Sucking the gas or liquid through the membrane filter. Adding a suitable immersion liquid makes the filter optically transparent for direct microscopical examination.	[4.35]
Adhesive plate specimen	Solid particles and liquid drops dispersed in gases	Particle sedimentation on glass plates covered with adhesive substance (petroleum jelly, glycerol etc.), removal of particles or examination directly on the plate.	[4.10, 4.36]
Paste smear, film specimen	Examination of ointments and pasty specimens that must not be dried	A thin film is smeared on to the specimen slide and examined without cover glass.	[4.10, 4.37]

Table 4.8, continued

Type of specimen	Applicability	Preparation	Reference
Replica	Close examination of surface details, morphology of easily decomposing substances (more frequently used in electron microscopy)	Grains are pressed into the viscous replica medium (e.g., polystyrene) and dissolved after replica medium has solidified. The negatives are vacuum-deposited (e.g., with germanium) and the replica medium dissolved (e.g. with ethylene dichloride). The positive replicas remain for examination.	[4.38] [4.38]
Semi-embedded fibre specimen	Fast examination of fibres	Fibres are stuck to specimen slide by means of transparent adhesive tape.	[4.39]
Evaporation specimen	Grain size and grain shape analysis	Dispersion of grains on specimen slide, evaporation of a thin metallic layer, removal of grains. Grain silhouettes will appear bright against black background.	

lattice changes the optical data of the test substance, giving rise to measuring errors, as can be proved by parallel tests using other media.

Mixtures of immersing media, made to continuously vary the refractive index, should have a consistent, low vapour pressure; otherwise, higher volatilization of one fraction may change the composition and, thus, the refractive index. Section 6.1.1 contains a list of suitable embedding media.

Loose grain specimens do not yield information on aggregate intergrowth conditions, nor on the pore volume or the internal structure of grains. Incrustation of grains, e.g. because of surface decomposition, is disturbing but can be obviated by crushing the grains.

Decomposing, aggressive or hygroscopic substances and high- and low-temperature samples can only be prepared as loose grain specimens. Sodium-dried nujol, silicone oils of various viscosities and polystyrene-xylene mixtures are frequently used as embedding media. Transparent ointments and greases also have good embedding properties. Silicon oil is applicable up to 360 °C. Still higher temperatures up to 1300 °C require either low-melting glasses or salt melts as embedding media. This involves the considerable hazard of a reaction between sample, specimen slide and immersion medium. Generally, high temperature work is done without cover glasses. Immersing media for low-temperature microscopy are silicone oils, alcohols, toluene, xylene etc.

Table 4.9. Application of microscopic specimen types in various fields of chemistry

Field	Loose grains	Thin section	Polished section	Microtome section	Microcell
Basic and intermediate inorganic products	×××	×	×	–	×××
Basic and intermediate organic products	×××	–	–	–	×××
Agrochemistry	×××	×	–	–	×××
Silicate chemistry	×××	×××	××	–	×
Catalysis	×	×××	×××	–	–
Organic high polymers	×	×	×	×××	×
Photochemistry	–	–	–	×××	–
Pharmaceutical chemistry	×××	×	–	–	×××
Study of raw materials and recycled materials	×××	××	××	–	×

× rarely used (special cases)
×× frequently used
××× generally used

Loose grain specimens to be examined in dark-field reflected light require neither cover glass nor embedding medium. The grains are suspended in methanol and dispersed on the specimen slide, after which the liquid evaporates. Particles are well distributed that way and can be examined immediately without further treatment. Should highly volatile liquids be used for immersing, the cover glass must be provided with a coat of some adhesive along the margin, with a small gap left free through which the immersing liquid enters by capillary action. A slight pressure on the cover glass removes air bubbles (if any).

Another version of the loose grain specimen is known as semi-immersed specimen. A thin film of adhesive holds the grains to the specimen slide; they must not sink into the adhesive, so as to keep their sides free of it. Sideways, the grains are surrounded by a liquid immersion medium. It may be interchanged with the grains staying stuck to their substrate. Adhesives used are epoxy resin, nail lacquer, cover glass cement, polystyrene-xylene mixtures, greases, gelatin etc. Small glass strips are cemented to the sides of the cover glass in order to avoid its direct resting on the grains.

Microtome section

The microtome section may be regarded a universal preparation for many investigations of organic high polymers in reflected or transmitted light. As a disadvantage, cutting relieves stresses in the material, which thus cannot be measured in a section. On many other types of objects, the microtome section is important for special-purpose examinations (Table 4.10). Soft, elastic or pasty specimens can be prepared at low temperatures using the cryostat microtome.

Table 4.10. Microtome techniques for different objects (acc. to the literature)

Type of object	Section without embedding		Section of object embedded in	
	Dry	Freezed	High polymers	Paraffin
Textile fibres, fabrics	×	×	×	×
Moulded fibre board	×	–	–	–
Solid and hard-foam plastics	×	×	×	–
Foils, films	×	×	×	×
Lacquer and paint layers	×	–	×	×
Rubber, elastomers	–	×	–	–
Powders	–	×	×	–
Pastes	–	×	–	–
Industrial raw materials	×	×	×	–
Carbons	×	–	×	–
Metals	×	–	–	–

The *trimming* stand serves for cutting out certain parts of a specimen in three dimensions.

Metals and composite metals can be cut with diamond knives [4.30, 4.43]. The applicability of a microtome knife depends on its shape as well as on its material (Table 4.11).

To obtain a good-quality section, the knife must be applied at the proper angle (Fig. 4.5).

Cutting methods

(1) Direct cutting without embedding
(2) Cutting of embedded specimens

Embedding media must not disturb the specimen. Suitable mountants are methacrylates, cast polyester resins, polyamide derivatives, epoxy resins, paraffin

Table 4.11. Applications of various microtome knives

Type of knife	Application
Metal knives	
Broad, thin knives	Soft, elastic material; direct cutting without pre-treatment
Short, strong-backed knives	Good cutting properties for the majority of objects
Plane-type knives	Extremely hard and tough materials, especially for large-area cuts
Glass knives	Plastics, fibres, foils
Diamond knives	Metals, hard composite materials

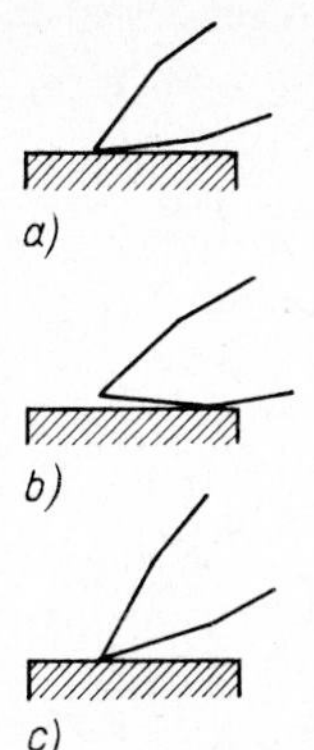

Figure 4.5. Microtomy
a) correct knife angle b) c) incorrect knife angles

and gelatin. For good, homogeneous sections, the specimen must be mounted on the microtome very securely. If direct clamping is not possible, the specimen is cemented to a supporting block by a fast-curing adhesive. Curling or fiberizing of dry-cut sections can be avoided by sticking a transparent adhesive foil to the surface created by the previous cut before cutting the next slice. For microscopy, the foil can be left on the specimen, whether this is embedded or not. Unembedded microtome sections are put on the specimen slide either dry or preferably with an embedding liquid (nujol, silicone, oil, water, glycerol). High-polymer sections may tend to warp if immersed in glycerol and to produce bubbles if embedded in canada balsam.

For fixation, the cover glass should be sealed around the edge with nail lacquer.

Section thicknesses are between 20 and 1μm on an average, or above or below that range for special purposes. As a rule, the actual thickness is greater than the theoretical value. The thinner a section is, the greater deviations will occur. For special problems, thin sections can be easily examined with reflected light, if they are placed on a mirror or similar highly reflective surface.

Permanent microtome section specimens of plastics can be prepared with glycerol gelatin or dammar; caedax, malinol or styrax should be used for plastics easily attacked by solvents [4.31, 4.32].

Etched microtome section

Minor differences in the specimen structure may not be clearly visible on the cut surface unless it is etched. Sections of *PVC* materials (including embedded *PVC* powders) appear structureless, while showing distinct features after treatment with solvents.

Ground thin sections

A cut aggregate plate ground to a thickness between 30 and 10 μm clearly reveals its internal structure and intergrowth conditions under the microscope, provided that at least part of the aggregate constituents are transparent at that

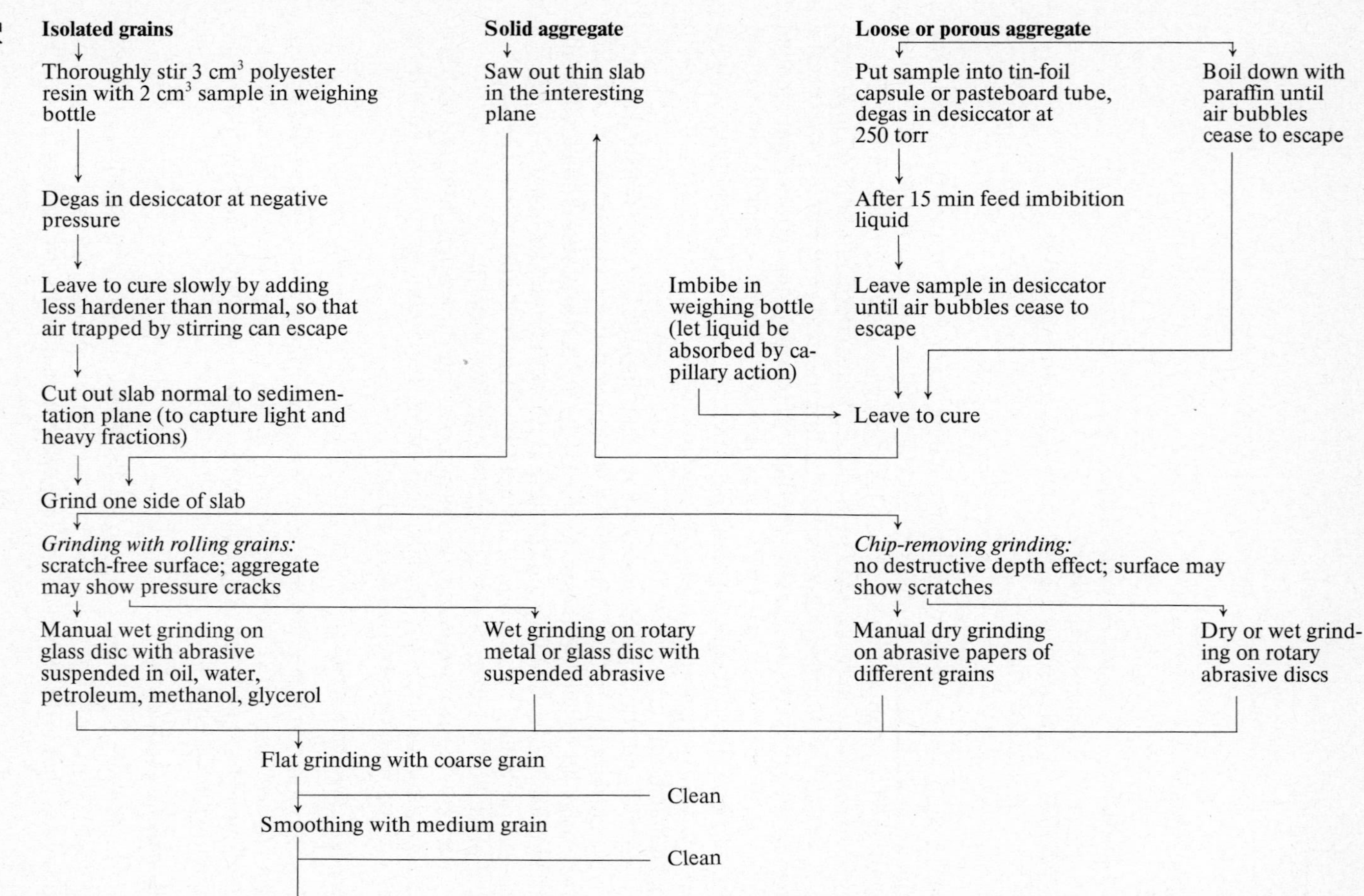
Isolated grains
Thoroughly stir 3 cm³ polyester resin with 2 cm³ sample in weighing bottle
Degas in desiccator at negative pressure
Leave to cure slowly by adding less hardener than normal, so that air trapped by stirring can escape
Cut out slab normal to sedimentation plane (to capture light and heavy fractions)
Solid aggregate
Saw out thin slab in the interesting plane
Loose or porous aggregate
Put sample into tin-foil capsule or pasteboard tube, degas in desiccator at 250 torr
Boil down with paraffin until air bubbles cease to escape
After 15 min feed imbibition liquid
Imbibe in weighing bottle (let liquid be absorbed by capillary action)
Leave sample in desiccator until air bubbles cease to escape
Leave to cure
Grind one side of slab
Grinding with rolling grains: scratch-free surface; aggregate may show pressure cracks
Chip-removing grinding: no destructive depth effect; surface may show scratches
Manual wet grinding on glass disc with abrasive suspended in oil, water, petroleum, methanol, glycerol
Wet grinding on rotary metal or glass disc with suspended abrasive
Manual dry grinding on abrasive papers of different grains
Dry or wet grinding on rotary abrasive discs
Flat grinding with coarse grain
Clean
Smoothing with medium grain
Clean

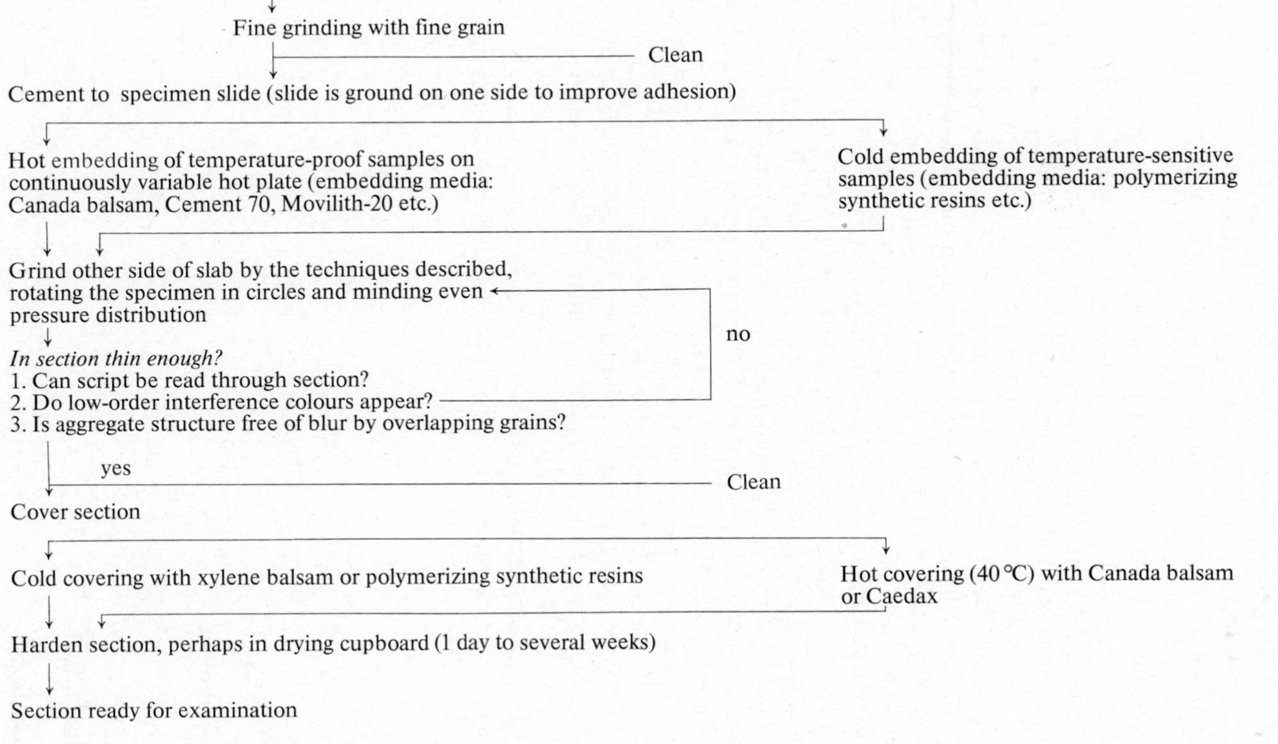

Chart 4.2. Preparation of thin sections

thickness. (Many aggregates of substances opaque to visible light can be investigated by transmission of infrared radiation.) If several sections cut along different planes normal or parallel to each other are studied in combination, the two-dimensional micro-images provide spatial information as well.

The preparation of thin sections (Charts 4.2, 4.27, 4.33) used to be relatively laborious, but is today facilitated by semi-automatic equipment (Fig. 4.6 a–d, see Annex).

Ground thin sections are very important for studying catalysts, solidified melts, sintered products, moulded granulates etc. Fine-grained aggregates must be ground to a thickness of 5 μm. Reliefless fine-grained sections must be ground with great care, especially if hardness differs greatly either within the aggregate or compared to the embedding substance. Grains will easily break away from the matrix or retain their rounded form as the softer material surrounding them is being ground off. The resulting relief shows dark fringes around the grains due to light reflection from the rounded edges, which makes exact grain size measurement difficult.

Porous or loose aggregates have to be solidified prior to thin grinding. This is effected by impregnation with a hardening agent [4.27, 4.33, 4.34, 4.44]. The specimen is first degassed in a desiccator at about 250 torr, after which the impregnating liquid is allowed to flow out of a dropping funnel into the specimen container, impregnating it from the bottom. If the funnel outlet tube protruding into the desiccator is offset at the lower end, it may be rotated for impregnating several specimens arranged side by side (Fig. 4.7). The specimens are left in the desiccator until no more air bubbles escape, which is after about 15 min. Care must be taken that volatile fractions of the impregnation medium are not exhausted along with specimen gas. Coloured or fluorescent impregnation media make specimen pores appear in rich contrast.

From the aggregate thus prepared, a slice of 2 or 3 mm thickness can be cut by means of a rock saw having a speed of 35 to 45 revolutions per second and cooled

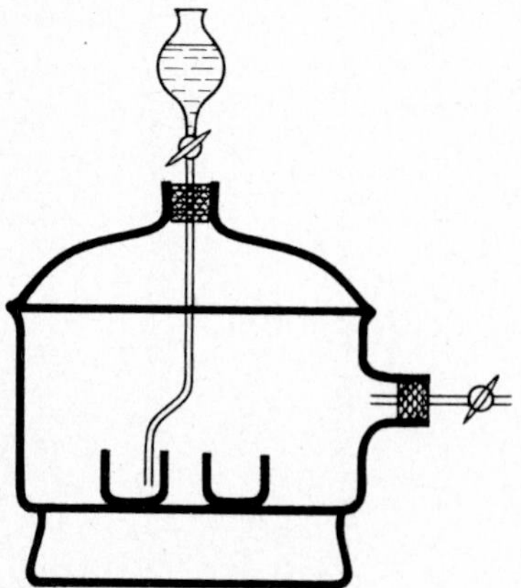

Figure 4.7. Hardening of loose aggregates by imbibition in the desiccator

Table 4.12. Refractive indices of some common embedding media for thin sections

Embedding medium (trade name)	Refractive index
Epoxy resins	1.47
Mowilith-20 (polyvinylacetate)	1.473
Pleximon 808 (ester of methacrylic acid)	1.50
Kollolith	1.53
Canada balsam	1.53–1.55
Cement-70	1.54–1.55
Caedax	1.55
Polyester, hard, type G (Schkopau)	1.55
Vestopol-H	1.56
Phthalate, type G (Schkopau)	1.569–1.572
Polystyrene BW, pellucid	1.588–1.589
Aroclor	1.64–1.67
Aroclor, sulphurized (up to 13%)	1.69
Hyrax	1.71

with petroleum (Fig. 4.6, Annex). Table 4.3 lists cutting discs for different specimen properties. The general rule is, *soft discs for hard specimens, and hard discs for soft specimens.*

The cement used for mounting the specimen on the slide and sealing the cover glass should have a refractive index that is either sufficiently different from that of the grains so as to allow rich-contrast presentation, or permits relative refractivity measurement (cf. 6.1.1). The cement must not attack the specimen (dissolution, decomposition) nor crystallize; it must be optically isotropic and colourless. Table 4.12 gives a survey of cements in present use.

Thin section grinding by several steps using abrasives of decreasing grain requires that the specimen be thoroughly cleaned between any two steps to prevent the coarser grains from entering the next finer fraction. Recommended cleaning agents are alcohol, carbon tetrachloride, xylene and toluene. Water-soluble salt aggregates are ground with oil, xylene or carbon tetrachloride suspensions. In that case it is advisable to use a dry box. Salt specimens may be ground by the following methods:
Flat grinding: abrasive paper, grain size 150
Coarse grinding: corundum of grain size 180, suspended in methanol
Fine grinding: abrasive paper, grain size 400/500

Recipes for impregnants to solidify loose or porous matter

(1) Polyester resin G (Schkopau)

Component	*Parts by weight*
Polyester G (Schkopau)	100
Monostyrene	10
Cyclohexanone peroxide paste	2
Cobalt accelerator	1

Curing at room temperature in a vacuum takes 1 to 2 days.

(2) Epoxy resins

Proved mixtures contain:

Component	*Parts by weight*
Epilox EGK 19 with hardener	3 to 4
Xylene (carbon tetrachloride, trichloroethane)	1
Epilox EGK 19 with hardener	5 to 10
Acetone	1
Epilox M 106 with hardener	4 to 5
Acetone	1
Epilox M 54 with hardener	1
Carbon tetrachloride	1

Epilox may also be added after the specimen has been impregnated with the solvent.

The addition of the diluents retards the curing process. The most favourable curing temperatures are between 20 and 70 °C. The latter temperature must not be set until 1 or 2 days after impregnation, otherwise the mixture will boil. Carbon tetrachloride and trichloroethylene cure at the lowest temperatures. Curing is greatly affected even by a slight water content of the specimen.

Polished sections

Polished sections are plane, preferably highly polished specimen surfaces made by cutting, sawing, grinding, etching or electrolytic treatment. Their examination provides information on constituent phases, distribution and intergrowth in the respective plane. Polished sections are particularly suited in studying solid aggregates that consist wholly or partly of opaque matter (coal, catalysts, ores, metals). In special cases, polished sections are used in examining transparent substances (high polymers).

Quantitative measurement of optical data is only possible for surfaces polished to mirror finish, and therefore restricted essentially to metals. For the majority of other specimens, polished sections only permit observation and measurement of microscope images themselves.

Polished sections of easily decomposing, hygroscopic, toxic or radioactive substances have to be made in suitable preparation boxes (cf. 4.2.2).

Charts 4.3 and 4.4 outline the sequence of operations in making polished sections [4.27, 4.33].

The specimens are best mounted in rings made by cutting lengths of 10 mm aluminium or plastics tubes of about 40 mm diameter. Epilox EGK 19 (10 parts) with hardener (1 part) is one suitable mounting cement. The plate below the rings should be greased or made of flexible *PVC* to allow easy removal after curing.

Abrasive grains must be rolling for rough grinding, fixed for fine grinding and polishing. If hard and soft phases occur side by side in a specimen, reliefless surfaces can be obtained if soft metal (lead-antimony alloys), polyethylene or wet basswood discs are used rather than such covered by polishing cloth. Generally, a separate disc is employed for each different grain size of the grinding and poli-

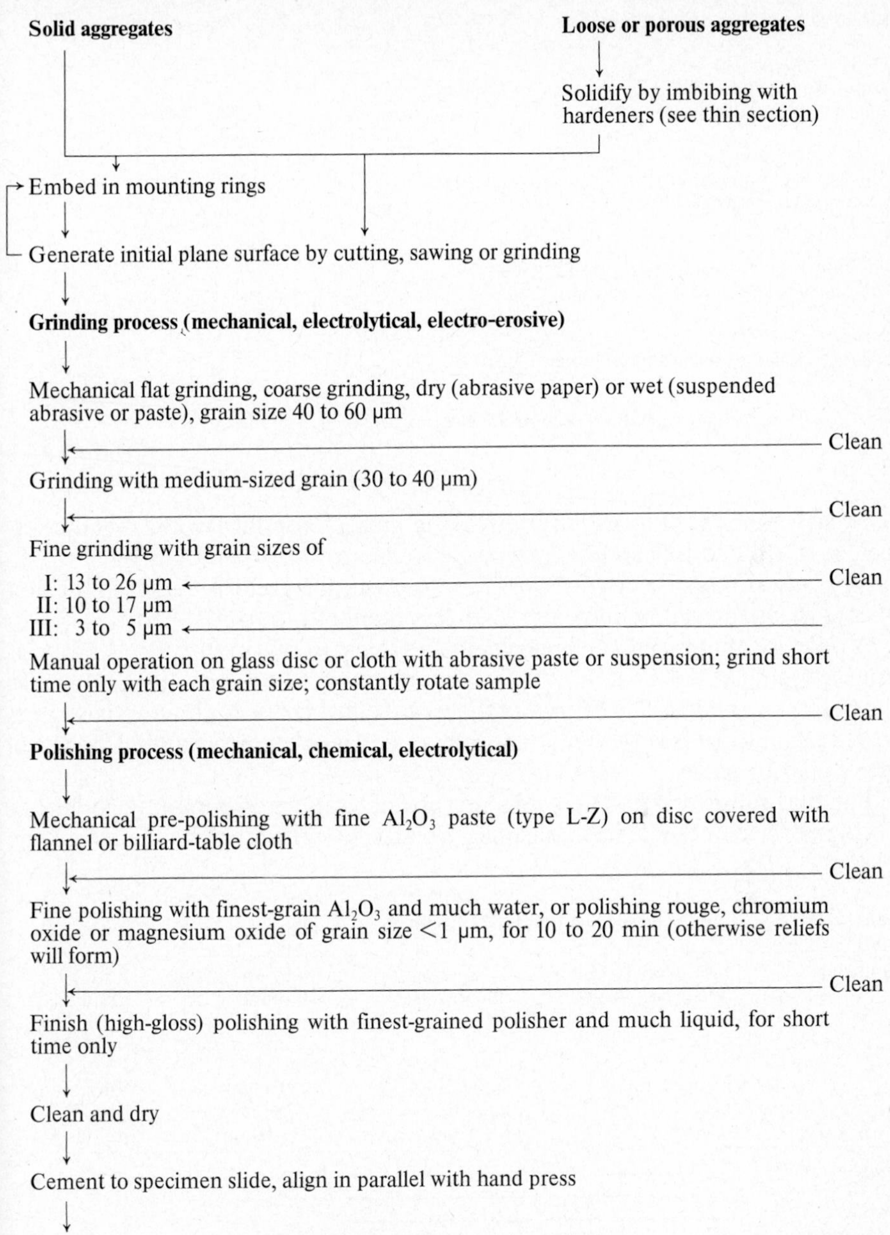

Chart 4.3. Preparation of polished sections

Cut out a smooth-faced sample

↓

Grind (wet method) with silicon carbide abrasive papers of decreasing grain (coarse grained to fine grained)

↓

Rough-polish on polishing disc (polishing leather) with diamond paste of 3 μm grain

↓

Fine-polish on polishing disc (polishing leather) with diamond paste of 0.25 μm grain

↓

Etch with reagents according to tables 6.36, 6.37

Chart 4.4. Preparation of polished sections of high polymers acc. to [7.227]

shing agents. A carrying over of grains to the subsequent stage would mar the surface; it is avoided by careful cleaning between any two operations by blowing, rinsing, ultrasonic or electrolytic treatment. Also, the grain size used in the first operation should not be too coarse, i.e. less than 0.01 mm.

A mixture of three parts of petroleum and one part of paraffin oil will be a versatile grinding liquid. Pastes generally have better grinding and polishing properties than loose grains. The fat base of the pastes effects better gliding and prevents surface cracks. Loose grains may easily by mixed with glycerol or ethylene glycol into a viscous paste.

For final polishing, however, polishing powder is more favourable (Table 4.13).

Compared to mechanical polishing, electrolytic polishing has the advantage

Table 4.13. Applicability of various grinding and polishing agents

Abrasive	Finest polishing, finish polishing	Fine polishing, fine grinding	Pre-polishing, rough polishing, fine grinding	Rough grinding
	Roughness height 0.1 μm	Roughness height 1–0.1 μm	Roughness height 1–10 μm	Roughness height 10–100 μm
MgO	———	———		
Fe_2O_3	———	———	———	
Cr_2O_3	———	———		
Corundum (synthetic)		———	———	———
Al_2O_3	———	———	———	———
SiO_2			———	———
B_4C			———	———
Diamond	———	———	———	———

that the surface is not subject to any pressure. The specimen forming the anode is submerged together with a cathode in a constantly circulating electrolyte.

Polished and etched sections

In various cases, structural details of polished sections become distinctly visible only after etching (Table 6.35). Apart from chemical etching using various solvents, cathodic etching has importance for metal examinations. Atoms are removed from the surface by bombarding it with positive ions of high energy.

Polished thin sections

A polished thin section is a combination of thin section and polished section. Employed for aggregates containing opaque as well as transparent grains, it permits examination of the opaque and strongly coloured parts by reflected light and the transparent and weakly coloured grains by transmitted light.

To make polished thin sections, the following operations are necessary:
(1) Make a polished section.
(2) Cement it to a specimen slide.
(3) Grind it to the normal thickness of thin sections (20 to 30 μm).
(4) Re-cement it to another specimen slide with the polished section facing up.
(5) Clean and repolish the surface.

4.3.2 Kinematic microscopy

Kinematic microscopy observes the changes of microscopic objects during physical and chemical processes. In most cases, it requires specimen preparation methods different from those of static microscopy. Above all, it is necessary to increase the distance between cover glass and specimen slide so as to form a specimen chamber that allows unhindered progress of the process observed.

For transmitted-light examination the specimen chamber is flat and only as deep as the useful transmission depth of the specimen allows. Specimen depth is limited by the requirement that the microscopical image must still be resolvable into its two-dimensional features without being disturbed by grains lying above (behind) each other. On the other hand, the chamber must be deep enough to exclude any influence of the chamber walls on the specimen (retardation of object movement or concentration equalization, checking of flows, pressure on the object, influence of interfacial forces). Between the two requirements, optimum chamber depths are specific for each particular magnification and must absolutely be observed in kinematic transmitted-light microscopy. These depths vary between one and ten millimetres. The shallower the chamber is, the higher is the object magnification attainable. Deep chambers, if placed on a microscope stage, require special, long focus condensers.

Specimen chambers for reflected-light work may be of any size, although the practical depth range is limited by the working distance provided by the objective. That range varies between a few centimetres and one millimetre, measured inwards from the specimen surface.

Specimen chambers must allow the process of interest to proceed without hindrance and to be observed. This requires the design features listed in Table 4.14.

The walls of specimen chambers with very small volumes must be made hydrophobic with a silicone oil emulsion, such as Aquaphob 2 (made by VEB Chemiewerk Nünchritz). Contrary to working instructions given elsewhere, the effect sets in immediately after the solvent has vaporized, without heat treatment being necessary.

Based on the general design principles outlined above, quite a variety of specimen chamber configurations for transmitted and reflected light are available (Table 4.15) [7.8, 7.64 to 7.67]. They all serve specific purposes; their general use as standard models is restricted. Systematic method studies have hardly been made so far.

Specimen chambers in use are designed for either horizontal or vertical ray paths. Illumination techniques with transmitted and reflected light include bright field, dark field and oblique light. Accessories permit the use of polarizing, phase and interference microscopy. Illumination possibilities are improved by relative rotation between specimen chamber and microscope, shifting the chamber normal to the ray path, or incorporation of prisms and mirrors for simultaneous examination of top, bottom and sides of the specimen. If the ray path is horizontal or if there is a flow in the chamber, the grain must be cemented to the wall or to the point of a thin specimen carrier, unless a turbulent current holds the grains dispersed in the liquid.

Specimen slide chambers and simple cells

Chambers made from specimen slides or configured as simple specimen cells permit simple kinematic examinations of single- or multiple-grain spec-

Table 4.14. Features of various types of specimen cells

Feature	Specimen slide chamber, simple cell	Flow-through cell	Micro-reaction cell
Three-dimensional specimen cavity	+	+	+
Controlled flow rate and thermostatting	–	+	+
Controlled heating and cooling	+	+	+
Thermometry	(+)	+	+
Optical path normal to glass walls	+	+	+
Micromanipulation	(+)	(+)	+
Microstirrer	(+)	(+)	+

Table 4.15. Preparation methods in kinematic microscopy

Type of specimen	Single grain	Aggregate	Object durable in dry state	Object hygroscopic or easily decomposed	Object temperature	
					High	Low
Specimen slide chambers, simple cells						
Uncovered slide	(×)	×	×	×	×	×
Raised cover glass	×	×	×	×	×	×
Ring-mounted specimen	×	×	×	×	×	×
Variable chamber	×	×	×	×	–	–
Suspended drop specimen	×	–	×	×	–	–
Specimen slide with ground-in cavity	×	×	×	×	×	×
Counting chamber cell	×	×	×	×	×	×
Microcone	×	–	×	×	–	–
Moist chamber	×	–	×	×	(×)	–
Flow-through cell	×	×	×	×	×	×
Microreaction cell	×	×	×	×	×	×

imens. Though of uncomplicated design (Fig. 4.8, Table 4.16), they have all essential features of a specimen chamber (Table 4.14). Ray paths through them are horizontal or vertical. They mainly serve for transmitted-light microscopy.

Microscopy at high or low temperatures requires a special way of preparation. In accordance with the design features of hot and cold stages (cf. 3.4), the uncovered specimen slide has proved a standard form for reflected and transmitted light, and the raised cover glass (ring specimen) for transmitted light.

A general problem of transmitted-light microscopy at high and low temperatures is presented by the embedding medium required to avoid the total-reflection fringes appearing around the grains in air. Suitable embedding liquids for the +20 to −100 °C range are easy to find (organic solvents, or the liquid phase of the system examined). In the lowest temperature range, freezing of the liquid phase about the grains also provides for an »embedding medium« whose refractive index is fairly usable.

In the 20 to 360 °C interval, temperature-resistant silicone oils can be used. If there is a possibility for a grain to be displaced by degassing due to decomposition, it should be fixed to the substrate by a thin film of high-temperature grease. At still higher temperatures, grains may be embedded in various salt and glass melts, such as $NaPO_3$ with an addition of 2% of NaF. It is always necessary, however, to make parallel examinations of large grains in air in order to eliminate side reactions between the specimen and the embedding medium. A more favourable way of studying solid aggregates at high or low temperatures is to use thin sections, where grain intergrowth replaces additional embedding media.

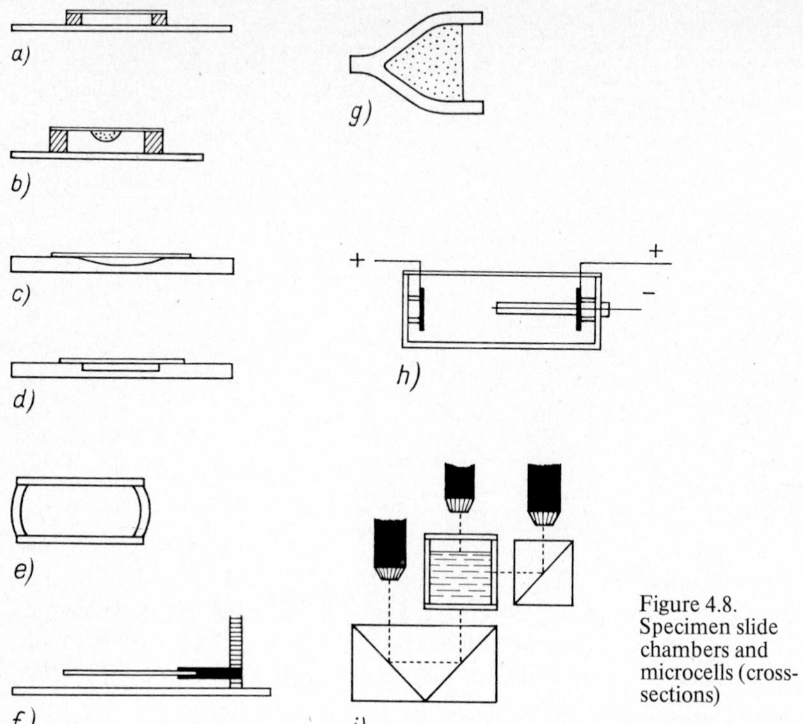

Figure 4.8. Specimen slide chambers and microcells (cross-sections)

a) raised cover glass, ring-mounted specimen b) hanging drop c) cavity slide (well slide) d) counting chamber cell e) flat capillary f) variable specimen slide chamber g) microcone h) microcell for electrochemical processes i) prism arrangement for simultaneous observation of processes in specimen cells from top, bottom and side k) humidity chamber (see Annex)

The effect of total-reflection fringes can be diminished also by selecting a suitable illumination technique (combined transmitted and reflected light).

Reflected-light examinations at varying temperatures can be carried out on polished sections and, above all, loose or briquetted grains.

A special form of specimen for variable-temperature microscopy is known as contact specimen (Chart 4.5). It permits the study of the changes of the solid and liquid phases in binary and ternary systems during melting and freezing on the hot and cold stage. Developed by Kofler [6.12] for the investigation of organic systems, it can also be employed for inorganic substances provided that the melts are not to viscous and form a distinct interface (cf. 6.4.4.2).

Specimen slides for high-temperature microscopy must be highly corrosion-proof. Corundum and fused quartz have become established for transmitted-light, as sheet platinum and special high-temperature materials have for reflected-light examination. Nevertheless, reactions with the substrate occur, e.g. in study-

Table 4.16. Design versions of specimen slide chambers and simple cells (cf. Fig. 4.8)

Description	Design principle	Application
Uncovered specimen slide	Insert in hot or cool stages, made of glass, fused quartz, corundum, platinum, special materials	Study of objects at high or low temperatures
Raised cover glass	Narrow glass supports at the cover glass edge confine the specimen space on the slide	
Ring-mounted specimen	Specimen space confined by a plastic ring between slide and cover glass	Qualitative studies of stationary liquid phase, lateral introduction of a micromanipulator is possible, thermostatting by placement on hot or cold stage
Variable chamber	A fixture adjustable in height holds cover glass at variable distance above slide	
Hanging-drop specimen	Drop of liquid suspended freely from the raised cover glass	For special purposes only (e.g., qualitative study of precipitation from liquid phase)
Cavity slide (well slide)	Usable with cover glass or without	Qualitative observation of crystallization, dissolution or decomposition of grains in liquids
Counting chamber cell	Exactly dimensioned cavity ground into specimen slide	Measurements requiring exact knowledge of liquid layer thickness in the filled chamber
Microcone	Conical space, open on one side, made by fusion-sealing and immediate drawing of capillaries (used in the humidity chamber)	Standard reaction vessel of chemical ultramicroanalysis; introduction of microstirrers, microelectrodes and micropipettes possible [6.191]
Humidity chamber	Transparent chamber with lateral opening for the micromanipulator; control of the gaseous phase in the chamber by evaporation of liquid entered by means of soaked wad of cottonwool	

ing salt melts. Such effects may in some cases be diminished by evaporation of a thin, transparent film of some heat-resistant material (as SiO) on to the specimen slide.

Flow-through cells [4.45 to 4.51]

Flow-through cells have a cross-section that ensures perpendicular light passage through both side walls (Fig. 4.9 a and b, Annex). They serve for observing grains fixed in the cell while a solution flows through, and for examining flowing suspensions. Reproducible conditions can be created by connection to a thermostat and installation of a temperature sensor.

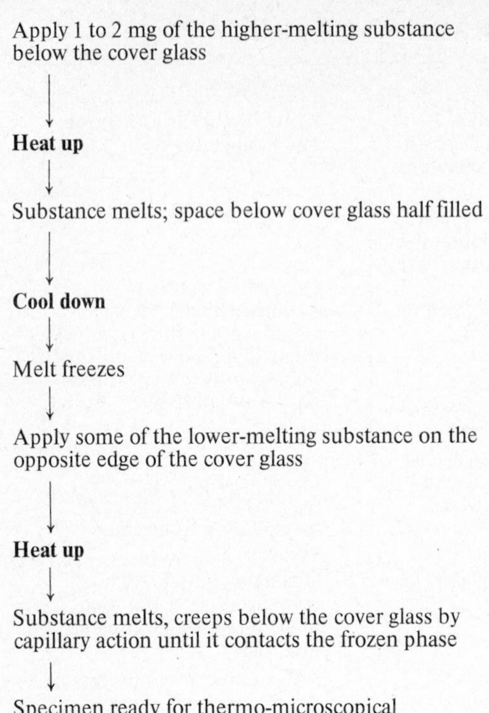

Chart 4.5. Preparation of a contact specimen for binary systems

The pump as well as the cross-sections of feed pipes and cell must be of adequate capacity to take up the expected flow rate of solid particles so as to avoid obstruction.

The cement used for joining the cell parts must meet very exacting demands. Epoxy resin, e.g., proved unfit for concentrated, flowing salt solutions. A seal not subject to attack is created by direct fusion between the glass parts of the cell.

Microreaction cells

Laboratory apparatus may be miniaturized and simplified to such an extent as to closely approximate microscopical conditions and to permit the investigation of physical or chemical processes under near-system conditions. By the subsequent variation of flow rate, stirring speed, concentration, temperature and grain size of the solid phases according to a complex test programme, information can be derived on the effect each variable has on the process. Two configurations are possible that differ by their degree of similitude to the comparison plant.

(1) Standardized microreaction cells

The similarity of these cells to the comparison plant is rather deformed as far as their component modules are concerned (stirrer, heating, cooling, valves), whereas a fairly close similarity of substance properties and process parameters is maintained. Standardized cells offer advantages for measuring the relationships between substance and process parameters, with equipment components remaining constant.

(2) Scaled-down plants

The plant approach does not, or not sufficiently, meet the requirements of microreaction cell design. Observation possibilities are limited (to reflected light and, in part only, transmitted light).

Microreaction cells made of 1 mm thick glass plates can be transilluminated vertically and horizontally (Fig. 4.9, Annex).

Temperature measurement is by thermistors. The flow inlet and outlet tubes lead to a thermostat, a feed or collecting tank, or a laboratory plant. Flow rate can be controlled by the pump or by hose clips, and measured by flowmeters. One or two rotary or vibration stirrers are employed, which must be of corrosion-proof material (e.g., plastics) if the liquids are aggressive, such as concentrated salt solutions. The stirrers may be arranged side by side or one above the other so as to correspond to the desired flow pattern, which can be modified by built-in additional obstacles. If the side walls are lined with a thin, transparent, conductive layer, the cell can be restistance-heated. The cell configuration must match the flow profile (Fig. 4.10).

Special apparatuses for investigating electrochemical processes, crystallization and dissolution, processes in textile chemistry and other special processes are described in the literature [4.45 to 4.51, 7.20, 7.64, 7.159].

Assessment of examination results

Sensible linear scale transformation down to the tiny dimensions of microreaction cells is, generally, impossible. Questions related to apparatus design cannot, therefore, be explored by the above methods, some special cases excepted.

Physical-chemical and process data may be obtained if a corresponding laboratory plant is used for comparison studies at the same time. In this case, the methods described supply significant information on the micropattern of the process of interest. Favourable factors are the easy creation of isothermal conditions, and the quickly reached concentration balance or thermodynamic equilibrium,

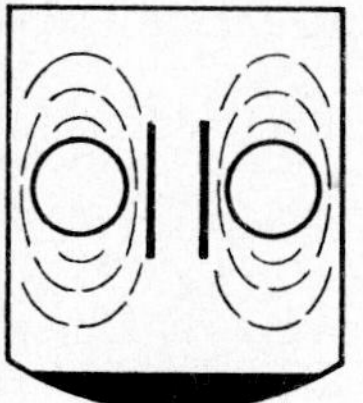
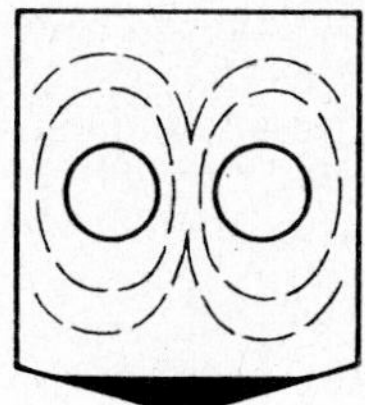
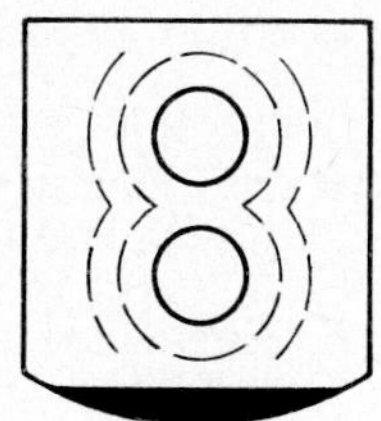
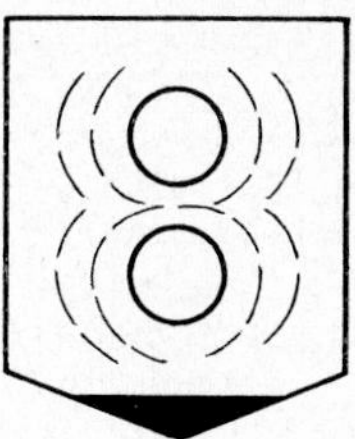

Figure 4.10. Microcells with stirrers for various flow profiles

respectively. Compared to this, large process plants have imperfect isothermal conditions, temperature gradients, concentration differences, slow or incomplete equilibrium setting, inhomogeneous distribution of reaction products, and additional design influences (occurrence of flow, turbulence, wall effects, »dead angles«, failure of diffusion influences etc.). In the interpretation of microscopical examination results, these special conditions must be considered just as much as a possibly inadequate ratio of particle diameter to cell dimensions.

References

[4.1] *Zettler, H.:* Moderne Methoden der anorganischen Analyse-Probenahme. Z. Erzbergbau Metallhüttenwes. 18 (1967) 165–171

[4.2] *Mulvany, J.:* Mikroprobenahme für chemische und instrumentelle Analysen, Lab. Pract. 15 (1966) 1409–1413

[4.3] *Gy, P.:* Probenahme fein- und feinstkörniger Erze. Freiberger Forschungshefte Reihe A, 415 (1967) 7–14

[4.4] *Hutschenreiter, W.:* Fehlerfortpflanzung bei der Probenahme. Bergakademie [Freiberg] 9 (1965) 537–542

[4.5] *Göll, G., and R. Helfricht:* Probenahme an körnigen Stoffen. Baustoffindustrie 15 (1972) 1, 23–28

[4.6] *Gehrke, C. W.,* et al.: Probenahme in Massendüngern. J. Assoc. off. analyt. Chemisto 50 (1967) 2, 382–392

[4.7] *Avy, A. P.,* et al.: Vergleich von Probenahmegeräten und -verfahren für Stäube. Staub-Reinhalt. Luft 27 (1967) 11, 469–480

[4.8] *Maddox, W. L.,* et al.: Automatischer Probenehmer zur Gewinnung bestimmter Proben der austretenden Substanz eines kontinuierlichen Analysators. Chem. Instruments 1 (1968) 1, 105–111

[4.9] *Diener, S., Kruhme, H., Rother, R., and K. Steinike:* Richtlinie zur Untersuchung von Körnerpräparaten. Wiss. Techn. Information des Zentr. geol. Instit. Berlin, Jg. 4 (1963) Sonderheft 3

[4.10] *Seyfarth, H.-H., and H.-H. Emons:* Probenvorbereitung und Präparation chemischer Objekte für die Durchlichtmikroskopie. Jenaer Rdsch. 4 (1970) 215–220

[4.11] *Sansoni, G.:* Zur Präparation von Aufbereitungsprodukten für lichtmikroskopische Untersuchungen. Z. angew. Geol. 14 (1968) 214–217

[4.12] *Wallace, S. R.:* Präparation von Mineralkörnern aus Dünnschliffen. Amer. Mineralogist 40 (1955) 927–931

[4.13] *Marshall, W. W., and W. E. Mayhead:* Herstellung von Analysenproben aus sehr hartem Material. J. Brit. Ceram. Soc. 5 (1968) 59–63

[4.14] *Eckhoff, R. K.:* A Rapid Method for Preparing a Random Distribution of Small Particles for Microscopy. Microscope 14 (1965) 11, 490

[4.15] *Brien, F. B.:* On Electrostatic and Magnetic Separation. Mining. Engng. 15 (1963) 105–106

[4.16] *Hofmann-Degen, K.:* Über ein neues Verfahren der Schlämmanalyse zur Untersuchung feinster pulverförmiger Stoffe. Zement-Kalk-Gips 5 (1960) 193–208

[4.17] Gross, W., and W. Haertel: Wie erhält man einwandfreie Ergebnisse von Schwimm- und Sinkanalysen feinster toniger Kohlenschlämme? Kohle und Erz 23 (1926) 343–352

[4.18] *Grunewald, V.:* Untersuchungen zur gravimetrischen Phasentrennung von Salzmineralien unter besonderer Berücksichtigung der Korngröße 0,1 mm. Bergakademie [Freiberg] 21 (1969) 2, 114–118
[4.19] *Vincent, H. C. G.:* Mineral Separation by an Electro-chemical Magnetic Method. Nature 167 (1951) 4261, 1074
[4.20] *D'ans, J., and R. Kühn:* Trennung von Salzgemengen der Kalisalzlagerstätten durch auswählende Auflösung und das Verhalten der Salzmineralien in einigen organischen mit Wasser mischbaren Flüssigkeiten. Berlin: Kali-Forschungsanst. für die Kaliindustrie 1945 (not published)
[4.21] *Svihla, G.:* Isolation of Radioactive Fallout Particles for Microscope Examination. Microscope 17 (1969) 1, 25
[4.22] *Otto, L.:* Der Mikromanipulator und seine Hilfsgeräte. Berlin: VEB Verlag Technik 1954
[4.23] *Hoppe, G.:* Ein pneumatisches Auslesegerät für kleine Partikel. Z. angew. Geol. 6 (1960) 515–516
[4.24] *Rumm, H., and R. Rumm:* Arbeiten mit und in Handschuh-Arbeitskästen (Glove-Boxes). GIT Fachz. Lab. 11 (1967) 6, 587–588
[4.25] *Tracey, M. W.: Lagerung feuchtigkeits- und luftempfindlicher Verbindungen. Chemist-Analyst 56 (1967) 1/2, 37*
[4.26] *Moeken, H.-H. P.,* et al.: Fernbedienungsanlage zur Analyse von Spaltproduktlösungen. Z. Radioanalyt. Chem. 1 (1968) 2, 113–112
[4.27] Sonderheft zur mikroskopischen Präparation. Wiss.-Techn. Information des Zentr. Geol. Instit. Berlin, Jg. 4 (1963) Sonderheft 10
[4.28] *Ludwig, G.:* Präparationstechnik für Aufbereitungsprodukte zur quantitativen erzmikroskopischen Untersuchung. Mineralium Deposita 1 (1966) 193–200
[4.29] *Bahrmann, E., and E. Urbach:* Zum Dickenoptimum der Deckgläser für Mikropräparate. Feingerätetechnik 13 (1964) 11, 502
[4.30] *Walter, F.:* Das Mikrotom. Leitfaden der Präparationstechnik und des Mikrotomschneidens. Ernst Leitz GmbH Wetzlar, Scharf's Druckerei 1961
[4.31] *Schrader, M., and U. Weber:* Über die Eignung verschiedener Einschlußmittel für Dauerpräparate von Kunststoffdünnschnitten. Mikroskopie 22 (1967) 7/8, 199–201
[4.32] *Nettelnstroth, K., Schmidt, G., and T. Loske:* Methoden zur leichteren und besseren Herstellung von textilen Mikrotomschnitten und Gießharzblöcken mit neuartigen Bettungsmitteln. Melliand Textilber. 41 (1960) 2, 133–135
[4.33] *Völker, H.:* Allgemeines über die Anfertigung von Dünnschliffen und Anschliffen. Z. f. Museumstechnik 13 (1967) 2, 155–169
[4.34] *Sinkankas, J.:* Hochdruck-Imprägnierung von porösen Materialien mit Epoxidharz für Dünnschnitt- und Mikrosondenanalyse. Amer. Mineralogist 53 (1968) 1/2, 339–342
[4.35] *Einbrodt, H. J., and K. H. Maier:* Staubmikroskopie mit Membranfiltern. Staub-Reinh. Luft (1954) 264–270
[4.36] *Kisser, J., and Z. Lehnert:* Die Herstellung von Abklatschpräparaten mit Hilfe der Kleband-Methode. Mikroskosmos 47 (1958) 165–168
[4.37] *Zeidler, W.:* Preparation of Thin Film Sections. Amer. Mineralogist 53 (1968) 9/10, 1773–1774
[4.38] *Bailey, G. W., and J. R. Ellis:* Particle Replication Techniques and Application. Microscope 14 (1965) 8, 306
[4.39] *Landgraf, R., and H. Keilhauer:* Ein neues Verfahren und Gerät zur rationellen Herstellung und sofortigen mikroskopischen Betrachtung von Faserstoffquerschnitten. Faserforsch. u. Textiltechn. 19 (1968) 2, 84–87
[4.40] BRD-Auslegeschrift 1 135 678 L 37 192 IXa/42h of 30. Aug. 1962, Kl. 42h, 15/03 *Poganski, S.:* Kühlbarer Objektträger für Lichtmikroskop

[4.41] *Owen, H. G.:* A Technique for the Preparation of Thin Sections for Thermal Treatment, together with their Remounting. Mineralog. Mag. 31 (1956) 232, 426–427
[4.42] *Peterfi, T.:* Die heizbare feuchte Kammer. Z. wiss. Mikroskopie (1927) 296–308
[4.43] *Walter, F.:* Mikrotomtechnik im Industrielaboratorium. Leitz-Mitt. Wiss. u. tech. V. (1970) 2, 45–56
[4.44] *Etienne, I., and J. Le Fournier:* Anwendung gefärbter synthetischer Harze für die Untersuchung von Speichereigenschaften in Dünnschliffen. Rev. Inst. franç. Pétrole, Pans 22 (1967) 4, 595–629 (French)
[4.45] *Petrow, T. G.:* Anordnung zur Untersuchung des Kristallwachstums unter dem Mikroskop. Kristallografija 2 (1967) 777 (Russian)
[4.46] *Montoriol-Pous, H., and H. M. Amigo:* Apparatur zur Untersuchung der Abhängigkeit des Habitus von der Kristallisationstemperatur an natürlichen Kristallen mit einer Abscheidungstemperatur aus der Lösung unter 100 °C. Rost Kristallov 7 (1967) 178–182
[4.47] *Schiemer, H. G.:* Beschreibung einer Zellkammer und Durchströmungseinrichtung zur Untersuchung von Gewebekulturen mit dem Interferenzmikroskop und UV-Mikrospektrographen. Mikroskopie 14 (1959) 91–99
[4.48] *Koulovskij, M. J.:* Einrichtung zur mikroskopischen Untersuchung des Kristallwachstums aus Lösungen. Kristallografija 3 (1958) 509–510 (Russian)
[4.49] *Oratovskij, V. I.:* Apparat zur Untersuchung des Kristallisationsprozesses bei erhöhten Temperaturen. Zavod. Labor. Moskva 30 (1967) 1, 118–119 (Russian)
[4.50] *Stadelmann, E.:* Eine einfache Durchströmungskammer.Zeitschr. wiss. Mikroskopie 64 (1959) 286–298
[4.51] *Reinert, G. G.:* Eine Durchflußkammer mit veränderlicher Kammertiefe. Zeiss-Nachr. 2 F. (1937) 121–125

5 *Image recording and image analysis*

5.1 Image recording

5.1.1 Photomicrography

To permit photographic recording of the object images observed through it, the microscope needs some design modifications. The eyepiece is replaced by a projection ocular. The photographic section proper consists of modular units varying by exposure metering technique, beam splitter design and camera type (Fig. 5.1, see Annex). There are two basic design concepts:

(1) Photomicroscopes

These are integrated apparatuses for automatic and serial photography of microscopic images at high quality.

(2) Camera attachments

They are accessories to be attached to research-grade microscopes if the images observed are to be recorded.

For a synopsis of photomicrographic techniques, see [5.1 to 5.4].

It is imperative in any photomicrographic work to exactly adjust Koehler illumination (Chart 5.1) in order to ensure an evenly lit field.

Methods to find the correct exposure time (shutter speed)

(1) Test exposures at stepped shutter speeds.
(2) Exposure meter in the ray path, especially for special exposures under difficult lighting conditions.
(3) Automatic exposure control for routine and serial photography; the camera shutter is controlled by the photocurrent.

The microscope's coarse focussing drive must be stiff enough to withstand the added weight of the camera attachment, or else the micrographs will be blurred. Vibrations or shocks during an exposure have the same effect.

When using a beam splitter (cf. Fig. 5.1), first adjust the eyepiece to focus on the double crosshairs before using the coarse and fine controls for focussing on the object. That way, the object image will appear in focus on the film, too.

The magnification of a micrograph can be identified by exposing a stage micrometer together with the object or subsequently. Principally, one should use a high-power objective and a low-power projection lens.

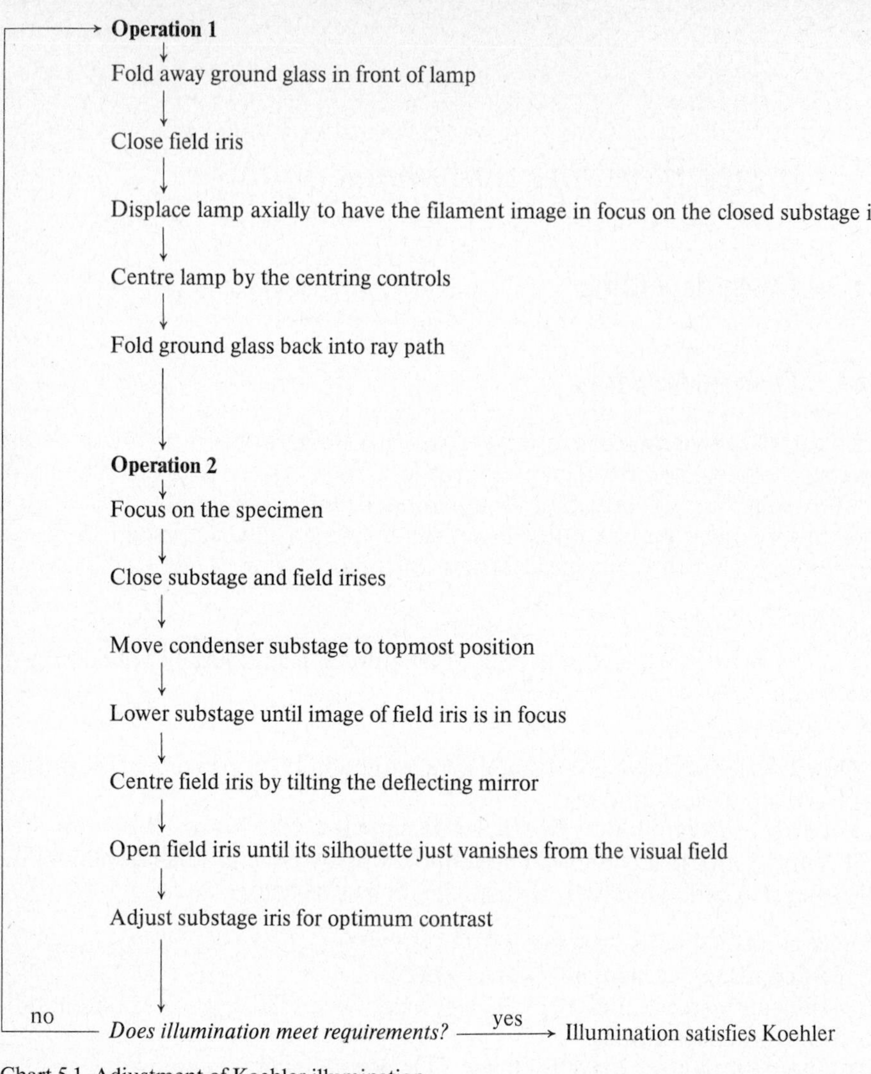

Chart 5.1. Adjustment of Koehler illumination

In black-and-white photomicrography, 15 DIN (25 ASA) is the most suitable film speed for most objects.

Parameters affecting the contrast range of a negative for a given object

(1) Gradation of the emulsion
(2) Developer composition and developing time (High developer concentration and long developing time produce negatives of higher contrast than those produced by greatly diluted developer and short time.)

(3) Wavelength of the exposure light (Generally, short-wave light produces weaker contrast than long-wave light.)

After development, even the densest parts of a negative should have some degree of transparency. Details in dark areas must show up well if the negative is held against a white paper background.

To reduce the influence of the diffuse halo (light scattered within the emulsion), one should employ thin-emulsion films and surface-acting developers. Tank and fine-grain developers are less suitable, as their action penetrates below the emulsion surface. Film grain is also important; emulsions of higher sensitivity have less definition than fine-grain thin-emulsion films.

The black-and-white rendition of coloured objects requires film material that ensures the transformation of colour tones into equivalent brightness levels (grey tones). Of the three film types available, none is capable of a completely adequate rendition of object hues corresponding to the eye's colour perception.

Unsensitized film:

Blue appears bright, whereas other hues do not.

Orthochromatic film:

Approximately corresponds to the eye's colour response. Sensitive to yellow and green. Blue is rendered too bright. There is no response to red.

Panchromatic film:

Sensitive to blue, yellow, green and red. Blue is rendered too bright.

Film materials for photomicrography should therefore be selected by consideration of colour distribution and contrast in the object and the size of the structures to be resolved. Table 5.1. assigns various film types to some selected object conditions.

A frequent task is to photograph dim objects, especially in dark-field, fluorescence and phase microscopy (Table 5.2). Here it is appropriate

(1) to use film of small frame size,
(2) to employ highly sensitive film,
(3) to use illuminants of high intensity (Table 5.3), inserting a heat-absorbing filter in case of thermolabile objects,
(4) to select the smallest possible magnification and maximum aperture,
(5) to remove from the ray path all unessential optical elements (filters, neutral density filters, beam splitters), and
(6) not to use zoom optics.

The photomicrography of moving or changing objects often calls for special devices. A micro-flash unit is indispensable for capturing fast-moving objects. Unblurred micrographs of particles in microcells, agitated at speeds of 500 to 600 r.p.m., can only be taken at exposure times between 1/5,000th and 1/10,000th second. As the light quantity received by the film cannot, in such cases, be control-

Table 5.1. Preferable film material for various microscopic objects

Microscopic object		Film material
Colour	Structure, brightness	
Black-white, grey-white or coloured, no dark reds	Numerous fine structural details	Orthochromatic, fine grain (10 and 15 DIN) with high resolving power
	Dim objects	Orthochromatic, coarse grain (20 and 28 DIN)
Red tones dominating		Panchromatic, highly sensitive to red

led by varying the duration of exposure, it is necessary to employ attenuation filters of different densities for the purpose. Where agitating speeds are slower so that no flash unit is required, note that shutter speeds between 1/500th and 1/2,000th second do not allow the use of A.C.-operated light sources, or else the camera's focal-plane shutter would depict the sequence of light pulses on the film as a sequence of alternating bright and dark strips.

To study changing objects, micrographs can be taken at a programmed sequence by means of an automatic timer. Combined with automatic exposure control, the timer releases the shutter at preselected time intervals.

5.1.2 Cinemicrography

Cinemicrography [5.5 to 5.9] records the course of motions or change processes in microscopical specimens. Depending on the specific purpose, cine cameras for different film gauges are available.

(1) 8mm and Super-8mm narrow-gauge (Fig. 5.2, see Annex)

Recording of processes for research and laboratory purposes. Not suitable for larger audiences. Trial exposures for larger-format recording.

(2) 16mm narrow-gauge

Table 5.2. Illuminance values in various microscopical techniques

Technique	Illuminance (lx)		
	Minimum	Maximum	Main range
Bright field	0.4	44	8–13
Phase contrast	0.2	3.5	0.5–1.3
Fluorescence	0.08	2	0.2
Polarization	0.02	44	2–4

Table 5.3. Illuminants for microscopical examinations, acc. to [5.3]

Description	Lamp voltage (V)	Lamp current (A)	Power input (W)	Luminance (1 cd/cm²)	Service life (h)
6V, 15W incandescent lamp	6	2.5	15	900	100
6V, 30W incandescent lamp	6	5	30	1,000	100
12V, 100W incandescent lamp	12	8.3	100	2,800	25
6V, 10W halogen lamp	6	1.6	10	2,000	100
12V, 50W halogen lamp	12	4.5	54	3,000	50
12V, 100W halogen lamp	12	8.3	100	3,200	50
D.C. carbon arc lamp	50	6	300	16,500	1 per pair of carbons
A.C. carbon arc lamp	50	10	500	10,500	
XBO-100 xenon arc lamp	14	7.1	100	9,000	120
XBO-101 xenon arc lamp	16	6.2	100	9,500	300
XBO-450 xenon arc lamp	18	25	450	35,000	2,000
HBO-50 mercury arc lamp	30	1.6	50	17,500	150
HBO-100 mercury arc lamp	18	5.2	100	85,000	150
HBO-200 mercury arc lamp	65	3.0	200	22,000	300

For research and demonstration tasks requiring higher-quality recording. Suitable for TV presentation. Individual frames can be analyzed.

(3) 35mm standard

Scientific cine-records satisfying highest quality standards, presentable to large audiences, with sound. High information content of the individual frame. Possibility to improve contrast photographically after recording.

The use of cine reflex cameras permits direct focussing of the microscopic image on the ground-glass screen (e.g., the Pentaflex 8, Fig. 5.2).

In projecting cine films, the maximum useful magnification is 400 times the size of the film frame. The projected image dimensions possible for the various film gauges are given below.

Film gauge	*Projected frame size*
35mm standard	6.40 × 8.80 m
16mm narrow-gauge	3.00 × 4.10 m
Super-8mm narrow-gauge	1.65 × 2.35 m
8 mm narrow-gauge	1.45 × 1.95 m

Table 5.4 lists the most common parameters for the different cine-film gauges.

Similar to photomicrography, the cinemicrographic and projection magnifications are found from the equations

$$M_{\text{frame}} = M_{\text{objective}} \cdot M_{\text{proj. ocular}} \cdot \textit{camera factor, and}$$

$$M_{\text{projected frame}} = M_{\text{objective}} \cdot M_{\text{proj. ocular}} \cdot \textit{camera factor} \cdot \textit{projection factor.}$$

Table 5.5 relates useful photographic magnifications to other exposure parameters.

It is often necessary in cinemicrography to decrease magnification in order to influence the exposure time per frame for dim objects. Adapters for narrow-gauge cine cameras with small scale factors (such as 0.5 in the JENOPTIK

Table 5.4. Film gauges and frequently used cine-projection parameters acc. to [5.5]

Film gauge	Normal frame rate (frames/s)	Frames per metre of film length	Film length per 1 s of projection time at normal frame rate (m/s)
35mm standard	24	52.5	0.456
16mm narrow-gauge	24	131	0.183
Super-8mm narrow gauge	18	234	0.077
8m narrow-gauge	16	261	0.061

Table 5.5. Cinemicrographic parameters and useful magnifications acc. to [5.5]

Film gauge	Frame size (mm)	Microscope objective	Projection oculars for the lower and upper limits of useful magnification	
35mm standard	16×22	4/0.16	×10	×20
		10/0.30	×8	×16
		40/0.95	×6.3	×12.5
		100/1.40	×4	×8
16mm	7.5×10.3	4/0.16	×5	×10
		10/0.30	×4	×8
		40/0.95	×3.2	×6.3
		100/1.40	×2	×4
Super-8mm	4.2×5.9	4/0.16	×2.5	×5
		10/0.30	×2	×4
		40/0.95	×1.6	×3.2
		100/1.40	×1	×2
8mm	3.6×4.9	4/0.16	×2	×4
		10/0.30	×1.6	×3.2
		40/0.95	×1.25	×2.5
		100/1.40	×0.8	×1.6

JENA GmbH equipment) offer a considerable gain in light, apart from the larger object field covered by the frame size.

The exposure time per frame is an important point to consider in cinemicrography, as it cannot be selected freely. Continuous film travel provided, the exposure time per frame is a function of the camera's frame rate and its sector aperture:

$$t_{\text{frame}} = \frac{\text{sector aperture angle}}{360° \cdot \text{frame rate}} \text{ (sec.)}$$

To control exposure time for a given photographic magnification, one can vary film speed, object illumination and objective N.A. The brightness of a frame increases with the square of demagnification (e.g., by using a lower-power projection ocular). At the same time, the object field covered increases.

The time scale is the ratio of the duration of a process to the duration of its presentation by the cine projector. n : 1 indicates that the rate of change is accelerated in projection *(time lapse)* , whereas 1 : n denotes the reduction of speed in projection *(time dilatation).* With extremely rapid or extremely slow processes, the time scale can alternatively be varied by the camera frame rate. High-speed cinemicrography, above all, is highly important for the examination of many technical processes (agitated crystallizers, fast decomposition processes). To make full use of the microscope's resolving power, it is imperative that

the resolving power of the film emulsion be two or three times that of the microscope.

If an 8mm narrow-gauge camera is employed, the emulsion may sometimes not be capable of resolving all details presented by the microscope.

5.1.3 Video techniques

Analogously to cinemicrography, microscopic images of moving or changing objects can be recorded on video tape in cassette recorders [5.7, 5.8]. The presentation of the record on a TV screen can also be optimized by time lapse, time dilatation or editing. Advantages of taped records over cine films are the possibilities of subsequent electronic magnification and image intensification to improve brightness or contrast.

As a drawback, the video-recording of dim objects involves a loss in contrast. On the average, TV image tubes require a minimum illumination of 5 lx (cf. Table. 5.2). Special-purpose tubes perform already at 0.1 lx.

To avoid contrast loss due to the line structure of the TV screen image, the resolution limit of the image tube, determined by its line spacing, must be below that of the microscope image. Whether this requirement is satisfied can easily be checked by recording a fine-structured test object.

The images of several cameras can be fed alternatingly to a single receiver by means of a camera selector switch. That way, the process itself and the correlated variation of measured data may be demonstrated at a time. Several receivers operated in parallel permit to present a process simultaneously as observed with polarized light, by phase contrast and interference contrast, for example.

Frequently, processes observed by interference or polarizing microscopy show characteristic changes of interference colours, which can be presented by colour video techniques.

A stage micrometer supplied with the equipment indicates the magnification of the screen image. Magnification may be varied continuously with a zoom system, although at the cost of image quality.

5.2 Image analysis

The extraction of (measurable) information from a microscopic image always involves some form of analyzing the image. The subject of analysis is the distribution of colour tones or grey levels and their controlled variation by the adjustments, or insertion into the ray path, of measuring elements. The measurement taken at a given point in the image can, at the same time, be related to the area surrounding that point, which is essential to the exact interpretation of the measured value. That makes microscopy distinctly different from other physical measuring techniques, which employ »blind« registration.

5.2.1 Visual image analysis

Visual analysis of the microscopic image or its photographic records is still the most important method of extracting information in microscopy. The human eye proves to be an extremely sensitive detector even of very faint effects. The subjective nature of the information sensed is, of course, a disadvantage.

The microscopic image itself lends itself to visual analysis where objects are static or changing slowly, whereas the micrograph also permits the study of rapid processes that the eye cannot resolve.

Methods of visually analyzing micrographs are outlined in Table 5.6 [5.9 to 5.12].

5.2.2 Automatic image analysis

Any quantity that can be registered microscopically can, in principle, be registered automatically. In recent years, automatic measurement has established itself especially in grain and petrofabric analysis and in kinematic microscopy [5.13 to 5.43].

Apart from routine serial examination in process control, it is especially the registration of parameter variations in fast processes under research that calls for automated microscopes. Such measurements can easily be resolved into constant, programmable steps.

Both visual and automatic image analysis register differences in absorption, reflection, scatter or refractive index. In automatic analysis, a photoelectric detector converts the light flux with its intensity variations into electrical signals.

The signals obtained in scanning an image carry two kinds of information:
(1) *optical information*, represented by the amplitude of the current pulse, and
(2) *morphological-geometrical information*, represented by pulse width or sequence.

Some microscopical quantities are not measurable directly, but indirectly by way of analogous secondary quantities. The melting point, for example, can be

Table 5.6. Analysis of photomicrographs

Analysis method	Application	Accuracy, error sources
Weighing method Cutting out interesting objects from micrograph; comparison of paper weights	Area analysis	Low accuracy despite high time consumption
Tracing of feature outlines from micrograph print on millimetre graph paper, or overprinting of a grid	Area analysis	Inaccuray of tracing; partly impaired contact between print and grid
Particle size analyzer acc. to Endter and Gebauer. Size-grading by 48 classes by variation of a circular reference area	Serial analyses of object sizes (classifying area analysis)	Accuracy insufficient for individual measurements, partly sufficient for large series
Use of *planimeter and integraph*	Accurate area analysis	Micrograph must be held entirely flat
Use of a low-power *measuring microscope*	Measurement of lengths and angles	Low setting accuracy if contrast is low and contours blurred. Measuring magnification depends on photographic resolution
Automatic length measurement with electronic control	Measurement of lengths	Very high measuring accuracy
Measuring microscope and equidensity methods	Photographic density measurement	Less subjective errors, better contour definition, higher contrast, less detail
Subjective densitometry (silver density, colour density) with microdensitometer and selective colour filters	Photographic density measurement	Subjective errors, possible structure differences between test area and reference area
Objective densitometry (silver or colour density): densitometer with photocell and electronic amplifier, selective colour filters	Photographic density measurement	Objective results; amplification susceptible to interference; uncertainty of electrical meter indication
Series of still micrographs taken at constant intervals, series of cine frames	Analysis of object changes by measuring areas, lengths, angles and densities in each frame	Advantage of visual inspection of measured object changes
Time-dilatation and time-lapse presentation of cine films	Assessment of the overall pattern of a process at a specified rate	

measured by the change of absorption of transmitted light by the specimen.

Potential error sources in automatic image scanning

(1) Errors due to the geometric-optical imaging properties of the objective. May be minimized by using high-qualitiy microscope optics.
(2) Errors due to the dependence of signal significance from object imaging. Effective only near the microscope's limit of resolution. Influenced by condenser aperture and by the nature, size and shape of objects.
(3) Errors due to the influence of inappropriate sample preparation (e.g., overlapping objects), grain intergrowth and incrustation, inhomogeneous grain structure and insufficient contrast between grain and matrix.

5.2.2.1 Differential automatic image analysis

The image is scanned at a grid of closely spaced points. By logic connection of the values measured at the points, information on the scanned object is obtained. Differential image analyzers may be grouped into several configurations depending on whether the object or the scanning light beam are moved.

(1) Moving object, stationary beam

This design concept is applicable, e.g., in the instantaneous testing of running threads, wires and foils, and flowing media in flow-through cells or agitated microreaction cells [5.24, 5.25]. The scanning stage also operates by that principle. It allows movement of very large surfaces past the scanning beam of a photometer in a raster fashion. An arrangement for automatic phase analysis in connection with a photometer attachment is described by Kroll [5.26].

(2) Moving beam (on observation side), stationary object

This is the most important arrangement in microscopic image analysis.

The following quantities can be established automatically, with digital output:

Phase fraction (area fraction)	Proportion of accumulated intercept lengths to total scan length
Mean grain size	Accumulated intercept lengths divided by number of phases hit
Specific grain surface	Number of scan line/phase intersections per mm of scan line
Form factor	Statistical quantity for a representative individual grain
Size distribution	Statistical registration of intercept length distribution
Degree of anisotropy	Ratio of mean linear phase dimensions before and after varying the scanning direction by a definite angle
Sequence (phase distribution)	Logic connection of phase boundaries counted on each scan line

Two design versions are known:

TV image analysis [5.27 to 5.31]

The microscope's intermediate image plane is occupied by the receiver surface of a TV camera, which scans the image line by line. Advantages are the high scanning frequency and the easy logic connection between the signals. The signal amplitude of the TV tube must be linear with illumination. Accurate photometric measurements are possible after calibration with standard specimens.

Photometer scanning [5.32]

A photometer on the observation side of the microscope has a scanning spot, which scans the intermediate image in a raster fashion (Fig. 5.3). Apart from an error due to rise time at the pulse edges, the duration of a signal is proportional to the length of the object. The signal height represents the luminance (absorbance, reflectance) of the object. Devices operating by that principle are *Kontron-Digiscan, Vickers Scanning Microdensitometer, Ameda* and others.

(3) Moving beam (on illumination side), stationary object [5.33 to 5.35]

In instruments of this *»flying spot« type,* the beam or scanning spot is moved on the illumination side. The main application is in the grain size analysis of disperse systems.

5.2.2.2 Integrating automatic image analysis

The desired quantity (intensity, absorbance, reflectance) is measured by integration over the entire visual field or a larger scanning area. Differences within that area are of no interest. Various arrangements are possible depending on the movement of object or scanning area.

(1) Moving object, stationary scanning area

This arrangement is employed for testing running threads, wires and foils and substances in agitated or flow-through cells, if a value that integrates many small objects over a large area is required.

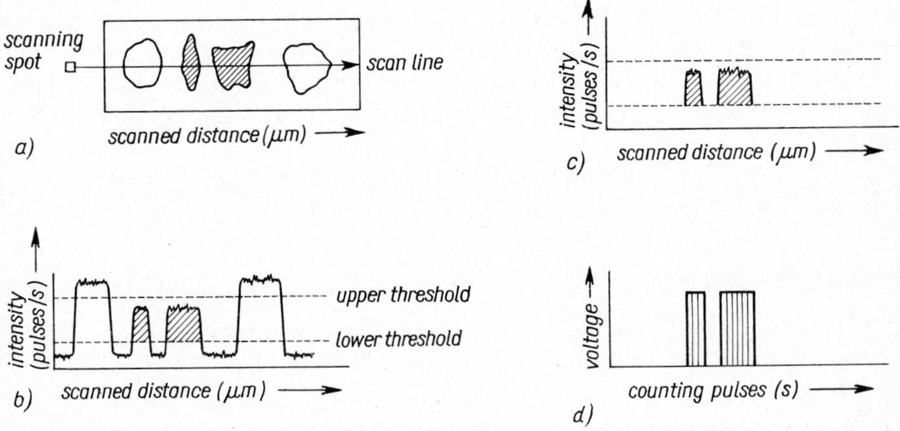

Figure 5.3. Automatic image analysis by photometric scanning
a) specimen features intercepting the scan line b) intensity distribution along scan line c) discriminated voltage d) counting pulses

(2) Moving beam, stationary object

The movement of a small scanning spot across the object (on the observation or illumination side of the microscope) provides a multitude of individual measurements, which are integrated into a value representative of the entire area. The integration may be computer-controlled according to a programme.

(3) Stationary but changing object, stationary scanning area

This is the most important arrangement in integrating automatic image analysis and is the easiest one to implement. Its main application is the measurement of object changes in kinematic microscopy.

Heterogeneous scanning area

Measurements using a heterogeneous scanning area are made in studying crystallization, dissolution and decomposition processes. The various microprocesses going on within the scanning area are integrated into a total value, whose change in time characterizes the overall course of the process. The simplest arrangement consists of a photometer attachment (e.g., the exposure meter section of a photomicrographic equipment), which registers brightness throughout the visual field. The set-up and its application in the study of crystallization and decomposition processes is described by Heide [7.3].

Fischer and Schramm [5.36] report on a special arrangement including a rotary analyzer. They studied the melting behaviour of high polymers in a temperature-controlled cell by measuring the light intensity through crossed nicols (Fig. 5.4). Temperature and light intensity are recorded simultaneously. Light losses due to absorption or scatter in anisotropic specimens affect the light intensity, but not the character of the recorded curves. With optically isotropic substances, the curves oscillate invariably between zero and a specimen-specific maximum; with anisotropic specimens they oscillate between a minimum and a maximum value.

A quantitative evaluation of the graphs yields the proportions of anisotropic and isotropic regions, their orientation and their change as a function of temperature.

A different set-up, especially for determining the integrated birefringence at high-polymer fibres by way of intensity measurements, is described by Schauler [5.37].

Homogeneous scanning area

Measurements with homogeneous scanning area include all studies of monocrystalline regions of uniform orientation. Two kinds of information may be obtained:

(1) Automatic measurement of physical properties of the crystal lattice

This includes techniques for the automatic measurement of birefringence [5.38, 5.39], extinction conditions on the universal stage [5.40], refractive index [5.41] and others.

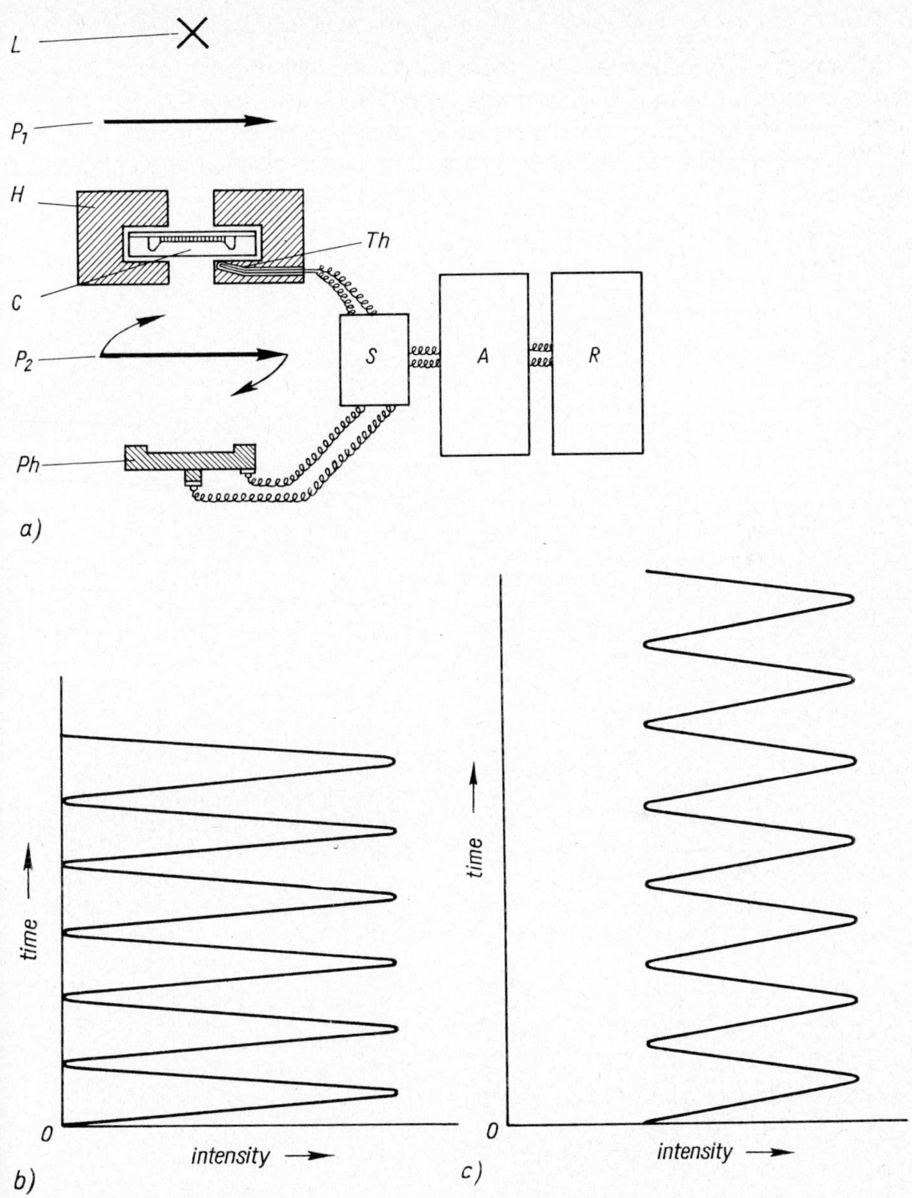

Figure 5.4. Intensity measurements acc. to Fischer and Schramm [5.36]
a) measuring set-up b) intensity curve for optically isotropic objects (not having changed during the measurement) c) intensity curve for optically anisotropic objects (not having changed during the measurement)

L light source P_1 polarizer H heating block C measuring cell Th thermocouple P_2 rotating analyzer Ph photometer S selector switch A amplifier R recorder

(2) Automatic measurement of the change of physical properties of the crystal lattice

Changes, e.g. of temperature on the hot and cold stage, result in changed specimen properties within the scanning area, which can be recorded against time. Corresponding equipment serves for measuring melting points (Fig. 5.5 [5.42]), polymorphic modification changes or lattice disintegration processes (Fig. 5.6).

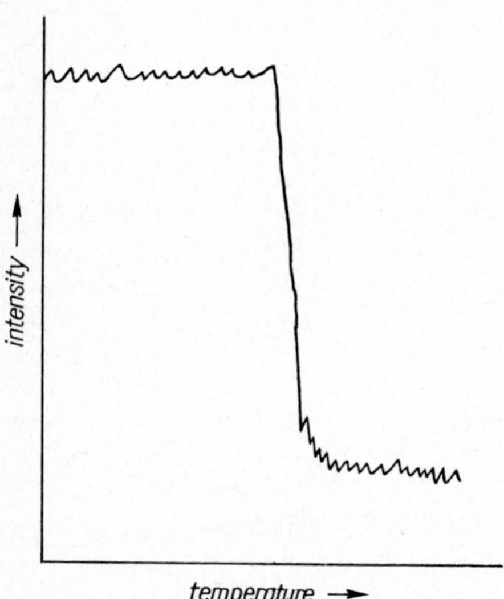

Figure 5.5. Melting point determination by automatic microscope-photometric intensity measurement

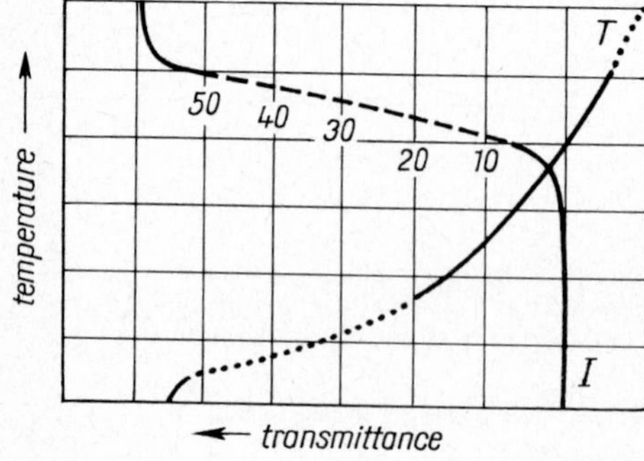

Figure 5.6. Study of phase change in the thermal decomposition of $CaSO_4 \cdot 2H_2O$, by microscopical photometry acc. to [7.169]

T temperature curve I intensity curve

References

[5.1] *Michel, K.:* Die Mikrophotographie. Wien - New York: Springer-Verlag 1967

[5.2] *Stade, G., and H. Staude:* Mikrophotographie. Leipzig: Akademische Verlagsgesellschaft Geest und Portig KG 1958

[5.3] *Bergner, J., Gelbke, E., and W. Mehliss:* Praktische Mikrophotographie. 2nd ed. Leipzig: VEB Fotokinoverlag 1973

[5.4] *Ludwig, C.:* Das Aufnahmematerial für die Schwarz-Weiß-Wiedergabe (Aus der mikrophotographischen Praxis). Leitz-Mitt. Wiss. u. Tech. 11 (1963) 6, 168–172

[5.5] *Weidel, G.:* Mikrokinematographie I und II. Bild und Ton 24 (1971) 1, 5–10 and 2, 37–50

[5.6] *Reumuth, H.:* Mikro-Kinematographie in Wissenschaft und Technik. Kino-Technik 62 (1960) 200–202

[5.7] *Robertson, J. A.:* Video Techniques Applied to Chemical Microscope. Mikroscope 16 (1968) 4, 305–310

[5.8] *Weidel, G.:* Fernsehmikroskopie – angewandtes Fernsehen mit dem Mikroskop. Bild und Ton 24 (1971) 4, 105–110

[5.9] *Diedrich, U.:* Methoden der Film- und Photoanalyse. Berlin: VEB Deutscher Verlag der Wissenschaften 1966

[5.10] *Schluge, H.:* Der Teilchengrößenanalysator nach Endter. Zeiß-Werkzeitschr. 7 (1959) 68–71

[5.11] *Greif, H.:* Ein photographisches Verfahren zur Messung der Volumenanteile inhomogener Stoffe. Feingerätetechnik 12 (1963) 157–159

[5.12] *Marsh, G. E.:* Measurement of Components of Thin Sections with the Planimeter. Amer. Mineralogist 23 (1938) 412

[5.13] *Wasmund, H.:* Mikrophotometrische Untersuchungsmethoden an feinen Teilchen. Veröffentlichungen Haus der Technik (Vulkan-Verlag) Essen (1970) 247, 76–88

[5.14] *Ernst, T., and H. Kohler:* Photoelektronische Messungen im Polarisationsmikroskop. Sitz. Ber. Bayr. Akad. Wiss. math.-nat. Kl. (1964) 93–96

[5.15] *Stroud, A.*, et al.: Kombination von Mikroskopen und Computern zur Analyse von Chromosomen. Microscope 15 (1967) 68–77

[5.16] *Welkowitz, W.:* Programming a Digital Computer for Cell Counting and Sizing. Rev. sci. Instruments 25 (1954) 1202

[5.17] *Plöckinger, E., and G. Dörfler:* Moderne Verfahren der stereometrischen Analyse. Radex-Rdsch. (1967) 3/4, 777–781

[5.18] *Petzow, G., and H. E. Exner* (Hrsg.): Quantitative Bildanalyse. 11 Vorträge vom III. Quantimet-Kolloquium, Sonderheft der »Praktischen Metallographie«. Stuttgart: Dr. Riederer-Verlag 1970

[5.19] *Nehls, R.:* Automatisches Messen, Zählen und Charakterisieren von Partikeln. Verfahrenstechnik 5 (1971) 12, 491–498

[5.20] *Rifkin, J. L.:* Verbindung eines Meßokulars mit einem Spannungsteiler für die kontinuierliche Aufzeichnung von gemessenen mikroskopischen Dimensionen. Analyt. Biochem. 24 (1968) 2, 192–196

[5.21] *Nassenstein, H.:* Die Praxis der automatischen Dispersoidanalyse, Chem.-Ing.-Techn. 29 (1957) 92–104

[5.22] *Walton, W. H.:* Automatic Counting of Microscopic Particles. Nature 169 (1952) 518

[5.23] *Kötter, K., and W. D. Langner:* Automatisierung von statistischen Reflexionsmessungen an Kohlen. Brennstoff-Chemie 42 (1961) 380–385

[5.24] *Barret, P. T.:* Polarisationsoptisches Verfahren zur Regelung der Gleichmäßigkeit der Eigenschaften von schmelzgesponnenen Fäden während des Spinnens. BRD-Auslegeschrift 1 097 167 (I 9324 IX/42h) of 12. 1. 1961, Kl. 42h 21, IKL G 01 j; n

[5.25] *West, P.:* Automatic, Non-contacting Measurement of Fine Filaments. Research (1971) April, 6–8

[5.26] *Kroll, J. M.:* Mikrophotometrische Bestimmung selektiv angefärbter Feldspate in keramischen Rohstoffen und Sanden. Leitz-Mitt. Wiss. u. Techn. IV (1967) 1/2, 19–23

[5.27] *Korn, M., and L. Rolf:* Korngrößenmeßtechnik mit dem Fernsehmikroskop Quantimet im Bereich von 40 μm. Aufbereit.-Techn. (1971) 154–160

[5.28] *Gerlach, H. D.:* Elektonische Hilfsmittel zur Automatisierung spannungsoptischer Messungen. Diss. Univ. Karlsruhe, F. f. Maschinenbau, 1968

[5.29] *Dudek, M. B.:* Die Anwendung des Quantimet in der Metallographie. Radex-Rdsch. (1967) 3/4, 790–794

[5.30] Quantitatives Fernsehmikroskop »Microscan« (Oberkochen-Zeiß-Siemens). Naturwiss. Rdsch. 22 (1969) 11, 507

[5.31] *Hofer, F.:* Erfahrungen beim Einsatz eines Fernsehmikroskops in einem Stahlwerk. Radex-Rdsch. (1967) 3/4, 794–805

[5.32] *Dörfler, G.:* Sequenzanalyse, ein neues Analysenverfahren zur quantitativen Ermittlung der Phasenverteilung von Festkörpern. Radex-Rdsch. (1967) 3/4, 782–785

[5.33] *Young, J. Z., and F. Roberts:* Ein Mikroskop, bei dem das Objekt mit einem Lichtstrahl abgetastet wird. Nature 167 (1951) 10/2, 231

[5.34] *Roberts, F., and J. Z. Young:* The Flying Spot Microscope. Proc. Inst. Elec. Engrs., part III A, 99 (1956) 747

[5.35] *Hackenberg, P.:* Korngrößenbestimmung mit dem Coulter Counter. Tonind. Ztg. 92 (1968) 12, 482–487

[5.36] *Fischer, K., and A. Schramm:* Optische Eigenschaften polykristalliner Systeme in Abhängigkeit von der Temperatur. Angew. Chem. 68 (1955) 12, 406–411

[5.37] *Schauler, W.:* Eine neue Methode zur Messung der Doppelbrechung mit Hilfe der Intensität des polarisierten Lichtes. Leitz-Mitt. Wiss. u. Techn. 5 (1970) 3/4, 95–98

[5.38] *Snell, C.:* Automatische Messung der Doppelbrechung. Nature 214 (1967) 5083, 78–79

[5.39] *Gahm, J.:* Quantitative polarisationsoptische Messungen mit Kompensatoren. Zeiß-Mitt. 3 (1964) 5, 153–192

[5.40] *Behr, H. J., Bonnke, H., and S. Voigt:* Ein neues Gerät zur automatischen Registrierung von ebenen und axialen Richtungselementen in mikroskopischen Korngefügen. Feingerätetechnik 17 (1968) 8, 360–363

[5.41] *Chromy, S.:* Photoelectric Apparatus for Refractive Index Determination by the Immersion Method. Amer. Mineralogist 54 (1969) 549–553

[5.42] *Kolb, A. K., Lee, C. L., and R. M. Trail:* Automatische mikroskopische Schmelzpunktsbestimmungsmethode. Analytic. Chem. 39 (1967) 10, 1206–1208

[5.43] *Dörfler, G.:* Ein flexibles System zur stereometrischen Analyse mittels Lichtmikroskop, Elektronenmikroskop und Mikrosonde. Praktische Metallographie VI (1969) 144–158

6 *Microscopical measurement of physical properties*

Microscopical studies of chemical compounds can resort to a large number of measurable quantities (Table 6.1). Tables 6.2 and 6.3 list the microscopical techniques and equipment required [6.1 to 6.21]. Depending on the problem, one selects the optimum property with regard to measuring technique and information content (Tables 6.4 and 6.5).

Optical properties serve, first of all, to characterize the structure of compounds, to analyze the phases of solid reactant mixtures, and to study the kinetics of physical-chemical processes on solid phases.

Morphological properties are important for many problems in process engineering, such as filterability, agglomeration behaviour, fine grain fraction and material strength. Microscopical volume analysis yields quantitative data on the chemical composition of solid reactants. Studies at high or low temperatures permit to investigate the behaviour of solid-solid, solid-liquid and solid-gas systems and to characterize the structural processes in these phase transitions.

Substances that are hygroscopic or tend to decompose require special preparation and measurement techniques. Table 6.6 surveys properties that can be measured on such substances by means of special methods.

Table 6.7 gives examples of minimum sample quantities required to measure microscopical properties.

6.1 Optical properties

6.1.1 Refractivity

The refractive index n of optically isotropic substances and the principal refractive indices n_o, n_e, n_α, n_β, n_γ rank among the most important quantities measured in chemical microscopy. In most cases they are both highly informative and easily determinable.

Special methods are required only for hygroscopic or easily decomposing compounds, reaction products of extremely fine grain, and some organic high polymers. Table 6.8 gives a survey of the various measuring methods and their fields of

Table 6.1. Microscopic properties and lattice symmetry

Microscopic property	Optically isotropic		Optically uniaxial			Optically biaxial		
	Amorphous	Cubic	Tetragonal	Hexagonal	Trigonal	Rhombic	Monoclinic	Triclinic
Refractive index	×	×	×	×	×	×	×	×
Birefringence	–	–	×	×	×	×	×	×
Optic axial angle	–	–	–	–	–	×	×	×
Optic sign	–	–	×	×	×	×	×	×
Extinction angle	–	–	×	×	×	×	×	×
Extinction	(×)	(×)	×	×	×	×	×	×
Absorption, colour, pleochroism	×	×	×	×	×	×	×	×
Dispersion of optical properties	×	×	×	×	×	×	×	×
Fluorescence	×	×	×	×	×	×	×	×
Reflectance	×	×	×	×	×	×	×	×
Internal reflections	×	×	×	×	×	×	×	×
Grain size	×	×	×	×	×	×	×	×
Thickness	×	×	×	×	×	×	×	×
Grain shape	×	×	×	×	×	×	×	×
Cleavability	–	×	×	×	×	×	×	×
Twinning	–	×	×	×	×	×	×	×
Surface texture	×	×	×	×	×	×	×	×
Inclusions, grain homogeneity	×	×	×	×	×	×	×	×
Volume fractions	×	×	×	×	×	×	×	×
Solubility, etching behaviour	×	×	×	×	×	×	×	×
Sublimation	×	×	×	×	×	×	×	×
Melting point	×	×	×	×	×	×	×	×
Decomposition point	×	×	×	×	×	×	×	×
Modification	(×)	×	×	×	×	×	×	×
Characteristic eutectics	×	×	×	×	×	×	×	×
Microhardness	×	×	×	×	×	×	×	×
Density	×	×	×	×	×	×	×	×
Conductivity	×	×	×	×	×	×	×	×
Temperature	×	×	×	×	×	×	×	×

Table 6.2. Measurement of microscopic properties by transmitted and reflected light

Microscopic property	Transmitted light			Reflected light	
	Orthoscopy		Conoscopy	Bright field	Dark field
	1 nicol	Crossed nicols			
Refractive index	×	(×)	–	–	–
Birefringence	×	×	×	–	–
Optic axial angle	(×)	×	×	–	–
Optic sign	(×)	×	×	–	–
Extinction angle	×	×	–	–	–
Extinction	–	×	–	–	–
Absorption, colour, pleochroism	×	–	–	–	×
Dispersion of optical properties	×	×	×	–	–
Fluorescence	×	–	–	×	×
Reflectance	–	–	–	×	–
Internal reflections	–	–	–	(×)	×
Grain size	×	–	–	×	×
Thickness	×	×	–	×	×
Grain shape	×	–	–	×	×
Cleavability	×	–	–	×	×
Twinning	×	×	–	×	×
Surface texture	×	–	–	×	×
Inclusion, grain homogeneity	×	(×)	–	×	×
Volume fractions	×	×	–	×	×
Solubility, etching behaviour	×	–	–	×	×
Sublimation	×	–	–	×	×
Melting point	×	×	–	×	×
Decomposition point	×	×	(×)	×	×
Modification	×	×	×	×	×
Characteristic eutectics	×	(×)	–	(×)	(×)
Microhardness	×	–	–	×	×
Density	×	–	–	×	–
Conductivity	–	–	–	×	–
Temperature	–	–	–	×	–

application [6.22 to 6.68]. Table 6.9 shows the average errors involved in some significant techniques.

The refractive indices of inorganic substances essentially cover the range of 1.5 to 1.6; those of organic compounds lie between 1.5 and 1.7 (Fig. 6.1).

6.1.1.1 Becke line measurement [6.22 to 6.34]

The Becke line is a thin light fringe at the boundary between a grain and the immersion liquid (Fig. 6.2). It is best observed on a grain specimen in the orthoscopic ray path using the lower polarizer, the highest possible magnification and a small condenser iris aperture.

Table 6.3. Equipment for determining microscopic properties

Microscopic property	Polarizing microscope for reflected and transmitted light	Phase contrast outfit	Interference microscope	Goniomicroscope, universal stage, spindle stage, Waldmann sphere	Cold stage, hot stage	Fluorescence outfit	Microhardness tester	Integrating stage, micrometer eyepiece	Microscope photometer, microscope spectrograph	Compensator
Refractive index	×	×	×	–	×	–	–	–	–	–
Birefringence	×	×	×	–	(×)	–	–	–	–	×
Optic axial angle	×	–	–	×	(×)	–	–	–	–	×
Optic sign	×	(×)	×	×	(×)	–	–	–	–	–
Extinction angle	×	–	–	×	×	–	–	–	–	–
Extinction	×	–	–	–	×	–	–	–	×	×
Absorption, colour, pleochroism	×	–	–	–	–	–	–	–	×	–
Dispersion of optical properties	×	×	(×)	×	×	–	–	–	–	×
Fluorescence	×	–	–	–	–	×	–	–	×	–
Reflectance	×	–	–	–	–	–	–	–	×	–
Internal reflections	×	–	–	–	–	–	–	–	(×)	–
Grain size	×	–	×	–	–	–	–	×	–	–
Thickness	×	–	×	(×)	–	–	–	–	–	–
Grain shape	×	–	–	–	–	–	–	–	–	–
Cleavability	×	–	–	×	–	–	–	–	–	–
Twinning	×	–	–	×	–	–	–	–	–	–
Surface texture	×	–	×	–	–	–	–	–	×	–
Inclusions, grain homogeneity	×	×	×	–	–	–	–	–	–	–
Volume fractions	×	×	–	–	–	–	–	×	×	–

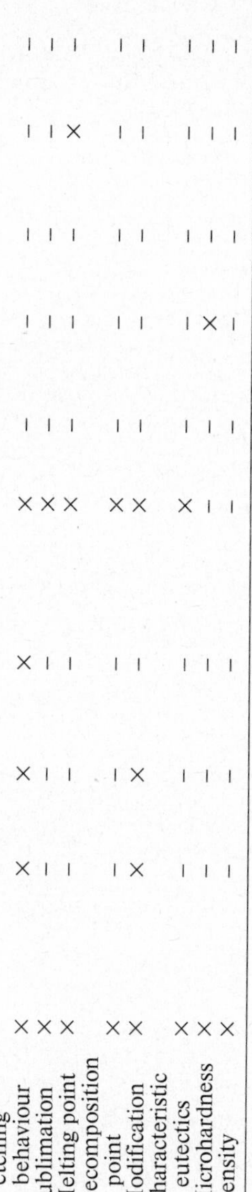

Solubility, etching behaviour	×	×	×	×	×	–	–	–	–	–
Sublimation	×	–	–	–	×	–	–	–	–	–
Melting point	×	–	–	–	×	–	–	–	×	–
Decomposition point	×	–	–	–	×	–	–	–	–	–
Modification	×	×	×	–	×	–	–	–	–	–
Characteristic eutectics	×	–	–	–	×	–	–	–	–	–
Microhardness	×	–	–	–	–	–	×	–	–	–
Density	×	–	–	–	–	–	–	–	–	–

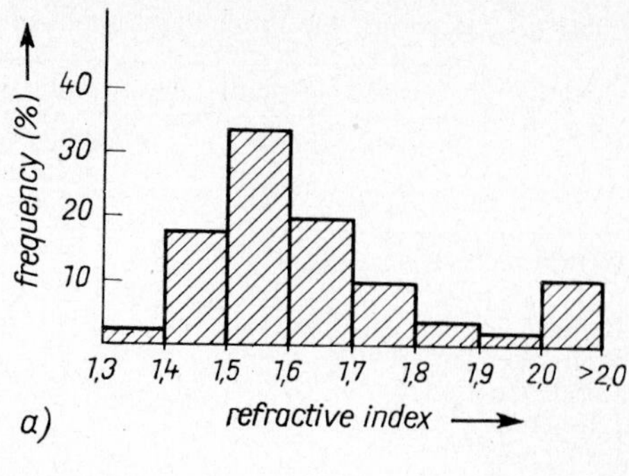

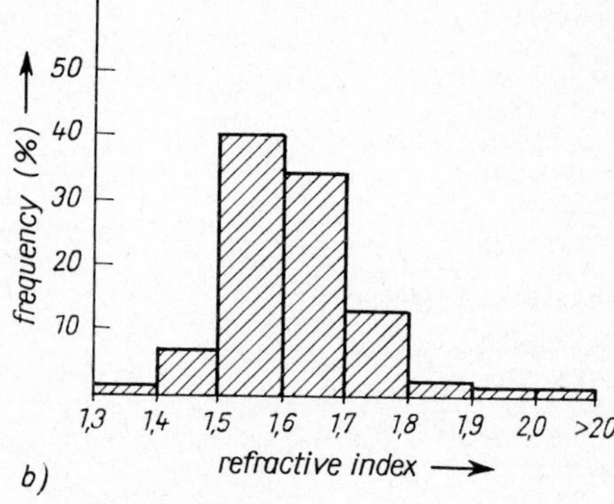

Figure 6.1. Frequency distribution of refractive indices, acc. to [7.102 and 7.103]

a) inorganic substances
b) organic substances

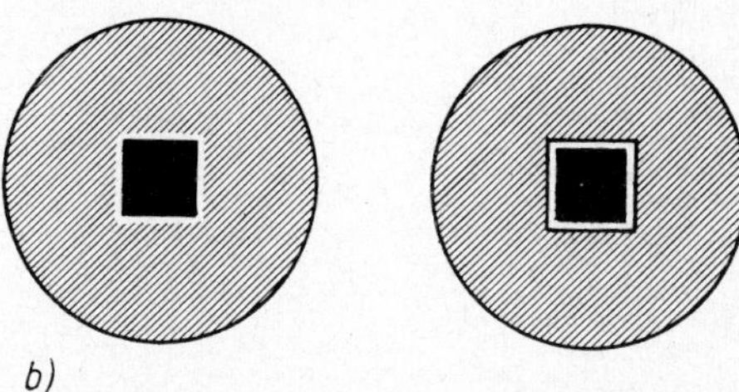

Figure 6.2. Becke line a) Becke line along the edges of paraffin crystals (see Annex) b) schematic representation of the Becke line outside and inside grain boundary c) halo around paraffin crystals seen through phase microscope (see Annex)

Table 6.4. Quantities applicable to problems in chemical microscopy *(transmitted light)*

Quantity	Type of analysis problem				
	Phase analysis	Lattice structure analysis	Petrofabric analysis	Morphological grain analysis	Surface analysis
Optical properties					
Refractive index	×	×	×	–	–
Birefringence	×	×	×	–	–
Extinction angle	×	×	–	–	–
Extinction	×	×	×	–	–
Pleochroism	×	×	–	–	–
Colour	×	×	–	–	–
Transparency	×	×	×	–	–
Optic sign	×	×	–	–	–
Optic axial angle	×	×	–	–	–
Dispersion of optical properties	×	×	–	–	–
Fluorescence	×	–	–	–	–
Circular polarization	–	×	–	–	–
Morphological properties					
Cleavability	×	×	×	×	×
Grain size	–	–	×	×	–
Thickness	–	–	×	×	–
Grain shape	–	×	×	×	–
Surface	–	–	–	×	×
Inclusions	–	×	×	×	×
Twinning	×	×	×	×	–
Volume fractions	–	–	×	–	–
Porosity	–	–	×	–	–
Grain homogeneity	(×)	×	–	×	×
Other substance-specific properties					
Solubility	×	×	–	–	–
Sublimation	×	×	–	–	–
Staining	×	–	–	–	–
Melting point	×	×	–	–	–
Decomposition point	×	×	–	–	–
Modification	×	×	–	–	–
Density	×	–	–	–	–
Characteristic eutectics	×	–	–	–	–

The technique is that of balancing step by step the light refraction caused by an immersing medium to that caused by the unknown substance, the Becke line being an aid in determining the residual difference in refractive index. As the microscope tube is raised, i.e. as the distance between specimen and objective increases, the light fringe moves into the medium having the greater refractive index. By operating the fine focussing control it is easy to ascertain the direction of travel. If several light fringes are visible, use the brightest one. If the refractive indices of grain and immersing liquid are equal, the grain is no longer seen against its background, if monochromatic light is used. With white light, coloured fringes appear, and grain contrast is greatly reduced, but the grain will not completely disappear in its background.

In case of equal refractive indices, the refractometer reading is that of the grain examined.

Balancing is effected by repeated changes of immersing liquids or by varying the proportion at which two liquids are mixed (Tables 6.10, 6.11).

Table 6.5. Quantities applicable to problems in chemical microscopy *(reflected light)*

Quantity	Type of analysis problem				
	Phase analysis	Lattice structure analysis	Petrofabric analysis	Morphological grain analysis	Surface analysis
Optical properties					
Reflectance	×	×	×	–	×
Absorbance	×	×	×	–	–
Gloss	×	×	×	–	×
Colour	×	×	×	–	–
Transparency	×	×	×	–	–
Fluorescence	×	×	×	–	–
Anisotropy	×	×	×	–	–
Bireflection	×	×	×	–	–
Internal reflections	×	×	(×)	–	–
Morphological properties					
Cleavability	×	×	×	×	×
Grain size	–	–	×	×	–
Thickness	–	–	×	×	–
Grain shape	(×)	(×)	×	×	×
Surface	–	–	×	×	×
Inclusions	–	–	×	×	×
Twinning	×	×	(×)	×	×
Volume fractions	–	–	×	–	×
Porosity	–	–	×	–	×
Grain homogeneity	(×)	–	(×)	×	×
Other substance-specific properties					
Solubility	×	(×)	(×)	–	×
Sublimation	×	×	–	–	×
Staining	×	–	×	–	×
Etching	×	×	×	–	×
Melting point	×	×	(×)	–	×
Decomposition point	×	×	(×)	–	×
Modification	×	×	–	–	×
Microhardness	×	×	(×)	–	×

Table 6.6. Properties measurable under special conditions

Microscopic property	Decomposing substances at 20°C	Temperatures above 20°C	Temperatures below 20°C
Refractive index	×	(×)	×
Birefringence	×	×	×
Extinction	×	×	×
Absorbance	×	×	×
Optic sign	×	×	×
Optic axial angle	×	(×)	×
Dispersion of optical properties	×	×	×
Reflectance	–	×	×
Melting point	–	×	×
Characteristic eutectic temperatures	–	×	×
Decomposition point	–	×	×
Temperature of polymorphous inversions	–	×	×
Sublimation	–	×	×
Etching	–	×	×
Solubility	×	×	×
Microhardness	–	(×)	×
Cleavability	×	×	×
Twinning	×	×	×

Table 6.7. Minimum object size and mass required to measure some important properties, acc. to [7.18]

Quantity	Size (μm)	Mass (g)
Melting point	5	10^{-10}
Solubility	1	10^{-12}
Density	20	10^{-8}
Refractive index	0.2	10^{-13}–10^{-14}
Principal refractive indices	5	10^{-10}
Fluorescence	1	10^{-12}
Microchemical analysis	1	10^{-12}

The technique is limited to grain sizes not smaller than 5 μm and refractive index differences down to the 4th decimal place. At steeply sloping grain edges, the direction of Becke line movement may be misrepresented if that movement is slower than the shift of the depth-of-focus limit.

Very small particles show light phenomena that sometimes defy exact determination. The whole particle may appear bright, especially if the focus is set somewhat above or below it.

The amount of refractive index difference between grain and immersing medium is also indicated by the »relief« of the grain, i.e. its contrast. Similar to surface contours, boundaries between grain and background may be distinct or

Table 6.8. Main methods of refractive index determination in solids

Principle	Method	Application
Refractometry	*Direct measurement* by placing crystals or specimens on the refractometer	Inapplicable for decomposing or hygroscopic compounds, requires large crystals
	Indirect measurement of refractive index of a solution of the substance	Refractive index determination in organic high polymers
Microscopical immersion methods, transmitted light	Becke line methods, phase contrast methods, bordering dark field method	Standard methods for most chemical compounds
	Interference microscopical methods	Organic high polymers, decomposing compounds
	Schröder-van der Kolk shadow method	Very large grains, no decomposing substances
	Tyndall effect method, Christianssen filter method	Ultrafine powders, submicroscopic particles
Microscopical methods without special immersion media	Calculation from universal stage methods, interference microscopical methods, Kofler's glass powder methods, Mozzherin's variascopic method, Duc de Chaulnes' method	Decomposing or hygroscopic compounds
Reflected light microscopy	Critical angle of total internal reflection, Brewster angle method	Polishable metals and metal alloys

Table 6.9. Mean errors in methods of refractive index determination

Method	Mean absolute error
Becke line	±0.001
Phase contrast	±0.0001
Interference	±0.0001
Dark field	±0.001
Temperature variation	±0.001
Wavelength variation	±0.001
Double variation (temperature, wavelength)	±0.0004
Glass powder method	±0.0003

ndistinct. A grain showing up in bold relief has a refractive index differing greatly from that of the immersing medium. If the grain hardly stands out against the medium, the difference in refractive index is correspondingly small. The difference in relief only represents the refractive index difference as such without indicating whether the grain is lower or higher in refractivity than the medium.

Chart 6.1 gives a survey of steps to be followed in refractive index determination.

Table 6.10. Important immersion media for refractive index determination (acc. to the literature)

Medium	n_D	Medium	n_D
Methyl alcohol	1.33	Cedarwood oil	1.516 1.51
Water	1.333 1.336	1,2-Dibromopropane	1.520
Ethyl ether	1.352	Collolith	1.520
Acetone	1.359	Iodoethane	1.522
Ethyl alcohol	1.36 1.362	Chlorobenzene	1.525 1.527 1.521
Hexane	1.375 1.39	Canada balsam	1.53
Propyl alcohol	1.385	1,2-Dibromoethane	1.536 1.539
Methyl butyrate	1.386	Methyl salicylate	1.537
Heptane	1.388 1.40 1.387	Fennel oil	1.538
Ethyl valerate	1.393	Creosote	1.54 1.542
Isobutyl alcohol	1.394	Eugenol (corr. to clove oil)	1.542 1.544
Octane	1.397	Bitter almond oil	(1.546)
Paraldehyde	1.398 1.404	o-Nitrotoluene	1.548 1.547
Amyl alcohol	1.40 1.409	Nitrobenzene	1.553 1.554 1.552
Ethyleneglycol monobutyl ether	1.42	Dimethylaniline	1.559
Ethyl monochloroacetate	1.423	Aniseed oil (anethol)	1.56 1.547 1.562
Isoamyl isovalerate	1.43	Caedax	1.56
Ethyl dichloroacetate	1.434	Monobromobenzene	1.562 1.56
1,2-Dibromoethane	1.44	Benzyl benzoate	1.569
Chloroform	1.446 1.45 1.443	Methyl benzoate	1.569
1,2-Dichloroethane	1.446	o-Toluidine	1.573 1.572
1,3-Dichloropropane	1.449	Cassia oil	1.586
Cyclohexanone	1.450	Aniline	1.586 1.587
Isoamyl sulphide	1.454	Bromoform	1.589 1.59 1.588
Lavender oil	1.461 1.463	Chloroaniline	1.592
Carbon tetrachloride	1.463 1.461 1.466	Benzaldehyde	1.60
Cyclohexanol	1.465	Cinnamon oil	1.602 1.605
Methyl thiocyanate	1.469	Cinnamaldehyde	1.615
Olive oil	1.47 1.468 1.476	Ceylon cinnamon oil	1.619
Cajeput oil	1.47	Iodobenzene	1.621 1.618
Glycerine	1.47 1.473 1.467	Cassis oil	1.621
Turpentine oil	1.471 1.472	Chinoline	1.624 1.62
Paraffin oil	1.475	Carbon disulphide	1.63 1.628 1.626
Almond oil	1.478	Diphenyl sulphide	1.635
Decalin	1.479	α-Chloronaphthalene	1.635 1.633 1.639
Castor oil	1.48 1.478 1.49	Carbon tetrabromide	1.639
Linseed oil	1.485	Phenyl isothiocyanate	1.651
Methyl fuorate	1.488	α-Bromonaphthalene	1.658 1.66 1.655
Dibutyl phthalate	1.49	o-Bromoiodobenzene	1.664
1,1,2,2,-Tetrachloroethane	1.494	α-Bromonaphthalene : Diiodomethane 2 : 1	1.68
p-Cymene	1.495 1.50	α-Bromonaphthalene : Diiodomethane 1 : 1	1.70
n-Xylene	1.497 1.50	Diiodomethane	1.74
Toluene	1.497	Diiodomethane : sulphur-saturated diiodomethane 1 : 1	1.76
Benzene	1.501 1.502 1.498	Diiodomethane, sulphur-saturated	1.78
Cedarwood oil	1.503	Tetraiodoethane	1.81
Pentachloroethane	1.504	$AsBr_3$ with 10% sulphur	1.81
Toluene	1.495		
Xylene	1.495 1.49 1.487		
Sandalwood oil	1.507 1.508		
1,2-Diiodoethane	1.513		
Anisole	1.515		

Table 6.10, continued

Medium	n_D	Medium	n_D
Phenyldiiodoarsine	1.84	$AsBr_3$, sulphur, As_2S_3, HgS (red)	2.03
Diiodomethane, sulphur, iodides	1.86 (1.87)	Diiodomethane, saturated with sulphur and phosphorus	2.06
Diiodomethane, phosphorus-saturated	1.94	Selenium	2.7
Sulphur	2.00	Selenium, arsenic selenide	3.17
$AsBr_3$, sulphur, As_2S_3, Se (black)	2.01		

Table 6.11. Mixtures of immersion media (acc. to the literature)

Refractive index range	Mixture
2.292 - 1.411	Perfluorotributylamine-chlorotrifluoroethylene
1.33 - 1.47	Water-glycerine
1.372 - 1.396	Ethyl acetate-amyl acetate
1.396 - 1.498	Amyl acetate-xylene
1.45 - 1.638	Kerosene (2 °C fraction)-α-chloronaphthalene
1.47 - 1.62	Glycerine-chinoline
1.482 - 1.658	Paraffin oil-α-bromonaphthalene
1.598 - 1.658	Bromoform-α-bromonaphthalene
1.62 - 1.658	Chinoline-α-bromonaphthalene
1.638 - 1.742	α-chloronaphthalene-diiodomethane
1.658 - 1.742	α-bromonaphthalene-diiodomethane
1.658 - 1.81	α-bromonaphthalene-10 % solution of sulphur in arsenic tribromide
1.68 - 2.10	Piperidine-antimony triiodide-arsenic triiodide
1.74 - 1.78	Diiodomethane-sulphur
1.74 - 1.78	Diiodomethane-arsenic tribromide
1.74 - 1.81	Diiodomethane-arsenic tribromide-sulphur
1.74 - 1.84	Diiodomethane-phenyldiiodoarsine
1.74 - 1.87	Diiodomethane-sulphur-iodides
1.74 - 1.94	Diiodomethane-phosphorus
1.74 - 2.07	Diiodomethane-sulphur-phosphorus
1.78 - 1.92	Arsenic tribromide-arsenic trisulphide
1.93 - 2.51	Sulphur-arsenic trisulphide
1.93 (2.0) - 2.92 (2.7)	Sulphur-selenium
2.42 - 2.78	Thallium bromide-thallium iodide
2.7 (2.92) - 3.17	Selenium-arsenic selenide

Determination of the principal refractive indices of optically anisotropic substances (Chart 6.2, Fig. 6.3)

(1) Statistical method

The principal refractive indices of optically anisotropic substances can be determined by ascertaining the n_α' and n_γ' values on many different grains with light transmitted at different directions.

The measurements are analyzed statistically by the following relations:

Optically uniaxial

n_o corr. to the n_γ' or n_α' value if this is constant in all measurements,
n_e corr. to the maximum or minimum of the n_α' or n_γ' value if this is different in all measurements.

Optically biaxial

n_γ corr. to the maximum n_γ' value measured,
n_α corr. to the minimum n_α' value measured,
n_β corr. to the mean of the minimum n_γ' value measured and the maximum n_α' value measured.

Grain specimen with grains of refractive index n_g, any immersion liquid of refractive index n_l.

↓

Determine the ratio n_g:n_l.
by Becke line, Schröder-van der Kolk effect, brightness relation between grain and background in phase contrast (monochromatic light) etc.

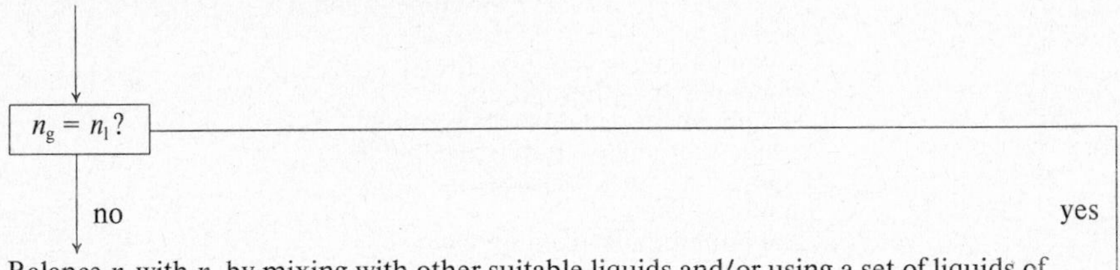

Balance n_l with n_g by mixing with other suitable liquids and/or using a set of liquids of graded n_l.

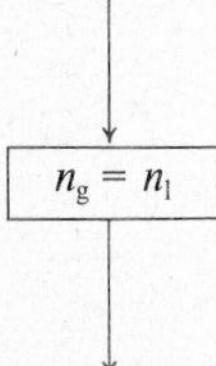

Measure n_l by refractometry to obtain n_g. ←

Chart 6.1. Immersion method of determining refractive indices of optically isotropic substances

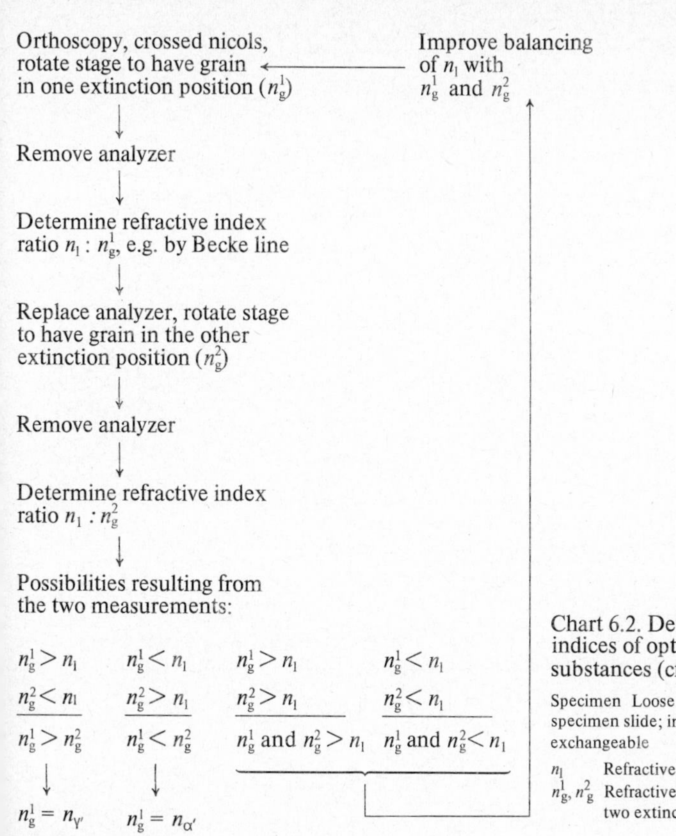

Chart 6.2. Determination of refractive indices of optically anisotropic substances (cf. Fig. 6.3)

Specimen Loose grains, permanently fixed to specimen slide; immersion liquid thus exchangeable

n_l Refractive index of immersion liquid

n_g^1, n_g^2 Refractive indices of a selected grain in the two extinction positions

The accuracy of the method increases with the number of different section planes examined.

(2) Determination at defined transmission directions

The exact determination of the principal refractive indices requires examination at defined transmission directions or of sections of defined orientation, respectively. Optically uniaxial substances require one special orientation of the optical indicatrix, whereas biaxial ones require two. Suitable crystals are selected by the interference figures seen in conoscopy. The order of interference colours observed in the orthoscopic ray path is another important aid in identifying transmission directions.

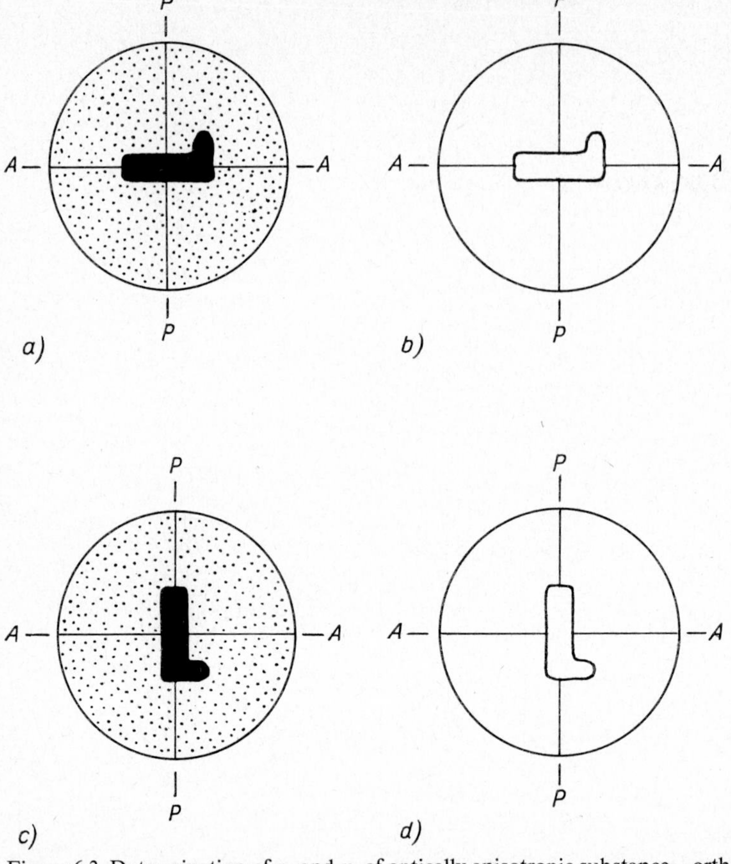

Figure 6.3. Determination of n_α and n_γ of optically anisotropic substance – orthoscopic procedure
a) setting the grain to extinction between crossed nicols b) determination of the relative refractive index difference $n_{grain} \gtrless n_{liquid}$ with the analyzer (A) out c) setting the grain to the other extinction position between crossed nicols (stage rotated through 90°) d) as b)

Measurement of n_o and n_e in uniaxial substances

The section must be parallel to the optic axis of the crystal. The interference figure must show a cross of wide, dark, rather blurred isogyres, with the four quadrants slightly brightened. The cross brightens up upon the slightest rotation of the stage. The microscope then being switched to the orthoscopic path, the section plane just found will be characterized by the highest-order interference colours. The two principal refractive indices are then found at the two extinction positions as described in Chart 6.2 (Fig. 6.3).

By determining the optic sign we obtain n_e and n_o.

Optically positive: greater value = n_e, smaller one = n_o;
optically negative: greater value = n_o, greater one = n_e.

Measurement of n_γ and n_α in biaxial substances

The section must be perpendicular to the optic normal, i.e. parallel to the optic axial plane. The interference figure is similar to the one described above for uniaxial specimens. The values of n_γ and n_α are determined according to Chart 6.2.

Measurement of n_β in biaxial substances

The refraction of light by a crystal is equal to n_β if transmission is parallel to an optic axis. In the conoscopic path, the vertex of the black hyperbola will then be at the centre of the crosshairs. Orthoscopically, the specimen appears optically isotropic if viewed between crossed nicols. (Thick specimens appear dark, or slightly brightened in case of slight inclination from the axial direction.) Refractometry according to Chart 6.1 yields n_β.

6.1.1.2 Phase contrast measurement [6.35 to 6.38, 7.54, 7.83]

The phase contrast technique is mainly suitable for the highly accurate measurement of refractive indices of finely grained specimens (1 to 50 μm).

The technique requires immersion, but is much more sensitive than bright-field methods with regard to balancing the refractive indices of grain and liquid. Grain sizes above 50 to 80 μm will cause disturbing effects that make measurement extremely difficult.

(1) Measurement by monochromatic light

A strong green filter of $\lambda_{max} = 520$ nm in the illuminating rays will suit most examinations.

Relationships for positive phase contrast:

Object darker (darker green) than background, *with bright fringe*	$n_{obj.} > n_{imm.liqu.}$
Object indiscernible against background	$n_{obj.} = n_{imm.liqu.}$
Object brighter (brighter green) than background, *with dark fringe*	$n_{obj.} < n_{imm.liqu.}$

The principal refractive indices are determined by the procedures given in Charts 6.1 and 6.2.

(2) Measurement by white light

Depending on how far the refractive indices of grain and liquid differ, various colour effects are observed (Table 6.12). With refractive indices being equal, these effects are clearly characteristic; at the same time, the grain contours more or less merge with the background.

Table 6.12. Object image variation in the phase microscope with varying refractive index differences between grain (n_g) and immersion liquid (n_l) (Liquid: mixture of cinnamaldehyde and diethyl oxalate)

Phase difference	Monochromatic light (green)	White light
$n_g \gg n_l$	Contrast reversal; total grain or thickest parts brighter than background	Sharp outline, grain surrounded by thin black line, pale colour
n_g clearly $> n_l$	Distinct contrast, dark grain, bright halo	Grain dark blue, halo less coloured, clear outline
n_g slightly $> n_l$	Weak contrast, grain of darker green than background, bright halo	Grain dark blue, coloured halo, blurred outline
$n_g = n_l$	Grain merged with background	Grain brilliant blue, coloured halo, blurred outline
n_g slightly $< n_l$	Weak contrast, grain of brighter green than background, dark halo	Grain light blue, coloured halo, blurred outline
n_g clearly $< n_l$	Distinct contrast, bright grain, dark halo	Grain light greyish blue, halo less coloured, clear outline
$n \ll n_l$	Contrast reversal; total grain or thickest parts darker than background	Sharp outline, grain surrounded by thin black line, pale colour

Table 6.13. High-dispersion immersing liquids, suitable for phase microscopy (acc. to the literature)

Immersion liquid	n_D (20°C)	$n_F - n_C$ (20°C)
Diiodomethane	1.742	0.0369
α-Iodonaphthalene	1.697	0.0368
α-Chloronaphthalene	1.638	0.0300
Cinnamaldehyde	1.621	0.0430
Ethyl cinnamate *(trans)*	1.558	0.0277
Ethyl salicylate	1.521	0.0210
Mineral oil	1.480	0.0125
Glyceryl triacetate	1.429	0.0072
Tributyl phosphate	1.422	0.0065
Diethyl oxalate	1.410	0.0077
Triethyl phosphate	1.405	0.0056

If suitable immersion media are selected, all grains having the same refractive index show the same colour. Thus, the fraction of a phase in the grain aggregate can be estimated at a glance or established automatically by photometry.

Immersion liquids for white light examinations should have as high a chromatic dispersion as possible (Table 6.13). A mixture of cinnamaldehyde (n_D = 1.6195) and oxalic acid ester (n_D = 1.4104) has been found useful for many substances. Table 6.14 lists the mixing proportions for obtaining various refractive indices within that interval.

The dispersion staining adopted by optically isotropic substances is independent from stage rotation, whereas anisotropic substances vary in colouration as

Refractive index	Proportion of cinnamaldehyde (% by vol.)	$n_F - n_C$ (20 °C)
1.409	0	0.008
1.420	5	0.009
1.431	10	0.011
1.440	15	0.013
1.451	20	0.014
1.462	25	0.016
1.473	30	0.018
1.482	35	0.020
1.494	40	0.022
1.505	45	0.023
1.514	50	0.025
1.525	55	0.027
1.536	60	0.029
1.547	65	0.030
1.558	70	0.032
1.567	75	0.034
1.578	80	0.036
1.590	85	0.038
1.600	90	0.039
1.610	95	0.041
1.621	100	0.043

Table 6.14. Mixing proportion, refractive index and dispersion of a cinnamaldehyde - diethyl oxalate mixture

the stage is rotated. To determine n_γ' and n_α', staining for the n_γ' and n_α' directions is effected successively by setting the grain to its two corresponding extinction positions in the respective section plane (Chart 6.2, Fig. 6.3).

The higher the birefringence ($n_\gamma' - n_\alpha'$), the more the refractive index in random sections will vary between the n_γ' and n_α' values as the stage is rotated. This means that the equality of refractive indices and, hence, dispersion staining, are restricted to narrow ranges of stage rotation. General rules are:

Low birefringence	All grains are stained equally well in any transmission direction.
High birefringence	The grain is stained within such a stage rotation angle only (i.e. angle between vibration directions in the grain and the polarizer) in which grain and liquid have equal n_D.

6.1.1.3 Universal stage measurements

Though less accurate than the immersion techniques described, refractive index determination using the universal stage often is the only possibility for the chemist to obtain physical data of decomposing or strongly hygroscopic substances. The decisive facility is that any immersion liquid or the parent solution itself may be used. Both the measurement and its analysis are rather laborious. The following paragraphs are only meant as a first introduction to the fundamentals of these techniques [6.39 to 6.45].

Measuring methods using the spindle stage largely correspond to the universal stage techniques. The smaller freedom of rotation makes spindle stage measurement even more time-consuming (Table 6.15).

Table 6.15. Special methods of refractive index determination and their applications

Method	Principle	Applications	References
Interference microscopy	Beam split up into 2 branches, separate modulation, interference between the recombined beams, analysis of the interference figures.	Refractive index measurement in organic high polymers and decomposing or hygroscopic compounds, precise measurements in fine powders, high-contrast presentation of phases of equal n_D in reaction products, presentation and measurement of refractive index gradients in solid and liquid media (e.g. in high polymer fibres and materials and in dissolution and crystallization processes in the liquid phase).	[3.2, 6.48–6.51]
Schröder-van der Kolk shadow method	Lateral insertion of an opaque stop into the tube above the objective (e.g. the outer edge of compensator slide). The grain will appear brighter on the side opposite the stop, if the grain has a higher refractive index than the liquid, and vice versa (Fig. 6.4).	Crystalline compounds and glasses of large grains, by immersion.	
Variation method	By varying the temperature and the light wavelength (Tables 6.17 and 6.18), the refractive indices of grain and immersion liquid can be balanced without changing the liquid.	Investigation of single-grain microsamples of decomposing substances that do not allow repeated change of immersion liquids. Fine adjustment after pre-selecting a liquid of nearly the same refractive index.	[6.53]
Spindle stage method	The grain is rotated within the liquid contained in the specimen cell to adjust special transmission directions permitting (favourably combined with the variation method) to determine principal refractive indices of immersed objects. (Change of liquids by fast cell	Determination of refractive indices of decomposing or hygroscopic compounds or of microsamples (single grains).	[6.54]

Table 6.15, continued

Method	Principle	Applications	References
	changer or double cell with crystal etalon as a micro-refractometer.)		
Glass powder method	Reversal of the immersion method principle. Anisotropic crystals are fused. The isotropic melt is characterized by one refractive index only, which is easier to be found than the principal refractive indices of anisotropic crystals. Measurement by strewn-in glass particles of a series of known refractive indices. Equality of refractive indices is set by the Becke line or by the behaviour in the phase microscope.	Applicable only for congruently melting crystals and melts. Fast method for testing organic crystalline compounds.	[6.55]
Bordering dark field method	In special dark-field beam arrangement only the light diffracted by the object contributes to the image. In white light, grain boundaries show characteristic colours (violet-blue) if grain and liquid have equal refractive indices.	Fore-runner of the phase contrast method to determine refractive index in fine powders.	[6.56]
Flow-through method after Kordes	The grain is contained in a chamber slowly passed by an immersion liquid. By varying the liquid's composition its refractive index can be balanced with that of the grain by the usual criteria.	Hygroscopic substances, single grains.	[6.57]
Tyndall effect method (after von Engelhardt)	The intensity of light scattered in suspensions corresponds to the refractive index difference between the liquid and solid particles. With refractive indices equal, intensity is at its minimum (Fig. 6.5).	Grain sizes below 1 µm; n_α and n_γ of optically anisotropic substances determinable.	[6.58, 6.59]

Table 6.15, continued

Method	Principle	Applications	References
Christiansen filter method	Based on light scatter measurement, this method avoids the error sources of the Tyndall effect method. The solid particles are densely packed so that they cannot change their relative positions during measurement. The specimens are held in small cells (Christiansen filters).		[6.60, 6.61]
Interferometer method	Computation of solid particle refractive indices from refractive indices of the suspension and the suspending liquid and the solid mass concentration of the suspension.	Grain sizes below 1 μm	[6.60]
Critical angle of total reflection	The microscope is used as a refractometer. *2 versions:* 1. Laterally incident light, special eyepiece, universal stage with highly refractive segment. 2. Transmitted light, refractive index balancing by temperature variation, measurement of the liquid refractive index at the critical angle with lateral illumination.	Single, smooth-faced crystals	[6.62]
Brewster angle method	Arrangement as in critical angle measurement by incident light	Transparent, optically isotropic substances only (e.g., glasses). Specimen surface must be homogeneous and well polished.	[6.63]
Duc de Chaulnes method	Computation of refractive index from thickness measurement with the fine focussing control.	Plane-parallel grains (limited accuracy).	[6.64]
Cherkassov method	Immersion method utilizing dispersion colour effects, requires parallel beam, special objective with stopped-down iris, filters removed, bright lamp.	Grain sizes above 20 μm	[6.65, 6.66]

Table 6.15., continued

Method	Principle	Applications	References
	Isotropic grains $n_{grain} > n_{liquid}$: grain blue, green $n_{grain} = n_{liquid}$: grain yellow $n_{grain} < n_{liquid}$: grain red **Anisotropic grains** Characteristic colour of grain margins *Sections normal to optic axes:* equal colour all around the grain edge (n_β) *Any other sections:* Edge colours varying with the ratio of n_γ' or n_α' to liquid. Blue edge: n_γ' ($n_{grain} > n_{liquid}$) Red edge: n_α' ($n_{grain} < n_{liquid}$) (Blue and red edges on opposite sides of the grain.)		
Varia- scopic method after Mozzherin	Various grain transmission directions are implemented by special illumination arrangement and shifting the grain within the visual field, so that it is transmitted at a different angle in each place. The place where refractive indices are equal can be measured by a coordinate system. This is a special version of the immersion method.	Grains and thin sections. The method is little known so far.	[2.46]
Refractive index deter- mination in solid embedding media	*Embedding media:* solidified melts or briquetted powders	Highly refractive substances; high-temperature microscopy	[6.67]
Indirect refractive index determination from a solution	The refractive index of a solid phase is determined from that of its solution in organic solvents. Solutions must be diluted, volume additivity provided. The concentration of the solution, and the partial volume and refractive index of the solvent must be known.	Organic high polymers, to avoid difficulties in preparation	[6.68]

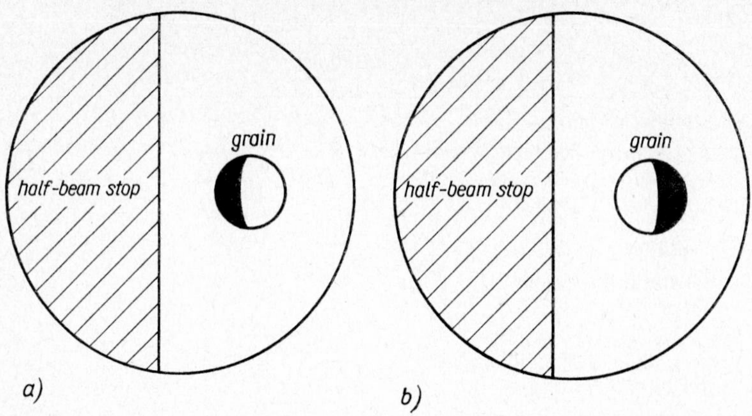

Figure 6.4. Schröder-van der Kolk method of refractive index determination (shadow or oblique illumination method)

a) grain shadow facing field shadow:
$n_{grain} > n_{liquid}$ b) grain shadow opposite field shadow: $n_{grain} < n_{liquid}$

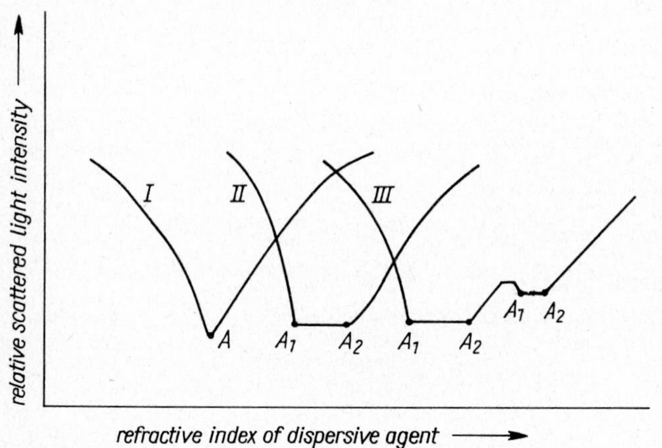

Figure 6.5. Refractive index of submicroscopic particles as a function of scattered light intensity, acc. to [6.58]

I optically isotropic substances
II optically anisotropic substances
III mixture of two optically anisotropic phases
A, A_1, A_2 n, or n_α and n_γ, of the substance examined

(1) Determining the mean refractive index of optically uniaxial substances [6.46]

A crystal to be measured by this method should be plane-parallel (thin section, platelet-shaped crystal in a grain specimen). Its optic axis must be parallel to the specimen plane (slide surface). This transmission direction is selected in the same way as described for n_0 and n_e determination (cf. 6.1.1.1). With the crystal on the universal stage, the A_4 (K) axis is rotated through 5° intervals and the retardation measured in each position (cf. 6.1.2). After reduction of the measured values to the specimen thicknesses corresponding to the various tilt angles, a mean refractive index is found by graphical determination or computation:

$\bar{n} = (n_e n_0)^{1/2}$.

By computation,

$$\bar{n} = n_s \left(\frac{Ri_1^2 \cdot \sin^2 i_1 - Ri_2^2 \cdot \sin^2 i_2}{Ri_1^2 - Ri_2^2} \right)^{1/2}$$

where Ri_1 and Ri_2 are the retardations at tilt angles i_1 and i_2 of the universal stage axes, and n_s is the refractive index of the segment used. Where birefringence and axial angle are small, the procedure can be applied to biaxial substances, with the triaxial optical indicatrix being treated like an ellipsoid of revolution:

Optical sign + $\quad n_\alpha \approx n_\beta \neq n_\gamma$

Optical sign − $\quad n_\alpha \neq n_\beta \approx n_\gamma$

2. *Determining the principal refractive indices and the optic axial angle 2V of optically biaxial substances* [2.11]

Basically, this method was developed by Fedorov [3.66]. It requires a section in which one optical plane of symmetry of the indicatrix is not too far inclined from normal to the specimen surface. The method permits to determine $2V$ to an accuracy of 0.1° and refractive indices to the third decimal place. The angle S (uncorrected half axial angle) is measured once each via n_γ and n_α, the refractive index of the segment (n_s) to be greater than that of the respective grain. That way, $2\,S\,n_\alpha$ and $2\,S\,n_\gamma$ are smaller than $2\,V n_\alpha$ and $2\,V n_\gamma$ and fall within the range of A4 (K) axis rotation.

Another technique reported by Fedorov employs the deviations in stereographic projection from the nominal 40° arc lengths between the optical axes of symmetry. The result is a mean refractive index of the object relative to the segment. It is not very accurate.

Quite a number of modifications of Fedorov's method using different arrangements (cf. 2.11) have been proposed by various authors. One successful method is to use a heated universal-stage refractometer with monochromatic light of variable wavelength (Emons double variation method).

Analysis of birefringence distribution relative to planes of symmetry acc. to Berek [2.11]

This method is based on the fact that birefringence is distributed symmetrically on either side of an optical plane of symmetry of the indicatrix. If the retardations reduced by the tilt angle show systematic deviations, the refractive indices of segment and grain differ. From the deviations, a mean refractive index can be derived by the equations

$$V n_\alpha + V n_\gamma = 90°$$
$$S n_\alpha + S n_\gamma < 90°$$

$$n_s \sin S n_\alpha = n_\beta \sin V n_\alpha$$
$$n_s \sin S n_\gamma = n_\beta \sin V n_\gamma$$
$$n_s \sin S = n_\beta \sin V$$

By substituting experimental values for n_β one obtains $V n_\alpha$ and $V n_\gamma$ angles whose sum is 90°. Thus, n_β and $2V$ are known.

In order to ascertain n_α and n_γ, first determine the approximate specimen thickness. The desired quantities result from the differences $n_\gamma - n_\beta$ (n_α parallel to the microscope axis) and $n_\beta - n_\alpha$ (n_γ parallel to the microscope axis).

Various attempts have been made to extend the applicability of the universal stage by combination with the immersion method. One favourable solution is provided by an „Index" accessory designed after Berek's universal stage refractometer. The refractive index of the immersion medium is balanced with that of the grain by temperature variation (hemisphere with heated ring). After switching from transmitted to diffusely reflected light, the refractive index can be determined from the position of the boundary of total reflection observed through a special eyepiece. With the grain oriented correspondingly, highly accurate values of n_α, n_β and n_γ are found in succession.

6.1.1.4 Special methods to determine refractive indices

In addition to the techniques described, there are a number of methods for special chemical applications. They are presented in Table 6.15.

Tables 6.16 and 6.17 supply data for low-temperature refractometry.

6.1.2 Birefringence

Birefringence Δn is an important parameter for characterizing the structures of both crystalline, optically anisotropic compounds and semicrystalline, organic high polymers and liquid crystals. Birefringence has to be determined indirectly through measurements of retardation and specimen thickness (cf. 2.3.1).

For measuring maximum birefringence ($n_\gamma - n_\alpha$), special sections are required, viz. normal to the optic axis in optically uniaxial, and normal to the optic axial plane in biaxial substances. For adjusting these section planes see 6.1.1.

This section only describes retardation measurements [6.69 to 6.79]. For specimen thickness measurement, see Table 6.27.

With polarizers crossed at 45° in the orthoscopic ray path, the amount of retardation can be estimated from the nature of the interference colours observed (Fig. 6.6), provided that grain thickness is known or grains of equal thickness or size are compared.

Table 6.16. Refractive indices of immersion liquids for low-temperature microscopy acc. to [6.26]

Liquid	Temperature (°C)														
	+20	+17	0	−9	−15	−20	−28	−40	−58	−60	−64	−80	−88	−95	−100
Ethyl iodide	1.514	1.520	1.526		1.532	1.538	1.540	1.549	1.559	1.561		1.572		1.584	1.586
Acetone-hexane mixture		1.372		1.381					1.404		1.412		1.423		
Ethyl iodide-diiodomethane mixtures															
95:5	1.522		1.533			1.545		1.556		1.567		1.579			1.591
90:10	1.532		1.543			1.555		1.566		1.577		1.578			1.601
Ethyl iodide-toluene mixtures															
51:49	1.501		1.512			1.523		1.534		1.545		1.556			1.567
55:45	1.503		1.514			1.525		1.535		1.546		1.557			1.568
63:27	1.505		1.516			1.527		1.537		1.548		1.559			1.570
66:33	1.506		1.517			1.528		1.538		1.549		1.560			1.571
76:24	1.507		1.518			1.523		1.539		1.550		1.561			1.572
84:16	1.508		1.517			1.528		1.538		1.549		1.559			1.569
92:8	1.514		1.525			1.536		1.547		1.558		1.569			1.580

Simple compensators facilitate correct interference colour determination. By insertion of a full-wave plate (1st order red) its fixed retardation of 550 nm is added to, or subtracted from, that of the grain, depending on the positions of n_γ and n_α of grain and compensator. The new retardation shows up by a changed interference colour; the nature of the colour change permits the initial retardation to be classified reliably (Fig. 6.6).

A quartz wedge compensator serves the same purpose, with the advantage of the comparison path difference being continuously variable. The microscope's condenser iris is narrowed down to permit easier recognition of the compensating effect. The interference colours produced by the wedge are seen more distinctly with the Bertrand lens swung in.

In a subparallel position [6.69], the full-wave compensator allows very small retardations to be discerned. From its initial position (n_γ aligned with the vertical crosshair or with the vibration direction of the polarized light), the compensator is tilted about 7° to the right or left by means of a lever. Birefringence will then show up by characteristic colour effects.

Retardation measurement [6.70 to 6.77]

Measurement is performed by compensating for zero retardation. This means that the retardation of the object is fully compensated by that introduced by tilting the compensator. This state is characterized by the object appearing dark against a bright background. Compensator types are diverse to suit special path difference ranges (Table 6.20).

The use of a half-shade plate increases the accuracy of compensation setting, especially where retardations are very small. With polarizers in exactly crossed position and without a specimen on the stage, the four quadrants of the plate are uniformly dark, same as with the compensator inserted in zero position. By turning the compensator in order to compensate for the specimen inserted, two opposite quadrants brighten up, whereas the other two get dark. The state of complete compensation is identified by the object being equally dark in the a and b quadrants.

Table 6.17. Dispersion of refractive indices of immersion liquids for low-temperature microscopy acc. to [6.26]

Liquid	Temperature	Light wavelength (nm)					
	in °C	486	520	555	589	620	656
Ethyl iodide	+17	1.529	1.524	1.520	1.517	1.515	1.513
	−15	1.541	1.536	1.533	1.529	1.526	1.524
	−28	1.550	1.544	1.540	1.537	1.534	1.532
	−58	1.570	1.564	1.559	1.556	1.553	1.551
	−95	1.595	1.589	1.586	1.580	1.575	1.575
Acetone-hexane mixture	+17	1.375	1.373	1.372	1.372	1.370	1.369
	−9	1.385	1.383	1.381	1.380	1.379	1.378
	−57	1.408	1.406	1.404	1.403	1.402	1.401
	−64	1.416	1.414	1.412	1.410	1.409	1.408
	−88	1.427	1.425	1.423	1.421	1.420	1.419

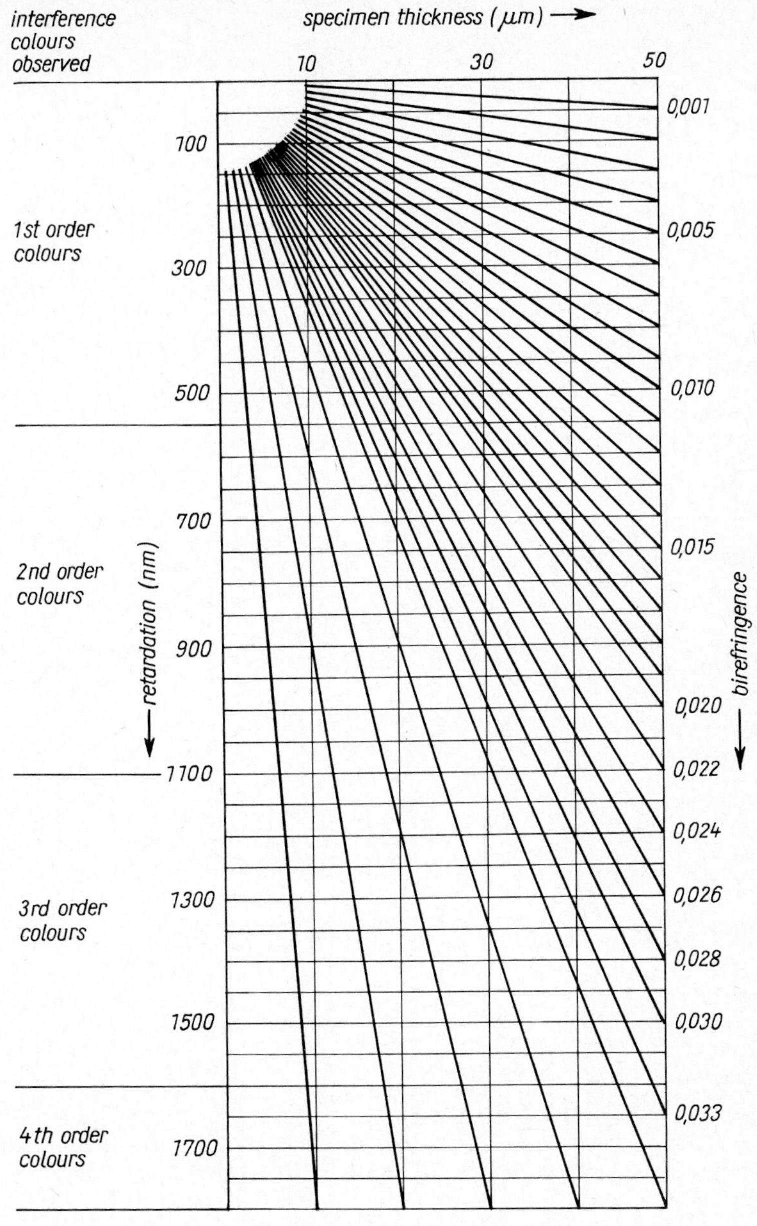

Figure 6.6. Retardation, birefringence and thickness of optically anisotropic substances as functions of interference colours observed

Table 6.18. Immersion liquids for wavelength/temperature double variation acc. to [7.3]

Liquid	Boilg. pt. (°C)	n_C (10 °C)	n_C (18 °C)	n_C (50 °C)	n_D (18 °C)	n_D (50 °C)	n_F (10 °C)	n_F (18 °C)	n_F (50 °C)	Temperature coefficient $(n/°C) \cdot 10^{-4}$
Ethyldiiodoarsine	-	1.808	-	1.777	-	-	1.834[1])	-	1.800[2])	8.1
Diiodomethane + Sulphur	-	1.775[3])	-	-	-	-	-	-	-	-
Diiodomethane	180	1.737	1.732	1.711	1.742	-	1.774	1.769	1.747	6.8
α-Iodonaphthalene	305	1.698	1.693	1.678	1.702	-	1.734	1.730	1.714	4.7
α-Iodonaphthalene + α-bromonaphthalene	-	1.675	1.669	1.652	1.6732[4])	1.6572	-	-	-	5.2
α-Bromonaphthalene + diiodomethane	-	-	-	-	1.664	-	1.687	1.682	1.665	5.2
Phenyl isothiocyanate	220	1.646	1.641	1.623	1.639	-	-	-	-	5.3
s-Tetrabromoethane	-	-	-	-	1.633[5])	-	1.660	-	1.640	4.7
α-Chloronaphthalene	259	1.629	-	1.611	1.620	-	1.644	1.638	1.620	5.5
Iodobenzene	188	1.618	1.613	1.596	-	-	1.628	-	1.607	5.0
Bromoform	149	-	1.593	-	-	-	1.608	-	1.586	5.5
Glycerol tribromohydrin	220	1.583	-	1.560	-	-	1.590	-	1.568	5.5
m-Chlorobromobenzene	196	1.570	-	1.550	1.573	-	-	1.589	-	5.1
o-Toluidine	200	-	1.567	-	-	-	1.573	-	1.552	5.0
o-Bromotoluene	181	1.555	-	1.534	1.553	-	-	-	-	4.8
Nitrobenzene	211	-	-	-	1.547	-	1.568	1.564	1.547	4.9
o-Nitrotoluene	222	1.544	1.541	1.525	1.5490[7])	1.5352	-	-	-	4.8
Clove oil	117[6])	-	-	-	1.539	-	-	-	-	5.6
1,2-Dibromoethane	131	-	-	-	-	-	1.551	-	1.533	4.6
Ethyl salicylate	233	1.519	-	1.501	1.520	-	-	1.528	-	5.4
1,2-Dibromopropane	-	-	1.516	-	1.504	-	-	-	-	4.9
Pentachloroethane	-	-	-	-	-	-	1.531	-	1.512	4.7
Ethyl benzoate	211	1.502	-	1.483	-	-	1.503	-	1.483	4.9
Cymene	176	1.490	-	1.471	-	-	-	-	-	-
Methyl thiocyanate	130	1.469	1.466	1.449	1.469	-	1.481	1.477	1.459	5.4
1,3-Dichloropropane	-	-	1.446	-	1.449	-	-	1.454	-	4.9
Petroleum + NM 15 silicone fluid	-	-	-	-	1.4390[8])	1.4270	-	-	-	4.0
Ethyl monochloro-acetate	-	-	1.420	-	1.423	-	-	1.428	-	4.9

[1]) n for $\lambda = 550$ nm
[2]) n for $\lambda = 550$ nm
[3]) n for $t = 25$ °C and $\lambda = 589.3$ nm
[4]) for $t = 20$ °C
[5]) n for $t = 24$ °C
[6]) Flash point
[7]) n for $t = 20$ °C
[8]) n for $t = 20$ °C

Table 6.19. Immersion liquids for the temperature variation method acc. to [2.4]

Liquid	Refractive index	Temperature coefficient of refraction
Diiodomethane	1.747	68
α-Iodonaphthalene	1.706	47
α-Iodonaphthalene + α-bromonaphthalene	1.681	47
α-Bromonaphthalene	1.657	48
Phenyl isothiocyanate	1.655	56
s-Tetrabromoethane	1.642	53
1,1,2,2-Tetrabromoethane	1.635	54
Iodobenzene	1.625	57
Chinoline	1.622	49
Bromoform	1.603	60
Aniline	1.582	52
o-Toluidine	1.577	51
Bromobenzene	1.560	54
1,2-Dibromoethane	1.555	60
Nitrobenzene	1.553	51
o-Nitrotoluene	1.551	49
1,2-Dibromoethane	1.543	56
Chlorobenzene	1.525	55
1,2-Dibromopropane	1.524	54
Anisole	1.515	51
1,3-Dibromopropane	1.513	48
Pentachloroethane	1.508	47
Methylfluorate	1.491	45
Methyl thiocyanate	1.473	54
Isoamyl sulphide	1.458	45
1,3-Dichloropropane	1.453	49
Ethyl dichloroacetate	1.441	47
Ethyl monochloroacetate	1.426	47

The light source should have a high light output that stays constant during a measurement. Visual measurements should be made in a dimly lit room. The microscope is focused optimally on the object in its initial position. The focus must not be changed during the compensation process. The microscope optics must not have stress-induced birefringence.

Charts 6.3 and 6.4 show the procedure of retardation measurement step by step.

Depending on the nature and condition of the specimen, it may be advisable to narrow down the condenser iris in order to achieve favourable contrast.

Dispersion of birefringence

The dependence of birefringence on light wavelength is another substance-specific quantity, to be employed, e.g., where compounds have almost equal opti-

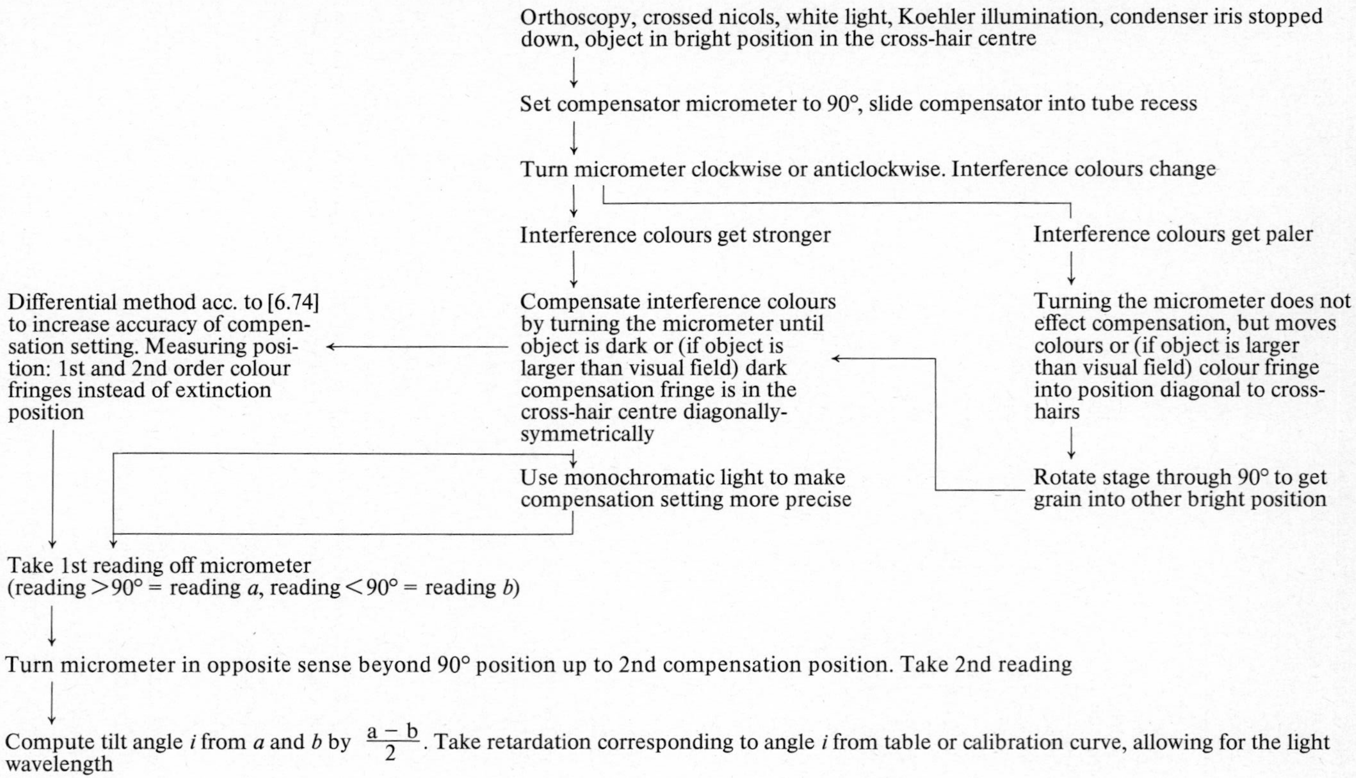

Orthoscopy, crossed nicols, white light, Koehler illumination, condenser iris stopped down, object in bright position in the cross-hair centre

↓

Set compensator micrometer to 90°, slide compensator into tube recess

↓

Turn micrometer clockwise or anticlockwise. Interference colours change

↓ (a) Interference colours get stronger

↓

Compensate interference colours by turning the micrometer until object is dark or (if object is larger than visual field) dark compensation fringe is in the cross-hair centre diagonally-symmetrically

→ (optional) Differential method acc. to [6.74] to increase accuracy of compensation setting. Measuring position: 1st and 2nd order colour fringes instead of extinction position

→ (optional) Use monochromatic light to make compensation setting more precise

↓ (b) Interference colours get paler

↓

Turning the micrometer does not effect compensation, but moves colours or (if object is larger than visual field) colour fringe into position diagonal to cross-hairs

↓

Rotate stage through 90° to get grain into other bright position (→ back to "Compensate interference colours")

↓

Take 1st reading off micrometer (reading $>90°$ = reading a, reading $<90°$ = reading b)

↓

Turn micrometer in opposite sense beyond 90° position up to 2nd compensation position. Take 2nd reading

↓

Compute tilt angle i from a and b by $\frac{a-b}{2}$. Take retardation corresponding to angle i from table or calibration curve, allowing for the light wavelength

Chart 6.3. Measurement of great retardations with rotary compensators 0 to 6λ and 0 to 130λ

Table 6.20. Methods of retardation measurement and their applications

Compensator	Application
Full-wave plate (1st order red) Quarter-wave plate (1st order grey) Quartz wedge, half-wave to **3λ**	Qualitative determination of retardation
Full-wave and quarter-wave plates in subparallel position	Qualitative determination of small retardations
Measuring compensator with calcite combination plate (Ehringhaus method)	0 – 133λ (range)
Measuring compensator with quartz combination plate (Ehringhaus method)	0 – 6λ (range)
Measuring compensator with quartz combination plate and Wright eyepiece	0 – 3λ (range)
Measuring compensator with azimuthal rotation (De Senarmont, Brace-Koehler, Bear-Schmitt and MacCullagh methods)	0 – 1λ 0 – λ/8 0 – λ/16 (range)
Compensation in retardation measurement by interference microscopy (cf. 3.2.2.)	Very small and very great retardations, depending on methods
Automatic retardation measurement	
1. Compensator rotated by motor and gears, linked to a line recorder 2. Intensity measurements without compensators	Very accurate measurement of great and small retardations

Orthoscopy, crossed nicols, white light, Koehler illumination, low condenser aperture

↓ → Accurately determine the 4 extinction positions of the compensator by the angle scale, as permanent references for the microscope

Move object into cross-hair centre in 45° position, insert compensator into tube recess, set compensator to any one extinction position

↓

Turn compensator to have object in maximum darkness

↓

Determine angle ϑ (always $<45°$) between this position and the nearest compensator reference point

↓

Retardation: $R_{compensator} \cdot \sin 2\vartheta$

($R_{compensator}$ = instrumental constant)

Chart 6.4. Brace-Koehler method of measuring small retardations using measuring compensators rotating in azimuth

cal data. The amount of that dispersion is characterized by the Ehringhaus number:

$$N = \left(\frac{(n_\gamma - n_\alpha)_D}{(n_\gamma - n_\alpha)_F - (n_\gamma - n_\alpha)_C} \right).$$

The dispersion is also evidenced by the occurrence of anomalous interference colours.

6.1.3 Optic axial angle

The axial angle $2V$ of the indicatrix of optically biaxial compounds (cf. 2.3.2) can be measured very accurately with any immersing medium. It is therefore a suitable parameter to characterize decomposing or hygroscopic substances. A change in $2V$ is an extremely sensitive indicator of slightest structural changes, a property that is utilized, e.g., for precise mixed crystal investigations or for detecting initial lattice changes.

The major methods of determining $2V$ in chemical compounds are outlined by Tables 6.21 and 6.22 [6.81 to 6.111, 6.19].

Table 6.21. Methods of optic axial angle measurement

Ray path	Method	Average error in degrees
Orthoscopic	Computation from the 3 principal indices	1 - 5
	Direct measurement with universal stage, spindle stage or Waldmann sphere	0.5 - 3
	Computation or graphical determination from indicatrix measurements	1 - 5
Conoscopic	Estimation from isogyre curvature	10
	Measurement in oriented sections	
	1. Sections normal to acute bisectrix, both axis emergence points visible	1 - 2
	2. Sections oblique to the acute bisectrix	1 - 5
	Direct measurement with universal stage, spindle stage or Waldmann sphere	0.1 - 3
	Computation or graphical determination from indicatrix measurement	1 - 5

Table 6.22. Accuracy of $2V$ measurements with the universal stage

Ray path	Measuring conditions	$2V$ range in degrees	Average error in degrees
Orthoscopic	**Low objective magnification** (most favourable condenser aperture; crystal and universal stage segments having refractive indices about 1.55)	0 - 60 60 - 90	0.2 - 0.4 0.5 - 1
	High objective magnification (least favourable condenser aperture; refractive indices of crystal and segment differing greatly)	0 - 60 60 - 90	2.5 5 - 6
Conoscopic	**Most favourable condenser aperture** (crystal and universal stage segments having refractive indices about 1.55)	0 - 60 60 - 90	0.1 - 0.2 0.4 - 0.6
	Least favourable condenser aperture (refractive indices of crystal and segments differing considerably)	60 - 90	0.5 - 1

6.1.3.1 Estimation from isogyre curvature

The dark band (isogyre) of an interference figure curves more strongly with smaller $2V$(Fig. 6.7). Observation and estimation are carried out with conoscopic rays and in the 45° position. Comparison with reference specimens of known $2V$ is of advantage.

The interference figures of very small objects can be evaluated if it is possible to occlude the environment by stopping down the field iris. Parts of other grains protruding into the field (or carried into it by stage rotation if objectives are decentred) will totally upset interference figures.

If grains are too small to permit stopping down to a uniformly extinguishing area, special techniques may help in some cases [6.80].

Frequently, the following *fast method* will be successful: Apply a drop of glycerol on the cover slip or directly on the specimen and quickly stir with a stick. A multitude of tiny air bubbles will be produced above the specimen. Put a cover slip on top and focus on the bubbles with orthoscopic rays. Slightly lower the body tube. Every bubble will then show an interference figure that corresponds to the optical behaviour of the object below it. The bubbles constitute lenses of high aperture and short focal distance; they represent the objective, whereas the microscope's Bertrand lens and eyepiece serve as an auxiliary microscope.

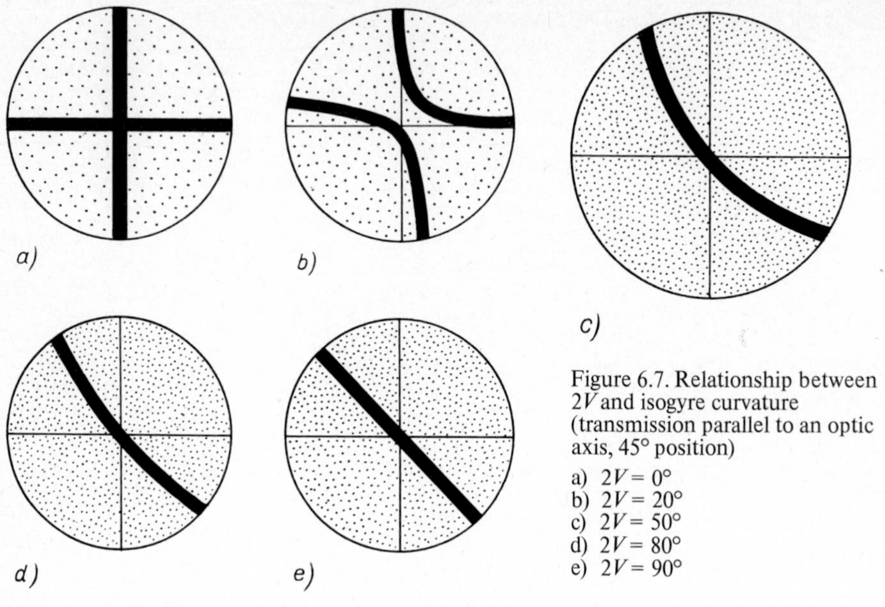

Figure 6.7. Relationship between $2V$ and isogyre curvature (transmission parallel to an optic axis, 45° position)

a) $2V = 0°$
b) $2V = 20°$
c) $2V = 50°$
d) $2V = 80°$
e) $2V = 90°$

6.1.3.2 Measurement in special section planes [2.4, 2.11, 6.82]

These methods of evaluating interference figures in the conoscopic ray path are primarily important in high- and low-temperature microscopy, where it is hardly possible to measure $2V$ by changing grain position.

Sections normal to the acute bisectrix of the indicatrix

Chapter 2.3.2 describes how to select these sections, in which the distance between the vertices of the hyperbolas in the 45° position correspond to the optic axial angle. The limit axial angle at which both vertices are still visible at the margin of the visual field at a given objective N.A. can be taken from Fig. 6.8. Angles larger than those cannot be determined by this method.

Depending on whether the Mallard constant [2.4, 2.11] is known, $2V$ can be found from the vertex distance by one of two procedures. The Mallard constant has to be determined separately for every objective-eyepiece combination by means of a standard specimen of known $2V$. Measurements will be inaccurate near the edge of the visual field, where the relation given below does not strictly apply.

(1) Mallard constant known

Computation is based on the relation

$$\sin V = \frac{K h_E}{n_\beta} = \frac{h_{1,2} K}{2 n_\beta}.$$

K Mallard constant

h_E distance of axis exit point from centre in eyepiece micrometer intervals
n_β mean principal refractive index of the substance
$h_{1,2}$ distance between the two axis exits in eyepiece micrometer intervals

(2) Mallard constant unknown

Computation is carried out by means of the *d*-value (Fig. 6.9):

$$d = \frac{2D}{2R} \qquad \sin V = \frac{dA}{n_\beta}.$$

A numerical aperture
n_β mean principal refractive index of the substance
$2R$ visual field diameter in eyepiece micrometer intervals
$2D$ distance between axis exits in eyepiece micrometer intervals

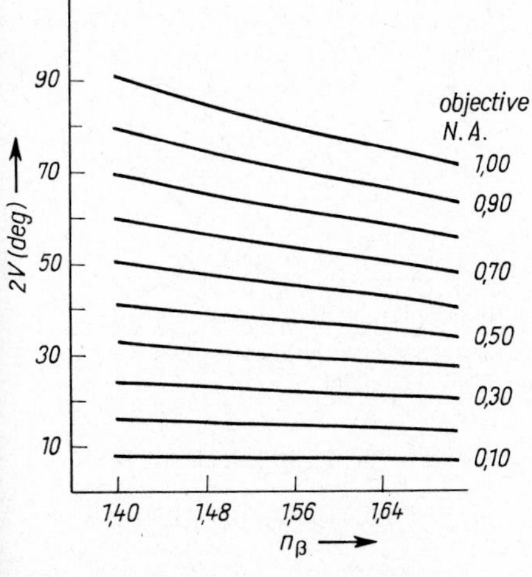

Figure 6.8. Limit optic axial angle $2V$ at which both vertices of the hyperbolic isogyre are still seen at the field margin (for dry objectives of a given N.A.), acc. to [7.114]

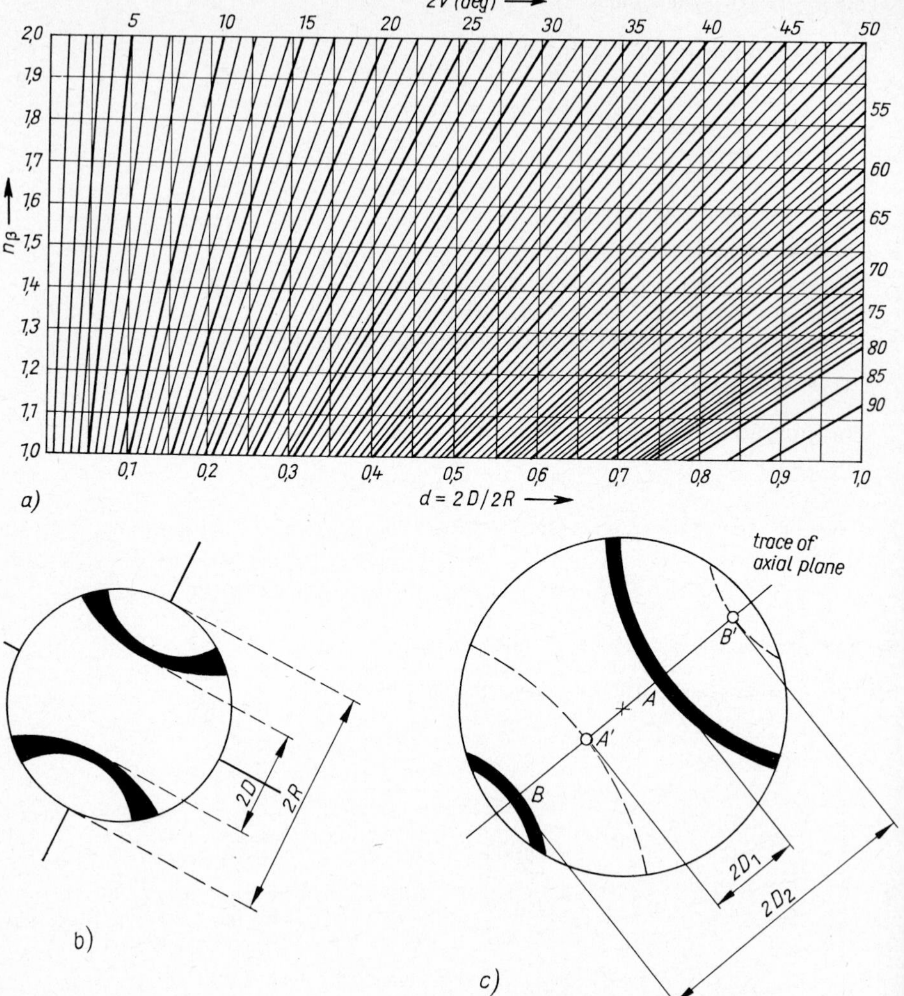

Figure 6.9. Determination of $2V$ by the measurement of conoscopic interference figures
a) nomogram acc. to [7.5] for transmission parallel to acute bisectrix b) key for a)
c) transmission oblique to acute bisectrix, but within optic axial plane

Sections normal to the optic axial plane (but not normal to the acute bisectrix)

Fig. 6.9 shows the interference pattern of such a section in 45° position. The relations applicable are

$2E = E_1 + E_2$
$\sin E_1 = D_1 K$
$\sin E_2 = D_2 K$
$\sin E = n_\beta \sin V$

D_1, D_2 distances of axis exits from centre
E half apparent axial angle
K Mallard constant

Sections with one isogyre only in the visual field (Rittmann's method [6.83])

By this method, the optic axial angle can be determined if the axis exit point is in the centre of the field. An auxiliary quantity Q is required:

$$Q = \frac{R_1}{R_2}.$$

R_1 minimum birefringence in the visual field
R_2 maximum birefringence

R_1 and R_2 are at the edge of the visual field along the intersection of the plane of symmetry with the interference pattern in 45° position.

If Q and n_β are known, $2V$ for an N.A. of 0.85 can be taken directly from the nomograph, Fig. 6.10.

6.1.3.3 Measurement with the universal stage

Universal stage measurement is the most favourable technique of optic axial angle determination, since the random transmission direction of the specimen is compensated by tilting the grain about axes that are normal to each other.

(1) 2V measurement by orthoscopic rays [3.53, 3.58, 3.60, 3.65, 3.69, 6.57 to 6.90, 6.97] (cf. Chart 6.5)

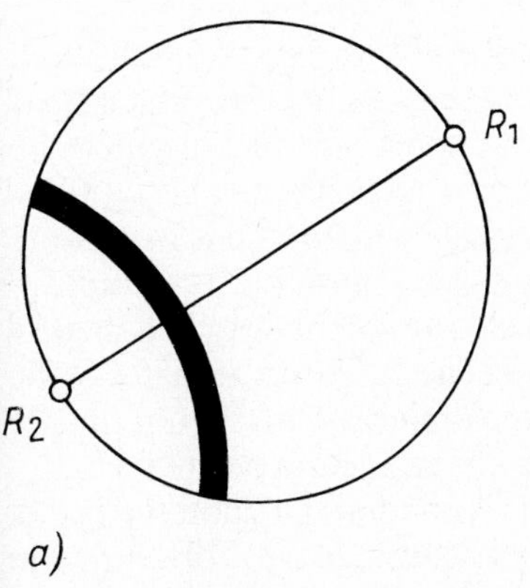

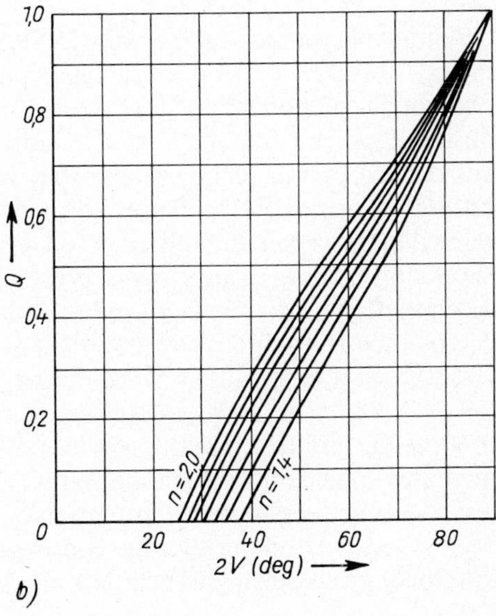

Figure 6.10. Determination of $2V$ acc. to Rittmann [6.83], with one hyperbola visible
a) key
b) nomogram for n_β between 1.4 and 2.0

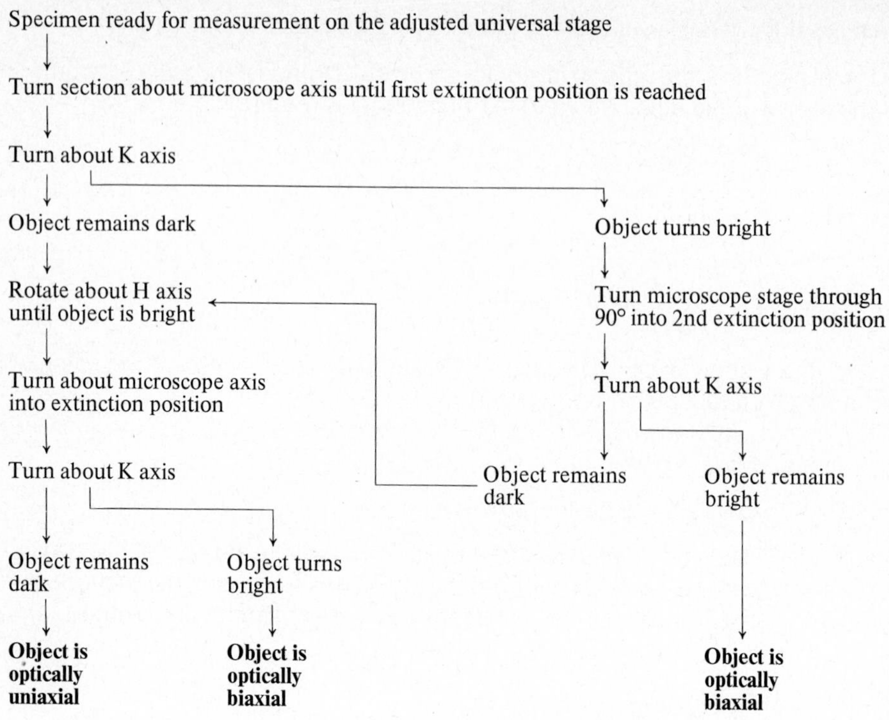

Chart 6.5. Identification of optically uniaxial and biaxial crystals on the universal stage

A grain of random section plane is tilted successively by turning the stage about its A_2 (H) and A_1 (N) axes. In each position it is necessary to check the influence of additional rotation about the A_4 (K) axis. The extinctions and brightenings observed permit the shape of the indicatrix to be concluded on. Being tilted about the A_4 axis, a section retains the extinct state set on it only if one of the optical planes of symmetry is normal to A_4. The indicatrix position thus found is related to the crystallographic axis system by the measurement of morphological elements (cleavage cracks, twins, crystal faces and edges). The coordinates of the measurements are then analyzed by stereographic projection in a Wulff net. $2V$ can be measured directly with the aid of the two 45° extinction positions.

An exact measurement of extinction positions is conditional upon the use of low-N.A. objective and condenser, and a bright light source. In case of extinction dispersion, monochromatic light will help. The accuracy of extinction setting increases considerably if compensators are employed. Orthoscopic methods are preferably applied to the examination of small grains or grain details that cannot be registered by universal stage conoscopy. Examples are the measurements of different optic axial angles at different areas of a mixed crystal of zonal structure.

Orthoscopic measurements are impeded by thin epitaxial growth layers on the grains and by high surface roughness.

(2) 2V measurement by conoscopic rays [3.64, 3.59]

In principle, the procedure corresponds to that described for orthoscopy. By subsequent rotation about the H and N axes, the optical plane of symmetry is adjusted normal to the K axis.

Dispersion effects or very small $2V$ require monochromatic light.

As an advantage, this method provides fast, easily apprehensible orientation about the shape and position of the indicatrix. Minimum grain sizes, though, have to be larger than those required for orthoscopic examination.

(3) 2V determination if only one or no axis is observable [2.11, 3.54, 3.66, 6.19, 6.93, 6.94, 7.5]

In some unfavourable section planes, only one or even none of the crystal optic axes may allow alignment with the microscope axis, even if all tilting facilities of the universal stage are utilized. In limit cases, the use of higher-refraction segments may help; the tilt angle must be carefully corrected (Fig. 6.11).

If it is possible to observe one axis, one measures the other parameters of the indicatrix and can then readily determine the position of the second axis, and thus the axial angle, from the stereographic projection of the measured values.

If no axis can be set, special procedures permit construction of the indicatrix in stereographic projection. Relying on mathematical relationships between the extinction directions and the positions of the optic axes (Biot-Fresnel law), these procedures are independent from the given initial section plane of the grain, but involve considerable arithmetics [6.20, 6.21, 6.91 to 6.94].

From the various approaches it may suffice to describe Berek's method [2.11]. It is applicable to the measurement of two optical symmetry axes of the indicatrix. The retardations in these directions, measured with a compensator and reduced to the true specimen thickness by multiplication by the cosine of the tilt angle, are substituted into the Mallard relation, yielding

$$\sin V_\gamma = \sqrt{\frac{n_\beta - n_\alpha}{n_\gamma - n_\alpha}}\,.$$

The Mallard relation is strictly applicable and provides accurate $2V$ values only if birefringence is small.

If only one principal path difference can be measured, a path difference measurement in an auxiliary direction of known orientation will be helpful.

Accuracy of 2V measurement on the universal stage (Table 6.22) [6.14, 6.98 to 6.101]

Measuring errors increase linearly with greater tilt angles and higher refractive index of the segments. Other influencing factors are the type of objective and the

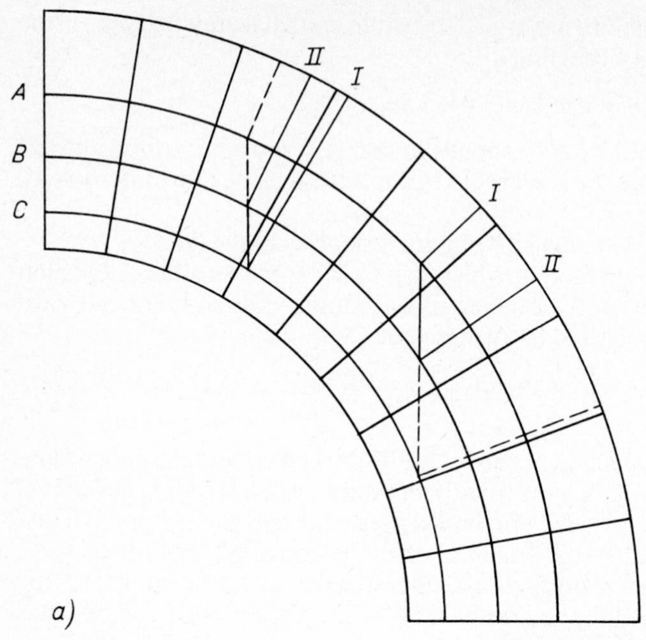

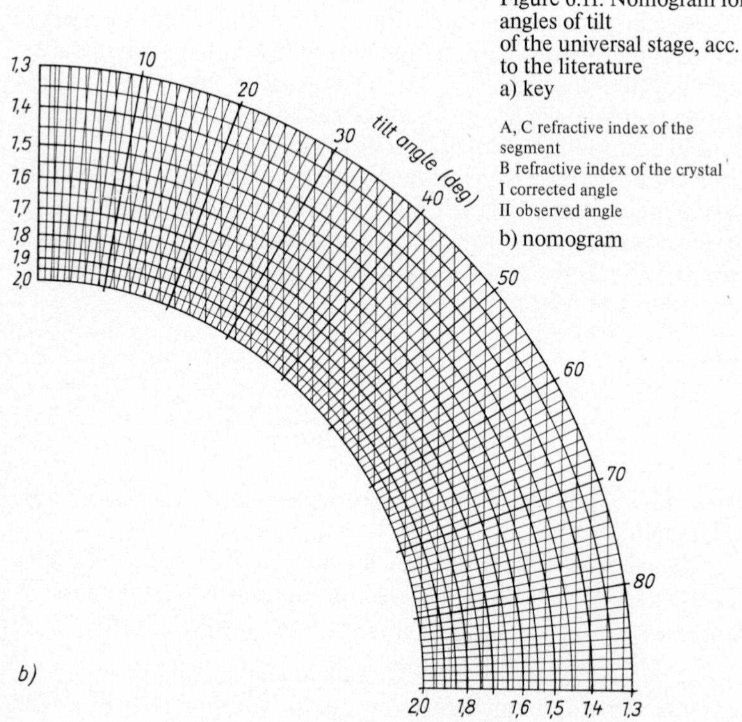

Figure 6.11. Nomogram for correcting angles of tilt of the universal stage, acc. to the literature
a) key

A, C refractive index of the segment
B refractive index of the crystal
I corrected angle
II observed angle

b) nomogram

condenser N. A. Errors in orthoscopic measurements are, in general, higher than those in conoscopic measurements. The conoscopic ray path permits greater accuracy in setting to axial directions. Greater axial angles involve greater errors. Errors due to less-than-optimum beam guiding (failure to adjust Koehler illumination) will show up if the light source image is observed with the Bertrand lens in and the stage inclined. The farther the light source image moves out, the greater will be the errors caused by the maladjustment.

Grain, segment and immersion medium should have approximately equal refractive indices. Substances having very high birefringence cause greater errors than low-birefringence substances.

(4) Measurements with the Waldmann sphere

The restricted rotary movement of the universal stage (120° to 130°) can be extended to about 270° if its two hemispherical segments are replaced by the hollow-glass Waldmann sphere. A special preparation technique permits examination of even small grains [3.101]. Turning the sphere within the stage fixture allows movement of the grain into the initial position most suitable for measurement. Measurement itself follows the universal stage methods described.

(5) Measurement with the spindle stage (cf. 3.2.1) [6.102 to 6.107, 6.18, 6.20, 6.21]

Methods of measuring $2V$ on the spindle stage largely correspond to universal stage methods. As a drawback, the spindle stage is more restricted in its facilities for positioning the grain, except in the case of tiltable designs (Table 3.5). There is a wide variety of methods to determine the indicatrix without direct measurement of both optic axes, e.g. by extinction measurements at various rotary positions of the grain.

6.1.3.4 Computation from the three principal refractive indices

The optic axial angle is mathematically related to the three principal refractive indices n_α, n_β and n_γ (or α, β, γ, respectively).

The literature records various equations for computing $2V$:

$$\sin V = \frac{n_\gamma}{n_\beta} \sqrt{\frac{n_\beta^2 - n_\alpha^2}{n_\gamma^2 - n_\alpha^2}} .$$

Optically positive

$$\sin^2 V = \frac{\gamma^2 (\beta^2 - \alpha^2)}{\beta^2 (\gamma^2 - \alpha^2)}$$

Optically negative

$$\sin^2 V = \frac{\alpha^2 (\gamma^2 - \beta^2)}{\beta^2 (\gamma^2 - \alpha^2)}$$

Optically positive

$$\cos^2 V = \frac{\alpha^2 (\gamma^2 - \beta^2)}{\beta^2 (\gamma^2 - \alpha^2)}$$

$$\tan^2 V = \frac{\gamma^2 (\beta^2 - \alpha^2)}{\alpha^2 (\gamma^2 - \beta^2)}$$

Optically negative

$$\cos^2 V = \frac{\gamma^2 (\beta^2 - \alpha^2)}{\beta^2 (\gamma^2 - \alpha^2)}$$

$$\tan^2 V = \frac{\alpha^2 (\gamma^2 - \beta^2)}{\gamma^2 (\beta^2 - \alpha^2)}$$

Approximation procedures:

$$\sin V = \sqrt{\frac{n_\beta - n_\alpha}{n_\gamma - n_\alpha}}$$

$$\tan V = \sqrt{\frac{n_\beta - n_\alpha}{n_\gamma - n_\beta}}$$

Optically positive

$$\sin^2 V = \frac{\beta - \alpha}{\gamma - \alpha}$$

$$\cos^2 V = \frac{\gamma - \beta}{\gamma - \alpha}$$

$$\tan^2 V = \frac{\beta - \alpha}{\gamma - \beta}$$

Optically negative

$$\sin^2 V = \frac{\gamma - \beta}{\gamma - \alpha}$$

$$\cos^2 V = \frac{\beta - \alpha}{\gamma - \alpha}$$

$$\tan^2 V = \frac{\gamma - \beta}{\beta - \alpha}$$

Various authors have provided graphical solutions to these computations (Fig 6.12) [6.108 to 6.111].

The accuracy of the computed axial angle $2V$ depends on the accuracy of refractive index determination. Computational errors are slightly different for varying values of $2V$ (Fig. 6.13).

6.1.3.5 Optic axial dispersion [2.11, 6.112]

In some substances, the optic axial angle varies markedly with wavelength if monochromatic light is used. White-light interference figures will show coloured areas next to the dark bands in the 45° position.

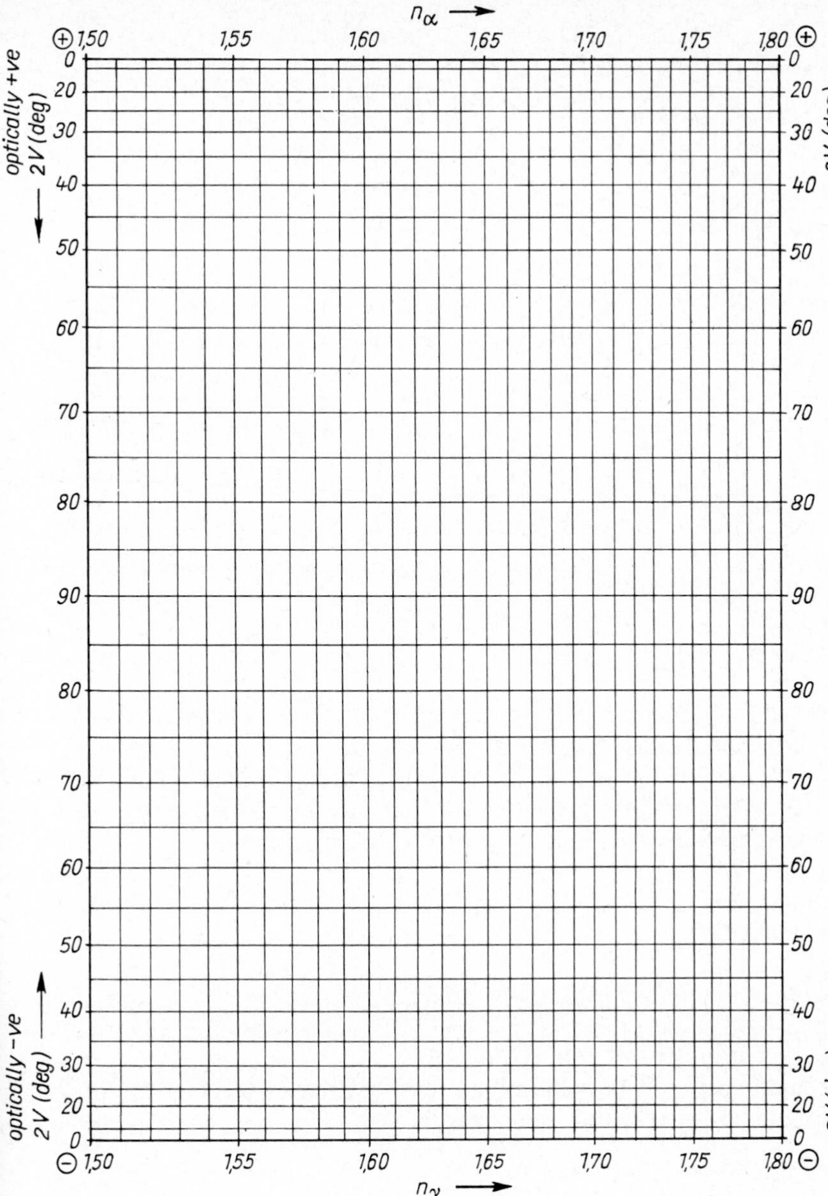

n_α →

⊕ 1,50 1,55 1,60 1,65 1,70 1,75 1,80 ⊕

optically +ve ← 2V (deg)

0 20 30 40 50 60 70 80 90 80 70 60 50 40 30 20 0

optically −ve → 2V (deg)

⊖ 1,50 1,55 1,60 1,65 1,70 1,75 1,80 ⊖

n_γ →

a)

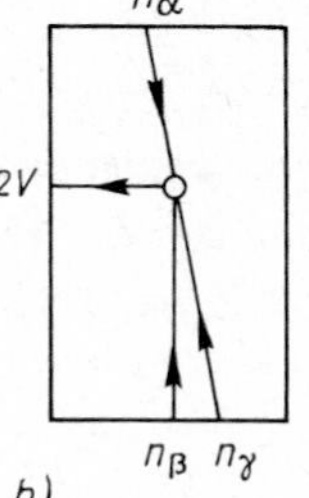

Figure 6.12. Graphical determination of $2V$ from the principal refractive indices, acc. to [7.104]

a) nomogram
b) key for determination

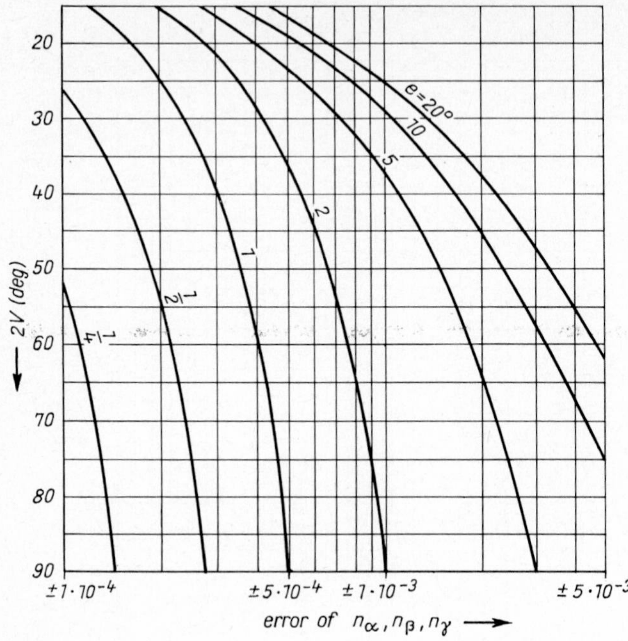

Figure 6.13. Nomogram for estimating errors in calculating $2V$ from the principal refractive indices, acc. to [7.5]
$2V = 2V' \pm e$
$2V'$ calculated optic axial angle

The colour distribution is a diagnostic that permits conclusion on crystal symmetry, provided that the dispersion is not too low.

Rhombic and monoclinic crystals show a regular colour distribution (Fig. 6.14). Triclinic crystals exhibit colour irregularities at the two dark bands.

In several substances, $2V$ varies greatly with temperature.

6.1.4 Optic sign

The optic sign characterizes the relation between the two typical parameters of the optical indicatrix of a substance (cf. 2.3.2). This relation is either *optically positive* or *optically negative* [6.112 to 6.115].

The optic sign can be determined from refractive index or birefringence measurements in the orthoscopic ray path.

Optically uniaxial substances: (cf. 6.1.1.1)

optically negative $n_0 > n_e$
optically positive $n_0 < n_e$

6.1.3.5 Optic axial dispersion [2.11, 6.112]
optically negative $(n_\beta - n_\alpha) > (n_\gamma - n_\beta)$
optically positive $(n_\beta - n_\alpha) < (n_\gamma - n_\beta)$

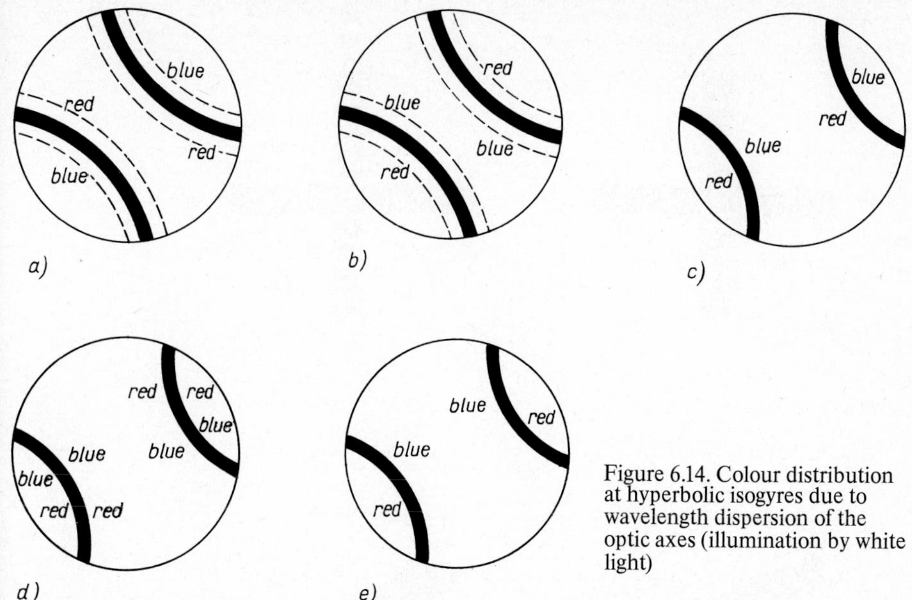

Figure 6.14. Colour distribution at hyperbolic isogyres due to wavelength dispersion of the optic axes (illumination by white light)

a) orthorhombic crystals, $2V$ red > blue b) orthorhombic crystals, $2V$ red < blue
c) monoclinic crystals, inclined dispersion d) monoclinic crystals, crossed dispersion
e) monoclinic crystals, horizontal dispersion

Determination by conoscopy requires interference figures of special sections (Tables 2.3 and 2.4). At least one of the isogyres must be visible. The figure is evaluated in the 45° position by checking the nature of colour distribution occurring when a full-wave compensator is inserted (Fig. 6.15).

6.1.5 Extinction and extinction angle

Homogeneous extinction will also occur in an aggregate of many submicroposition is a very sensitive indicator of uniform orientation and structure of the lattice in this region. Changes in the lattice, e.g. initial decomposition at the beginning of fusion, caused by mechanical strain or phase changes in the solid state, will show up very early by clearly inhomogeneous extinction. If lattice change intensifies, extinction becomes undulous or speckled; finally, the region can no longer be set evenly dark.

Object regions of uniform lattice structure but varying orientation exhibit uniform extinction, although at different stage rotation angles.

Homogeneous extinction will also occur in an aggregate of many submicroscopical monocrystal regions, if it is uniformly oriented (e.g., high polymers). Observed conoscopically, these aggregates supply interference figures that depict the symmetry conditions of their structural orientation.

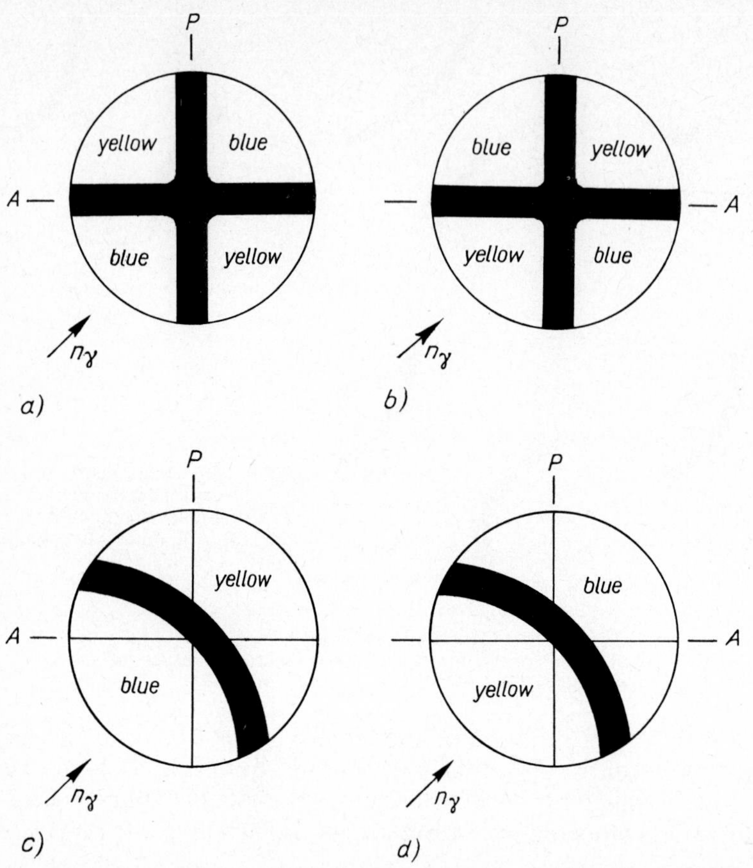

Figure 6.15. Colour distribution in the interference figure after inserting a full-wave (1st order red) compensator for determining optic sign. (The vibration direction n_γ of the compensator is shown at bottom left.) a) optically uniaxial, positive b) optically uniaxial, negative c) optically biaxial, positive d) optically biaxial, negative

If the grains are crystal-shaped, it is possible, with a polarizer inserted, to measure the angle between crystal edges and the eyepiece crosshairs after extinction setting with crossed polarizers (Chart.6.6). This angle characterizes the coincidence between principal crystallographic directions and optical vibration directions and, if sections of defined orientation are observed, the coincidence between the crystallographical and optical coordinate systems. Depending on the size of the angle, the extinction is straight, symmetric or oblique (Fig. 6.16, Table 6.23), cf. [6.116].

The accuracy of extinction setting for a grain can be increased by the use of high-intensity illuminators, setting to 1st-order red (Fig. 6.6) instead of darkness and observing the sensitive colour change, and the use of a half-shade plate. For a

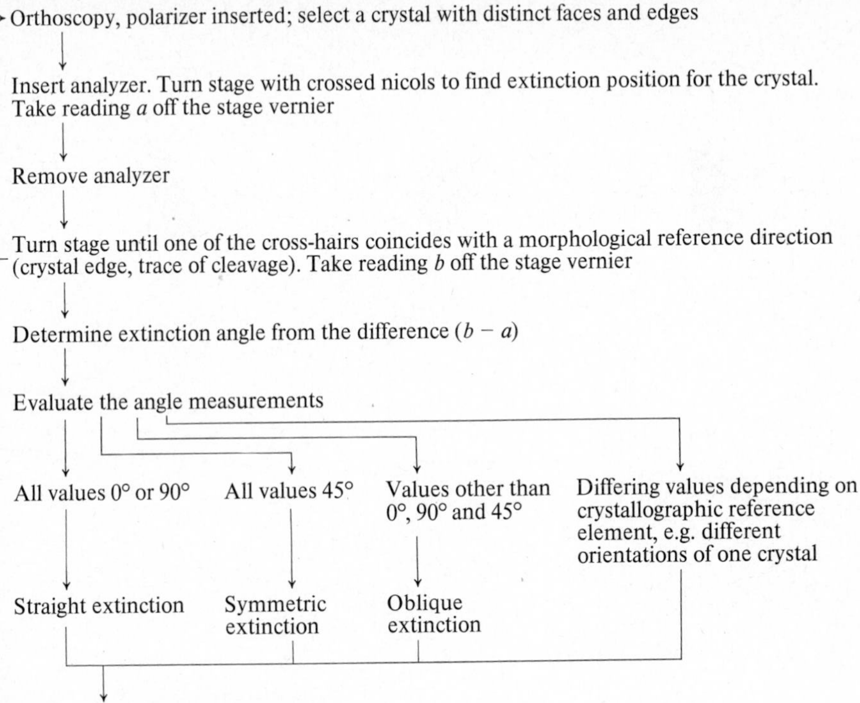

Chart 6.6. Measurement of the extinction angle of crystals

check, one of the polarizers is slightly rotated off the crossed position. If the extinction setting is correct, the slight rotation causes the grain and its environment to appear equally bright.

In transmission directions having oblique extinction (Table 6.23), extinction dispersion may occur. With monochromatic light, the homogeneous dark position will then be observed at different stage rotation angles depending on the wavelength. With white light, the grain appears coloured, and no dark position can be set.

6.1.6 Absorption and pleochroism

The spectral absorbance of a coloured object is an important parameter of coloured compounds, where other quantities often defy measurement. Absorbance also allows to draw important conclusions on the molecular structure of a substance [6.117 to 6.126].

Modern instrumentation (Table 3.10) permits various measurements on microscopical objects to be performed:

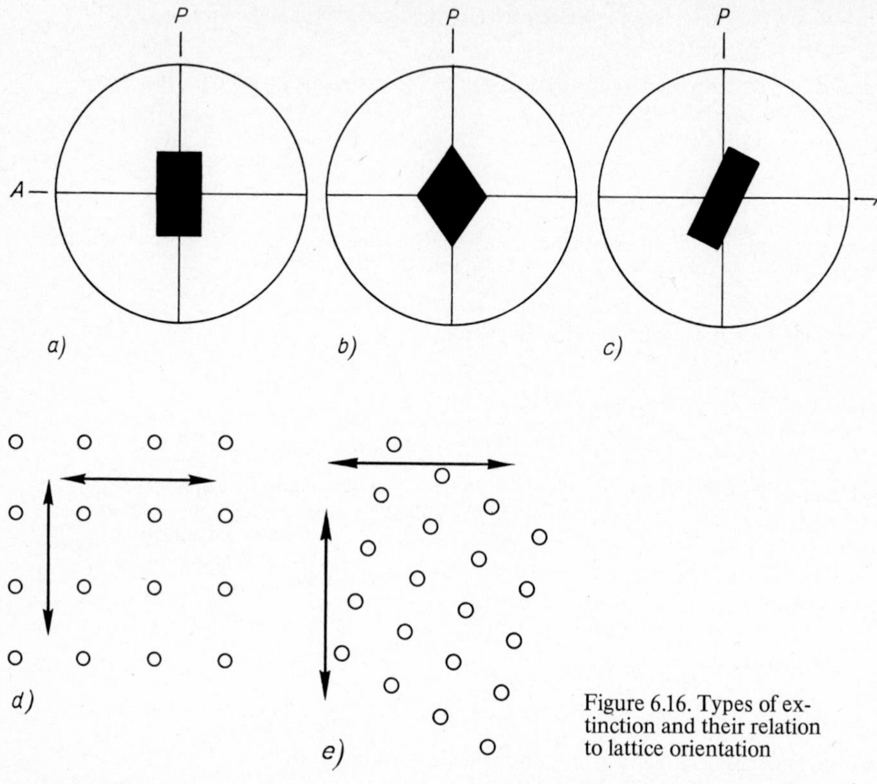

Figure 6.16. Types of extinction and their relation to lattice orientation

a) straight extinction b) symmetrical extinction c) inclined extinction d) interpretation of straight extinction and e) of inclined extinction by the orientation of light vibration in the lattice

(1) Recording of absorption spectra over a range from about 200 to 800 nm
(2) Absorbance measurements on object details at a selected wavelength in that range
(3) Recording of absorption profiles of inhomogeneous specimens along a scan line
(4) Recording of fluorescence spectra
(5) Reflectance measurements within a range of about 360 to 800 nm

Such measurements are made by transmitted (1., 2., 3., 4.) or reflected light (4., 5.).

Absorbance measurements register the ratio of the intensities of the light received, and the light transmitted, by the specimen. That ratio can be measured direct or through a derived quantity. It can be registered visually, photographically by microdensitometry, or electronically.

Compared to macro-techniques, *microscopical absorbance measurement* has some *specific features:*

(1) As an advantage, small individual grains or selected spots of an inhomogeneous specimen can be tested.

(2) Thickness variations of microscopic specimens have a disturbing effect on quantitative analysis. Grain inhomogeneities such as cleavage cracks, inclusions etc. cause error by additional scatter, diffraction or reflection.

(3) The fine pencil of light required for the measurement calls for special illumination techniques.

(4) As specimens are rather thin, substances must have sufficiently high absorption coefficients in order to be identified.

In a plot of spectral absorption, the positions and heights of absorption maxima can be evaluated.

In optically anisotropic crystals, absorption is different in varying lattice directions, a phenomenon known as *pleochroism.* Observed orthoscopically through a transmitted-light microscope with inserted polarizer, pleochroism shows up as a colour change of the object when the stage is being rotated.

Quantitatively, the pleochroism of a substance can be ascertained by measurements of spectral absorption along the coordinate axes of the indicatrix in sections of defined orientation.

Table 6.23. Types of extinction in the various crystal systems

Crystal system	Frequency of extinction types for statistically distributed transmission directions	Extinction exhibited by specific faces	
Cubic	Permanent darkness during rotation of the microscope stage		
Tetragonal		(001)	Permanent darkness during stage rotation
Trigonal Hexagonal	Extinction predominantly straight or symmetrical, sometimes oblique	(100), (010), (110), (hki0) etc.	Straight extinction
		(111), (101), (hkil)	Symmetric extinction
Orthorhombic		(010), (100), (001), (110) etc.	Straight or symmetric extinction
		(hkl)	Oblique extinction
Monoclinic	Extinction frequently oblique, sometimes straight	(100), (001) etc.	Straight extinction
		(010), (hkl)	Oblique extinction
Triclinic	Extinction always oblique (extinction angles may be very small)		

In reflected light, the colours of microscopic objects are best recognizable with dark-field illumination. The brightest grain will determine the apparent brightness of the others. Polished sections of transparent substances will appear dark.

The colour impression depends on the surrounding colour. A characteristic hue may appear dull and is easily misinterpreted if it is surrounded by similar hues, whereas its characteristic quality is emphasized in contrast with its complementary colour. Colours appear different under observation with dry and immersion objectives, the immersion liquid tending to enhance the colour impression.

Internal reflections

Grain boundaries, cleavage cracks and other grain inhomogeneities cause internal reflections under high-intensity incident light. This is explained by the fact that even highly absorptive substances are somewhat transparent if they are thin enough. The colours of such internal reflections are the true mineral colours. They are best absorbed with incident light, crossed polarizers and immersion objectives. In anisotropic substances, the reflections and their boundary lines frequently appear doubled.

While coloured microscopical objects can be analyzed by their natural spectral absorption, colourless substances may be stained artificially [6.124, 6.125]. Such selective staining may serve to add contrast to structured objects or to identify individual phases by their specific staining behaviour. Compared to modern, exact measuring techniques, staining methods have little importance, as they involve too many errors caused by the different natures, concentrations and action times of the dyes, influence of other substances on solubility, different interfacial forces, and different structures and compositions of the substances to be studied.

6.1.7 Other optical properties

In special cases, optical properties other than the quantities described may serve to characterize chemical compounds. Principles and methods of measuring these properties are outlined in Table 6.24.

6.2 Morphological properties

To characterize solid compounds morphologically, one requires information on grain shape, grain size, grain surface, and mechanical homogeneity (cracks, pores). These properties have a decisive influence on the reaction and pre-

Table 6.24. Microscopical detection of fluorescence, reflectance and optical activity

Optical property	Measuring principle	Methods	References
Fluorescence	UV light excites some objects to emit longer-wave radiation. Rare earths and carbons, e.g., emit object-specific primary fluorescence. Other objects may be impregnated with fluorochromic dyes to obtain secondary fluorescence. The latter is also used to study adsorption processes and aggregate structures (e.g., catalysts).	Examination by reflected and transmitted light. The exciting UV light does not contribute to optical imaging; it is eliminated by barrier filters. Thus, fluorescent objects appear coloured against a dark background. Fluorescence light has low intensity; in case of transmitted-light excitation it is attenuated by the specimen, so that a strong illuminator (*HBO* lamp) is required. Fluorescence excited by incident light reaches the objective directly and unattenuated.	[6.126 to 6.130]
Reflectance	Measurement by microscope photometry; calibration with standard specimens.	Measurement requires a highly polished specimen surface oriented exactly normal to the microscope axis, low but constant condenser aperture, determination of the background stray-light on a non-reflecting specimen, and stopping off the environment of the object.	[6.131 to 6.136]
Bireflection (random sections)	Optically anisotropic substances in random sections exhibit 2 direction-specific reflectance values, R_γ' (maximum) and R_α' (minimum), analogous to the refraction of transmitted light.		
Bireflection (definite sections)	*Optically uniaxial:* 2 principal reflection coefficients R_o and R_e. *Optically biaxial:* 3 principal reflection coefficients R_γ, R_β, R_α.	Suitable standard specimens are various ores, metal films evaporated on to glass substrates, quartz, and special glasses. For exact measurements, object and standard should not differ much in reflection.	
Optical activity	The right-hand and left-hand circularly polarized component rays of light passing an object may differ in refractive index *(optical rotation),* absorption *(circular dichroism)* or both *(Cotton effect),* caused	Transmitted-light measurements. **Orthoscopic procedure** Measurement of the angle of rotation by which the extinction position deviates from 0° with monochromatic light.	[2.12]

Table 6.24., continued

Optical property	Measuring principle	Methods	References
	by asymmetric lattice structure and/or asymmetric molecular structure. Comparative measurements on specimen solutions permit a decision.	Measurements require a strong effect in the specimen (e.g., organic high polymers); they are comparable only if specimen thicknesses are standardized. Wavelength variation yields *rotary dispersion.* **Conoscopic procedure** In case of strong effects, characteristic interference figures appear, such as central brightening of the cross of isogyres in optically uniaxial substances.	

cipitation behaviour and the quality parameters of industrial products. Measurements of grain size reduction permit registering the course of physical-technical processes in quantitative terms, such as the growth of a reaction layer, the dissolution or crystallization of solid phases etc.

For comprehensive accounts, see the literature [6.137 to 6.155].

6.2.1 Grain shape

The outer shape of isolated small objects varies widely. Of all methods, microscopy is the only way to directly observe and register grain shapes.

In a microscopic image, the shape of a polygonal grain can be positively identified only if its diameter is larger than the product of its number of corners and lateral resolution (about 0.5 μm). For example, an observer will recognize the shape of a hexagon only if its diameter is larger than 3 μm. Grain sizes below 1 μm permit distinction only between fibrous or long-columnar and isometric shapes, and no further distinction is possible by means of optical microscopy.

The two-dimensional microscope image shows the grain silhouette in the respective projection. To allow observation from all sides facilities must be provided for rotating the grain. One method is to embed the grain into a highly viscous medium and to shift the cover slip so that the grain rolls over. Mounting in a Waldmann sphere is a more accurate, but more time-consuming method.

(1) Regular shapes

Crystallizates occurring in a liquid phase very frequently have a tabular, acicular, cubic, skeletal or dendritic shape. The wealth of these forms can be registered and described by relatively simple crystallographic methods, such as by the Miller indices [2.6, 2.12]. These indices characterize the tracht of the crystals, i.e. the totality of a crystal's faces. The crystal habit (e.g., acicular, tabular, isometric) indicates distortions of the faces during growth.

In general, one cannot correlate characteristic crystal shapes with certain substances, as the crystal shape is influenced by many factors such as temperature and pressure during formation, concentration, solution components, trace impurities, polymorpous modifications etc. Nevertheless, a symmetry analysis of the crystal shape observed, preferably by measurement of the angles between faces and those between edges, provide useful hints on the crystal system or on lattice symmetry. From crystal configuration, information may also be derived about the relationship between cristallographic and optical principal directions (cf. 2.2).

(2) Irregular shapes

Irregular shapes are typical for non-crystalline substances; in crystalline compounds they occur if the formation of crystal shapes is prevented by the way the process is controlled.

Rough classification of irregular grain shapes

Sub-isometric
Sub-elongated
Sub-tabular
Irregular concave (e.g., vitreous clinker)
Rounded-to-elliptic (e.g., polymerization products)
Conical-fibrous (e.g., spherulites)
Intergrown aggregates

The shape of any grain can be described in quantitative terms by various parameters, viz.

(1) the ratio of grain axis lengths a : b : c intercepted by the enveloping rectangular parallelepiped;
(2) the grain shape coefficient ε (ratio of the surface area S_{sph} of a sphere of equal volume and the actual grain surface area S; $\varepsilon = S_{sph}/S$);
(3) the grain shape coefficient ε' rather than ε, as the true surface area is difficult to determine (ratio of the circumference C_c of a circle of equal area and the circumference C of the largest cross-section area; $\varepsilon' = C_c/C$);
(4) the ratio between length and width (to describe fibres and needles); and
(5) the ratio between thickness and the mean diameter of the table silhouette (for tabular shapes).

The degree of grain roundness varies within wide limits (Fig. 6.17). It is an overall term for a grain's surface area and surface configuration.

The examination of porous grains is a special case [6.137, 6.138]. A combination of thin section and polished section observations will yield information on pore

Figure 6.17. Varying degrees of roundness of grains seen through the microscope

size, pore distribution and permeability. Cavities can be detected better if the specimens are impregnated with coloured or fluorescent liquids or with immersion media of defined refractive index [6.141].

6.2.2 Grain size and thickness

The microscopical determination of grain size or thickness, apart from holographic methods, relies on length measurement in the two-dimensional image. Measurements in three dimensions require the grain to be revolved.

Direct grain measurement is possible by microscopical or electron-microscopical methods only. All other methods of establishing grain size are indirect (Fig. 6.18).

(1) Grain size measurement [6.142 to 6.150, 6.153, 6.154]

For spheres, the term of grain size is exactly defined. For all other grain shapes, it can be expressed approximatively by various parameters or diameter difinitions (Table 6.25, Fig. 6.19). Comparable results of grain size analyses would require equal diameter definitions.

Microscopical grain size determination primarily covers the range of 1 to 40 μm, which adjoins to the range of standard sieve analysis.

The practical lower limit of microscopical grain size measurement is influenced by the kind of illumination and the difference in refractive index between the grain and its surrounding medium. The limits of visibility are

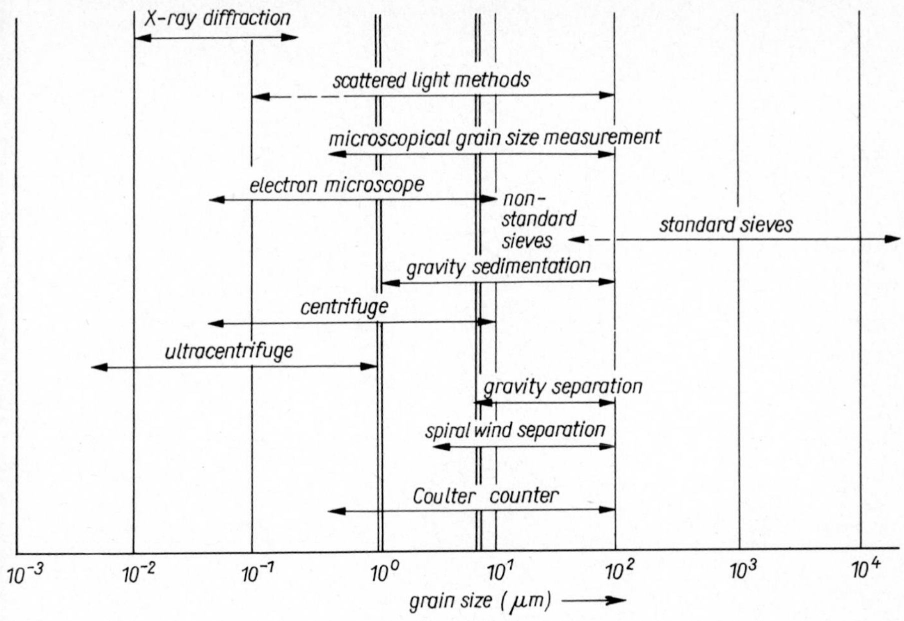

Figure 6.18. Range of application of microscopical grain size measurement as compared to other methods

Table 6.25. Some grain diameter definitions (cf. Fig. 6.19)

Term	Definition	Microscopical measurement
Diameter of a circle of equal area	Diameter of a circle having an area equal to that confined by the grain outline, observed normal to the grain's bearing surface	Endter particle analyzer, using micrographs
Feret's diameter	Normal distance between parallel tangents to opposite grain sides	Micrometer eyepiece, statistical evaluation of many measurements
Random linear intercept	Length intercepted by grain outline of any of parallel scan lines of constant spacing	Statistical measurement of many intercepts of parallel lines by automatic scanning, to obtain the »mean linear intercept.«
Krumbein's diameter	Longest possible intercept between grain outlines	Micrometer eyepiece; visual measurement if grains are few
Martin's diameter	Grain diameter along the line dividing it into two equal areas	Micrometer eyepiece; visual measurement if grains are few

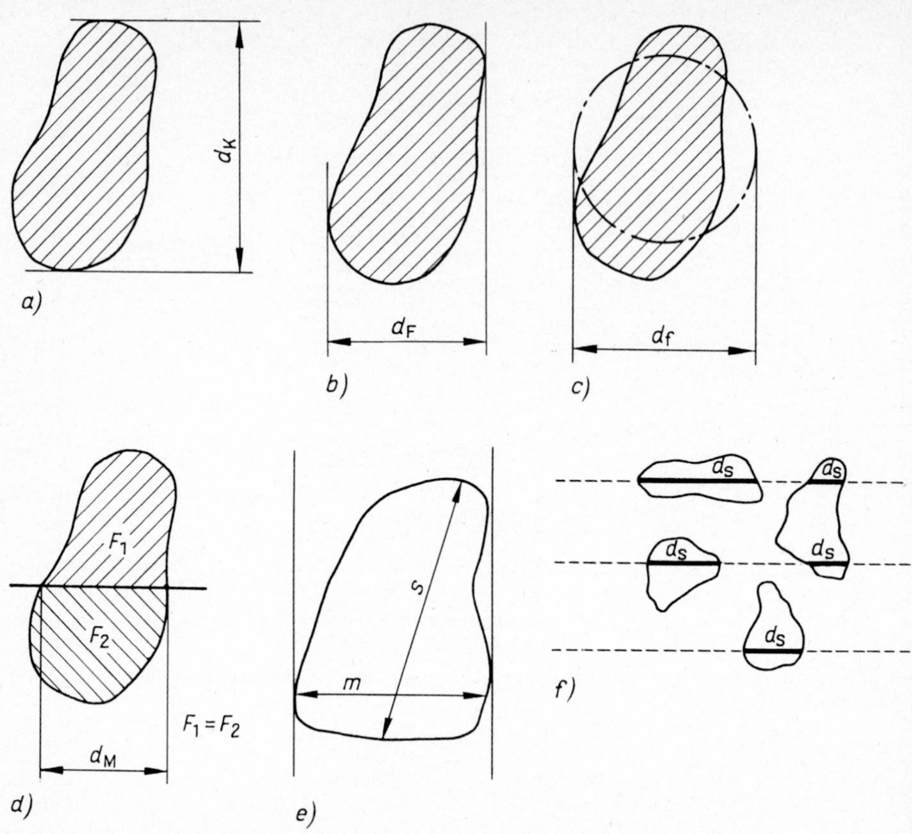

Figure 6.19. Various diameter definitions used in grain size measurement
a) Krumbein's diameter (d_K) b) Feret's diameter (d_F) c) diameter of the circle of equal area (d_f) d) Martin's diameter (d_M) e) arithmetical mean diameter = (m + s)/2 f) random horizontal intercept (d_s)

0.2 µm for bright-field microscopy,
0.1 µm for dark-field microscopy, and
0.01 µm for ultra-microscopy.

Table 6.26 lists various techniques of microscopical grain size measurement; its range of application is compared with other methods in Fig. 6.18.

The results of microscopical grain size analyses are influenced by the kind of specimen preparation.

In thin sections and polished sections, grains are cut at random, whereas in a loose grain specimen the grain settles on the specimen slide so that its two-dimensional silhouette area is the largest possible. Therefore, grain sizes ascertained

Table 6.26. Methods of microscopical grain size determination

Principle	Visual methods	Automatic methods
Area measurement in micrographs	Planimeter method, weighing method, Endter particle analyzer	Automatic image analysis (phase integrator), Nassenstein disperso-meter, Mullard particle size analyzer
Area measurement in the microscopical image	Net-ruled micrometer eyepieces	Oscillating-mirror measuring instruments
Length measurement in the microscopical image	Screw micrometer eyepiece, integrating spindle stage	
Point count in the microscopical image	Henning eyepiece, point counter	Flying-spot microscope, automatic image analysis (phase integrator)

from loose grains are always larger than those measured on thin sections or analyzed by sieving or sedimentation (Fig. 6.20).

In some cases, grain size data are ambiguous. Primary grain size as the dimension of a domain of uniform lattice orientation may not be identical with secondary grain size as a morphological quantity, such as in firmly intergrown grains. Microscopical methods yield information on both primary and secondary grain sizes. Grain intergrowth and coarse lattice dislocations show up in orthoscopy by different extinction conditions of the various uniformly oriented crystal domains. Such examinations are important, e.g., for assessing the firmness of grains of fertilizers and other powder chemicals.

In recent years, the emergence of holography has provided new possibilities for the direct, three-dimensional ascertainment of the spatial grain shape. First results obtained with grains dispersed in gases or liquids in technical reactors have been published. The method is very important where growth or dissolution processes in stirred laboratory crystallizers have to be watched, or where flow processes in polymer melts have to be represented in model processing plants.

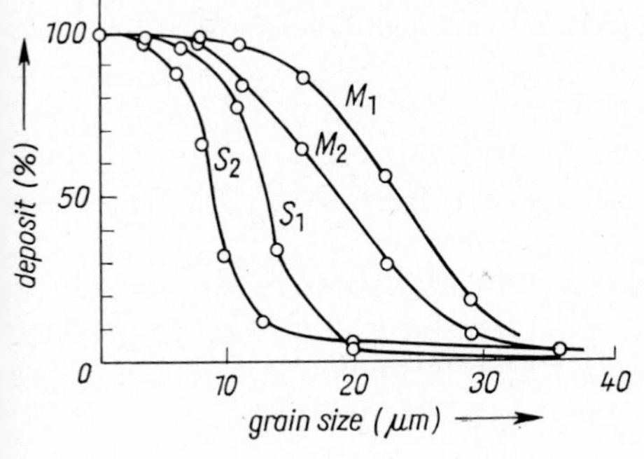

Figure 6.20. Microscopical grain size determination (loose grains) compared to sedimentation analysis, performed on two model SiO_2 dusts

S_1, S_2 results of sedimentation analysis
M_1, M_2 results of microscopical measurement

Grain size measurement on decomposing substances and at high or low temperatures

Exact measurements of grain sizes of decomposing or hygroscopic compounds can only be made by microscopy. The same is true for measurements below normal room temperature. In most of these cases, loose grains are the most suitable kind of specimen, especially if contained in specimen slide chambers. Preferably, the substance is immersed in the liquid phase of the formation system (mother solution). Aggregation and surface growth on grains by secondary formation may increase the apparent grain size, but the error is easily detected in microscopical examination. Grain size becomes a problematic term if the original grain has partially or wholly decomposed. By cautiously crushing the »aggregate« grains, one can identify the size of the individual grains under the microscope. For example, grains of $MgCl_2 \cdot HCl \cdot 7H_20$, originally sized 200 to 300 µm, consisted of an aggregate of $MgCl_2 \cdot 6H_2O$ grains of 5 to 10 µm each after decomposition.

Grain size measurements in the high-temperature range [6.150] involve some additional difficulties. Thermal etching on polished specimen surfaces decreases the visible grain size. Volume changes in phase transformation do not show up adequately in a polished section. Thin films of newly-formed substance (such as oxides) may cover the grains visible at the surface.

Grain size on specimens that cannot be polished is measurable only with immersion (cf. 4.3.1) because of total reflection effects.

(2) Thickness measurement (Table 6.27) [6.151, 6.152]

Measuring the thickness of films is an important task in photochemistry, high-polymer foils, and coated materials. The thickness of a layer passed by a light beam is significant in determining birefringence from path difference measurements (cf. 6.1.2).

6.2.3 Surface

Under the microscope, specimen surfaces can be characterized by their visual appearance as well as by quantitative measurement. Possibilities include

(1) observation of the surface image: surface growth, etching pits, inclusions, surface texture (step growth, spiral growth, cleavage cracks and other lines intersecting the surface), roughness tests;

(2) quantitative analysis of the surface image: distribution of etching pits, measurement of vicinal faces and etching surfaces, etc.;

(3) quantitative measurement of surface properties: reflectivity, film thickness, gloss of foils etc.

Surface examination requires reflected light. In addition to bright-field, dark-field and mixed light techniques, lateral illumination is employed to produce high-contrast images. In special cases, additional information is provided by the replica technique or by thin metal films evaporated in a vacuum.

Table 6.27. Methods of thickness measurement of microscopical specimens

Instrument	Principle	Application
Caliper	Mechanical	Thin slices, microtomed sections etc. (rough measurement)
Optimeter	Mechanical, with optical readout (contactor pin acting on hinged mirror)	Thin slices, microtomed sections etc.
Microscope fine focusing control	Difference between focussing on top and bottom sides of plane-parallel grains (transmitted light, high-power objective)	Thin sections, thin plates of transparent material
Rotation of grain embedded in viscous medium	Micrometer eyepiece measurement after rolling the grain into desired position by shifting the cover slip	Transparent and opaque grains
Waldmann sphere	Embedding acc. to [3.105], micrometer eyepiece measurement after turning into desired position	Transparent and opaque grains, thin plates, microtomed sections
Compensator and universal stage	Retardation measurement, if birefringence is known	Transparent substances
Interference microscope	Retardation measurement, if birefringence is known	Transparent substances
	Differential measurement of retardation at grooves (reflected light)	Layers on a substrate

Surface measurement by reflected light uses interference, phase contrast and ellipsometric techniques.

(1) Interference and phase microscopy

Depending on the wavelength employed, the lateral resolving power of light microscopical techniques is about 0.2 μm, whereas the vertical resolution obtained by phase and interference methods may be a few angstrom units under favourable conditions [7.46]. Practically, resolution in depth is influenced by many factors such as fine surface texture, the refractive index or reflectivity of the material, scattered light, and objective aperture. Phase and interference techniques reveal, e.g., the layer-by-layer growth of crystal faces and thus permit conclusion on the mechanism of the growth process.

(2) Ellipsometry

Ellipsometry is based on the change in the state of polarization of reflected light. It is a method of measuring the thickness and refractive index of thin films on surfaces, with main applications in the testing of metal and semiconductor surfaces. Measurable films may be as thin as a discontinuous single-atom layer (»submonolayer«). Practical uses are the examination of absorption processes on such surfaces and of surface layer growth due to chemical reaction with gases or electrolytes [6.155].

6.3 Volume and distribution analysis

In a microscopical specimen of heterogeneous composition, the distribution or the colume fractions of the various constituents can be determinded by suitable counting methods, provided that the constituents clearly differ by appearance. These differences can be enhanced by special preparation such as selective etching or dissolving, and special measuring techniques incolving pahse or interference contrast.

The differences between countable features in the microscopical image are the basis also for the automatic measurement of their volume fractions (cf. 5.2.2).

Some chemical applications

Different phases in solid mixtures of raw materials, intermediates and reaction products in chemical engineering processes
Phase distribution in pressed granulates, materials and ceramic products
Distribution of etching pits on monocrystal faces
Manufacturing faults (flakes) on high polymer foils
Impurities on film surfaces
Distribution analysis in aggregates by radioactive doping and microautoradiography

Microscopical volume analysis is superior to other methods especially where very small regions have to be analyzed, where mixtures contain decomposing substances, or where the phases involved are chemically similar to each other (cf. 7.1).

Semi-quantitative determinations can be made with the aid of estimation charts (Fig. 6.21) [6.156].

6.3.1 Point counter (counting at constant intervals)

In point counter, the specimen is indexed from point to point at a constant interval specified to correspond to the mean sizes and spacings of the features of interest. The points to be counted are those in the centre of the eyepiece crosshairs; counting is followed by statistical analysis.

Where grain sizes differ greatly, grains within certain size classes are counted in separate runs at suitable magnifications.

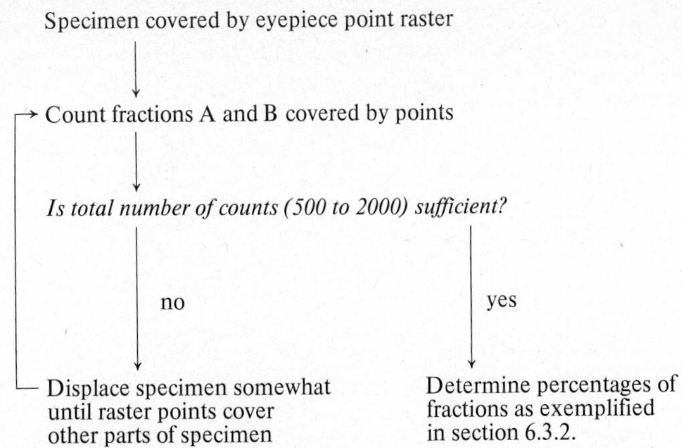

Chart 6.7. Quantitative analysis of a two-phase mixture, using the point counting eyepiece

The number of points counted within the areas of a structural constituent is proportional to the total area and, thus, to the volume occupied by that constituent, provided that a sufficient number of points are counted.

Instrumental setups for counting may include either integrating eyepieces or integrating stages. Charts 6.7 and 6.8 (see page 170) present the respective procedures [6.157 to 6.165].

6.3.2 Counting with no constant intervals

For special assignments it may be more appropriate not to count at constant intervals, but to use one of three methods directly related to the grains (features) [6.157 to 6.165].

(1) Grain count

This method requires features of approximately equal size, i.e. a narrow grain size fraction, and grains well isolated from each other. A mechanical stage attachment moves the specimen about the stage in two directions normal to each other and parallel to the eyepiece crosshairs. Counting starts in one of the top corners of the specimen. All grains passing the crosshair centre are recorded by their respective kinds. The accuracy of results increases with narrower grain size interval and with the number of grains having isometric shapes. Thus, the volume error will become insignificant (cf. Table 6.28).

The percentages by weight of the various features (phases) are derived from the percentages by weight of the density fractions and their percentage composition.

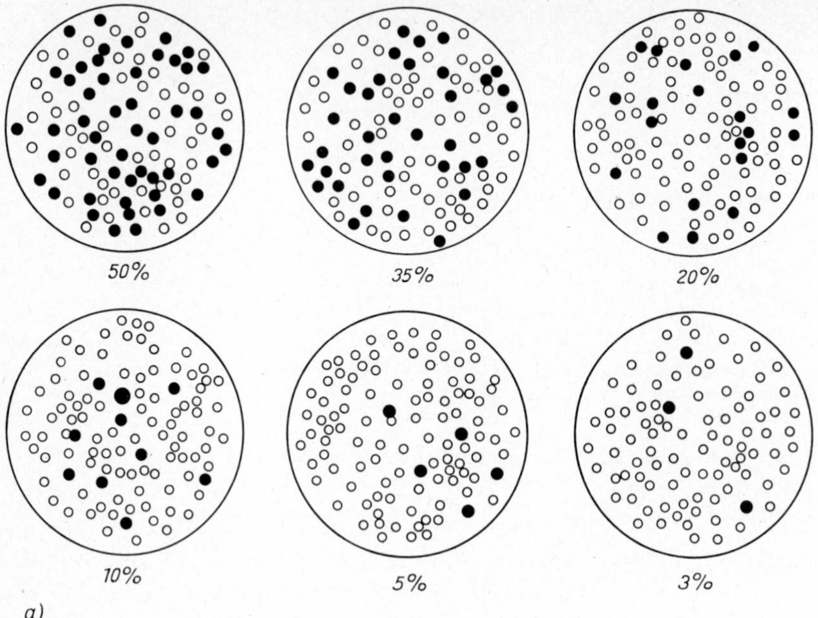

Figure 6.21. Semi-quantitative estimation of volume fractions
a) in a two-phase mixture
b) in the homogeneous background

Calculation example for quantitative microscopical phase analysis of a mixture of three phases A, B and C

Phase i	a_i	ρ_i	$a_i\rho_i$	*mass–%*
A	23	2	46	19
B	37	2	74	31
C	40	3	120	50
i	100		240	100

a_i grain count (point count, intercept length) of a phase
ρ_i density

$$mass{-}\% = \frac{a_i\rho_i \cdot 100}{\sum_{i=1}^{n} a_i\rho_i}$$

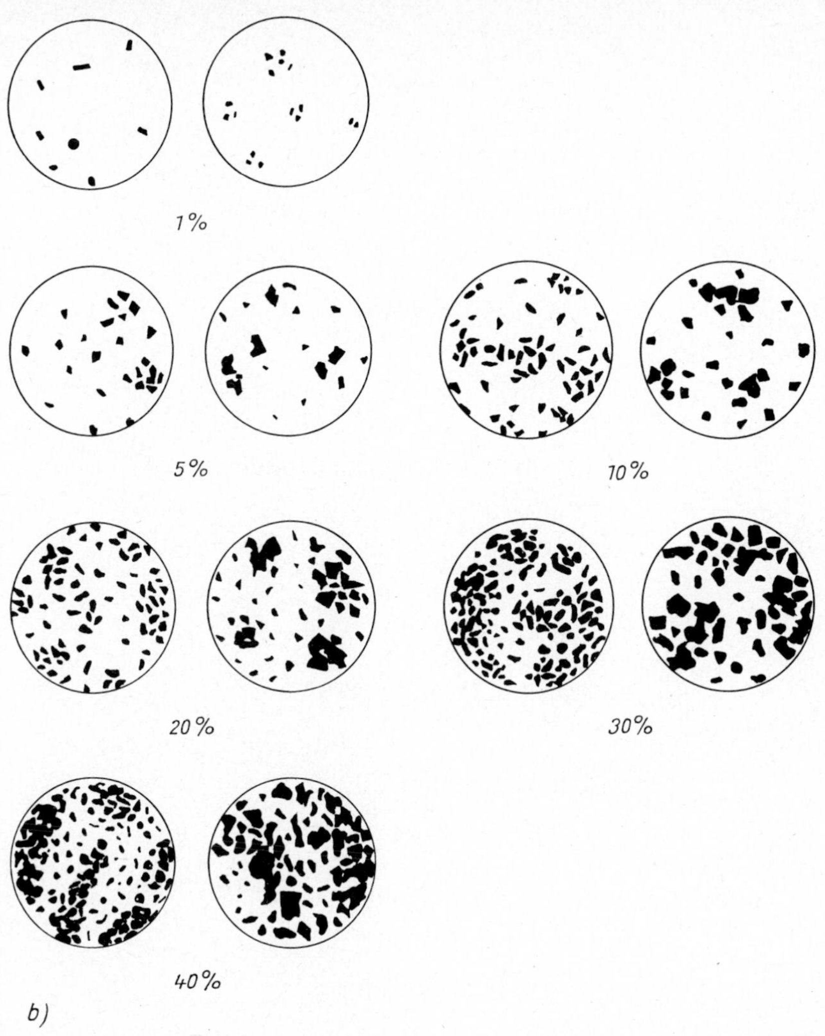

(2) Linear analysis

The method makes use of the statistical relationship between grain diameters or linear (chordal) intercepts and grain section areas, which are proportional to volume.

Linear analysis is especially applicable to specimens having widely differing grain sizes. This method is more laborious than the others.

Given a high density of measuring points, linear analysis may be automated by differential image analysis techniques (cf. 5.2.2).

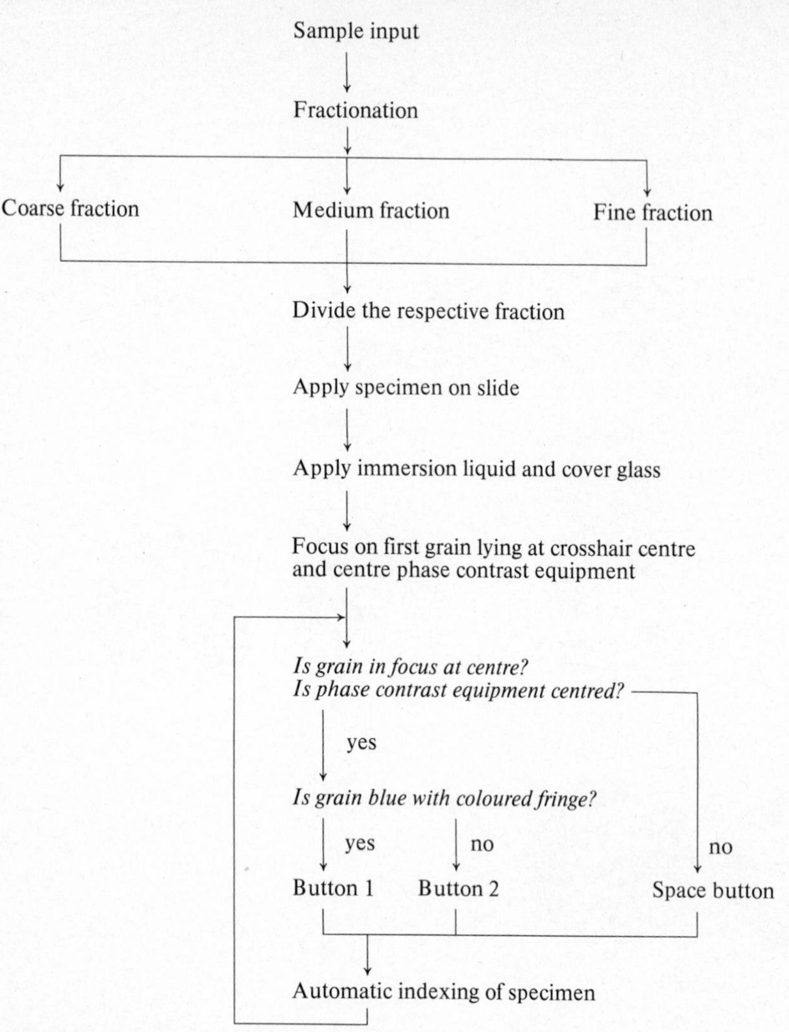

Chart 6.8. Quantitative phase analysis of a specimen, using the Eltinor integrator (to be repeated for each phase, with the respective phase stained)

(3) Area analysis

The volume fractions of the different phases are found from the corresponding area fractions in the microscopical image, which are determined either by microscopical planimetry or by means of automatic area integration techniques (cf. 5.2.2). If the distances between measuring points are very small relative to grain size, the method is hardly any different from point counting. Area analysis as a method of volume analysis is required especially if the specimen contains only few grains (objects) of interest.

Table 6.28. Errors in microscopical volume analysis

Type of error	Influence on result	Possible elimination
Sampling error	Sometimes extreme	Thorough mixing, statistical sample division
Volume error	Rounded grains: small Irregular or tabular grains: great	Grain form correction factor, counting within narrow size fractions
Crushing error	Small to medium	Reproducible preparation conditions in serial analyses, grain-preserving crushing, observation of all size classes
Error due to intergrowth	Small to medium	Material must be fully dispersed
Diagnosis error	Small to medium	Careful identification, unambiguous criteria
Counting error	Mostly small	Automatic counting
Errors in converting vol.% into mass-%	Small	Precise knowledge of densities, no grain intergrowth, no inclusions

6.3.3 Errors of microscopical volume analysis

Table 6.28 lists various errors that may occur in microscopical volume analysis [6.166 to 6.180].

Fig. 6.22 shows how the error varies with the number of points in point counter methods.

In general, the error is less for thin and microtomed sections than for loose grains and polished sections, because of the difference in registering fine and coarse grains. That is why grain specimens should be specially prepared, i.e. embedded in a solidifying substance. A slice should then be cut normal to the plane of sedimentation with an abrasive cutter.

Volume errors result from deviations between area and volume percentages where grains are not spherical. Practically it is impossible to quantify this relationship in mathematical terms, but the error influence may be corrected by factors determined empirically. If the grains of a phase have similar shapes, each phase may be corrected by the respective grain shape factor.

Main corrective measures [6.175 to 6.180]

Splitting up the volume influence into a fractionating factor and an average diameter;

combining the corrections for volume influence and density into one factor by determination of average grain weight. (Accuracy increases only if grain shapes differ widely.)

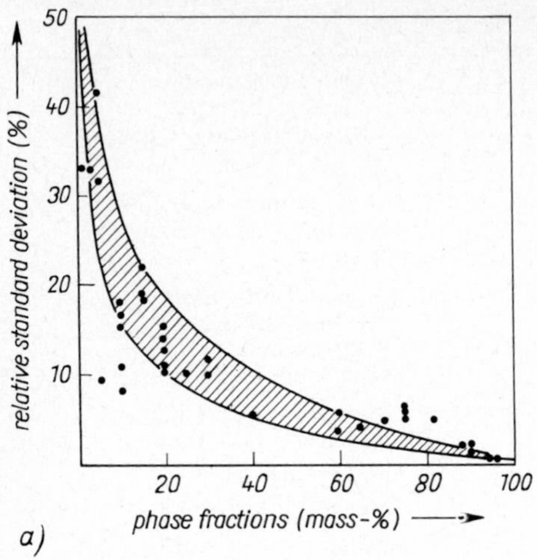

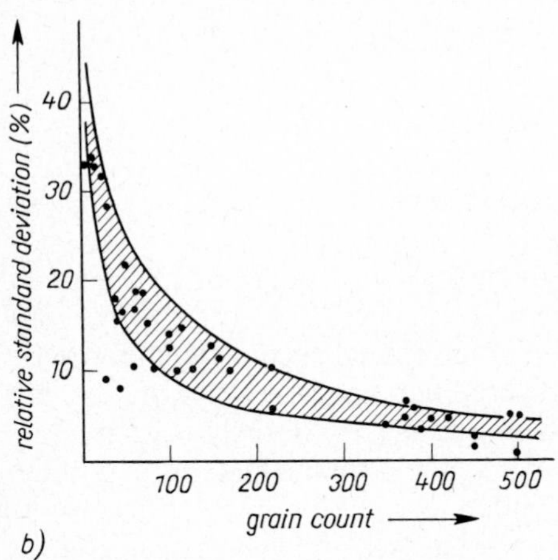

Figure 6.22. Error magnitude as a function of the number of measuring points
a) range of relative standard deviation vs. fraction of a phase in the total volume, and
b) vs. number of grains counted, for particle counting methods, acc. to [6.171] c) nomogram for determining absolute errors, and d) for determining relative errors, in point counting methods, acc. to [7.5]

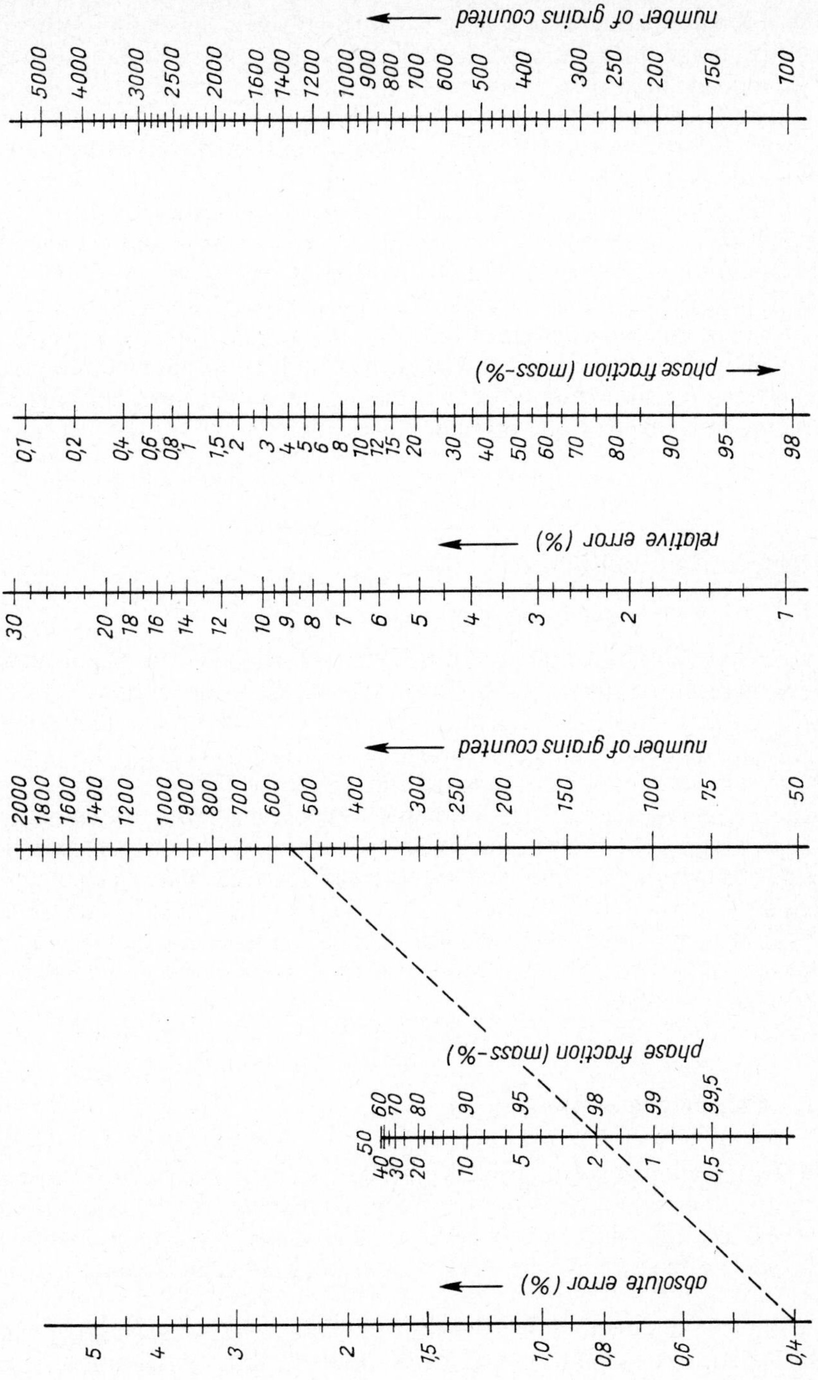

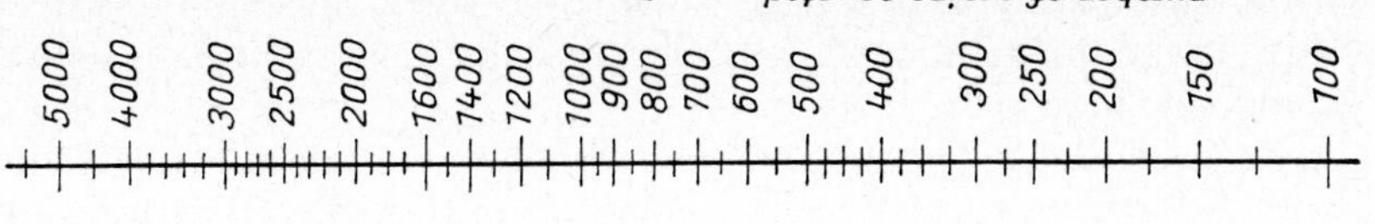

phase fraction (mass-%)
0,1 0,2 0,4 0,6 0,8 1 1,5 2 3 4 5 6 8 10 12 15 20 30 40 50 60 70 80 90 95 98

relative error (%)
30 20 18 16 14 12 10 9 8 7 6 5 4 3 2 1

d)

number of grains counted
2000 1800 1600 1400 1200 1000 900 800 700 600 500 400 300 250 200 150 100 75 50

phase fraction (mass-%)
50 60 70 80 90 95 98 99 99,5
40 30 20 10 5 2 1 0,5

absolute error (%)
5 4 3 2 1,5 1,0 0,8 0,6 0,4

c)

The influence of the volume error is particularly great in solid mixtures containing cubic and thin tabular phases and in unstatistical distributions such as localized individual grains.

Where transparent and opaque grains are intergrown, the counts are generally excessive for the opaque fraction. Here, too, correction factors may be introduced.

One difficult problem in volume analysis is to determine the fractions of secondary and trace constituents. The results become less reliable as the quantity of material increases and the content of the phase of interest decreases. It is of decisive importance to obtain a representative specimen. Despite intense mixing, the specimen inhomogeneity due to widely differing densities, grain sizes and grain shapes of the individual phases may be a multiple of the content of secondary and trace constituents in any part of the specimen. This error can be compensated only by taking a greater number of specimen.

6.4 Thermal methods

The microscopical exploration of the thermal behaviour of chemical compounds on the hot and cold stage belongs to those chemical methods having the richest tradition. Until today, the microscopic determination of melting point on the hot stage for identification, purity testing, and system behaviour testing of organic crystalline compounds is an established method in pharmacy and chemistry. The automation of such measurements by recording the intensity variation of the crystals (cf. 5.2.2) largely excludes subjective errors while retaining the advantages of microscopy, i.e. simultaneous observation of the phase change.

In a similar way, the processes of other thermal changes in chemical compounds can be measured. Microscopical techniques are steadily gaining in importance especially for kinetic measurements of decomposition processes and inversions in the solid phase.

6.4.1 Polymorphous inversion

The temperature of polymorphous inversions shows up in the curve recording an optical parameter vs. temperature. Within the stability range of a phase, optical properties change continuously whereas the phase transformation point is marked by a discontinuity (Fig. 6.23). For examples, see the literature [6.181 to 6.183, 7.163, 6.187].

The transformation point is determined by *direct measurement of an optical parameter* during heating-up or cooling-down. Preferably, this parameter is bire-

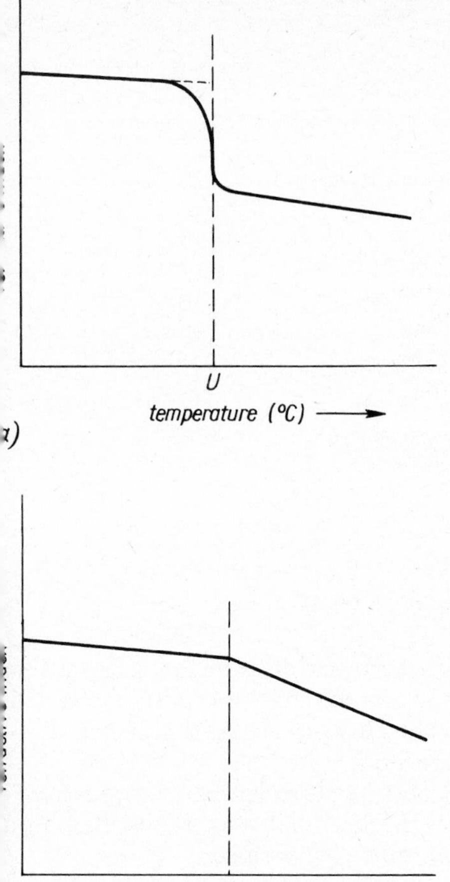

Figure 6.23. Change of refractive index due to polymorphous inversion, acc. to [6.183]
a) 1st-order inversion (α–β inversion)
b) 2nd-order inversions (order-disorder inversions)
U inversion temperature

ringence (retardation measurement), extinction position, optic axial angle or refractive index. *Indirect determination* is possible from the optical data of both modifications and their temperature coefficients.

Phase transition is a time-dependent process, so that heating rates should be very slow.

Tables 6.4, 6.5 and 6.29 list the various optical properties whose changes serve for establishing the transformation point. Exact determination is possible with transmitted light and in some cases, with reflected light (for polished sections). Opaque specimens that cannot be polished require reflected-light darkfield examination; the effects they show (if any) at the transformation point are not very conclusive.

Table 6.29. Microscopical criteria in polymorphic phase transitions

Phenomenon observed	Example
Twinning	$(K,Na)AlSi_2O_6$
Colour change	Yellow to red in mercury compounds
Transformation of spherulites into structureless crystals of uniform extinction, or vice versa	Cholesteryl acetate
Anomalous birefringence	$Mg_6(Cl_2B_{14}O_{26})$
Partition into panels	$(K,Na)AlSi_2O_6$
Clouds, reams, cracks	Paraffines
Undulatory extinction	$Mg_6(Cl_2B_{14}O_{26})$
Grain aggregates, increasing or decreasing grain size	Observed in most transitions
Changed extinction behaviour of crystals due to changed lattice orientation	

6.4.2 Temperature of decomposition (cf. 7.3.2.2)

The hot and cold stage microscope also permits to measure the decomposition point (or interval) of a substance; this point is indicated by a discontinuity in the variation of optical properties (cf. 6.4.1). This change is always irreversible, in contrast to enantiotropic polymorphous inversions.

Decomposition processes can best be observed if the specimen is immersed in silicone or paraffin oil. Bubbles caused by the separation of a gaseous phase are an important criterion in identifying polymorphous inversions.

The heating rate to be chosen is determined by the time dependence of the decomposition process. Too fast a rate may cause the false appearance of a decomposition interval, with the temperature rising before the process is completed.

At extremely slow heating rates, the temperature behaviour of decomposition shows steps (cf. 7.3.2.2).

6.4.3 Fusion temperature (melting point)

Melting point determination is the longest-known method in thermal microscopy. The melting point varies markedly with purity. Even slight impurities will change the fusion temperature by several degrees [6.184 to 6.187, 7.96]. The melting points of many substances are listed in tables [7.106, 7.109].

The physically analogous freezing point may be found to be considerably lower because of metastable supercooling. Supercooling can be ended earlier by means of seed crystals.

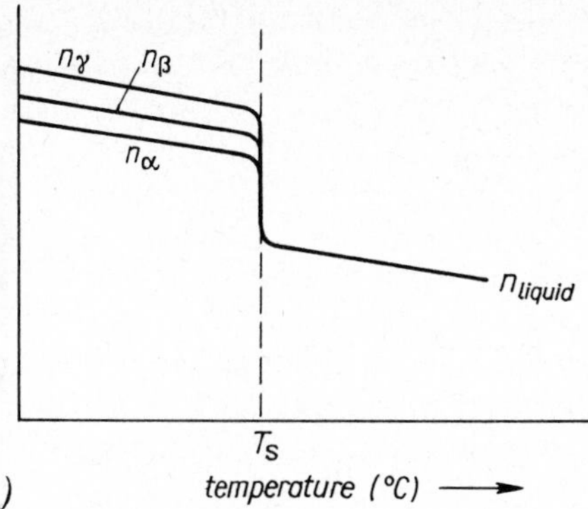

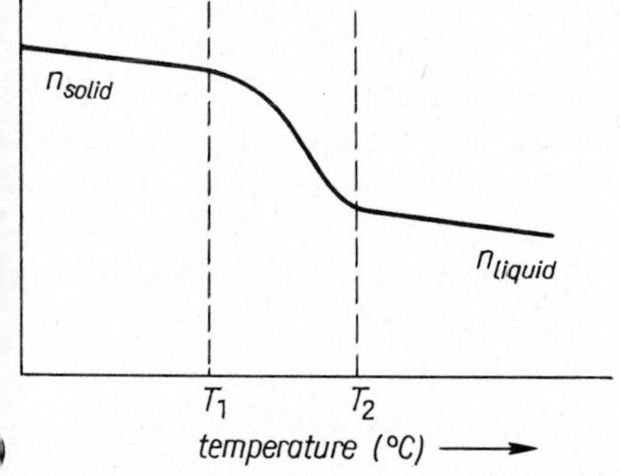

Figure 6.25. Change of refractive index during melting
a) crystalline solids
b) amorphous solids
T_s melting point
T_1–T_2 melting interval

At the melting point, the optical properties of a substance change just as dis-nctly as its outer state (Fig. 6.24 a to f, see Annex). The three principal refractive ıdices of an optically biaxial crystal, e.g., will be replaced by the uniform refrac-ve index of the melt (Fig. 6.25).

The temperature curves of birefringence, optic axial angle, the characteristic xtinction angles, reflectance, transmittance and absorbance are discontinuous, ɔo.

The temperature curve of light transmittance is employed in automatic melting oint recording by means of a photometer attachment to the microscope [6.185, .186].

Non-crystalline solids have no defined melting point but a melting interval glasses, waxes, organic high polymers). In this range, the temperature curve of efractive index, e.g., only bends downward (Fig. 6.25). An apparent melting ıterval in a crystalline substance is caused by impurities.

If the melting point of a substance is unknown, its approximate value may be ɔund by a coarse test at a relatively fast heating rate of 5 to 10 deg Centigrade/min.

The subsequent fine test should then be made at heating rates of 0.1 to 2 deg/mi Rounding of the crystal edges indicates that the melting interval has bee reached (Fig. 6.24).

If the end of the melting process is not clearly identifiable, the temperatur should be varied about the probable melting point by some degrees and kept co stant for some time. Any crystallites still present will then grow and becom visible. Several repetitions of that procedure at intervals of 0.3 deg permit th exact melting point to be defined.

If the sample is contaminated by substances having different melting point the melting phenomena may well be superposed to each other. During heatin e.g., a few higher-melting impurity crystallites may remain solid even after th main constituent has completely molten.

Microfusion studies are often affected by effects whose origin one must know i order not to draw wrong conclusions.

Volume contraction in phase transition may cause shrinkage cracks. Ga bubbles may occur when gases dissolved in a liquid phase are set free or in case c decomposition. In some delicate substances, the pressure of directly applie cover slips may give rise to mechanical twinning. In some cases, the grain and th liquid phase differ so little in refractive index that the grains appear only by phas or interference contrast techniques.

Table 7.32 classifies the crystallization phenomena observed during freezin

Mixed crystal fusion

Mixed crystals have a melting interval. Between the temperature ranges of th solid and liquid phases there is a range where both occur side by side, with the composition continuously changing. The temperature dependence of that ratio i easy to observe with the microscope. The composition variation of the initially se parated crystals is indicated by redeposition, dissolution and crystallization pro cesses. Curved, open annular and arabesque crystals may occur, spirall interwined in part, same as dendrite-like shapes. Simultaneously, the inte ference colour (retardation) changes, and smooth crystals appear grainy. Homo geneous areas may develop clouds and reams.

A fast heating or cooling rate will lead to zoning, due to incomplete concen tration balancing.

Zones of equal composition are identified clearly by their equal optical prope ties in a row of measurements across the crystal (Fig. 6.26).

The melting point of decomposing substances

There are two main methods to determine the melting point of an easily de composing substance, viz.

(1) Direct determination, with the substance immersed in nujol or silicone oil b means of a preparation box. The immersion medium prevents externa inoculation effects.

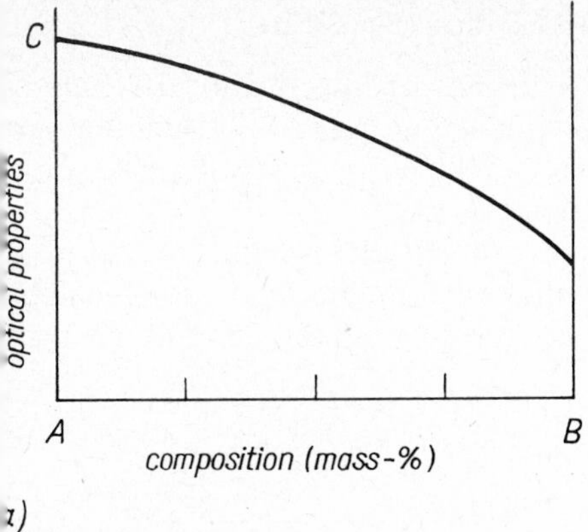

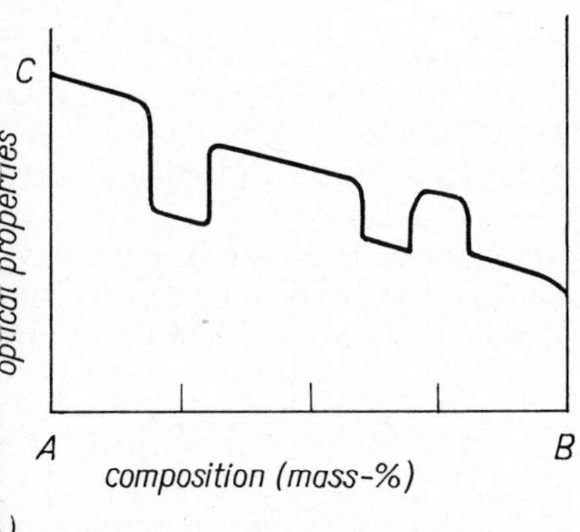

Figure 6.26. Variation of mixed crystal composition (A, B) represented by the variation of optical properties (C) (schematic)
a) continuous change
b) zoning

(2) Indirect determination, with the substance mixed with another substance to make it more durable (a frequent effect). The melting point of the pure substance can be extrapolated.

6.4.4 Examination of »solid-liquid« systems

The observation and recording of fusion and freezing processes on the hot and cold stage microscope yields important information on the recording of melting and freezing curves of »solid-liquid« systems.

In systems containing solvents, the isothermal and polythermal existence ranges can be found by the microscopical identification of the sediments of graded isothermal batches of different concentration.

6.4.4.1 The temperature behaviour of graded mixtures (Fig. 6.27)

This method requires graded mixtures with known molar fraction. In examining their melting and freezing behaviour one would record the first and last appearance of the solid or liquid phases, resp., at the various concentration ratios, from which the phase diagram can be derived directly.

The results obtained from a mixture do not by themselves identify the nature of a system. Binary systems, e.g., whether infinitely miscible or not at all, mostly show the same general response of specimen change to temperature changes from T_1 to T_4:

T_1: liquid phase only,
T_2: last solid residue in liquid phase,
T_3: last liquid residue in solid phase,
T_4: solid phase only.

A decision whether a true eutectic has been formed by fast crystallization or fusion of the eutectic mixture, and whether mixed crystals have been formed in case of these processes starting and ending at a slow rate, can be made after detailed observations only.

The specimens for such system examinations are thoroughly mixed fine powders of the initial substances (*powder method*). To improve homogeneity, it is advantageous to melt and re-freeze the mixture between specimen slide and cover slip (*crystal film method*).

6.4.4.2 Contact method

The contact method was developed by Kofler especially for organic crystal studies using the hot-stage microscope [6.12, 6.188], but it can be used analogously with other substances and other temperature ranges. The examination of a contact specimen (Chart 4.5, Tables 6.30 and 6.31) yields ample information on the nature and thermal data of binary and ternary systems.

The principle underlying the contact method is the observation of changes in the mixing zone of two substances as the temperature is varied. The specimen represented by the graph in Table 6.31, e.g., when being heated up, first shows the eutectic temperature T_1 in the mixing zone, then the melting point of substance A at T_2 in the left half of the specimen, and finally the melting point of substance B in the right half of the specimen.

If a compound occurs in a binary system, a contact specimen permits the measurement of

the melting point of substance A,
the melting point of substance B,
the melting point of the compound AB,
the eutectic temperature A/AB, and
the eutectic temperature B/AB.

The transitions appear as sharp fronts. Phase reactions show up by fused fronts. Tables 6.30 and 6.31 record the appearance of contact specimens and the measurable data for the most important cases of binary systems.

In order to obtain purely by fusion methods quantitative information on characteristic system points (maxima and minima positions of the fusion curve), Kofler combined the contact method with the glass powder method for determining the refractive indices of melts (cf. 6.1.1).

The basic idea is that the temperature required to give a melt a particular refractive index varies linearly with concentration. If one adds glass fragments of refractive index 1.540 to a contact specimen, the temperatures lie on a straight line, where the melts of the various mixtures have this refractive index. In monochromatic light, the glass grain will then be invisible. Fig. 6.28 illustrates this relationship.

From the temperature-vs.-refractive index curve (I), the mixing ratio of the eutectic can be found by measuring the temperature at the point of eutectic composition in the contact specimen. More complex systems can be measured quantitatively in the same way.

With the contact method, mixed crystals are easily detected. Both phases are brought into contact, and the higher-melting phase is inoculated should freezing not set in in time.

During further cooling down, the isomorphous growth of one crystal kind in the melt of the other kind can be followed clearly.

If melts are low in viscosity, contact specimens of inorganic crystals frequently

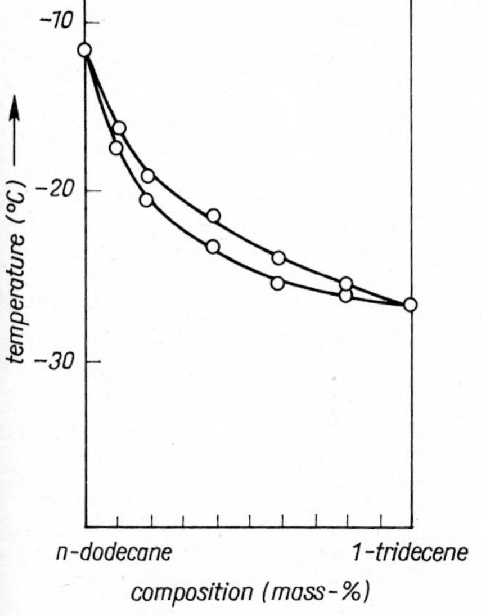

Figure 6.27. Microscopical determination of the melting and freezing curve in an n-dodecane/1-tridecene system

Table 6.30. Contact method in binary system with mixed crystal formation

System	Freezing points of all mixtures are between those of the pure components	Freezing curve goes through a maximum	Freezing curve goes through a minimum	Freezing curve with transition point	Freezing point with eutectic point
Schematic illustration of contact specimen	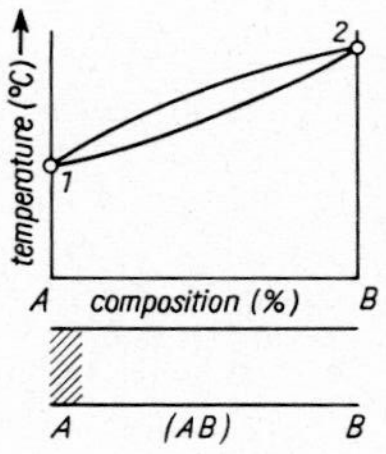 A (AB) B	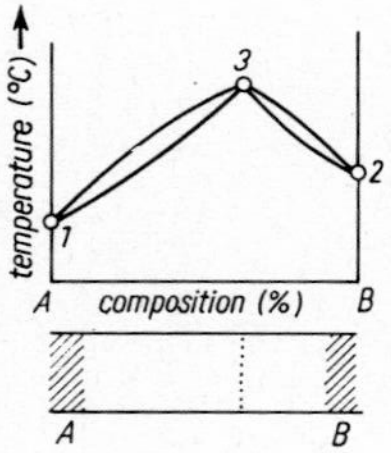 A B	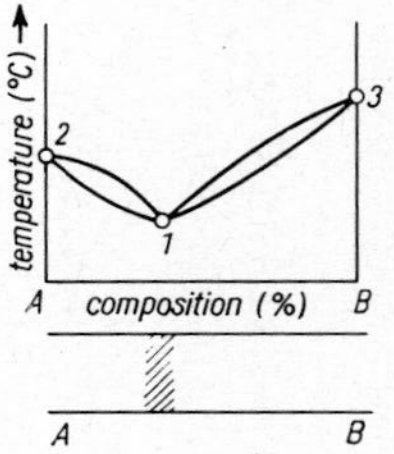 A B	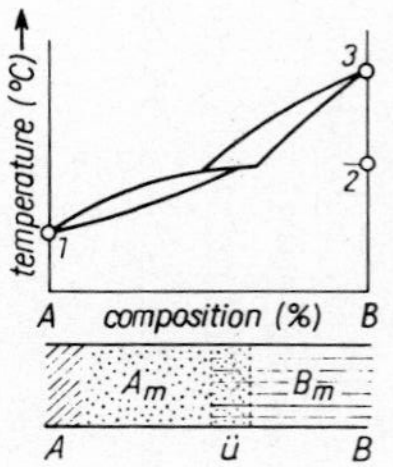 A_m B_m A ü B	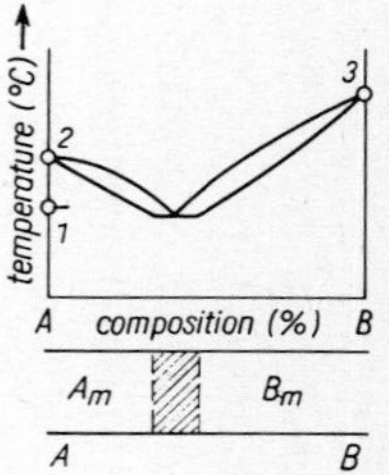 A_m B_m A B
Border zone in contact specimen		Stripe of mixed crystals, corresp. to maximum	Stripe of mixed crystals, corresp. to minimum	Overlap zone ü between mixed crystal types A (A_m) and B (B_m)	Border zone between mixed crystals A (A_m) and B (B_m)

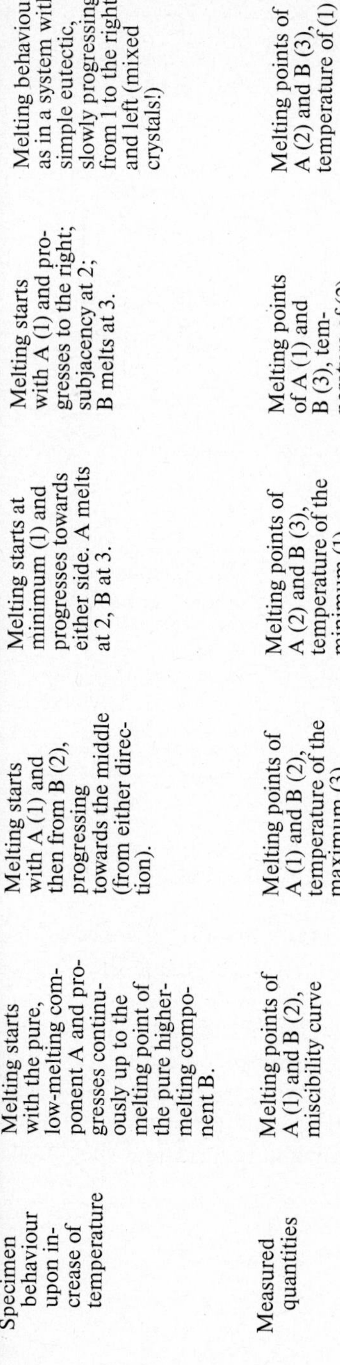

Specimen behaviour upon increase of temperature	Melting starts with the pure, low-melting component A and progresses continuously up to the melting point of the pure higher-melting component B.	Melting starts with A (1) and then from B (2), progressing towards the middle (from either direction).	Melting starts at minimum (1) and progresses towards either side. A melts at 2, B at 3.	Melting starts with A (1) and progresses to the right; subjacency at 2; B melts at 3.	Melting behaviour as in a system with simple eutectic, slowly progressing from 1 to the right and left (mixed crystals!)
Measured quantities	Melting points of A (1) and B (2), miscibility curve	Melting points of A (1) and B (2), temperature of the maximum (3)	Melting points of A (2) and B (3), temperature of the minimum (1)	Melting points of A (1) and B (3), temperature of (2)	Melting points of A (2) and B (3), temperature of (1)

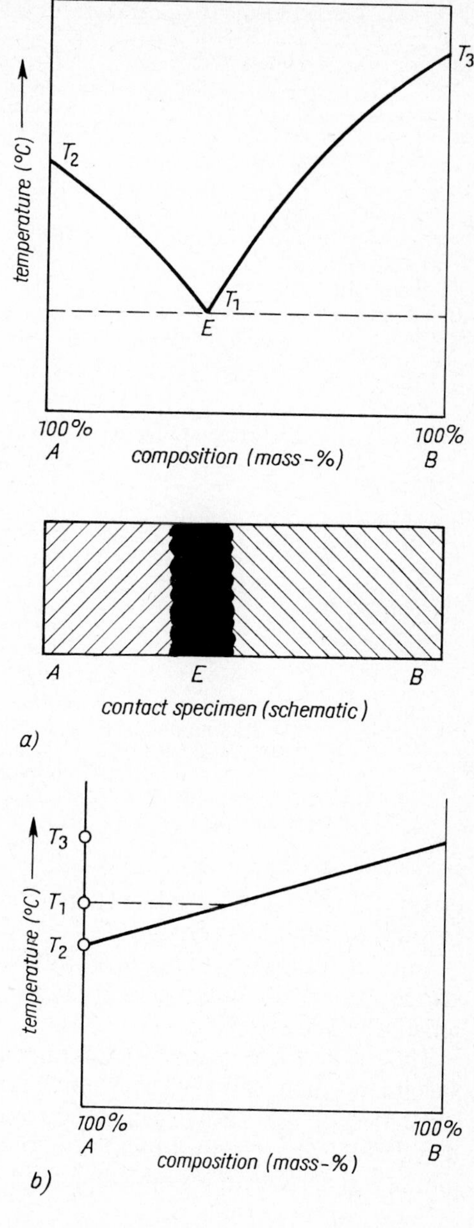

Figure 6.28. Contact method (schematic)
a) simple two-component system, with schematic illustration of the contact specimen
E eutectic mixture
b) combination with glass powder method

Table 6.31. Contact method in binary systems without mixed crystal formation

System	Simple eutectic	Dystectic, compound with congruent melting point	Peritectic, compound with incongruent melting point
Schematic illustration of contact specimen	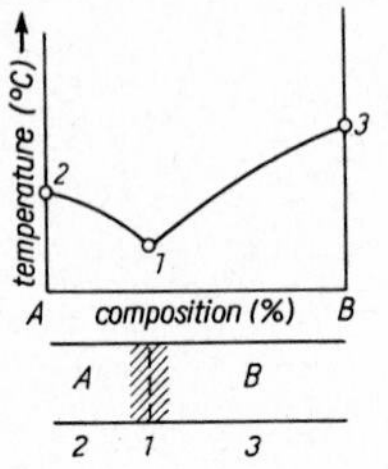 temperature (°C) ↑; A composition (%) B; points 2, 1, 3; specimen: A, B; 2 1 3	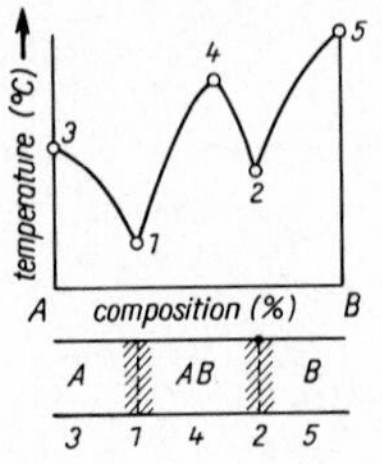 temperature (°C) ↑; A composition (%) B; points 3, 1, 4, 2, 5; specimen: A, AB, B; 3 1 4 2 5	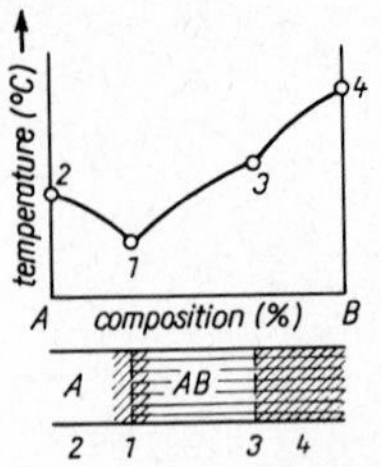 temperature (°C) ↑; A composition (%) B; points 2, 1, 3, 4; specimen: A, AB; 2 1 3 4
Border zone in contact specimen	Eutectic composition	Specimen divided into three by two eutectic zones 1 and 2	Between 3 and 4, A is superposed on compound AB
Specimen behaviour upon increase of temperature	Melting starts in the border zone (1), then A (2) melts, followed by A (3).	Melting starts in the eutectic border zone (1), then progresses to (2), later to phase A (3), the compound AB (4), and ends with phase B (5).	Melting starts in the eutectic zone (1), then progresses to phase A (2). At 3, the compound AB disappears, producing A and a melt; phase B melts at 4.
Measured quantities	Eutectic temperature (1), melting points of A (2) and B (3)	Eutectic temperatures 1 and 2, melting points of A (3), AB (4), and B (5).	Eutectic temperature 1, peritectic 3, melting points A (2) and B (4).

do not exhibit a distinct contact zone. Examinations by this method are difficult because of imperfect contact.

One application of the contact method to the identification of organic substances (especially pharmaceuticals) is the technique of measuring the characteristical eutectic [6.12].

The substance to be identified is mixed with a reference standard. The contact specimen will quickly reveal the temperature of the common eutectic. The initial substance can now easily be identified by comparative measurements. Table 6.32 lists several specimens of pharmaceutical importance having identical melting points and demonstrates how they can be distinguished in a contact specimen.

6.4.5 Microsublimation

Sublimation under the microscope is significant for the following problems:
(1) Purification or separation before microscopy (mainly organic substances);

Table 6.32. Characteristic eutectic temperatures for identifying organic substances of equal melting point, acc. to [6.12 and 7.2]

Substance	Melting point (°C)	Eutectic temperature (°C) with	
		Benzil	Acetanilide
m-Bromobenzyl-piridyl-diethyl-ethylene-diamino-maleate	108 - 110	80	68
Fructose	100 - 110	94	98
Fluoranthene	110	64	85
Rhotan	110	66	88
Acridine	110	67	80
Acetotoluidide	110	78	75
0,0-Diphenol	110	65	48
Anisoin	110	78	89
Ocrin	110	67	28
Methylbutyl-ketone-2,4-dinitro-phenylhydrazine	110	74	88

(2) Measurement of the sublimation temperatures of compounds; studying the temperature behaviour of sublimating substances; and
(3) Examination of crystal growth from the gaseous phase.

Microsublimation can be observed by transmitted or reflected light, both preferably at low power. For qualitative work, a raised cover slip with specimen slide on the hot stage will serve the purpose. Special apparatus available for sublimation in vacuum [6.12] permits one to examine specimen and standard simultaneously for comparison. The solid phases formed are identified by measuring their optical properties.

6.4.6 Thermometry with the microscope

In recent years, some facilities have become known that permit thermometry on microscopic objects by registration of the heat radiation emitted by them [6.189]. Applications are the temperature measurement of spun threads or of microelectronic devices. The exact calibration of the radiometric apparatus is still a problem.

6.5 Ultramicroanalysis [6.190 to 6.192, 6.200 to 6.208]

The term of ultramicroanalysis denotes the application of modified chemical analysis methods in miniaturized apparatus under microscope observation (Tables 6.33, 6.34). If the micromanipulator is employed as the main tool, some authors refer to the technique as micrurgy.

The fundamental methods of ultramicroanalysis were published by Benedetti-Pichler and Kirk in 1937. The principal idea is that smallest substance quantities can be chemically analyzed if reactant concentrations are made sufficiently high by radically diminishing the sample volume. That is to say that any reduction of the sample solution volume permits one to detect ever smaller substance quantities (Table 6.35).

If solution volumes are drastically reduced to 10^{-2} to 10^{-5} ml, reactant concentrations even in case of substance quantities below 1 µg can be kept within conventional limits; thus, many assaying techniques approved in macro- and microanalysis may principally be transferred to microgramme quantities. That way it is possible to chemically analyze substance quantities at the very limit of naked-eye visibility. As the sample volume is reduced, the threshold of detectability decreases. Ultramicroanalysis is employed where sample quantities available are extremely low; thus it is different from trace analysis, which seeks to detect very low contents in considerably larger sample quantities.

Although ultramicroanalysis is a genuine micro-method requiring minute sub-

Table 6.33. Comparison of ultramicroanalysis with other methods of chemical analysis, acc. to [6.191]

Analysis method	Solution volume (ml)	Substance weight (mg)	Detection threshold (mg)
Macro methods	2	100	10^{-2}
Micro methods	0.5 - 0.05	10 - 0.5	10^{-5}
Ultamicroanalysis	10^{-3} - 10^{-6}	10^{-3}	10^{-7}

Table 6.34. Surface-to-volume relationships in macro- and ultramicroanalysis, acc. to [6.202]

Analysis method	Vessel dimensions (cm)	Wetted vessel area (cm^2)	Solution volume (cm^3)	Quotient surface area/ volume (cm^{-1})
Macro methods	r = 2.5 h = 3	67	59	1.13
Ultramicro methods	r = 0.05 h = 0.2	0.07	0.0015	47

Table 6.35. Units and symbols commonly employed in ultramicroanalysis

Length	1 m	$= 10^3$ mm $= 10^6$ µm
Mass (weight)	1 g	$= 10^3$ mg $= 10^6$ µg $= 10^9$ ng ($= 10^6$ γ) ($= 10^6$ E*))
Volume	1 l	$= 10^3$ ml $= 10^6$ µl $= 10^9$ nl ($= 10^6$ λ $= 10^9$ mλ) ($= 10^6$ mm^3)
Equivalent	1 Eq	$= 10^3$ mEq $= 10^6$ µEq ($= 10^6$ ε)

*) E = Emich, unit used in the literature to honour the microchemist, F. Emich.

stance quantities, trace analyses may also be used in simple cases if the main component can be readily separated.

Ultramicroanalysis is further important where optical data have to be correlated to chemical composition in order to identify unknown substances.

Equipment

Microscope

Ultramicroanalytical operations are carried out on the specimen stage of a reflected-light microscope, sometimes with a transmitted-light accessory outfit. Volumes are determined by means of an eyepiece screw micrometer.

Micromanipulator

Because of the delicacy of the equipment and the working under the microscope, it is necessary to use a micromanipulator. Many different types are available (Fig. 4.3a). It is essential that the unit should have

(a) a vertical motion through ± 20 mm,
(b) a horizontal motion through ± 25 mm towards and away from the sample,
(c) a facility to incline the stage by 15° to 20°, and
(d) a facility to swivel the entire system about its axis.

Reaction vessels

The sample is contained in a »microcone«, a thin-walled capillary tube of 1 to 3 mm length, whose bore tapers down into a solid rod. The configuration corresponds to that of the pointed tube used in microanalysis. Microcones are produced easily by the extension of glass capillaries; they can hold between 0.2 and 3 µl of analysis solution depending on their diameter. Microcones, same as any other glassware used in ultramicroanalysis, must be made hydrophobic. To do this, the cleaned glass surface is coated with a thin film of an organic silicon compound, e.g. by immersion in a 3% solution of CH_3SiCl_3 in carbon tetrachloride and subsequent heating to about 130 °C for one hour. The film thus applied is resistant to aqueous salt solutions and diluted acids, but less so to alkaline solutions (Table 6.36).

Table 6.36. The stability of a methylchlorosilane film at room temperature, acc. to [6.202]

Solution concentration	HCl	HNO_3	H_2SO_4	NaOH
	Time in hours			
0.1 N	15	15	15	2
0.5 N	15	25	15	1
1.0 N	3	1.5	1.5	0.5
3.0 N	1	1	1	0.25
6.0 N	0.5	0.5	0.5	-

Hydrophobing of the glassware with organic silicon compounds has some important benefits for ultramicroanalysis, viz.

(1) vessels need not be flushed,
(2) titration is more accurate,
(3) salts do not creep along vessel walls,
(4) rate of evaporation off low volumes is distinctly less because miniscuses are flattened,
(5) reading of capillary volumes is more accurate, and
(6) substance losses due to adsorption by vessel walls are considerably reduced.

During the analysis, the microcones are held horizontally in holes drilled into the end face of a block of plastics material. The solutions will not run out thanks to surface tension.

Humidity chamber

To avoid evaporation losses, the holder with the microcones is contained in a humidity chamber. It consists of a box made of perspex slabs open at the front, and with removable bottom and cover glass plates. The holder is surrounded by moistened cottonwool, which maintains a damp atmosphere inside the chamber. Concentrated salt solutions should not be kept in the chamber too long, as they will dilute themselves by isothermal distillation. Protruding into the chamber through a hole in the side is a shiftable, electrically heated platinum wire loop (powered at 1 to 6 volts). If pushed below the microcone, the loop serves to evaporate and heat the sample solution.

Pipetting equipment

Piston pipettes are employed for conveying smallest liquid quantities. The piston, a length of wire actuated by a screw, displaces a barrier liquid (decocted water) reaching up to a glass capillary tightly connected to the drive screw by means of a rubber stuffing box. This capillary receives the sample. Its bore diameter is between 0.2 and 1 mm, and its sample end tapers down into a capillary point 5 to 10 mm long and 10 to 100 μm wide. The liquid to be pipetted must be sucked into the capillary so that it is separated from the barrier liquid by an air bubble of 20 to 30 mm length.

Centrifuges

Sediments obtained during analytical procedures must be centrifuged so that they are collected in the point of the microcone. Suitable are fast-starting, fast-stopping centrifuges with horizontal rotors. A special holder is required to receive the microcones.

Measuring capillaries

Measuring capillaries are needed for precisely measuring small liquid volumes and for calibrating the burettes. They are thin-walled glass capillaries of circular cross-section and 20 to 30 mm length. Their diameters, ranging from 0.2 to 1.5 mm, can be measured accurately with the eyepiece screw micrometer. For measurement, the capillary is held in a horizontal position inside the moist chamber. A surplus of the liquid to be measured is taken in. The column of liquid in the hydrophobic capillary is limited by entirely plane, well-defined surfaces. The volume of the sample taken can be computed from the length of the liquid column before and after the analysis sample has been removed.

Burettes

The burette most used in ultramicroanalysis is the micrometer burette. A wire piston whose advance is checked on a dial gauge displaces a corresponding amount of barrier liquid and, thus, an equal amount of titration solution from the attached glass capillary. After calibration with measuring capillaries, readings obtained may be accurate within 0.07 nl ($= 7 \cdot 10^{-8}$ ml).

Evaporation and concentration apparatus

To evaporate and concentrate solutions, ultramicroanalysis uses a set of small teflon or quartz vessels, some of which are conical. For ease of manipulation, they have handling rods. As the evaporation of solvent progresses, one transfers the sample to ever smaller vessels until the desired volume has been reached. The last stage of evaporation should invariably be made in the microcone below the microscope, the volume evaporated from the microcone to be replenished with a pipette.

Stirrer

A mechanical stirrer is useful for titration. The vibrations of a vibrator are transferred to the microcone or to a part immersed in the liquid (pipette, electrodes).

Electrodes

Electrodes are metal wires 0.1 to 0.05 mm thick and up to 10 mm long. They are attached, by melting or soldering, to well-insulated copper wires of 0.5 to 1 mm thickness clamped into the micromanipulator. Each electrode may be contained in a glass capillary.

Techniques (Table 6.37)

At its present state, ultramicroanalysis normally permits quantification of substance quantities of 10^{-9} g, but in special cases the limit is even lower. The standard deviation of a single measurement is less than 10 %. Electrochemically controlled techniques such as coulometry are capable of determining quantities as small as 10^{-11} g (using not the microcone but a micro-droplet suspended under paraffin). The use of electron microscopy and micromanipulators specially designed for it allows to attain sensitivities between 10^{-15} and 10^{-19} g, which approach the theoretical limit of 10^{-20} g.

Gravimetry

Gravimetry, although a classical method of chemical analysis, has gained little importance in ultramicroanalysis, mainly because of the lack of balances capable of weighing the minute substance quantities involved.

Commercially available silica torsion microbalances, which are rather expensive, are practically useful for quantities down to 10^{-6} g.

Table 6.37. Survey of ultramicroanalytical techniques, acc. to [6.200]

Description	Least substance quantity analyzable	Remarks
Separation techniques		To enrich the substance for the following determination methods:
	10^{-6} - 10^{-9} g	Ion exchange
	10^{-6} - 10^{-9} g	Extraction (especially for subsequent photometry)
	10^{-3} - 10^{-5} ml	Distillation, sublimation
	10^{-6} - 10^{-9} g	Electrolysis
Qualitative analysis	10^{-6} - 10^{-9} g	Analogous to the separation process of classical chemical analysis
Gravimetry	10^{-5} - 10^{-6} g	High experimental outlay; limited by the quality of ultramicrobalances
Colorimetry Photometry	10^{-7} - 10^{-10} g	Microscope photometer required
Volumetry	10^{-6} - 10^{-9} g	Universal application, as small volume differences are easy to measure.
Electrochemical procedures	10^{-7} - 10^{-10} g	Possibilities: amperometry, potentiometry, pH testing, conductometry, dead-stop titration, Coulometry

The few reports published yet on gravimetric ultramicroanalysis state substance weights between 10^{-4} and 10^{-5} g. Anyhow, it is easier to determine small volume differences than small weight differences, so that there is little probability that the technique will find widespread use.

Colorimetry and spectrophotometry

Techniques employing colorimetry and spectrophotometry are widely applied to ultramicroanalysis. They are fast and highly sensitive. The main problems involved are the design of an appropriate narrow-gauge capillary cell, its bubble-free filling, the modification of existing spectrophotomectric equipment, and error-free colour formation in smallest volumes.

Visual colorimetric determinations can be made without much equipment outlay, by comparing the colour densities of reference and sample solutions contained in capillaries of 10 to 30 mm length and 0.2 to 0.5 mm width arranged vertically under the microscope. According to [6.201] this technique is suited to detect $2 \cdot 10^{-12}$ g of chromium qualitatively and to quantify 10^{-10} g of that element.

For spectrophotometry, a technique of absorbance measurement is reported [6.201] that uses capillaries of 1 to 2 mm diameter, arranged horizontally in the transmitted-light beam of a microscope photometer. The capillaries serving as sample cells are held in a plastics holder placed on the specimen stage. The sample and reference solutions are filled into one and the same capillary, separated by a narrow air gap. The solution surfaces are well-defined planes. The capillary ends are closed by water membranes.

The use of round cells is subject to the following conditions, viz.

- objective power $\geqq$ 10x;
- constant field size throughout a series of measurements;
- constant position of the focal plane during a series of measurements; it is expedient to focus on to the upper (objective-side) liquid-to-glass interface;
- measurement in the middle of the capillary (lengthwise);
- the same capillary to be used for a measurement and for recording the calibration curve.

That way, absorbance measurements of aqueous solutions can be made with volumes of 0.2 µl having optical path lengths of 1 mm. Detection thresholds are between 0.5 and 5 ng depending on the absorptivity of the substance.

Volumetry

Volumetric techniques have a key position in quantitative ultramicroanalysis. They are accurate, require little time and little equipment and are largely similar to familiar methods in macro- and microanalysis. As the substance quantity is found from the amount of standard solution added (i.e. there is no concentration measurement during the determination), the very low evaporation losses that might occur are of no consequence. The sensitivity and reproducibility of such techniques are functions of the accuracy to which the reaction end point is determined. In general, the change of a physico-chemical parameter is made use of as an indication of end point. In the ultramicro range, concentration-dependent parameters can at present be determined with sufficient accuracy if substance concentrations are not less than 10^{-4} or 10^{-5} mol/l. Especially in volumetric work, sensitivity and reproducibility depend on the smallest possible experimental volume, as the assay of extremely small substance quantities requires that the analysis volume be reduced as much as possible. For quantitative work, the lower limit of the analysis volume is 10^{-4} ml now.

Volumetric titrations with colour indicators call for high indicator concentration, because of the small path length across the microcone. As a consequence, indicator errors may occur that restrict determinable substance quantities to not less than about 10^{-7} g. In some cases, special techniques requiring greater experimental outlay permit access to the submicrogramme range as well.

Electrochemical techniques (Table 6.38)

Electrochemical techniques do not involve the problems of end point titration in volumetric analysis. Potentiometric, amperometric or conductometric titrations cover even the nanogramme range (less than 10^{-8} g of substance) without difficulty. The important point is to have a sample cell of optimum design. Many suggestions for this have been reported [6.202].

Apart from electrochemically indicated titrations, ultramicroanalyses increasingly use techniques that do not require the addition of a standard solution, such as coulometry. The advantage of such methods is that the volume remains unchanged during the procedure; volumetry is replaced by the measurement of the quantity of electricity, which frequently is simpler and more accurate.

In coulometric titration, the required reagent is produced electrolytically. Its quantity and, thus, that of the substance to be determined can be registered by measurements of current and time. It is possible today to assay substance quantities in the nanogramme range, and this does not even exhaust the potentialities of the method.

One very accurate method also applicable to the ng range is dissolution coulometry, which uses a constant potential to determine the quantity of precipitated substance due to excess current by ascertaining the quantity of electricity required to redissolve the precipitated substance.

Separation procedures

Electrolytic separation

Metals can be separated from solutions by electrolysis, a simple and fast procedure. As electrode surfaces in ultramicroanalysis are small, it is necessary to maintain a current density of 0.10 to 0.01 A/cm^2 (10^{-4} to 10^{-5} A for a D.C. voltage of 4 to 6 volts). Preferably, the electrodes are made of platinum wire about 0.1 to 0.2 mm thick.

Substances that do not deposit on platinum electrodes are separated by means

Table 6.38. Electrolytic separation of substances (acc. to the literature)

Electrode		Electrolyte		Electrolytic conditions			Electrolytic product	
Diameter (cm)	Surface (cm^2)	Composition	Metal concentration	Voltage (V)	Current (A)	Duration (min)	Composition	Quantity (gr)
Platinum								
0.02	10^{-4}	$CuSO_4$	$8 \cdot 10^{-3}$	2.2	10^{-4}	20	Cu	$8 \cdot 10^{-6}$
0.02	10^{-4}	$CuSO_4$, $NiSO_4$ also with $FeSO_4$ each	$1 \cdot 10^{-3}$ $3 \cdot 10^{-4}$	2.2	10^{-4}	10	Cu	$5 \cdot 10^{-7}$
0.02	10^{-4}	$Pb(NO_3)_2$	$1 \cdot 10^{-3}$	2.7	10^{-4}	10 – 15	PbO_2	$5 \cdot 10^{-7}$
Mercury								
Cathode surface 2×10^{-3}		$FeSO_4$ $(NH_4)NO_3$ (H_2SO_4)	$3 \cdot 10^{-4}$ $2 \cdot 10^{-4}$	3.4 – 3.8	$3.4 - 3.8 \cdot 10^{-4}$	15 – 20		*Precipitated at the cathode:* Fe $1.5 \cdot 10^{-7}$ *Remaining in solution:* V $1.5 \cdot 10^{-7}$

of mercury. Under ultramicroanalytical conditions, this calls for a specially designed electrolysis cell. Because of the intensive gassing frequently occurring, the cell mostly is a glass capillary open at both ends, into which a platinum fork supporting a mercury drop is introduced.

Precipitation

Precipitation is the easiest method of separation. A suitable reagent is added to the solution contained in the microcone, and the sediment is separated by centrifugation. Frequently, it becomes visible only after centrifugation, because of the small volume. Before immersing the pipette mouth into the solution, one should generate a slight excess pressure in the pipette so that the reagent leaks out as soon as the pipette touches the sample, and the pipette mouth cannot be blocked by adhesive action.

Ion exchange

This method uses columns consisting of capillaries between 30 and 40 mm long and 2 to 3 mm thick, each provided with a small fritted glass filter at the bottom end. The capillary is filled with finely ground resin and the solution added on top into the flaring upper end. The various eluate fractions are collected in microconi. The column is arranged outside the humidity chamber.

Simple exchange operations may also be performed directly in a microcone partly filled with finest-grained resin powder.

Extraction

Extractions at an ultramicro scale require some experimental skill, because the organic liquids used as extraction agents are highly volatile. The humidity chamber must therefore be saturated with the respective solvent vapour. For extraction, both the aqueous and organic phases are fusion-sealed into a capillary and centrifuged. The organic phase may also be added to the solution as a drop suspended from the pipette mouth, which is moved to and fro through the aqueous phase several times.

Subsequently, the joined extracts are collected in a second cone and evaporated.

Distillation and sublimation

Another microcone serves as a receiving and cooling vessel. Its diameter just fits over the cone containing the sample. A microheater is cautiously applied, and the receiving vessel centrifuged after the process has finished. The receiving vessel may be cooled additionally by carbon-dioxide ice.

Special method: Determination of density of small particles

The density of a substance (the quotient of its mass and its volume) is a substance-specific constant, which is easy to ascertain and can contribute to characterizing the substance. According to a procedure reported by Helbig [6.203], the density of particles down to a volume of 0.1 mm^3 can be determined with ultramicroanalytical techniques in a measuring capillary. The procedure is based on the classical displacement method, by which the sample volume is found from the volume of liquid displaced by it. One end of the measuring capillary is fusion-sealed. The capillary is filled with a liquid (e.g., water) and fixed to the microscope horizontally, e.i. at right angles with the ray path; after this, the specimen grain is inserted. Its volume is found from the shifting of the liquid level.

For determining the specimen weight, balances of appropriate accuracy (sensitive to ± 1 μg) are required.

If the density of a particle is less than 1 g/cm^3, the sample is centrifuged to the surface of the liquid and the capillary rotated so that the liquid level is above the particle. Upon slight knocking the particle will immerse into the liquid. The capillary forces prevent the water from leaking out.

Maximum deviations of individual measurements from the mean are between ± 0.5 and ± 2 %. Volume measurement with the measuring capillary is affected by average measuring errors of ± 0.3 %. Mass determination must be accurate to ± 0.1 %.

Microcrystallography [6.190] as a special technique of qualitative analysis has lost its importance.

The detection sensitivity of many ultramicroanalytical techniques can be increased by subsequent identification of the reaction products by means of measuring optical properties. Identification by typical crystal shapes alone is not always unique [6.208].

Limits and errors

Theoretically, it should be possible to analyze substance quantities of any degree of smallness by means of corresponding reductions of the solution volume. The minimum quantity is found from statistical considerations on the number of reacting particles required, which is about 10^{-20} g. Owing to various disturbing effects such as increased solubility of smaller crystals, oversaturation and surface effects, the minimum substance quantity practically ascertainable at present is 10^{-10} g.

It is essential for this analytical technique that dust-free conditions are carefully maintained. A single dust particle may considerably affect the analysis result. The purity of the reagents is highly influential, too.

A list of possible errors in volumetric ultramicroanalysis is given in [6.191] (see Table 6.39).

Table 6.39. Analysis error possible in volumetric ultramicro-analysis, acc. to [6.191]

Operation	Chemical errors[1])	Error in end point titration	Instrumental and reading error	Gross errors[2])
Preparation of analysis	b	d	c	a
Specimen dimensions	b	d	c	a
Determination operation	b	d	c	a
End point indication	a	b	c	a
Computation of results	d	d	d	a

a error not estimable
b error partly estimable
c error estimable
d free from errors

[1]) The author understands chemical errors to be influences of side reactions, incomplete reaction process, interface influences, etc.

[2]) Gross errors originate from contamination by smallest dust particles, uncontrolled loss of substance etc.

6.6 Other physical properties

6.6.1 Solubility and etching behaviour

Microscopical investigations of the solubility and etching behaviour of solid phases provide a variety of information, viz.
(1) Separation of unwanted phases in solid mixtures (Table 6.40)
(2) Identification of phases (Table 6.41)
(3) Behaviour of solid compounds against certain solvents
(4) Mechanism and kinetics of dissolution processes
(5) Visualization of structural differences on surfaces (Table 6.42)
(6) Detection of lattice defects (Table 6.43) [6.194]

Microscopical structure examination after etching the specimen is suited, above all, for coarsely grained substances, thus bordering favourably on the range of X-ray techniques.

Investigation into solubility and etching is subject to rather complex conditions, which make the interpretation and comparison of results difficult.

Table 6.40. Phase separation of technical crude potash salts by selective dissolution, acc. to [7.39]

Existing phases	Phase separated by dissolution	Solution employed
$K_2SO_4 \cdot MgSO_4 \cdot 2\ CaSO_4 \cdot 2\ H_2O$ (polyhalite) $KCl \cdot MgSO_4 \cdot 3\ H_2O$ (kainite)	$KCl \cdot MgSO_4 \cdot 3\ H_2O$	20 °C: 126.02 g $MgSO_4 \cdot 7\ H_2O$ 17.5 g K_2SO_4 100 g H_2O
NaCl (rock salt) KCl (sylvine)	KCl	20 °C: saturated NaCl solution
NaCl (rock salt) $KCl \cdot MgCl_2 \cdot 6\ H_2O$ (carnallite)	$KCl \cdot MgCl_2 \cdot 6\ H_2O$	20 °C: saturated NaCl solution
KCl (sylvine) NaCl (rock salt)	NaCl	20 °C: saturated KCl solution
$3\ K_2SO_4 \cdot Na_2SO_4$ (glaserite) KCl (sylvine)	KCl	25 °C: 12.42 g KCl 4.35 g K_2SO_4 100 g H_2O
$KCl \cdot MgSO_4 \cdot 3\ H_2O$ (kainite) KCl (sylvine)	KCl	35 °C: 68.92 g KCl 12.64 g $MgCl_2 \cdot 6\ H_2O$ 31.33 g $MgSO_4 \cdot 7\ H_2O$ 100 g H_2O

The solubility and etching behaviour of objects under examination is greatly influenced by various factors, viz.

concentration of the solvent or etchant,
duration of action,
temperature,
influence of foreign substances on solubility composition and structure of the specimen substance, position of the crystal face tested, swelling processes and side reactions.

Errors may result from imperfectly formed faces which make for inaccurate measurement, unrepeatable etching conditions, and different progress of etching in different specimens. It is essential that the etchant should act in one spot and that its action is not face-selective. The etched faces must be clearly defined crystallographically (no vicinal faces).

The investigation into the solubility and etching behaviour requires some instrumental outlay.

Table 6.41. Microscopical identification of clinker phases in cement by selective etching, acc. to [6.193]

Clinker phase	**Etchant** H_2O	1ml HNO_3+ 100 ml ethyl or amyl alcohol	HF 1. concentrated 2. diluted 1:10 3. vaporized	10% KOH in H_2O	8 ml NaOH (10%) + 2 ml Na_2HPO_4 (10%)	10 ml n-oxalic acid + 90 ml ethyl alcohol (95%)	0.4 or 1.0% alcoholic solution of borax	Mixture of glacial acetic acid: ethyl alcohol (1:100)
	Etching temperature							
	RT	RT	RT	30 °C	50–55 °C	RT	RT	RT
	Etching time							
	5 – 15 s	2 – 15 s	1.2 – 3 s 2.5 – 10 s 3.10 – 20 s	15 s	60 s	5 – 15 s	10 min	2 – 5 s
Alite ($3\,CaO \cdot SiO_2$)	T	TS	(T)	–	T	–	T	S
Belite ($2\,CaO \cdot SiO_2$)	S	TS	T	–	–	–	–	(S)
Brownmillerite ($4\,CaO \cdot Al_2O_3 \cdot Fe_2O_3$)	–	S	–	–	T	–	–	–
Aluminate ($3\,CaO \cdot Al_2O_3$)	T	–	–	T	T	T	–	–
Calcium oxide (CaO)	T	T	T	–	T	–	T	S
Iron monoxide (FeO)	–	T	T	–	–	–	–	–
Periclase (MgO)	not attacked by etchants							
Potassium sulfate (K_2SO_4)	etchants act very rapidly							

Etching behaviour

T tarnish etching
(T) weak tarnish etching
S structural etching
(S) weak structural etching
RT room temperature

Table 6.42. Etchants for polyethylene, acc. to [7.227]

Etchant	Temperature (°C)	Etching time (min)
Heptane	65	1 - 3
Benzene	65	3 - 6
Xylene	70	0.25 - 1
Benzine	40 - 60	5
Chloroform	55	1 - 2
Carbon tetrachloride	70	2

Table 6.43. Examples for etching techniques used in microscopical detection of edge and screw dislocations in metals (acc. to the literature)

Metal, alloy	Etching conditions	Face observed
Ag	Chemical etching *Etching time:* 5 to 45 s *Etchant:* 2.5 parts NH_4OH (concentr.) + 1 part H_2O_2 (30%)	(111), (100)
Cu	Chemical etching *Etchant:*20 ml saturated solution of $FeCl_3$ + 20 ml HCl + 5 ml CH_3COOH + 5 to 10 drops Br_2	(111)
	Etching time: 10 to 60 s *Etchant:* conc. aqueous solution of HBr with varying quantities of $CuBr_2$	(111)
Fe + 3% Si	Electrolytic etching at abt. 20 to 30 mA/cm^2, annealing at abt. 160 °C during 15 min. *Etchant:* Morris electrolyte for steel polishing; 133 ml glacial acetic acid + 25 g CrO_3 + 7 ml H_2O	All faces, also suitable for polycrystalline samples
	Chemical etching *Etchant:* 1 part conc. NH_4OH solution + 2 parts $CuSO_4$ solution (25%)	(112), (100)
Zn	Chemical etching *Etching time:* abt. 15 s *Etchant:* 4 ml HNO_3 (conc.) per 100 ml H_2O	(0110)

Evaporating solvent or etchant must not act on the microscope objective. For direct observation of the process, specimens are best contained in specimen slide chambers and microcells.

The solvent may have the following effects on the specimen, viz.

(1) The grain does not change.
(2) The grain gets dissolved, etched, stained, and tarnished.
(3) Some gas bubbles develop.
(4) The grain dissolves under vehement gas development.

These effects provide information about the dissolution mechanism (cf. 7.3.4.2) and help to distinguish between the solution properties of the compounds to be compared.

6.6.2 Microhardness

The microhardness of a solid denotes the resistance offered by the lattice to an intruding body as long as the lattice structure remains intact, with the substance deforming in an elastic and/or plastic way. If the penetrating body destructs the structure, the indentation size bears no meaningful relation to the original structure. Microhardness is a theoretically complex quantity, which is difficult to compute, but in many cases it permits to characterize fine structural differences in solids.

Microhardness tests require specimens with smooth surfaces.

The microhardness H_m of a material is the ratio of the penetration force to the surface of the indentation of the diagonal d:

$$H_m = \frac{P}{d^2}.$$

P penetration force
d diagonal

The indentation can be inspected by reflected or transmitted light. Of the various penetration bodies, the Vickers pyramid and Hanemann's design have proved to be the most favourable devices. The straight-lined edges of the indentation are easier to measure than the circular ones of spherical or conical indentations [6.195, 6.196].

Microhardness may be measured by three different techniques, viz.

(1) Penetrating equipment separate from the microscope (Vickers equipment): microscopical measurement of the indentation
(2) Penetrating body fixed to the front lens of the objective (Hanemann technique): microscopical selection of the spot of interest, subsequent measurement of the indentation

(3) Penetrating body fixed to the front lens, permitting transmitted light observation and measurement during the penetration process under pressure (arrangement reported by Müller [6.197]).

The results of the three techniques are not readily comparable with each other. For microhardness tests it is essential that

(1) each test should consist of at least three indentations each with three different test loads, and
(2) if an indentation in an anisotropic material is distorted, measuring the diagonal is not sufficient for characterizing the substance. Each edge of the indentation must be measured individually.

Müller's arrangement is preferred for testing organic high polymers, whereas the Vickers and Hanemann techniques are employed for less elastic materials.

The results of microhardness tests are useful, above all, for characterizing non-crystalline objects (glasses, resins, high polymers). Polycrystalline material of high density yields good statistical hardness values. Microhardness testing is unsuitable for measuring the structural strength of aggregates.

In making polished sections, even slight hardness differences cause a relief, which shows up by a line of light (Schneiderhöhn line). This line changes its position as the microscope tube is raised or lowered. The rule is that *with the body tube being raised, the line invariably moves from the harder to the softer phase.*

6.6.3 Cleavage, twinning, density, conductivity

Table 6.44 provides a synopsis of the observation of, and information obtained from, these microscopical properties.

Table 6.44. Microscopical determination of cleavability, twinning, density and conductivity

Property	Observation method	Evaluation	References
Cleavability (partition of a crystal along lattice planes of weak bonding force)	Medium magnifications, transmitted or reflected light, orthoscopy, one polarizer, narrow condenser aperture. Subjective classification of cleavage cracks by rectilinearity, parallelism and distinctness	Indication to crystal structure in grains without external crystal shape, important reference faces for correlating optical indicatrix to lattice structure, statistical information from observation of many crystals of same substance, crystallographic indexing of cleavage faces (requires measurement of larger, well-formed crystals)	[2.6]
Twinning (regular intergrowth of lattices of homogeneous crystals)	Medium magnifications, orthoscopy; one polarizer (characteristic, reentrant angles on crystals); crossed polarizers (different extinction positions of the two crystals as the stage is rotated); exact measurement of twins with universal stage	Frequent occurrence of twinning in some substances is an additional characteristic. Measurement of twins provides hints on crystal structure.	[2.6]
Density	Flotation method in a microcell, indication of direction of movement (and thus, of density difference) by direction of particle deviation after setting to cross-hair centre; horizontal beam on optical bench, transmitted light, density of liquid found from refractive index measurements	With the density and the microscopically determined symmetry of small crystals known, powder X-ray diffraction patterns can be indexed.	[6.198, 6.199]
Conductivity	Special sensor moved along smooth specimen surface, reflected light; recording a conductivity profile along the scanned line	Phase boundaries cause leaps in the conductivity curve.	

References

[6.1] *Tatarskij, V. B.:* Kristalloptika i immersionnyj metod issledovanija mineralov. Moskva: Izd. »Nedra« (1965)

[6.2] *Fiedler, H. J.:* Die Untersuchung der Böden. Bd. 2, Laboruntersuchung. Dresden: Verlag Theodor Steinkopff 1965

[6.3] *McCrone, W. C., Draftz, R. G., and J. G. Delly:* Supplement to the »Particle Atlas«. Michigan Ann Arbor Science Publ. Inc. 1969

[6.4] *McCrone, W. C.:* Ultramikroanalyse mit dem Mikroskop. Adv. Analyt. Chem. Instrument (1968) 7, 1–40

[6.5] *Zussmann, J.:* Physical Methods in Determinative Mineralogy. New York: Academic Press 1967

[6.6] *Zaumseil, G.:* Methoden und Geräte zur Ausmessung von Gefügebestandteilen. Feingerätetechnik 3 (1954) 19–26

[6.7] *Berl, W. G.:* Physical Methods in Chemical Analysis, vol. 1. New York: Academic Press 1950

[6.8] *Cameron, E. N.:* Apparatus and Techniques for the Measurement of Certain Optical Properties of Ore Minerals in Reflected Light. Econ. Geol. 52 (1957) 3, 252–268

[6.9] *Ehrenberg, H.:* Die mineralogisch-mikroskopische Untersuchung feinster Aufbereitungsprodukte. Z. Erzbergbau Metallhüttenwesen 4 (1951) 285–293

[6.10] *Schumann, H.:* Neuere optische Methoden für die Bestimmung von Schwermineralien. Erdöl u. Kohle 4 (1951) 684–687

[6.11] *Jones, F. T.:* Fusion Techniques in Chemical Microscopy. Microscope 16 (1968) 1, 37

[6.12] *Kofler, L., Kofler, A., and M. Brandstätter:* Thermo-Mikromethoden zur Kennzeichnung organischer Stoffe und Stoffgemische. Weinheim: Verlag Chemie 1954

[6.13] *McCrone, W. C.:* Fusion Methods in Chemical Microscopy. New York, London: Interscience Publishers Inc. 1957

[6.14] *Piller, H.:* Bemerkungen über den Einfluß der Segmentgröße und Präparatdicke auf die Genauigkeit bei der Messung von Neigungswinkeln mit dem U-Tisch. Mikroskopie 12 (1957) 5/6, 166–174

[6.15] *Saenz de, J. M., and J. C. Tessore:* A Directional Constant Derived from the Optical Indicatrix and its Application in Quantitative Conoscopy. Schweiz. Mineralog. Petrogr. Mitt. 48 (1968) 2, 471–485

[6.16] *Schumann, A.:* über die optische Orientierung durchsichtiger Kristalle. Z. angew. Physik 1 (1949) 8, 343

[6.17] *Joel, N.:* The Use of the Gnomonic Projection in the Determination of the Optical Indicatrix of Crystals. Mineralog. Mag. 213 (1951) 602

[6.18] *Saylor, C. P.:* Accurate Microscopical Determination of Optical Properties on Small Crystal in *Barer, R., and V. R. Cosslett:* Advances in Optical and Electron Microscopy, vol. 1. New York: Academic Press Inc. 1966, 41–76

[6.19] *Wenban-Smith, A. K.:* Computerprogramm zur Bestimmung optischer Parameter sowie optischer und kristallographisch-optischer Vorzugsrichtungen direkt aus U-Tisch-Ablesungen. Canad. Mineralogist 9 (1967) 2, 269–270

[6.20] *Joel, N., and F. E. Tocher:* Conical Extinction Curves: a New Universal Stage Technique. Mineralog. Mag. 265 (1964) 853–867

[6.21] *Joel, N., and J. Garaycochea:* The »Extinction Curve« in the Investigation of the Optical Indicatrix. Acta Crystallogr. 10 (1957) 399–456

[6.22] *Bokij, G. B.:* Immersionnyj metod. Moskva: Izd. MGU 1948

[6.23] *Nakovnik, N. I.:* Immersionnyj metod v primenii k petrografičeskim šlifom. 3-e Izd. Moskva: Gos. Geol. Tech. Izdat. 1957

[6.24] *Mozžerin, J. V.:* Statističeskij immersionnyj metod. Zap. vses. mineralog. Obšč. 2, 94 (1965) 516–529
[6.25] *Tatarskij, V. B.:* Vervollkommnete Immersionspräparate. Zap. Xses. mineralog. Obšč. 2, 95 (1966) 614–616 (Russian)
[6.26] *Bokij, G. B., Zurinov, G. G., Sokol, V. I., and V. Z. Kologjažnyj:* Immersionnye židkosti dlja kristallooptičeskich izmerenij pri nizkich temperaturach. Ž. neorg. Chim. VI (1961) 8, 1754–1758
[6.27] *Bokij, G. B., and V. I. Sokol:* Refraktometričeskaja charakteristika kristallogidratov chloridov litija i natrija. Ž. strukturnoj Chim. 5 (1964) 4, 594–597
[6.28] *Weaver, C. F., and T. N. McVay:* Immersion Oil with Indices of Refraction from 1,292–1,411. Amer. Mineralogist 45 (1960) 469
[6.29] *Cargille, J. J.:* Der richtige Gebrauch von Immersionsöl für das Mikroskop. Lab. Pract. 14 (1965) 11, 1289–1294
[6.30] *Bacanov, S. S., and E. D. Ruckin:* Hochlichtbrechende Immersionsflüssigkeiten. Zap. Vses. mineralog. Obšč. 2, 96 (1967) 3, 355–356 (Russian)
[6.31] *Meyrowitz, R.:* Solvents and Solutions for the Preparation of Immersion Liquids of High Index of Refraction. Amer. Mineralogist 41 (1956) 49–59
[6.32] *Lembeck, R.:* Einige Vorschläge zur Verbesserung der Technik zur Bestimmung der Brechungsindizes nach der Immersionsmethode. Staub 21 (1961) 4, 183–186
[6.33] *Hafner, H. C., and J. L. Rood:* Notwendige Vorsichtsmaßnahmen für Präzisionsmessungen des Brechungsindex. Mater. res. Bull. 2 (1967) 2, 303–309
[6.34] Über den Einfluß verschiedener Flüssigkeiten auf den Brechungsindex von Tonmineralien. Z. Kristallogr. 95 A (1936) 464
[6.35] *Emmons, R. C., and R. M. Gates:* The Use of Becke-line Colour in Refractive Index Determination. Amer. Mineralogist 33 (1948) 9/10, 612–618
[6.36] *Gillberg, M.:* The Error Caused by Inexact Orientation in the Determination of Refractive Indices of Minerals by the Immersion Method. Ark. Mineralog. Geol. 2 (1960) 509
[6.37] *Schmidt, K. G.:* Neues Arbeitsblatt zur Brechzahl- und Dispersionsbestimmung staubförmiger Mineralien. Staub 18 (1958) 247
[6.38] *Bertoldi, G.:* Phasenkontrast in der Mineralogie als Hilfsmittel bei der Brechzahlbestimmung mittels Immersionsmethoden. Neues Jb. Mineralog. Mh. 93 (1959) 10–18
[6.39] *Wilcox, R. E.:* Immersion Liquids of Relatively Stong Dispersion in the Low Refractive Index Range (1,46–15,2). Amer. Mineralogist 49 (1964) 5/6, 683–688
[6.40] *Brown, K. M.:* Dispersion Staining. Microscope and Crystal Front 13 (1962) 311 and 14 (1963) 39
[6.41] *McCrone, W. C., and A. Teetsov:* Measuring the Refractive Indices of Subnanogram Particles. Microscope 17 (1969) 1, 83
[6.42] *Girault, J. P.:* A New Method for Measuring the Refractive Indices of Minerals. Amer. Mineralogist 35 (1950) 421
[6.43] *Lindberg. M. L.:* Measurement of the Alpha Index of Refraction in Minerals. Amer. Mineralogist 31 (1946) 317
[6.44] *Swift, P. M.:* The Indirect Determination of Beta Index of Refraction and 2 V. Amer. Mineralogist 39 (1954) 838
[6.45] *Rosenfeld, J. L.:* Determination of All Principal Indices of Refraction on Difficultly Oriented Minerals by Direct Measurement. Amer. Mineralogist 35 (1950) 9/10, 902–905
[6.46] *Mosebach, R.:* Zur Bestimmung einer definierten mittleren Lichtbrechung optisch einachsiger Minerale im Dünnschliff. Heidelberger Beitr. Mineralog. Petrogr. 4 (1954) 269–287

[6.47] *Lichačev, A.:* Opyt izmerenija pokazatelej prelomlenija mineralov v immersii na fedorovskom stolike. Moskva: Trudy NJGRJ, vyp. 62, 1964

[6.48] *Ebina, A., and S. Shionova:* Einfache Methode zur Bestimmung eines genauen Brechungsindex mit einem Interferenzverfahren. Rev. sci. Instruments 36 (1965) 941-943

[6.49] *Schulz, G.:* Brechungsindexbestimmung im Interferenzmikroskop am Beispiel des CdS. Naturwissenschaften 41 (1954) 525

[6.50] *Sabatier, G.:* Interferometrische Methode zur Messung der Hauptindizes kleiner Kristalle. Bull. Soc. franç. Mineralog. Cristallogr. 88 (1965) 404-412 (French)

[6.51] *Wendelov, L. M., Wallin, L. E., and S. E. Gustafuson:* Messung des Brechungsindex von geschmolzenen Salzen mit der Wellenfrontschnitt-Interferometrie. Z. Naturforsch. 22 (1967) 8, 1180-1184

[6.52] *Canit, J. C., Berger, D., and M. Billardon:* Polarimetrische Messungen von Brechungsindexänderungen. Optica Acta 13 (1966) 255-270

[6.53] *Selm, R.:* Ein einfaches Verfahren zur Bestimmung der Lichtbrechung von Mineralien in Körnerpräparaten. Geologie 6 (1956) 540

[6.54] *Wilcox, R. E.:* Use of the Spindle Stage for Determination of Principal Indices of Refraction of Crystal Fragments. Amer. Mineralogist 44 (1959) 1272-1293

[6.55] *Brandstätter-Kuhnert, M.:* Versuche zur Verwendung der Phasenkontrastmikroskopie bei der Glaspulvermethode nach L. Kofler. Mikroskopie 17 (1962) 15-20

[6.56] *Dodge, N. B.:* The Dark-Field Colour Immersion Method. Amer. Mineralogist 33 (1948) 9/10, 541-549

[6.57] *Kordes, E., Vogel, W., and B. Rackow:* Über eine Hilfsapparatur zur Messung der Lichtbrechung von hygroskopischen Pulvern. Z. anorg. allg. Chem. 268 (1952) 4-6, 236

[6.58] *Engelhardt v., W.:* Eine Methode zur Bestimmung der Lichtbrechung submikroskopischer Teilchen. ZBl. Mineralog. A (1937) 103 and (1938) 212

[6.59] *Correns, C. W.:* Bestimmung des Brechungsexponenten in Gemengen feinkörniger Mineralien und von Kolloiden. Fortschr. Mineralog. Petrogr. 14 (1930) 232-233

[6.60] *Schwarz, F., and G. Hilbig:* Brechzahlbestimmung disperser Systeme. Silikattechnik 18 (1967) 12, 375-377

[6.61] *Heller, W.:* The Determination of Refractive Indices of Colloidal Particles by Means of a New Mixture Rule or from Measurement of Light Scattering. Physic. rev. 68 (1945) 5-10

[6.62] *Butterill, J. D., and E. H. Nickel:* Einfluß von Oberflächeneigenschaften auf die Bestimmung des Brechungsindex durch die Brewster-Winkelmethode. Amer. Mineralogist 52 (1967) 7/8, 1247-1250

[6.63] *Schumann, H.:* Zur Bestimmung des Brechungsquotienten von nicht opaken Mineralsubstanzen mit Hilfe des Polarisationswinkels. Nachr. Akad. Wiss. Göttingen, II. math.-physik. Kl. 8 (1960) 167-172

[6.64] *Miller, A,:* Bestimmung des Brechungsindex nach der Methode von Le Duc de Chaulnes. J. opt. Soc. America 58 (1968) 3, 428

[6.65] *Feklitčev, V. G.:* Opyt primenenija metoda »fokalnogo ekranirovanija« pri immersionnom opredelenii pokasatelej prelomlenija mineralov. Moskva: Trudy IMGRE, vyp. 18 1963

[6.66] *Čerkasov, J. A.:* Teodolitnodisperionnyj metod izmerenija pokazatelej prelomlenija. Novye metody v mineralogii i petrografii i rezultaty ich primenenija. Moskva: Gosgeotechizdat 1963, 48-52

[6.67] *Ruschkin, E. D., Batsanov, S. S., and Y. J. Vesnin:* Messung der Lichtbrechung von Kristallpulvern in festen Immersionsmitteln. Soviet Phys. Cryst. 9 (1965) 632-634 (Russian)

[6.68] *Wigand, G., and H. J. Veith:* Über die Bestimmung des Brechungsindex von

Polymeren aus Messungen ihres Brechungsinkrements. Plaste u. Kautschuk 16 (1969) 9, 671–675

[6.69] *Laves, F., and T. Ernst:* Die Sichtbarmachung des Charakters (+ oder −) äußerst schwacher Doppelbrechungseffekte. Naturwissenschaften 31 (1943) 5/6, 68–69

[6.70] *Mosebach, R.:* Das Messen optischer Gangunterschiede mit Drehkompensatoren. Heidelberger Beitr. Mineralog. Petrogr. 1 (1949) 5/6, 515

[6.71] *Rittmann, A.:* On the Determination of Birefringence in the Case of High Retardations. Bull. Fac. Sci. Univ. Alexandria 1 (1951) 1

[6.72] *Pfeiffer, H. H.:* Über Meßunsicherheit und Empfindlichkeit von Kompensatormessungen an biologischen Objekten. Optik 5 (1949) 217

[6.73] *Mosebach, R.:* Methodisch wichtige Beziehungen zwischen Gangunterschieden und optischen Konstanten in Dünnschliffen anisotroper Mineralien. Fortschr. Mineralog. 28 (1951) 50–57

[6.74] *Mosebach, R.:* Eine Differenzmethode zur Erhöhung der Meßgenauigkeit und Erweiterung des Meßbereiches normaler Drehkompensatoren. Heidelberger Beitr. Mineralog. Petrogr. 2 (1949) 167–172

[6.75] *Mosebach, R.:* Ein einfaches Verfahren zur Erhöhung der Meßgenauigkeit kleiner optischer Gangunterschiede
Heidelberger Beitr. Mineralog. Petrogr. 2 (1949) 132

[6.76] *Rath, R.:* Fehler bei Gangunterschiedsmessungen mit Berek-Kompensatoren. Mikroskopie 12 (1958) 9/10, 327-345 and 14 (1959) 3/4, 75–85

[6.77] *Wales, J. L. S., and H. Janeschitz-Kriegl:* Apparatus for the Measurement of the Flow Birefringence of Polymer Melts. J. polymer. Sci. A 2 (1967) 5, 781–790

[6.78] *Kuznecov, E. A.:* O dispersii dvuprelomlenija. Moskva: Izv. Vyst. uc. zav. (1959) 1, 60–67

[6.79] *Winchell, A. N., and W. B. Meck:* Birefringence-Dispersion Ratio as a Diagnostic. Amer. Mineralogist 32 (1947) 336

[6.80] *Mozžerin, J.,* et al.: Methodik der Konoskopie feiner Körner bei der mikroskopischen Untersuchung von Schliffen und ihre Anwendung bei der Arbeit nach dem U-Tischverfahren. Zap. vses. mineralog. Obsc. 2, 85 (1965) 434–435

[6.81] *Bonnke, H., and A. Lehr:* Neue Methoden zur Achsenwinkelmessung zweiachsiger Kristalle. Ber. Geol. Ges. DDR 10 (1965) 5, 643

[6.82] *Bergner, J.:* Zur Messung des Achsenwinkels optisch zweiachsiger Kristalle mit Hilfe des Polarisationsmikroskops. Feingerätetechnik 6 (1957) 103–108

[6.83] *Rittmann, A.:* Metodo del quoziente caratteristico del ritardi por la determinazione in diretta di 2 V. Rend. Soc. Mineral. Ital. 3 (1946) 221

[6.84] *Tröger, W. E.:* Ein neues Nomogramm zur Bestimmung des optischen Achsenwinkels. Heidelberger Beitr. Mineralog. Petrogr. 3 (1952) 44

[6.85] *Koritnig, S.:* Ein Nomogramm zur Bestimmung der veränderlichen Lichtbrechungsquotienten in beliebigen Schnitten optisch ein- und zweiachsiger Kristalle sowie zur Bestimmung des Achsenwinkels 2 V. Heidelberger Beitr. Mineralog. Petrogr. 1 (1948) 471

[6.86] *Tobi, A. C.:* A Chart for Measurement of Optic Axial Angles. Amer. Mineralogist 41 (1956) 516–519

[6.87] *Roy, N. N.:* Modifizierter Drehtisch, der die direkte Messung von 2 V erlaubt. Amer. Mineralogist 50 (1965) 1441–1449

[6.88] *Parker, R. L.:* A Stereographic Construction for Determining Optic Axial Angles. Amer. Mineralogist 41 (1956) 935–939

[6.89] *Munro, M.:* The Measurement of Large Optic Axial Angles with the Universal Stage. Mineralog. Mag. 35 (1966) 273, 763–769 and 34 (1965) 272, 656–660

[6.90] *Ehlers, E. G.:* Vereinfachte Methode zur 2V-Bestimmung mit Hilfe 3- und 4achsiger Tische. Mineralog. Mag. 35 (1966) 275, 958–962 and 36 (1967) 278, 299–300

[6.91] *Joel, N.:* Determination of the Optic Axes and 2 V; Electronic Computation from Extinctions Data. Mineralog. Mag. 35 (1965) 270, 412–417
[6.92] *Joel, N.:* Stereographic Construction for Determining the Optical Axes of a Biaxial Crystal Directly from a few Extinction Measurements. Mineralog. Mag. 33 (1964) 769–779
[6.93] *Tocher, F. E.:* A New Universal Stage Technique for the Determination of 2 V when neither Optic Axis is Directly Observable. Mineralog. Mag. 33 (1964) 266, 1038–1054
[6.94] *Tocher, F. E.:* Extinction and 2 V: A Simple Stereographic Solution of General Application. Amer. Mineralogist 49 (1964) 1622–1630
[6.95] *Vincent, H. C. G.:* Projector for Interference Figures and for Direct Determination of 2 V. Mineralog. Mag. 30 (1955) 229, 666
[6.96] *Mosebach, R.:* Über refraktometrische Messungen und Achsenwinkelbestimmungen mit Hilfe des mehrachsigen Drehtisches. Fortschr. Mineralog. Petrog. 27 (1950) 35
[6.97] *Mosebach, R.:* Eine Methode zur Bestimmung des wahren Winkels der optischen Achsen und des mittleren Hauptbrechungsquotienten zweiachsiger Mineralien im Dünnschliff. Heidelberger Beitr. Mineralog. Petrogr. 2 (1951) 432–436
[6.98] *Munro, M.:* Errors in the Measurement of 2 V with the Universal Stage. Amer. Mineralogist 48 (1963) 308
[6.99] *Tobi, A. C.:* Comments on »Errors in the Measurement of 2 V with the Universal Stage« by M. Munro. Amer. Mineralogist 49 (1964) 812
[6.100] *Wright, H.:* Bestimmung der Indikatrixorientierung und von 2 V mit Hilfe des Drehtisches: Eine Warnung und ein Test. Amer. Mineralogist 51 (1966) 19–24
[6.101] *Fairbairn, H. W., and T. Podolsky:* Notes on Precision and Acuracy of Optical Angle Determination with the Universal Stage. Amer. Mineralogist 36 (1951) 823–832
[6.102] *Villarroel, H., and N. Joel:* Extinction Curves and Equivibration Curves Obtained with a Variable-Axis-Spindle-Stage. Mineralog. Mag. 36 (1967) 277, 127–130
[6.103] *Garaycochea, J., and O. Wittke:* Determination of the Optic Angle from the Extinction Curve of a Single Crystal Mounted on a Stage. Acta Crystallogr. 17 (1964) 183–189
[6.104] *Wilcox, R. E., and G. A. Jzett:* Optic Angle Determined Conoscopically on the Spindle Stage. I. Micrometer ocular. Amer. Mineralogist 53 (1968) 1/2, 269–277
[6.105] *Noble, D.:* Optic Angle Determined Conoscopically on the Spindle Stage. II. Selected Rotation Method. Amer. Mineralogist 53 (1968) 1/2, 278–283
[6.106] *Wilcox, R. W.:* Optic Angle Determination on the Spindle Stage. Bull. Geol. Soc. Amer. 71 (1960) 2003
[6.107] *Noble, D. C.:* A Rapid Conoscopic Method for Measurement of 2 V on the Spindle Stage. Amer. Mineralogist 50 (1965) 180–185
[6.108] *Gravenor, C. P.:* A Graphical Simplification of the Relationship Between 2 V and N_Y, and N_Z. Amer. Mineralogist 36 (1951) 162
[6.109] *Wright, F. E.:* Computation of the Optic Axial Angle from the Three Principal Refraction Indices. Amer. Mineralogist 36 (1951) 543–556
[6.110] *Waldmann, H.:* Über eine graphische Auswertung der Achsenwinkelgleichung. Schweiz. mineralog. petrogr. Mitt. 25 (1945) 327
[6.111] *McAndrew, J.:* Relation of Optical Axial Angle with the Three Principal Refractive Indices. Amer. Mineralogist 48 (1963) 1265–1285
[6.112] *Bryant, W. M. D.:* Axial Dispersion with Change in Sign. J. Amer. chem. Soc. 65 (1943) 96
[6.113] *Lane, J. H., and H. T. U. Smith:* Graphic Method of Determining the Optic

Sign and True Angle from Refractive Indices of Biaxial Minerals. Amer Mineralogist 23 (1938) 457
[6.114] *Meixner, H.:* Die Starksche Methode zur Bestimmung des optischen Zonen charakters an keilrandigen Kristallen mittels Berek-Kompensator. Schweiz mineralog. petrogr. Mitt. 32 (1952) 2, 348
[6.115] *Ernst, E.:* Zur Bestimmung des optischen Charakters doppelbrechende Kristalle im polarisierten Licht. Neues Jb. Mineralog. Beil. Bd. 64, A (1931 47–65
[6.116] *Tunell, G.:* Two Definitions of Positive and Negative Extinction Angles in the Plagioclase Feldspars. Amer. Mineralogist 38 (1953) 404–411
[6.117] *Mandarine, J. A.:* Absorption and Pleochroism: Two Much-Neglected Optica Properties of Crystals. Amer. Mineralogist 44 (1959) 65–77
[6.118] *Whetty, E. T.:* The Microspectroscope in Mineralogy. Smiths Misc. Coll. 6 (1915) 5, 16 H
[6.119] *Gabler, F.:* Microphotometry and Its Application. Microscope 15 (1966) 3,85–9
[6.120] *Exner, G., and W. Schreiber:* Mikrophotometrische Untersuchungen an Modell präparaten in: Optik und Spektroskopie aller Wellenlängen. Berlin: Akademie Verlag 1962 (236–245)
[6.121] *Herrmann, R.:* Physikalische Grundlagen der Mikroskop-Photometrie. Z. analyt Chem. 252 (1970) 2/3, 81–82
[6.122] *Hansen, G.:* Energetische Grenzen der Mikrospektralphotometrie. Zeiß-Mitt. (1961) A 117–124
[6.123] *Girin, O. P.,* et al.: Notwendigkeit der Korrektur der beobachteten Absorp tionsspektren fester Körper bei der Untersuchung ihrer Mikrocharakteristiken Optika u Spektroskopija 25 (1968) 3, 438–440 (Russian)
[6.124] Anfärben von Mineralien. Literaturzusammenstellung Nr. 82. Arbeitsgruppe Dokumentation des Forschungsinstitutes für Aufbereitung. Freiberg 1960
[6.125] *Fay, W.:* Zusammenstellung und Durchführung sämtlicher in der Mineralogie üblichen und möglichen Mineralanfärbmethoden. Bergakademie Freiberg Meldearbeit 1962
[6.126] *Parker, C. A.:* Spektrophosphorimeter Microscopy – An Extension of Fluorescence Microscopy. Analyst 94 (1969) 1116, 161–177
[6.127] *Haitinger, M.:* Fluoreszenzmikroskopie. Berlin: Akademie-Verlag 1959
[6.128] *Haitinger, M.:* Die Fluoreszenzanalyse in der Mikrochemie. Wien: Verlag Haim 1937
[6.129] *Winkelman, J., and J. Grossmann:* Quantitative Fluoreszenzanalyse in undurchsichtigen Suspensionen durch Auflichtoptik. Anal. Chem. 93 (1967) 8, 1007–1009
[6.130] *Long, J. V. P., and S. O. Agrell:* Die Kathoden-Lumineszenz von Mineralien im Dünnschliff. Mineral. Mag. 34 (1965) 318–326
[6.131] *Ehrenberg. H.:* Reflexionsmessungen in der Erzmikroskopie. Eine Übersicht über die Methoden und neueren Instrumente. Z. wiss. Mikroskopie 66 (1964) 32–44
[6.132] *Trojer, F.:* Reflexionsmessungen in der Mikroskopie hüttenmännischer Produkte. Berg- u. Hüttenmänn. Mh. 107 (1962) 33–39
[6.133] *Kötter, K.:* Die mikroskopische Reflexionsmessung mit dem Fotomultiplier und ihre Anwendung auf Kohleuntersuchungen. Brennstoff-Chem. 41 (1960) 263–272
[6.134] *Mitsche, R., and H. Scheidl:* Die Anwendung der Mikroreflexionsmessung in der Metallographie. Berg- u. Hüttenmänn. Mh. 109 (1964) 82–85
[6.135] *Knosp, H.:* Messung der Mikroreflexion und der optischen Konstanten von Metallen und Hartstoffen. Z. Metallkunde 60 (1969) 6, 526–531
[6.136] *Feinleib, J., and B. Feldmann:* Absolute Reflexionsmessungen an kleinen Proben

bei tiefen Temperaturen. Rev. sci. Instruments 38 (1967) 1, 32–33

.137] *Hennig, A.:* Bestimmung der Oberflächen beliebig geformter Körper. Mikroskopie 11 (1965) 1–20

.138] *Watznauer, A., Bünger, K., and B. Knebel:* Beitrag zur räumlichen Darstellung des Porenraumes von Sedimentgesteinen. Z. angew. Geologie 11 (1965) 2, 86

.139] *Friehmelt, E.:* Beitrag zur mikroskopischen Untersuchung der räumlichen Struktur von Koks, Brikett und ähnlichem Material. Erdöl u. Kohle 20 (1967) 4, 289–292

.140] *Perez-Rosales, C.:* Eine vereinfachte Methode zur Bestimmung der spezifischen Operfläche. J. Petrol. Technol. 19 (1967) 8, 1081–1084

.141] *Graber, D. G.:* Interferenzmikroskopische Untersuchung poröser Materialien. Microscope 15 (1967) 43–50

.142] *Winkler, K.:* Zur Methodik der mikroskopischen Korngrößenanalyse. Diplomarbeit Universität Jena, Mineralogisches Institut, 1968

.143] *Šimecek, J.:* Mikroskopische Bestimmung der Korngrößenverteilung. Staub-Reinhalt. Luft 26 (1966) 162–167

.144] *Tuma, J.:* Zusammenhang zwischen Korngrößenbestimmung nach der Sedimentationsmethode und nach der mikroskopischen Methode. Staub-Reinhalt. Luft 26 (1966) 5, 211–241

.145] *Exner, H. E.:* Mikroskopische Bestimmung von räumlichen Korngrößenverteilungen in undurchsichtigen Stoffen. Prakt. Metallogr. 3 (1966) 334–341

.146] *Ludwig, G.:* Fehler bei der Korngrößenbestimmung in Anschliffen von Lockerprodukten. Jenaer Rdsch. 6 (1961) 187–190

.147] *Goldsmith, P. L.:* Berechnung der wahren Teilchengrößenverteilung durch die in einem Querschnitt beobachteten Größen. Brit. J. appl. Physics 8 (1967) 6, 813–830

.148] *Guljaev, A. M.:* O točnosti aprodelenija dliny volokna po metodu Glagoleva. Alma-Ata: Trudy kasach. Nauk. Yssl. Inst. Miner. Syroja, vyp. 3 (1960) 137–138

.149] *Smith, R. E.:* Grain Size Measurement in Thin Section. J. Sediment Petrology 36 (1966) 3, 841–843

.150] *Jeglitsch, F., and A. Kulmburg:* Über die direkte Bestimmung der Korngröße mit dem Hochtemperatur-Mikroskop. Radex-Rdsch. (1967) 3/4, 679–694

.151] *Goldstein, D. J.:* Schichtdickenmessung mit dem Interferenzmikroskop. Nature 213 (1967) 5074, 386–387

.152] *Palatnik, L. S.,* et al.: Methoden zur Stärkemessung dünner Filme. Zav. Labor 34 (1968) 3, 288–297 and 2, 180–190 (Russian)

.153] *Ligten van, R. F.:* Holographic Microscopy and Small Particles. Microscope 16 (1968) 4, 349

.154] *Fourney, M. E.,* et al.: Bestimmung der Teilchengröße und -geschwindigkeit von Aerosolen mit Hilfe der Holographie. Rev. scient. Instruments (1969) 2, 205–213

.155] *Beckmann, K. H.:* Methoden und Ergebnisse optischer Untersuchungen an Halbleitergrenzflächen. Angew. Chemie 80 (1968) 6, 213–224

.156] *Dennison, J. M.:* Zuverlässigkeit von visuellen Schätzungen von Kornmengen. J. Sediment. Petrology 36 (1966) 1, 81–89

.157] *Schucharot, E.:* Das Integrationsverfahren in der mikroskopischen Technik. Handbuch der Mikroskopie in der Technik, Bd. 1, Teil 1. Frankfurt/Main: Umschau-Verlag 1957, 565–588

.158] *Mrazek, J., and K. H. Richter:* Morphometrische Kornanalyse unter Anwendung elektromechanischer elektronischer Gerätetechnik. Geologie 17 (1968) 6/7, 835–846

.159] *Chayes, F.:* Petrographic Modal Analysis. New York: John Wiley and Sons 1956

.160] *Rath, R., and D. Spoerel:* Eignung der Zählverfahren zur Bestimmung des

Oberflächenanteils im Falle nicht normal verteilter Komponenten. Mikrosko- pie 22 (1968) 9/10, 252–265

[6.161] *Krause, W., and W. Lanitz:* Beitrag zur quantitativen Mineralanalyse von Pul verpräparaten unter dem Mikroskop. Silikattechnik 17 (1966) 2, 50–52

[6.162] *Greeman, N. N.:* The Mechanical Analysis of Sediments from Thin Sectior Data. J. Geology 59 (1951) 447–462

[6.163] *Fricke, P.:* Anwendung des Punktzählgerätes Eltinor zur Teilchendichtebestim mung. Kristall u. Technik 3 (1968) 1, K1–K4

[6.164] *Greif, H.:* Volumenanteilbestimmung an Planschliffen nach dem Punktzählver fahren. Silikattechnik 16 (1965) 4, 121–124 and 15 (1964) 5, 166–167

[6.165] *Behringer, C. A.:* Untersuchungen zur quantitativen Mikroskopie von Drogen pulvern. Dissertation TH Zürich 1963, Nr. 3343

[6.166] *Bayly, M. B.:* Errors in Point Counter Analysis. Am. Mineralog. 45 (1960) 447

[6.167] *Bradshaw, P. M. D.:* Measurement of the Modal Composition of a Graniti Rock Powder by Point-Counting, Infrared Spectroscopy and X-Ray Diffraction Mineralog. Mag. 36 (1967) 277, 94–101

[6.168] *Bandemer, H.:* Die für eine vorgegebene Genauigkeit notwendige Punktzah beim Point-Counter-Verfahren. Z. angew. Geol. 11 (1965) 3, 147–148

[6.169] *Plas van der, L., and A. C. Tobi:* Eine Tafel zur Abschätzung der Genauigkei von Point-Counter-Ergebnissen. Amer. J. Sci. 263 (1965) 1, 87–90

[6.170] *Mehnert, K. R., Willgallis, A., Kallies, B., Meier, W., and B. Schüler:* Die Ge nauigkeit der Kornzählanalyse zur Bestimmung des quantitativen Mineral bestandes von kompakten Gesteinen granitähnlicher Zusammensetzung. Beitr Mineralog. Petrogr. 6 (1959) 203–218

[6.171] *Böhm, G., and E. Bader:* Zur Genauigkeit der kristalloptischen Analyse vor Kalirohsalzen. Mitt. Kaliind. Sondershausen (1966) 1, 15–18

[6.172] *Chayes, F., and H. W. Fairbairn:* A Test of the Precision of Thin-Sectior Analysis by Point-Counter. Amer. Mineralogist 36 (1951) 704–712

[6.173] *Kalsbeek, F.:* Bemerkung zur Zuverlässigkeit von Punktzählanalysen. Neues Jb Mineralog. Mh. (1969) 1, 1–6

[6.174] *Solomon, M.:* Counting and Sampling Errors in Modal Analysis by Poin Counter. J. Petrol. 4 (1963) 367

[6.175] *Bernhardt, C., and K. Hennig:* Methode zur Bestimmung der Gewichtsprozent von Glimmern im Granit für die Baustoffpraxis. Z. angew. Geol. 6 (1960) 9 459–460

[6.176] *Kellagher, R. C., and F. J. Flanagan:* A Comparison of Two Methods for Con verting Grain Counts to Weight Percent Composition. J. Sediment Petro logy 26 (1956) 222–227

[6.177] *Chayes, F.:* The Relation Between Area and Volume in Micrometric Analysis Mineralog. Mag. 30 (1953) 221, 147

[6.178] *Chayes, F.:* A Correction Factor for Specific Gravity and Volume Difference in Fragment Analysis. Econ. Geol. 41 (1946) 749–760

[6.179] *Elliott, R. B.:* An Effect of Depth of Focus on Micrometric Analysis. Mineralog Mag. 31 (1965) 232, 272–275

[6.180] *Chyes, P.:* The Holmes Effect and the Lower Limit of Modal Analysis Mineralog. Mag. 31 (1956) 232, 272–275

[6.181] *Iwasaki Hiroshi* et al.: Refractive Indices of $LiTaO_3$ and High Temperatures Jap. J. appl. Physics 7 (1968) 9, 185–186

[6.182] *Yamago Taketoshi:* Temperature Dependence of Principal Refractive Indices o Orthorhombic KNO_3. Jap. J. appl. Physics 7 (1968) 9, 1056–1058

[6.183] *Heide, K.:* Dynamische thermische Analysenmethoden. Leipzig: VEB Deut scher Verlag für Grundstoffindustrie 1979

[6.184] *McCrone, W. C., and M. Bayard:* Mikroschmelzpunkte. Microchem. J. 1

(1966) 97-104
[6.185] *Jucker, H., and H. Suter:* Neue Methoden und Geräte der Schmelzpunktbestimmung. Fortschr. chem. Forsch. 11 (1969) 3, 430-472
[6.186] *Vaughan, H. P.:* Photometric Recording of Single Crystal-Melting Values and Other Fusion Phenomena. Microscope 17 (1969) 71-76
[6.187] *Reese, D. R., Nordberg, P. N., Eriksen, S. P., and J. V. Swintosky:* Technique for Studying Thermal Induced Phase Transitions. J. Pharm. Sci. 50 (1961) 177-178
[6.188] *Martinek, A.:* Die Kontaktmethode zur Mikrothermoanalyse nach A. Kofler. GIT-Fachz. Lab. 11 (1967) 4, 229-236
[6.189] Kontaktlose T-Messung an integrierten Schaltungen mit dem Mikrothermoskop. Feinwerktechnik 75 (1971) 5, 228-229
[6.190] *Alimarin, J. P., and M. N. Petrikova:* Anorganische Ultramikroanalyse. Berlin: VEB Deutscher Verlag der Wissenschaften 1962
[6.191] *Helbig, W.:* Untersuchungen über die Verwendbarkeit und das Leistungsvermögen elektroanalytischer Verfahren in der quantitativen Ultramikroanalyse. Habilitationsarbeit, TU Dresden 1968
[6.192] *Scullion, H. J.:* Reversible Complex Formation as an Identification Aid in Chemical Microscopy. Microscope 16 (1968) 1, 12
[6.193] *Sansoni, G.:* Zum Festigkeitsverhalten von PZ-Klinkern in Abhängigkeit von der Porosität und Struktur der Klinker. Freib. Forsch.-H. (1975) A 531, 209-220
[6.194] *Beckert, M., and H. Klemm:* Handbuch der metallographischen Ätzverfahren. Leipzig: VEB Deutscher Verlag für Grundstoffindustrie 1962
[6.195] *Lebedeva, S. I.:* Opredelenie mikrotverdosti mineralov. Moskva: Izd. AN SSSR 1963
[6.196] *Petzold, A.:* Untersuchungen zur Prüfung der Mikroeindruckhärte von Glas. Wiss. Ztschr. Hochsch. Architektur Bauwes. Weimar 13 (1966) 6, 635
[6.197] *Müller, K.:* Anwendung einer neuentwickelten Kleinlast-Härtemeßmethode auf der Basis des Vickers-Verfahrens. Kunststoffe 60 (1970) 4, 265-273
[6.198] *Shaub, B. M.:* Gebrauch des Mikroskopes zur Bestimmung des spezifischen Gewichts an kleinen Mineralkörnern. Amer. Mineralogist 44 (1959) 890-891
[6.199] *Franklin, F. A.,* et al.: Bestimmung der Dichten von verschiedenen meteorischen, vulkanischen und vom Eis stammenden Kügelchen. J. geophys. Res. 72 (1967) 10, 2543-2546
[6.200] *Helbig, W.:* Die Ultramikroanalytik - ein hocheffektives Verfahren der forensischen Chemie. Kriminalistik u. forens. Wissenschaften (1971) 5, 7-20
[6.201] *Helbig. W.:* Versuche zur Spektralphotometrie von Mikro- und Submikroliterproben. Kriminalistik u. forens. Wissenschaften (1972) 7, 59-66
[6.202] *Helbig, W.:* Ultramikroanalyse. Moderne chemische Methoden in der Klinik. 2nd ed. Suppl. Leipzig: VEB Georg Thieme Verlag 1968
[6.203] *Helbig, W.:* Beitrag zur orientierenden Dichtebestimmung an kleinen Partikeln. Kriminalistik u. forens. Wissenschaften (1976) 25, 31-38
[6.204] *Kirk, P. L.:* Quantitative Ultramicroanalysis. New York: McGraw Hill 1950
[6.205] *Schramm,.H. P.:* Zum Einsatz ultramikroanalytischer Analysenmethoden bei der Qualitätskontrolle von Halbleiterwerkstoffen. Diss. TU Dresden 1971
[6.206] *Korenman, I. M.:* Količestvennyj mikrochemičeskij analiz. Moskva: Goschimizdat 1949
[6.207] *Benedetti-Pichler, A. A.:* Qualitative Mikroanalyse. Monographien aus dem Gebiet der qualitativen Mikroanalyse, Bd. 1 und 2. Wien: Springer-Verlag 1958 and 1964
[6.208] *Keune, H., Seyfarth, H.-H., and D. Mielsch:* Zur polarisationsmikroskopischen Diagnostizierung der Fällungsprodukte mikrochemisch-analytischer Nachweisreaktionen. Wiss. Zeitschr. PH Erfurt-Mühlh. math.-nat. R., 13 (1977) 1, 7-12

7 *Application examples*

Applications of microscopical methods in chemistry are numerous and highly diversified [7.1 to 7.68]. They may be grouped into five types (Table 7.1). Microscopically identifiable properties and selected applications for each type are shown in Tables 6.4 and 7.2.

Grain size measurements and grain shape determination provide direct information on the problem examined, whereas other measurements have to be analyzed to extract the full information content. For example, the measured refractive index of a grain may be analyzed for different purposes, viz.

- identification and distinction from other grains,
- ascertaining a change in substance or structure by comparison with other refractive indices measured in the sample,
- information on the packing density of the structure as a whole or in the different spatial directions,
- visualization of structural differences within a crystal, e.g. zoning, or
- contrasting the grain in mixtures and aggregates.

Besides static microscopical methods, techniques of kinematic microscopy are gaining importance in chemical investigations [7.64].

Table 7.1. Types of analysis by microscopical methods in chemistry

Analysis type	Static microscopy	Kinematic microscopy
Phase analysis	Phase identification, phase composition of solid mixtures	Change of phase composition in physico-chemical processes
Microstructure analysis	Information on microstructure	Change of microstructure in physico-chemical processes
Fabric analysis	Measurement of texture orientation, intergrowth, porosity and distribution	Change of texture caused by physico-chemical processes
Grain analysis	Measurement of grain size and grain shape	Change of grain size and grain shape caused by physico-chemical processes
Surface analysis	Quantitative characterization of surfaces	Change of surface condition caused by physico-chemical processes

Table 7.2. Application of microscopical methods in various branches of chemical industry and research

Field of application	Phase analysis	Microstructure analysis	Grain analysis	Fabric analysis	Surface analysis
Basic and intermediate inorganic products	Production control analysis of solid raw materials, intermediate and end products Identification of side products and contaminants Constituents of solid wastes, industrial dusts, corrosion products Solid fraction in sewage Formation of new phases in wear reactions on reactor walls Compounds in system studies	Isomorphism and polymorphism of compounds Mechanism and kinetics of decomposition reactions Crystallization and dissolution processes Flocculation in sewage Relative orientation of electrodeposits and electrodes Structural disintegration of diaphragm material in electro-chemistry	Grain size changes in sintering processes Grain size and grain shape in relation to filtration behaviour, agglomeration of powders, bulk density, free flowing Bubble growth in electrolytic processes	Layer growth on electrodes in electrochemical processes Texture of deposits and corrosion on battery electrodes Relation between strength and grain texture in electrodes and compressed granulates Phase distribution in multiphase granulates	Corrosion processes on surfaces Thickness of thin films Quantitation of etching pit density Roughness of surfaces Grain orientation on polycrystalline surfaces Surface reactions with gaseous phases Surface of electrodes in electrodeposition
Agrochemistry	Production control of solid raw materials, intermediate and end products	Mechanism and kinetics of decomposition in agrochemicals	Influence of process parameters on grain size and grain shape, and their influence on free flowing,	Relation between texture and strength of compressed granulates	

Table 7.2, continued

Field of application	Phase analysis	Microstructure analysis	Grain analysis	Fabric analysis	Surface analysis
	Identification of side products and contaminations	Behaviour of technical dissolution and crystallization processes.	aggregation behaviour, and bulk density	Phase distribution in multiphase granulates	
	Determination of effective isomers or polymorphous modifications in insecticides and herbicides; e.g. isomerism of benzene hexachloride *(Lindan)*			Degree of intergrowth in crude salts	
Silicate chemistry (building materials, ceramics, glass)	Production control of solid raw materials, intermediate and end products	Reaction mechanism and kinetics of the hydration of silicates, aluminates and ferrites	Growth rate in the crystallization of glasses	Relation between texture and strenght in building materials and ceramics	Surface configuration of glasses, ceramics and building materials
	New formed phases in cement solidification and in technical calcination and baking processes	Behaviour of crystallization process in glasses	Change of grain size in sintering processes		
			Relation between strength and optimum grain size range in building materials		
	Newly formed crystalline phases in glasses	Mixed crystal formation and polymorphism in silicates, aluminates and ferrites			

Table 7.2, continued

Field of application	Phase analysis	Microstructure analysis	Grain analysis	Fabric analysis	Surface analysis
Catalysis	Products newly formed from consumed catalysts	Structural characterization of catalytically active substances	Monitoring and optimizing the process of catalyst production (precipitation, granulation)	Change of catalyst texture in its consumption	Study of catalyst surfaces
		Structural features of the carrier material		Relation between strength and texture in compressed granulates	
		Structure of molecular sieves and hydroxides, and dehydration studies		Volume of macropores, permeability	
		Structural changes of Al_2O_3 and alumino-silicate carriers when other components enter the lattice		Phase distribution in multiphase granulates	
Basic and intermediate organic products	Detection of cellulose in coal	Emulgation process in oils as a function of emulgator type and concentration	Optimum droplet size for liquid sprayers with different nozzle geometries		
	Carbon fraction in dust				
	Identification of compounds in residues of carbolic oil distillation	System studies on crystals, liquids and crystalline compounds	Influence of setting point depressants on grain size and grain shape of paraffins		
	Identification of effective modifications in organic explosives	Carbon structure under different production conditions			

Table 7.2, continued

Field of application	Phase analysis	Microstructure analysis	Grain analysis	Fabric analysis	Surface analysis
		Flocculation processes in industrial sewage Observation of nitration and acetylation of cellulose through change of birefringence			
Pharmaceutical chemistry	Identification of steroid hormones, barbiturates and sulphonamides Detection of trace impurities	Mixed crystals formation and polymorphic modification in drugs Decomposition of drugs – determination of shelf life	Grain size of powders, particle size in ointments and pastes	Homogeneity of grain and phase distribution in tablets and ointments Relation between the texture and strength of tablets	
Organic high polymers	Identification of textile fibres, pigment particles, impurities, inclusions, additives, decomposition products	Dissolution of PVC grains in gelatinizing Decomposition process of stabilizers Effects of various plasticizer contents in PVC Decomposition and melting behaviour during temperature increase	Grain studies in polymerization processes Relation between size and distribution of latex particles, and the permeability and protective properties of coats of paint Morphology of PVC powder grains and the gelatinization properties of PVC pastes	Additives distribution analysis to optimize dosage and mix Relation between texture and cooling schedule Texture distribution across the material and its relation to strength	Surface structure of PVC grains Surface contamination of foils Relation between surface structure and gloss effects Surface changes by ageing and corrosion

Table 7.2, continued

Field of application	Phase analysis	Microstructure analysis	Grain analysis	Fabric analysis	Surface analysis
		Structural changes during injection moulding, extrusion, spinning, stretching	Particle morphology and cohesion, and their influence on homogeneity and strength of compressed powders	Formation conditions of spherulitic textures, textures of copolymers and welded seams	Fibre and cloth surfaces and their finishing
					Coated surfaces
		Relation between chain branching and birefringence in high- and low-pressure polyethylene		Influence of distribution and orientation of fibres on strength of fibre-reinforced materials	
		Relation between structure and cooling schedule		Influence of texture on dissolution	
		Photoelasticity of isotropic high polymers (epoxy resin casting)			
		Process of polymerization of solid phases			
		Structure and flocculation of sols, gels and suspensions in polymerization			

Tables 7.3, 7.4 and 7.5 give a survey of the diversified applicabilities and the measuring quantities and methods employed.

Sections 7.1 to 7.4 deal with particular cases, including generalizations of the methods employed. The examples demonstrate that the success of present-day microscopical methods is still greatly depending on the expert selection of suitable methods of preparation, measurement and data analysis.

From the many procedures described, the user will have to select for himself the technique that suits his problem best. Chart 7.1 provides a guideline for the general approach, referring to various sections of this book.

7.1 Phase analysis

Analyzing the phases of solid mixtures is one of the main applications of chemical microscopy.

The result of a chemical analysis does not always indicate the phase composition of a sample. Where phases differ little in chemical composition, analytical techniques have to be based on specific, readily measurable physical properties. Preferred applications of various physical methods to phase analysis and the position held by microscopical techniques are found on chart 7.2 [7.69 to 7.114].

Chart 7.5 describes the general procedure of microscopic phase analysis; some

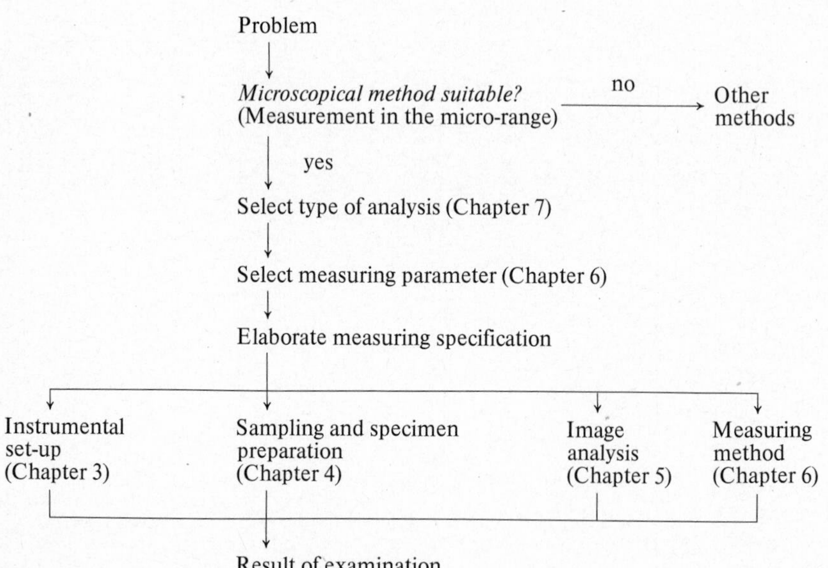

Chart 7.1. Elaborating a method sheet for microscopical analysis

Table 7.3. Fields of application of kinematic chemical microscopy

Phase analysis	Microstructure analysis	Grain analysis	Fabric analysis	Surface analysis
Mechanism of physico-chemical processes				
Kinds and volume fractions of solid reaction products of individual phases of a crystallization sequence or individual stages of dissolution, at t_n	*Solid/solid changes:* polymorphous transformations, mixed crystals formation, unmixing, structural changes of organic high polymers and in solid-state reactions, diffusion, swelling and addition processes	*Processes:* recrystallization, crystallization, dissolution, etching, melting, sublimation, grain size change in chemical reactions or by deformation (working, machining, granulation)	*Processes:* deformation by working, machining or granulation, bonding, solidification, transformation by chemical reactions, recrystallization, crystallization, selective dissolution	*Processes:* etching, dissolution, epitaxial growth, vacuum deposition, mechanical working, corrosion, decomposition, ageing, reduction and oxidation
Detection of unstable, ephemeral solid intermediate stages of processes				
Partial processes and side reactions	*Solid/liquid and solid/gaseous changes:* structural changes in dissolution, etching, crystallization, sublimation, and in reactions with liquid or gaseous phases			
Relation between process sequence and the variation of influencing parameters	*Liquid/liquid changes:* structural changes in liquid crystals and in colloid systems			
Determination of single- and multi-component systems				
Kinetics of physico-chemical processes				
Study of the relation between degree of conversion and yield, study of reaction rates, determination of reaction order and of activation energy				

Table 7.4. Possible micro-kinematic measurements

Number of grains involved	Measuring principle	Parameters measured
Several grains in visual field involved	Direct measurement of quantity of initial or new product vs. time in a two-dimensional image	Point integration, photographic area analysis, measurement of integral light transmittance
Single grain involved	Direct measurement of quantity change vs. time in a two-dimensional image	Photographic area analysis, measurement of integral light transmission
	Indirect measurement of quantity change vs. time by measuring a corresponding parameter in a two-dimensional image	Variation of birefringence, refractive index, axial angle and light transmittance

Table 7.5. Methods of kinematic microscopy

Method	Characteristic	Example
Discontinuous	Process reconstructed from observations made step-by-step	
With several batches	Examination of final stage	Analysis of precipitates after evaporation experiments differing in time
With one batch	Sampling during process	Evaporation with sampling at intervals
Continuous	Direct observation of process, continuous direct registration of data *Advantages:* One batch is sufficient. Substance tested remains in the reaction chamber. Connection to computer possible.	Measurement of retardation changes during heating on hot stage
In microreaction cells		Crystallization in stirred cells
By incorporation of respective apparatus into microscope ray path		Measurement of spun thread running past

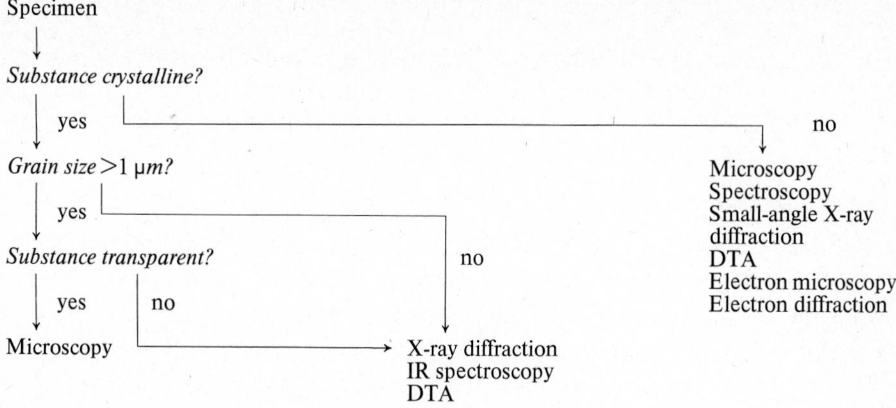

Chart 7.2. Physical methods applied to phase analysis of solid mixtures

of the steps may be simplified depending on the particular measuring conditions involved. By its inherent qualities (cf. chapter 1), microscopy has a number of preferred applications in phase analysis.

(1) Phase analysis of complex solid mixtures and assessment of secondary and trace constituents

Example (1): Analysis of precipitates in potash salt processing

Short dwell times in the processing plant cause wide differences in the paragenesis of precipitates due to failure to reach equilibrium.

Transmitted-light microscopy

Orthoscopy: Loose grain specimen, phase contrast outfit, immersion, refractive index measurement by Becke line method, or staining contrast in the phase microscope (cf. 6.1.1 and 6.1.2).
Conoscopy: Determination of optic sign with compensator, semi-quantitative determination of axial angle: in difficult cases measurement on universal stage (cf. 6.1.3.3).

The parameters measured permit one to positively distinguish all phases present. Quantity analysis is carried out on the phase microscope with semi-automatic integrating stage (Fig. 7.1, see Annex, cf. 6.3).

Example (2): Analysis of fuel ashes and roasting residues

Component phases in these substances outnumber 10 in most specimens, since the materials employed are of heterogeneous composition and are subjected to varied thermal influences in the respective plant. Specimens may differ widely as to the number of phases involved, depending also on grain size.

Transmitted-light microscopy

Loose grain specimen, micromanipulator selection of phases of interest (cf. 4.2.1); determination of refractive index, birefringence, optic sign and axial angle by phase or interference microscopy, universal stage, spindle stage, compensator, integrating stage.

Reflected-light microscopy

Polished section, measurement of absorption, reflection and interior reflexes through reflected-light microscope (bright- and dark-field paths).

Example (3): Analysis of matter newly formed in used-up catalysts.
Polished section, measurement of absorption and reflection by bright-field reflected-light microscopy; combination with microprobe or laser microspectral analysis.

(2) Analyzing phases of close chemical similarity

The various *reaction products formed during cement consolidation* are a typical example. They can be clearly distinguished in the mixture by their different optical constants, a possibility that chemical analysis fails to provide.

$3CaO \cdot Al_2O_3 \cdot nH_2O$	$3CaO \cdot Al_2O_3 \cdot 12H_2O$	$2CaO \cdot Al_2O_3 \cdot 7H_2O$
$3CaO \cdot Al_2O_3 \cdot 34H_2O$	$2CaO \cdot Al_2O_3 \cdot 9H_2O$	$3CaO \cdot Al_2O_3 \cdot 6H_2O$
$3CaO \cdot Al_2O_3 \cdot 18H_2O$	$4CaO \cdot Al_2O_3 \cdot 8.5H_2O$	$2CaO \cdot Al_2O_3 \cdot 5H_2O$
$4CaO \cdot Al_2O_3 \cdot 14H_2O$	$3CaO \cdot Al_2O_3 \cdot 8H_2O$	$CaO \cdot Al_2O_3 \cdot H_2O$
$4CaO \cdot Al_2O_3 \cdot 12H_2O$	$3CaO \cdot Al_2O_3 \cdot 7H_2O$	$3CaO \cdot Al_2O_3 \cdot 0.5H_2O$

Other examples are the *discrimination between paraffins of different numbers of carbon atoms* (Table 7.6) *or different polymorphic modification,* and the *identification of mixed phases in mixed paraffin-olefine crystals* (Fig. 7.2).

The pharmaceutically important *barbiturates* are hardly distinguishable by IR spectroscopy. Fast and positive identification is afforded by microthermal analysis, a microscopical phase-analytical method (Table 7.7).

Similar cases are provided by the analysis of various inorganic hydrates and hydroxides. Here, too, the microscopical measurement of optical constants permits a clear distinction between different hydration stages in a mixture (Tables 7.8 and 7.9).

(3) Analysis of easily decomposing or hygroscopic substances (cf. 7.1.4)

Same as in all physical methods of phase analysis, the unknown substance is identified in two steps:

I. Measurement of substance-specific microscopic properties (Table 6.1)
II. Comparison of measured data with reference values of all phases chemically possible in the system

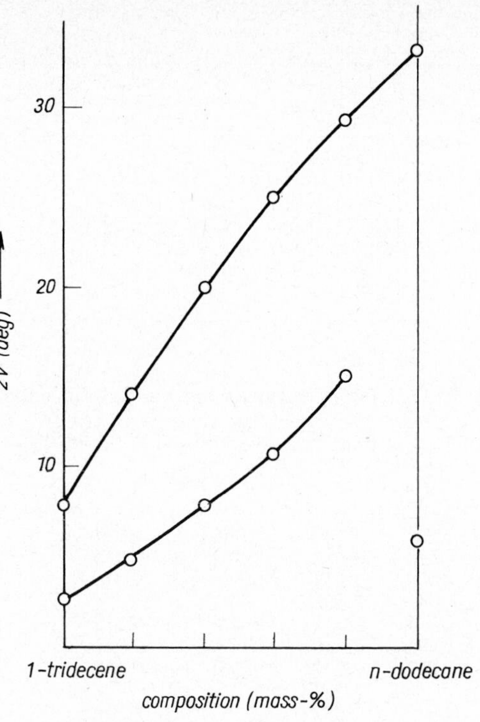

Figure 7.2. Distinction of modifications of mixed paraffin-olefin crystals by axial angle measurement

Table 7.6. Optical constants of paraffins depending on number of carbon atoms, acc. to [7.103]

Composition	Refractive index n_o	n_e	Birefringence
$C_{24}H_{50}$	1.4970	1.5420	0.0450
$C_{25}H_{52}$	1.4985	1.5465	0.0480
$C_{26}H_{54}$	1.5022	1.5487	0.0465
$C_{27}H_{56}$	1.5042	1.5508	0.0466

Table 7.7. Melting points of barbiturates differing under microscopical investigation though having identical absorption spectra, acc. to [7.2]

Substance	Melting point (°C)
1. Butethal	127
2. Aprobarbital	141
3. Ethyl-(3-methylbutyl-) barbituric acid	157
4. Butabarbital	165
5. Cyclobarbital	173
6. Barbital	190

Table 7.8. Optical constants of magnesium sulphate hydrates, acc. to the literature and authors' measurements

Hydrate	Optic sign	Axial angle (deg)	n_α	n_β	n_γ	Birefringence
$MgSO_4 \cdot H_2O$	+	57	1.523	1.525	1.596	0.063
$MgSO_4 \cdot {}^5/_4H_2O$		–	1.512	1.530	1.538	0.026
$MgSO_4 \cdot 3H_2O$	–	65	1.495	1.497	1.498	0.003
$MgSO_4 \cdot 4H_2O$	+	50	1.490	1.491	1.497	0.007
$MgSO_4 \cdot 5H_2O$	–	45	1.482	1.492	1.493	0.011
$MgSO_4 \cdot 6H_2O$	–	38	1.426	1.453	1.456	0.030
$MgSO_4 \cdot 7H_2O$	–	51	1.432	1.455	1.461	0.029

Table 7.9. Optical constants of aluminium hydroxides differing in composition, acc. to the literature

Composition		Optic sign	Axial angle (deg)	n_α	n_β	n_γ	Birefringence
AlO(OH)	I	+	84	1.702	1.722	1.750	0.048
	II	+	80	1.646	1.652	1.661	0.015
	III	–	3	1.649	1.649	1.665	0.016
	IV	amorphous	–	–	1.565	–	–
	V	?	?	1.645	?	1.655	0.010
$Al(OH)_3$	I	+	0	1.566	1.566	1.587	0.021
	II	+	0	1.577	1.577	1.595	0.018
	III	amorphous?	–	–	1.583	–	–
	IV	amorphous?	–	–	1.550	–	–

Depending on how far the microscopic properties of all possible phases are known, there are two ways of procedure.

Reference values known

In this case, reference data for the substances concerned are available in tabulated form in reference books [7.102 to 7.109] or specialized reports, for direct comparison with the data measured.

Reference values partly or completely unknown

The required reference data have to be established either before or during specimen analysis.

Reference data establishment before analysis

All phases likely to occur in the system have to be synthesized separately and measured under the microscope. These studies will also reveal which of the microscopical parameters are useful for identifying the unknown phase.

Reference data establishment during analysis

Where the phases of interest cannot be synthesized, reference values have to be derived from analysing the specimen itself. In general, there are two ways of correlating measured microscopical data of an unknown phase to its chemical composition:

(1) Isolation of individual grains of the phases A, B, C etc. under the microscope, using the micromanipulator, followed by quantitative chemical ultramicro-analysis (cf. 6.5).

(2) Deriving the correlation from the course of phase change (A, B, C etc.) in series of specimens, allowing for the respective microscopically determined volume fractions due to the chemical conditions in the system. One should mind the possible interference of impurities or secondary formations.

To determine the number of phases occurring in the specimen substance, the specimen is subjected, grain by grain, to various measurements using instruments selected according to Table 6.3. Errors may result where two phases happen to have closely similar constants (e.g., optic axial angles). To eliminate such errors, several optical constants should be measured on each grain.

A tentative work sheet should tabulate the measured data and give approximate volume estimates for the phases termed A, B, C etc. Table 7.10 is an example, in which a comparison of axial angles reveals a distinct course between specimen 1 and specimen 10.

Specimens 1, 3 and 4 in Table 7.10 consisted of a single phase; 2, 5 and 8 were almost homogeneous, whereas the rest contained several phases. After these relative data were found, complete evaluation was provided by full chemical analyses of the specimens.

Table 7.10. Work sheet for statistical evaluation of axial angle ($2V$) measurements in solid mixtures of unknown phase composition

Specimen No.	Phase A $+2V = 80°$	Phase B $+2V = 40°$	Phase C $-2V = 40°$	Phase D $-2V = 10°$
1	×	–	–	–
2	×	–	–	× ?
3	×	–	–	–
4	×	–	–	–
5	×	–	× ?	–
6	× (?)	–	×	× (traces)
7	–	×	×	–
8	–	×	×	–
9	–	×	×	× (traces)
10	–	×	×	–

7.1.1 Phase analysis by refractive index measurements

The principal refractive indices readily lend themselves to the phase analysis of solid mixtures, as they can be measured conveniently and accurately. Phases

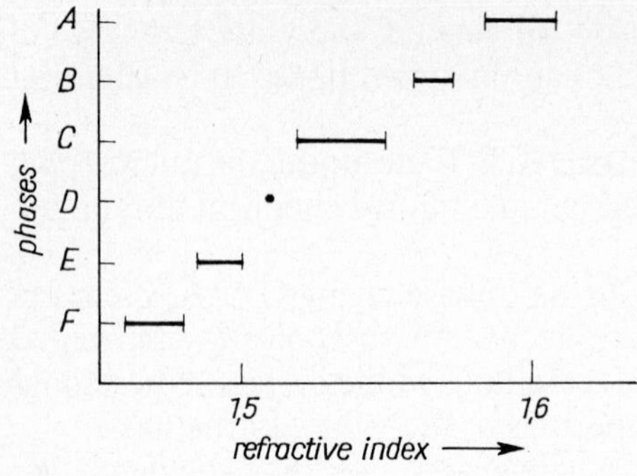

Figure 7.3. Phase analysis of solid mixtures by refractometry

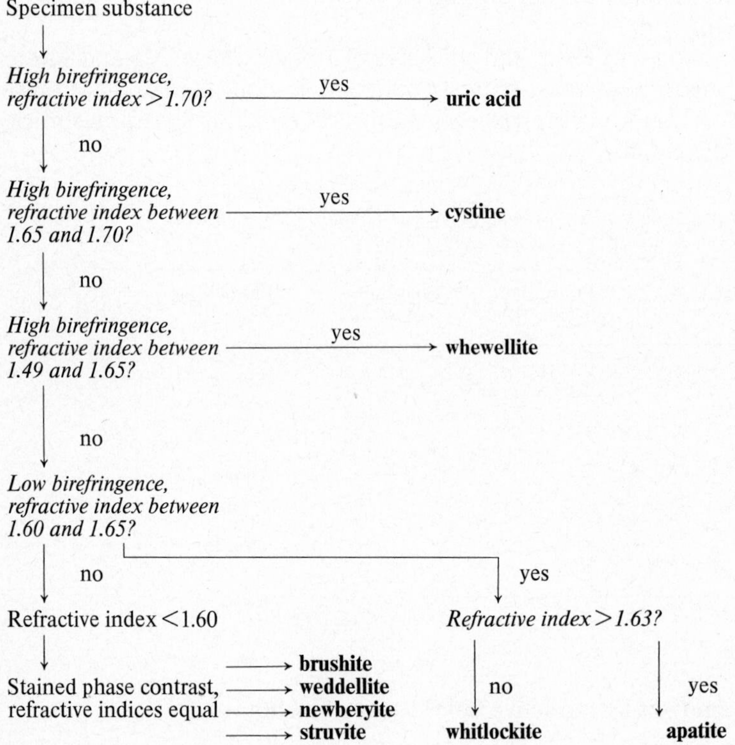

Chart 7.3. Procedure in microscopical routine determination of main constituents of kidney stones

showing measurable differences of refractive index can easily be detected in succession, if a suitable immersion medium is selected. The refractive index of the medium should either lie between the refractive index ranges of the individual phases or be equal to the $n_\beta(n_o$-)value of the respective phase (Fig. 7.3, chart 7.3, [7.110]).

Conditions are more complex in substances whose refractive index ranges overlap or coincide. Various overlaps are then possible, permitting different analytical results [7.112, 7.114].

Depending on the kind of overlap and on the immersion liquid used, overlap between phases A and B may establish one of three cases:

Phases A and B are quantitatively distinguishable without statistical methods (Fig. 7.4);

phases A and B are quantitatively detectable by statistical methods (Fig. 7.4); or

phases A and B are not detectable side by side (use other quantities measurable by microscopy).

The last case applies only if the phases have all principal refractive indices and the optic sign in common.

Fig. 7.4 demonstrates the principle of investigation by statistical methods. The extent and kind of overlap determine the ratio between phase-specific and non-evaluable n_α'-and n_γ'-values and, thus, the percentage of evaluable n_α'- and n_γ'-values [7.114]. n_γ'-values specific for phase A are greater than the refractive index

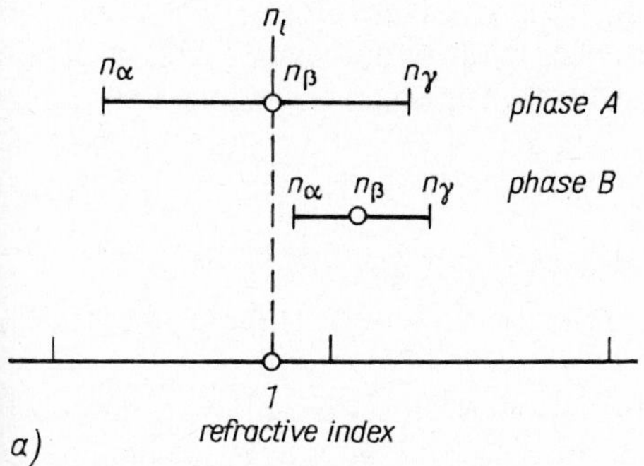

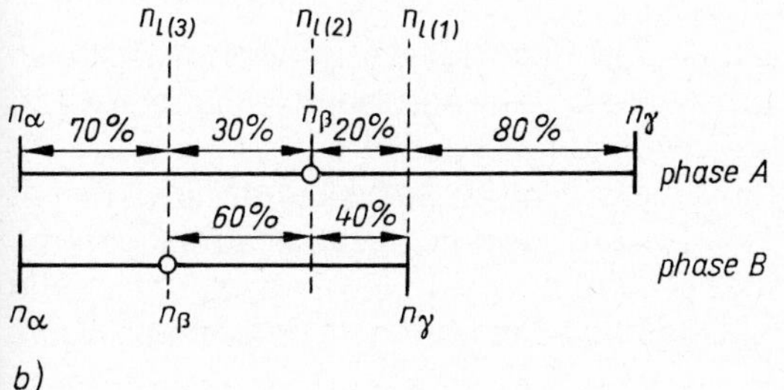

Figure 7.4. Phase analysis of solid mixtures having overlapping refractive index ranges
a) phases A and B quantifiable without statistical methods
b) phases A and B quantifiable by statistical methods

of liquid 1 ($n_{l(1)}$). If it is assumed that all transmission directions of the indicatrix have equal statistical frequencies, with a great number of isometric crystals measured, the number of grains having $n_\gamma' > n_{l(1)}$ corresponds to about 80 % of the phase A fraction in the mixture of A and B. The fraction of phase A can then be found from this relation and the total number of grains analysed; the balance will be fraction B.

Algorithm to Fig. 7.4b

80 % A	$= n_1$
100 % A	$= x$
$n - x$	$= y$
n	= 100 %
x	= % A of the mixture
100 − % A	= % B of the mixture

n number of A or B grains analyzed at $n_\gamma' \gtrless n_{l(1)}$
n_1 number of A grains analyzed at $n_\gamma' > n_{l(1)}$
x number of grains in phase A
y number of grains in phase B

To verify results, similar calculations can be made using the other relations and the liquids $n_{l(2)}$ and $n_{l(3)}$.

Chart 7.4 shows the sequence of steps to be taken in the example illustrated by Fig. 7.4b.

These theoretical relations constitute rigorous statistical patterns only if the number of grains analyzed is very large and if the grains are of spherical shape, i.e. if all transmission directions are represented with equal shares. Where grain shapes deviate from an ideal sphere, certain transmission directions will occur more frequently than others.

Depending on the degree of deviation from the sphere, there are two groups of grain shapes:

(1) quasi-isometric grains without a clearly preferred direction, and

(2) grains showing one or two clearly preferred directions (tabular or prolate shape).

For grains of the first group, deviation from an even statistical distribution of transmission directions is negligible if a large number of grains are analyzed. Grains of the second group cannot be evaluated for the present purpose unless special allowance is made for that deviation.

Crushing the grains or mounting them together with glass powder will change their initial orientation favourably in some cases. The non-isometric grains will be tipped in different directions by the irregular glass grains.

The number of second-group grains measured should be a multiple of that of first-group grains in order to yield useful results. Another factor of influence is the width of the basic phase-specific range of n_α' or n_γ' (in Fig. 7.4), e.g., $n_\gamma' > n_{l(1)}$. The smaller the interval is, the more grains should be measured to obtain statistically valid information.

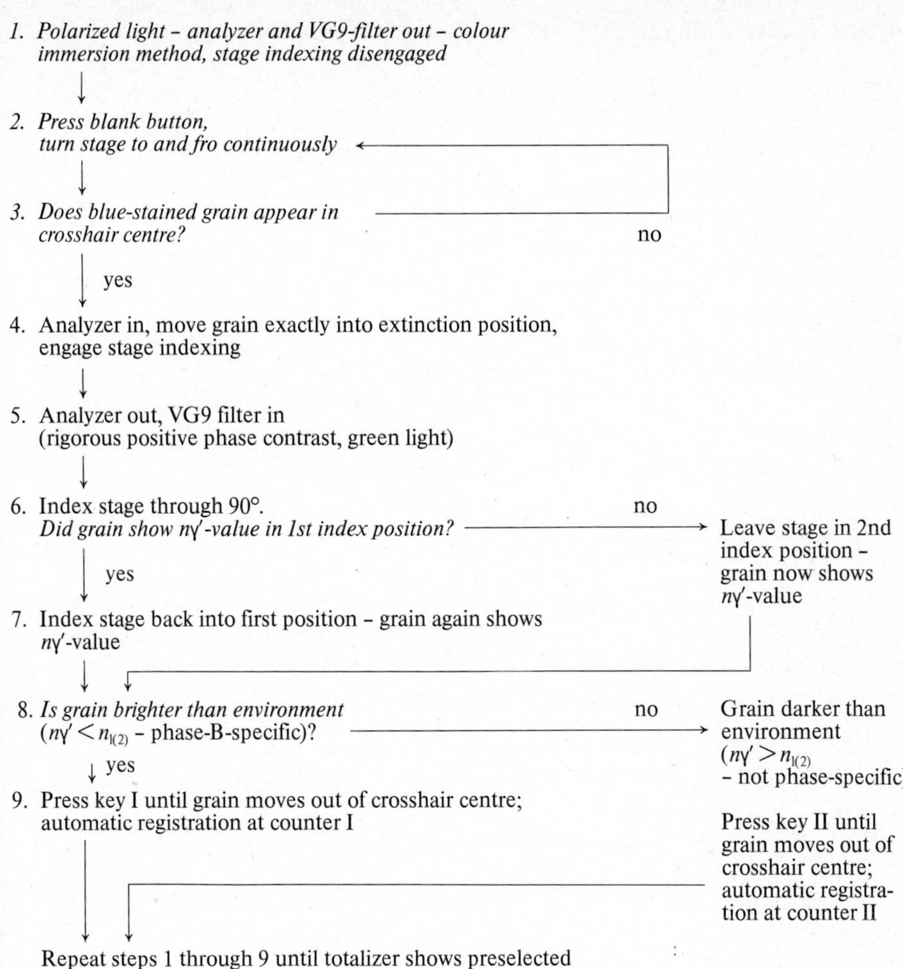

Chart 7.4. Schematic measuring procedure for identifying a combination group (A, B) and quantitative determination of phases A and B by a sequence of steps (Fig. 7.4, immersion liquid $n_{l(2)}$)

This technique is of practical importance, e.g., in the analysis of paragenetic K_2SO_4 and $3K_2SO_4 \cdot Na_2SO_4$ precipitates, a process of technological significance in the potash industry (Fig. 7.4).

Analytical evaluation is generally very difficult, if not impossible, where more than two phases at a time have overlapping refractive index ranges. Even then, qualitative or statistical-quantitative information can be obtained in favourable cases by resolving the overlap combination into groups of two and analyzing each pair in succession.

7.1.2 Determining fractions of different origin in a specimen

This problem is encountered in the investigation of technological dust emissions in industrial conurbations, where the share of different emitters in settled dust is of interest to pollution control authorities. If each component emission in a specimen is a homogeneous substance, each fraction represents one kind of emitter. If, however, individual emissions consist of several phases, or if several emissions have phases in common, no immediate correlation is possible.

The shares of component emissions in a dust specimen can be assessed in two steps by means of emission-specific indicator constituents:

(I) Assessing the percentage of the respective indicator constituent in the dust originating from a particular emitter yields the conversion relation: n % of indicator constituent = 100 % of the respective dust.

(II) The percentages of indicator constituents in the dust specimens, determined by phase analysis, are then converted into emitter shares.

The indicator constituents used should satisfy the following requirements:

(1) They should be specific for the respective emitter and should not, or in no appreciable quantities, occur in other dusts.

(2) Their share in the total composition of the settled dust specimens should be as large and as constant as possible. Trace fractions, though specific, will yield qualitative information only.

(3) Quantitative phase analysis should not be complicated or inaccurate.

Table 7.11 exemplifies the procedure in a chemical industry application [7.113].

7.1.3 Analysis of secondary and trace constituents

To determine small volume fractions in solid mixtures, microscopical methods are particularly successful. Here, a »grain-by-grain« approach is superior to other phase analytical methods.

There are three procedures, depending on the specimen properties (Table 7.12), [7.111].

Analysis after enrichment (Chart 7.5, see pages 234 and 235)

Analysis in situ is necessary especially in case of surface growth on grains, inclusions in grains, interparticle filling in aggregates etc. A surface-grown grain can be contrasted efficiently against the substrate by immersion in a liquid of suitable refractive index. If the grain of interest overlaps with other objects in the light path, optical data measurement is rather difficult.

A phase whose volume fraction is not too small can be tested, together with the surrounding material, by an additional measuring technique (e.g. powder roentgenography). The reflection pattern is identified by subtracting the known reflections of the surrounding matter.

Table 7.11. Diagnostically significant data of dust species sampled in four different places

Dust group	Dust species	Indicator constituent	Chief method of determination	Characteristic feature	Mode of calculation
I	Pyrite raw dust	Pyrite	Microscopic (fraction 1 to 70 μm)	Lustrous, light-yellow grains	85% pyrite in 100% pyrite raw dust
II	Roasting residues	Hematite (slag spherules)	Microscopic (fraction 1 to 70 μm)	Red internal reflexes, round shape	85% hematite (10% slag spherules) in 100% roasting dust
III	Lignite dust	Lignite fractions			In a 100 to 300 deg temperature interval, 60% ignition loss in 100% lignite dust
	LHT)* coke dust	*LHT* coke fraction	Fractionated determination of ignition loss	Temperature-dependent, characteristic ignition loss	In a 300 to 400 deg and 400 to 500 deg temperature interval each, 30% ignition loss in 100% LHT coke dust
	Soot dust	Soot			In a 300 to 400 deg temperature interval, 35% ignition loss in 100% soot dust
IV	Power station ash	Slag spherules	Microscopic (fraction 1 to 70 μm)	Round shape	50% slag spherules in 100% power station ash
		Calcium ferrite (?)	X-rays	$d = 2.384$ Å	Semi-quantitative evaluation of relative intensity
	Steam locomotive ash	Slag spherules	Microscopic (fraction 1 to 70 μm)	Round shape	25% slag spherules in 100% locomotive ash
		Anorthite (?)	X-rays	$d = 3.20$ Å	Semi-quantitative evaluation of relative intensities
	Coal residues	(Balance to total locomotive dust)			35% C in 100% locomotive ash

Table 7.11, continued

Dust group	Dust species	Indicator constituent	Chief method of determination	Characteristic feature	Mode of calculation
V	Abraded particles and atmo- spheric dust	None	Indirect determination		100% − (I% + II% + III% + IV%) = V%

*) LHT: lignite high-temperature coking process

Table 7.12. Possibilities to analyze secondary and trace constituents

Sample quantity available	Method of analysis
Much	Enriching secondary and trace constituents into main constituents
Little	Isolation of the interesting grain
Little	In situ (direct within the sample)

Identification of an isolated grain

(1) Microscopical examination

A single grain can be identified by any property that permits microscopical measurement (Table 6.1). If the grain has no special transmission direction, measuring facilities are extended considerably by repositioning with the universal stage, spindle stage or Waldmann sphere, or by shifting the cover slip over a highly viscous immersion medium.

A combination of polarizing, interference and phase contrast techniques is favourable (Chart 7.6, see pages 236 and 237).

The grain is fixed to the needle point of a spindle stage. By rotation, grain thickness can be measured accurately through the microscope. This permits interference microscopy, as retardations can be measured and analyzed (Table 7.13). The indicatrix of the grain is found by combination with refractive index measurement by the wavelength-temperature double variation method if a hot and cold stage is used (Fig. 3.5), and with spindle stage measurements at different angles of rotation.

(2) X-ray diffractometry

In some adverse cases, a grain cannot be fully identified by its microscopical properties alone. X-ray diffractometry offers a favourable complementation.

Table 7.13. Main methods of interference microscopy of objects fixed by adhesion

Interference method	Type of specimen
Shearing method with complete image splitting	Small grains, any grain shape
Fringe method	Larger objects of smooth surface, preferably of tabular shape

Sample preparation, Debye-Scherrer roentgenography and its graphical evaluation have been described elsewhere [7.111].

As the lattice planes available for reflecting the X-rays are few, the pattern obtained does not show continuous-density interference rings but spots lying along such rings. The pattern is comparable to that produced by an unaligned monocrystal and recorded with a rotary-crystal camera. Kittrick and Hope measure the Bragg angle ϑ_{hkl} by projecting the spots on to a family of curves that represent the course of the Debye-Scherrer rings on the film, and extrapolating the spots to the equator. The measuring error involved is relatively large, though; it may lead to misinterpretation where compounds having higher lattice parameters are to be identified. To use precision instruments for measuring the film would not improve things either, since the position of a spot has to be established by two measurements, whose errors would add up, viz.

(1) the distance of the diffraction spot projected to the equator from the spot produced by the direct ray, and
(2) the distance of the diffraction spot normal to the equator.

The sheet-film method, with only one distance measurement required, increases the accuracy of ϑ_{hkl} determination.

All interferences are on concentric circles whose radiuses are equal to the distance r of the spots from the direct ray spot. Each circle corresponds to a ring in a common Debye-Scherrer pattern and, thus, to a certain Bragg angle ϑ_{hkl} to be determined in the transmission mode:

$$\tan 2\vartheta_{hkl} = \frac{r}{A}$$

$$\vartheta_{hkl} = \frac{90}{\pi} \arctan \frac{r}{A} .$$

The distance A between film and specimen can be determined very precisely if the crystal is glued to the end of a thin wire. The spots produced by the metal serve for calibration. By varying A, the resolving power of the technique can be changed to suit the specimen type.

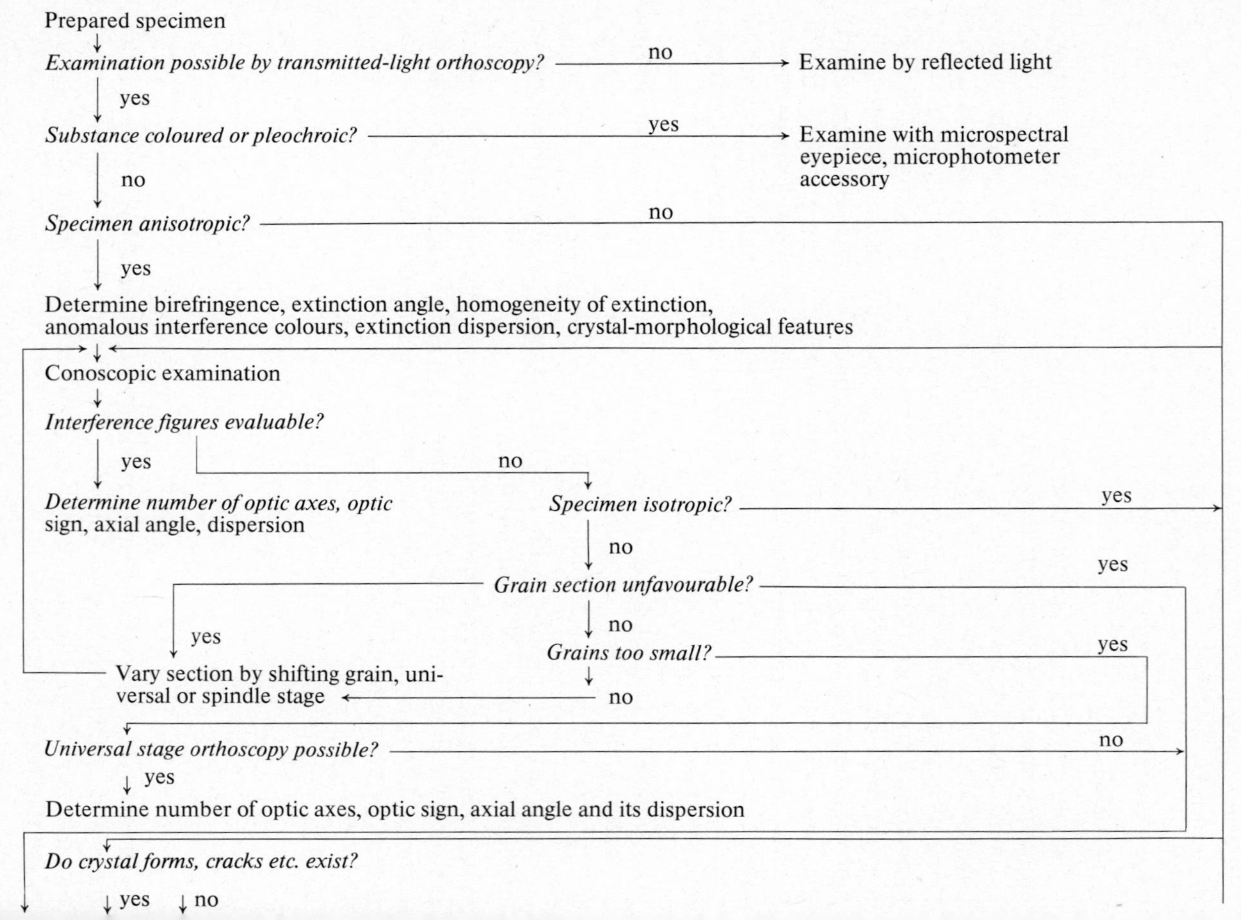
Prepared specimen
Examination possible by transmitted-light orthoscopy?
no
Examine by reflected light
yes
Substance coloured or pleochroic?
yes
Examine with microspectral eyepiece, microphotometer accessory
no
Specimen anisotropic?
no
yes
Determine birefringence, extinction angle, homogeneity of extinction, anomalous interference colours, extinction dispersion, crystal-morphological features
Conoscopic examination
Interference figures evaluable?
yes
no
Determine number of optic axes, optic sign, axial angle, dispersion
Specimen isotropic?
yes
no
Grain section unfavourable?
yes
yes
no
Grains too small?
yes
Vary section by shifting grain, universal or spindle stage
no
Universal stage orthoscopy possible?
no
yes
Determine number of optic axes, optic sign, axial angle and its dispersion
Do crystal forms, cracks etc. exist?
yes
no

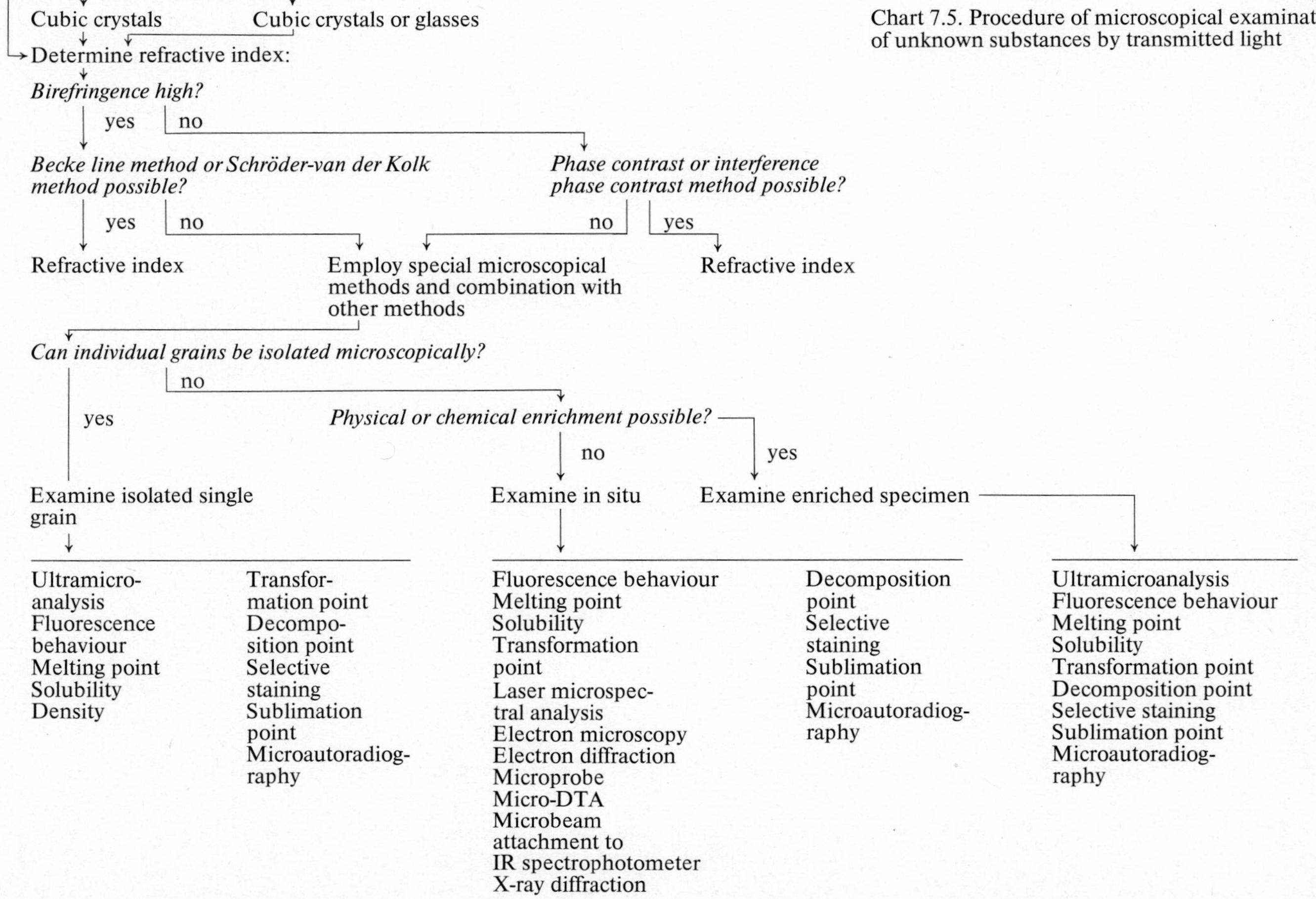

Chart 7.5. Procedure of microscopical examination of unknown substances by transmitted light

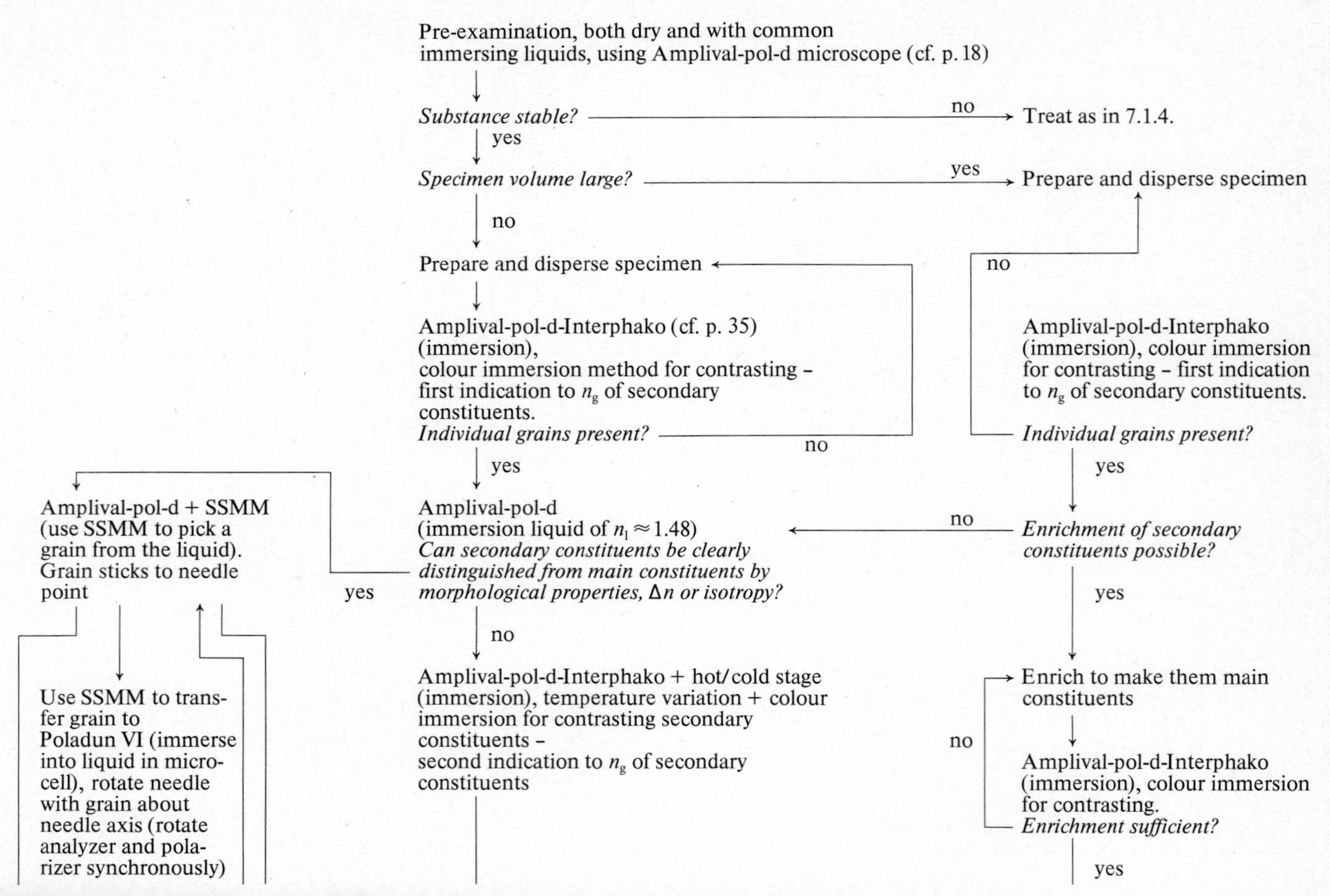
Pre-examination, both dry and with common immersing liquids, using Amplival-pol-d microscope (cf. p. 18)
Substance stable?
no
Treat as in 7.1.4.
yes
Specimen volume large?
yes
Prepare and disperse specimen
no
Prepare and disperse specimen
Amplival-pol-d-Interphako (cf. p. 35) (immersion), colour immersion method for contrasting – first indication to n_g of secondary constituents.
Individual grains present?
no
yes
Amplival-pol-d-Interphako (immersion), colour immersion for contrasting – first indication to n_g of secondary constituents.
Individual grains present?
no
yes
Amplival-pol-d (immersion liquid of $n_l \approx 1.48$)
Can secondary constituents be clearly distinguished from main constituents by morphological properties, Δn or isotropy?
yes
Amplival-pol-d + SSMM (use SSMM to pick a grain from the liquid). Grain sticks to needle point
Use SSMM to transfer grain to Poladun VI (immerse into liquid in micro-cell), rotate needle with grain about needle axis (rotate analyzer and polarizer synchronously)
no
Amplival-pol-d-Interphako + hot/cold stage (immersion), temperature variation + colour immersion for contrasting secondary constituents – second indication to n_g of secondary constituents
Enrichment of secondary constituents possible?
no
yes
Enrich to make them main constituents
Amplival-pol-d-Interphako (immersion), colour immersion for contrasting.
Enrichment sufficient?
no
yes

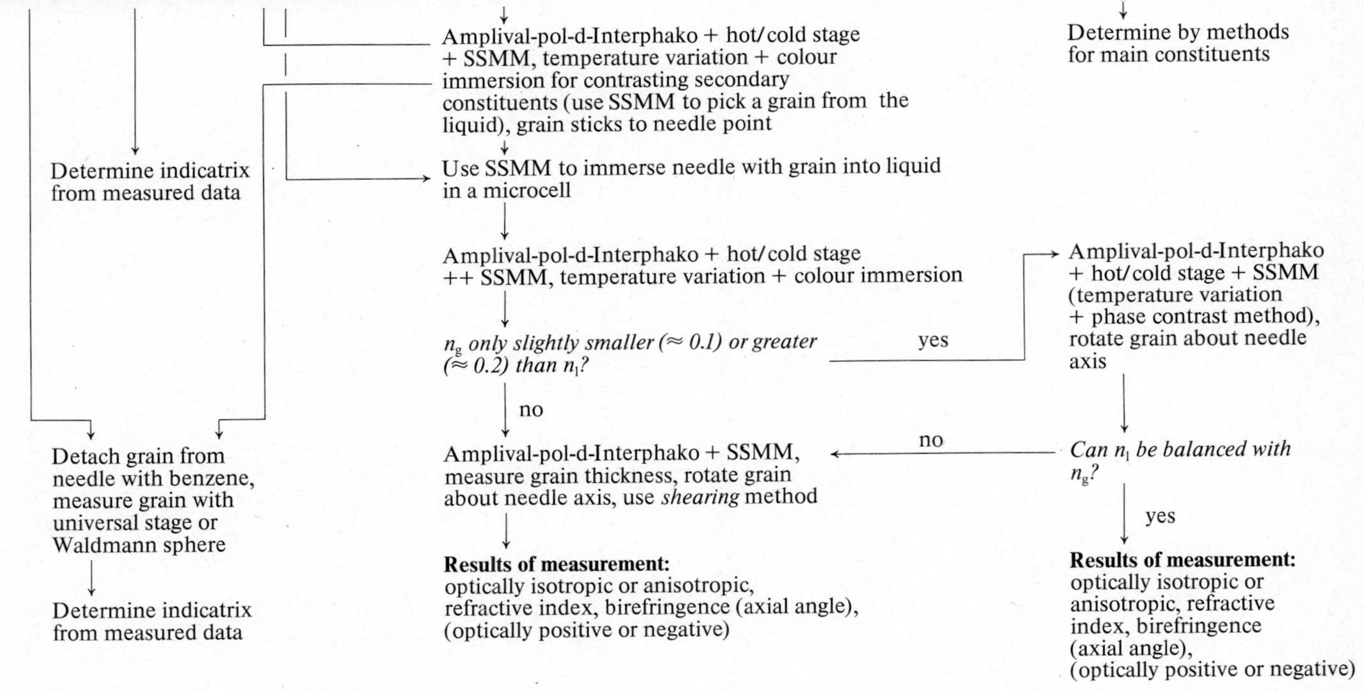

Chart 7.6. Methods of preparation and microscopical examination for identifying secondary constituents, acc. to [3.103] and [7.114]

SSMM = spindle stage/micromanipulator (cf. Table 3.5)

$n_g = n_{grain}$; $n_l = n_{liquid}$

The accuracy of measuring the spot distance *r* essentially depends on the exactness of defining the direct ray spot and the size of the diffraction spots. The average of the measurements of all spots of equal ϑ_{hkl} will provide a higher accuracy of *r*.

Primary ray diaphragm diameters between 0.2 and 0.1 mm have proved particularly suitable.

The method allows only a limited range of interference angles. In a transmission pattern of film diameter *D*,

$$\tan 2\vartheta_{hkl} = \frac{D}{2A}.$$

A compromise has to be found between resolution and angle range depending on the specimen to be examined.

(3) Ultramicroanalysis (cf. 6.5)

The chemical analysis of a grain permits correlation of the microscopical and diffractometric measurements with a definite chemical compound. One should note that the substance used for fixing the grain to the needle may degrade analysis results.

7.1.4 Special requirements for phase analyses of hygroscopic and decomposing substances

The main difficulty in analysing such substances is sample preparation (cf. 4.2.2). The adverse properties demand differently prepared specimens and, thus, different measuring techniques. For the properties of such substances that can be determined microscopically, and for the general prodecure, see Table 6.5 and Chart 7.7. Preparation and measurement are performed best in an integrated box-microscope unit (Fig. 7.5, see Annex). Its preparation section can be cooled with CO_2 or liquid N_2. The microscope has a cold stage and a micromanipulator arranged on the side.

Very often, surface decomposition interferes with microscopical observation. Even slight incrustation of grains with newly formed matter will be disturbing. An overlay of 1 % of the new phase is sufficient to prevent the exact measurement of birefringence and to make interference colours indistinct [3.105]. In the box, such grains can be crushed to obtain fresh fracture surfaces.

Despite those difficulties, microscopy is the only method where specimen decomposition can be detected visually during examination and the resulting adulteration of measurements obviated.

Example: Identification of the compound $KCl \cdot MgCl_2 \cdot HCl \cdot 7H_2O$ by measuring the optic axial angle

Preparation under an HCl atmosphere in a corrosion-proof PVC box, and transfer of selected grains into a Waldmann sphere. *Filling:* Polystyrene-xylene

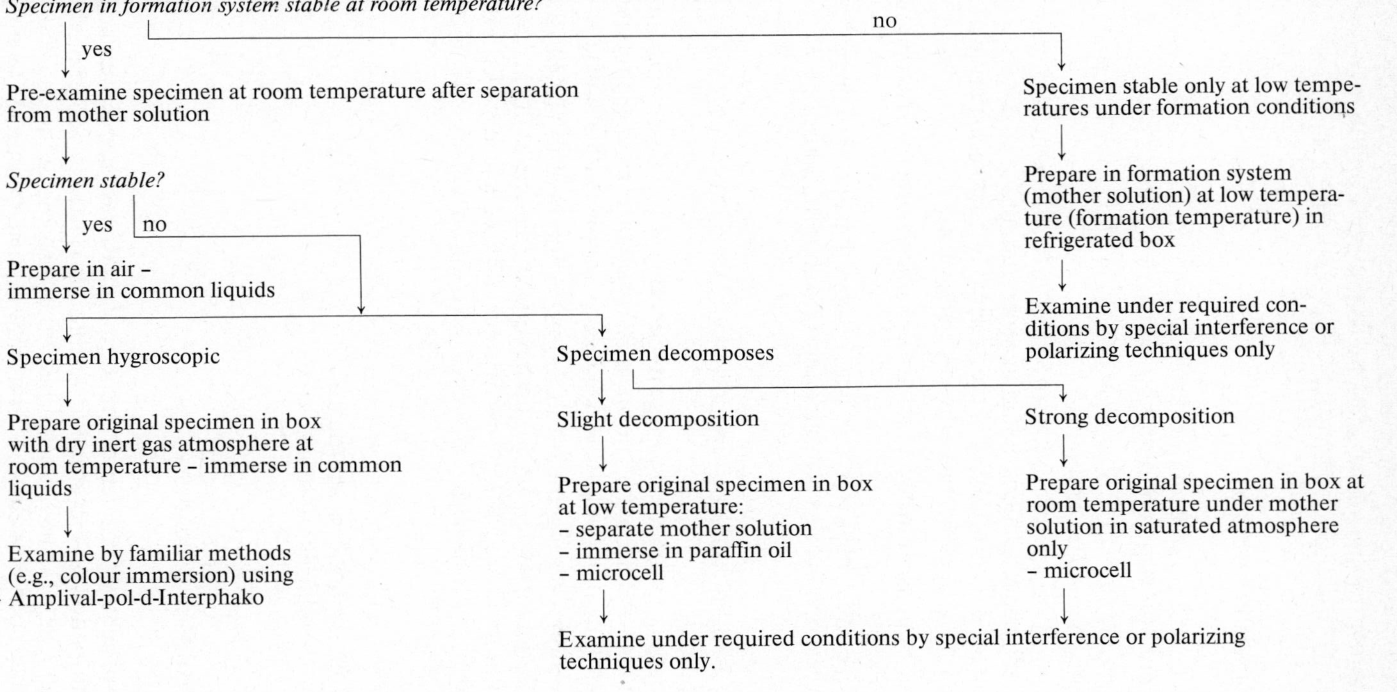

Chart 7.7. Specimen preparation and examination methods for unstable substances

mixture. Measurement of optic axial angle by conoscopy using the universal stage (see 6.1.3.2).

7.1.5 Phase analysis in processes

The change of a solid phase during a physical-chemical process can be measured by discontinuous or continuous procedures (cf. Table 7.5).

Example (1): Discontinuous study of precipitate during isothermal evaporation experiments on potash industry solutions
Parameters of interest: Temperatures between 50 and 90 °C, different evaporation times, different batch concentrations, addition of seed grains.

For the microscopical determination of the precipitate, various experimental series are run in parallel, the above parameters being varied.

Example (2): Study of phase transformations in ceramic baking, calcinating and combusting processes.

The newly formed substances will react with one another and with unconverted initial substances, depending on heating rate and concentrations.

Possible methods:

(1) *discontinuous* assessment of the products of various batches heated to different temperatures;
(2) *continuous* examination on the hot-stage microscope, direct observation and measurement of the behaviour of the raw materials under heating; photomicrographic recording, photometric registration of changing optical data.

Example (3): Continuous measurement of the phase composition change in the incongruent dissolution of $KCl \cdot MgCl_2 \cdot 6H_2O$ into KCl and solution
Methods: Microcell, transmitted-light microscopy, orthoscopic rays, crossed polarizers, automatic photometric recording of the intensity change due to the reduction of phases.

Basis of measurement:

$KCl \cdot MgCl_2 \cdot 6H_2O$ – optically anisotropic,
KCl – optically isotropic.

7.2 Microstructure analysis

The interaction of light with the lattice structure of solids, registered by microscopy, conveys information on that structure (Table 7.14) [7.115 to 7.163]. As the wavelength of visible light is much greater than the dimensions of unit cells and lattice elements, only such elementary effects show up that accumulate by mul-

Table 7.14. Lattice structure information obtainable from microscopical observations

Microscopical property	Information on the microstructure of chemical compounds
Refractive index	Anisotropy of packing density
Birefringence	Spatial pre-order state in liquid phases; lattice anisotropy of solids
Birefringence differing by sectors, but uniform extinction	Anomalous mixed crystals
Extinction angle	Lattice symmetry (crystal system)
Type of extinction	
1. homogeneous	Uniform lattice array, homogeneous lattice configuration
2. incomplete (undulatory or speckled)	Lattice imperfection
Colour, pleochroism, absorption spectrum	Spatial structure of lattice elements (molecules, ions), chemical bonds within and between the elements, spatial array of elements in the lattice
Transparency, turbidity	Submicroscopic interstitial impurities, solid phase separation
Optic axial angle	Form of indicatrix, expression of lattice anisotropy, anomalous biaxiality by submicroscopic twin lamellae or lattice imperfections affecting a large space
Dispersion of optical properties	Lattice symmetry (crystal system)
Circular polarization	Asymmetric arrangement of lattice elements, molecular structure of asymmetric elements
Cleavability	Bonding anisotropy in the lattice
Zoning	Mixed crystal formation
Sectorial growth	Growth inhomogeneity
Melting and decomposition temperatures	Strength and disintegration behaviour of the structure

tiple repetition within the lattice. Such effects are the substitution of lattice elements, variations of unit cell size etc. Inhomogeneities and lattice defects larger than 0.2 μm are directly visible and measurable through the microscope.

Some limited information may be obtained on the spatial arrangement of the lattice elements (molecules, ions) themselves. The configuration of the molecule, repeated in the lattice, influences the light passing through it and thus the optical data measured.

For interpreting the measurements, one needs a model concept of the structure in order to estimate or calculate the optical data to be expected; these have to be compared to the measured values. Quantitative model calculations for various organic and inorganic structures, e.g. polyamides, complex compounds, phyllo- and inosilicates, permitted the successful interpretation of refractive indices, birefringences and optic axial angles [7.146 to 7.148].

The limited information yielded by statical microscopical measurements may be enhanced considerably

(1) by comparing the data measured for several compounds (e.g. mixed crystals), and

(2) by kinematic-microscopical relative measurements of the change of optical data in case of structural changes – optical parameters react very sensitively to even slightest changes of structure.

The dextro- and laevorotatory forms of optically active substances have all optical data in common except for their optical activity, which is measurable with the microscope (Table 7.15).

Optical property	d-form	l-fom
Axial angle	86°	86°
n_α	1.5477	1.5477
n_β	1.5804	1.5805
n_γ	1.6191	1.6192
Birefringence	0.0714	0.0715

Table 7.15. Identical optical constants (except for optical rotation) of d- and l-forms of the monohydrate of 2-amino-3-carbamylpropionic acid, acc. to [7.103]

7.2.1 Structural interpretation of the optical indicatrix

The refractive index characterizes the amount of retardation to which a light wave passing a substance is subjected. That retardation can be considered as a differential back-shifting of the wave phase at the lattice planes (Fig. 7.6). The shift is caused by interference with weak elementary waves produced at the lattice planes.

In accordance with this model, birefringence is caused by the phase backshift being different at lattice planes normal to each other.

The form and size of the optical indicatrix is a coded representation of the structural state encountered by the light passing the crystal lattice in different spatial directions.

The light refraction in a particular direction is the resultant of three influences, viz.

(1) ionic polarization and its interaction with the field of light waves,

(2) spatial form and arrangement of lattice elements,

(3) number of atoms per unit volume (packing density).

The interpretation of refractive indices in a particular lattice direction has to allow for these three factors separately. Depending on the structure, one factor may clearly prevail, or the effects of several factors may either compensate or amplify each other.

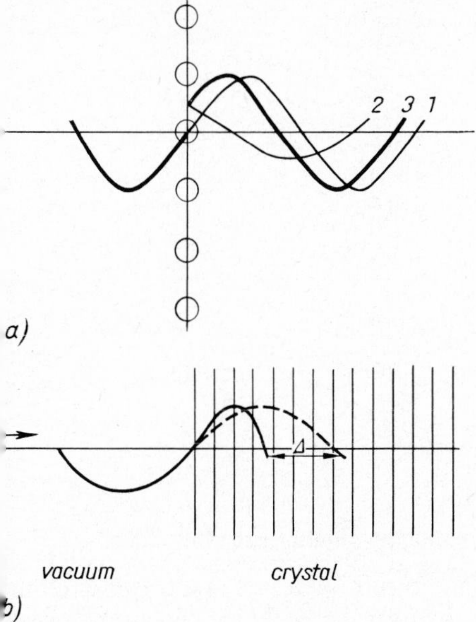

Figure 7.6. Interpretation of refraction in solids by phase retardation of the incident wave (acc. to [2.3])
a) at a row of points N
1 original wave 2 secondary wave 3 resultant wave
b) at the lattice planes
Δ phase retardation

Refractive index variation in hydrates of different water content (Fig. 7.7)

The increase of water molecules in the structure reduces the packing density and changes the lattice configuration. Refractive index and birefringence decrease. A similar effect is observed in a comparison between various modifications of a compound (Table 7.16).

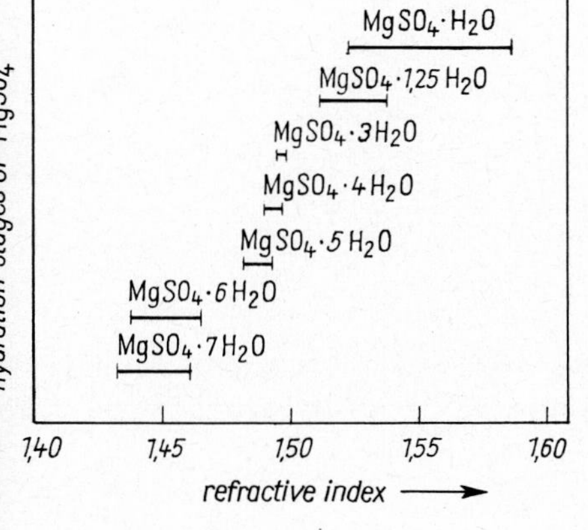

Figure 7.7. Refractive index variation in hydrates of a compound, exemplified by $MgSO_4$ hydrates

Table 7.16. Densities and refractive indices of various polymorphous modifications of a compound, acc. to the literature

Compound	Polymorphs (mineralogical names)	Density (g/cm^3)	Mean refractive index
SiO_2	Low tridymite	2.26	1.471
	Low christobalite	2.32	1.486
	Low quartz	2.65	1.549
$Al_2O_3 \cdot SiO_2$	Andalusite	3.15	1.639
	Sillimanite	3.23	1.666
	Disthene	3.6	1.720
$CaCO_3$	Calcite	2.72	1.572
	Aragonite	2.94	1.632
TiO_2	Anatase	3.84	2.524
	Brookite	3.95	2.637
	Rutile	4.24	2.760
$LiAlSiO_4$	High eucryptite	2.31	1.522
	Low eucryptite	2.63	1.580

Refractive index variation by replacing the cation by another one (Fig. 7.8)

Here, the change in refractive index and birefringence is due to the different ionic polarization of the sodium and calcium ions. Table 7.17 shows an analogous influence.

Refractive index in layer structures

Light rays incident normal to layer structures oscillate in the plane of the layers, they are characterized by low velocity or high refractive index. Light waves normal to these have a low refractive index. These two direction-specific extrema are responsible for the high birefringence and the optically negative character of layer structures. If the structural conditions are similar in various directions within the layer plane, the resulting indicatrix is optically uniaxial. Structural differences within the layer plane show up by a biaxial indicatrix.

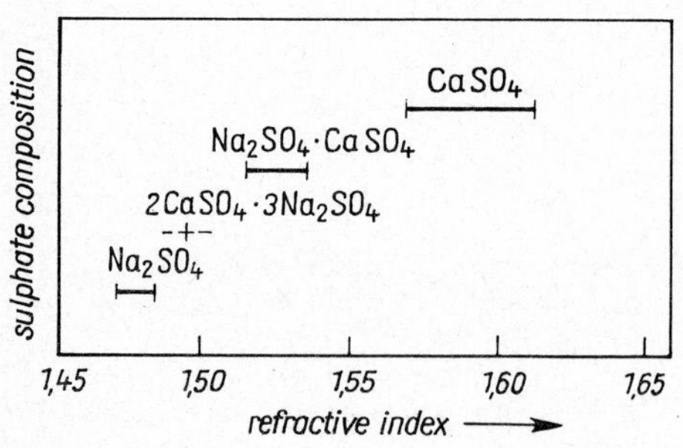

Figure 7.8. Refractive index variation by cation substitution, exemplified by Ca-Na sulphates

Table 7.17. Refractive indices of sodium halides

Compound	Refractive index
NaF	1.336
NaCl	1.544
NaBr	1.641
NaI	1.774

Structures having a layer-like arrangement of planar complex anions (CO_3^{2-}, NO_3^-, ClO_3^-) behave like true layer structures, showing high negative birefringence, too.

In $CaCO_3$ (calcite), birefringence is essentially a function of the dipole interaction of the oxygen atoms, whereas the influence of the carbon atoms is negligible and the effect of calcium is independent from the direction of the electrical field.

The polarization of each oxygen atom increases under the influence of the other ones if the electrical field acts in the direction of the planar array of the CO_3 group (0001). In the [0001] direction, polarization decreases.

Increased polarization means stronger oscillation of the atoms in the field, stronger secondary waves and, thus, a considerably greater retardation of the primary wave (Fig. 7.6).

Exceptions from such simple relationships are found in some uniaxial layer structures of various hydroxides. $Ca(OH)_2$, $Mn(OH)_2$ and $Cd(OH)_2$ show normal negative birefringence, whereas $Mg(OH)_2$ and γ-$Al(OH)_3$ are positive. A field extending normal to the layer plane causes a cylindrical polarization of the hydroxyl ions, generating a strong dipole moment normal to the layer, and thus n_e becomes greater than n_o. This special type of polarization of the hydroxyl ion counteracts the negative birefringence caused by the layer structure and may, if its influence prevails, result in an optically positive behaviour.

Structural inhomogeneity in addition compounds

A determination of the optical data of $MgSO_4 \cdot 2CH_3OH$ will reveal variations of birefringence and optic axial angle within individual crystals [7.170] (Fig. 7.9). Microscopically determined activation energy values of the decomposition reaction (cf. 7.3.2.2) indicate differences between individual crystals as well as crystal domains.

The compound contains the solvate molecule in the cavities of a loosely packed framework of SO_4 tetrahedron elements and Mg ions. By microscopy and X-ray diffractometry, the $MgSO_4$ host lattice was found to have a cubic symmetry. This is reduced by the entrance of the solvate molecule to an orthorhombic or monoclinic (optical) symmetry, the dimensional change of the cubic unit cell being hardly measurable.

The reason is the statistically irregular filling and discharging of the lattice cavities with guest molecules having a polar configuration.

As there are no stoichiometric relationships throughout the lattice of all grains, it is a problem to specify a chemical formula for that compound. The specification

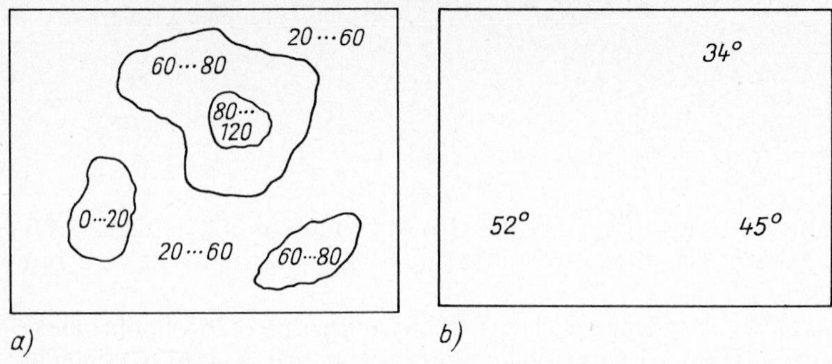

Figure 7.9. Variations of axial angle and birefringence within isometric $MgSO_4 \cdot 2CH_3OH$ crystals, shown for oblong tabular monocrystals
a) retardations within the monocrystal (nm)
b) axial angles within the monocrystal

of the solvate content of the precipitates constitutes a mean value of the filling of the host lattice for all grains of the sample substance.

7.2.2 Mixed crystals

Mixed crystals are distinguished by the fact that their optical properties vary continuously with their composition. Even the slightest difference in chemical composition produces distinctly measurable changes in refractive index, birefringence, optic axial angle and/or reflectance (Figs. 7.10 to 7.12, Tables 7.18 and 7.19) [7.128 to 7.132].

Zonal structure, caused by fast mixed-crystal formation without concentration balancing, is revealed conspicuously by the zonal arrangement of optical parameters in the microscopical image (Fig. 6.26).

A combination of microscopical and X-ray measurements may throw some light on the structures of mixed crystals. In studying hydrated Na–Ca sulphate, e.g., we found a partial formation of mixed crystals with a ratio of $CaSO_4 : Na_2SO_4$ between 1 : 1.5 and 1 : 1.8 [7.131]. Fig. 7.10 shows the variation of optical data with chemical composition and its relationship to the rhombic unit cell dimension determined by X-ray diffractometry.

Analogously to the three principal refractive indices changing in the same direction, the lattice constants at first vary in equal directions, but then b_0 and c_0 reverse their behaviour. In parallel with that, the volume of the unit cell at first reduces slightly, followed by a slight increase with different edge dimensions. The reversing point coincides with a reversal of the optical sign of the structure, i.e. a reversal of the quantities $(n_\gamma - n_\beta) : (n_\beta - n_\alpha)$ and of the optic axial angle $2V$.

The important point is that the optical properties change markedly, while the unit cell dimension varies within such narrow limits as to be practically constant.

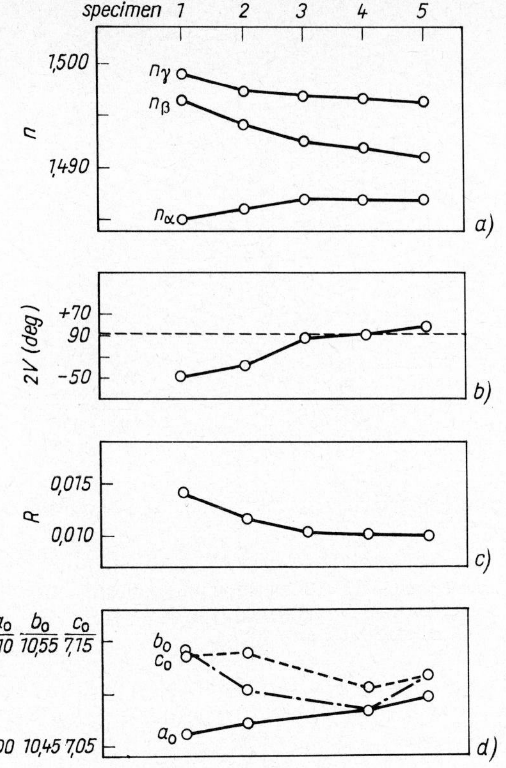

Figure 7.10. Variation of optical data and lattice constants with composition in partially mixed crystals of hydrated Ca-Na sulphates
a) principal refractive indices $n_\alpha, n_\beta, n_\gamma$
b) optic axial angle $2V$
c) birefringence
d) lattice constants a_0, b_0, c_0

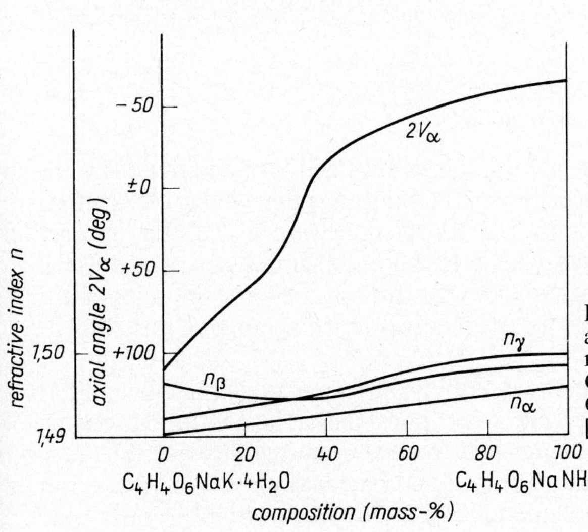

Figure 7.11. Variation of axial angle and principal refractive indices in mixed crystals of $C_4H_4O_6NaK \cdot 4H_2O$ and $C_4H_4O_6NaNH_4 \cdot 4H_2O$ (acc. to [7.103])

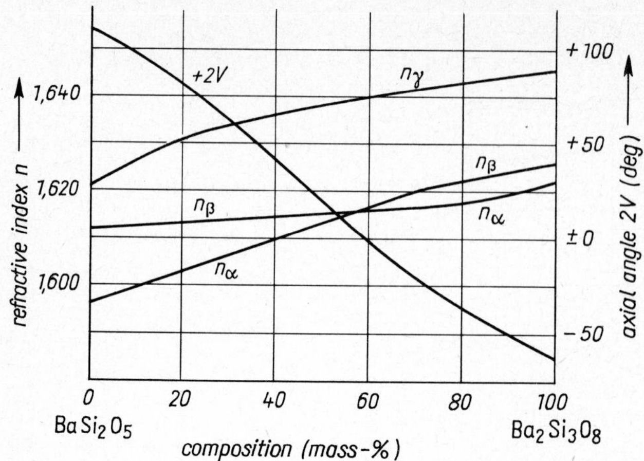

Figure 7.12. Variation of axial angle and principal refractive indices in mixed crystals of $BaSi_2O_5$ and $Ba_2Si_3O_8$ (acc. to [7.102])

Table 7.18. Composition and optical constants in mixed $(K,Na)_3$ $Na(SO_4)_2$ crystals, acc. to [7.102]

Na : K	n_o	n_e	$n_o - n_e$
4Na : K	1.485	1.490	0.005
Na : K	1.490	1.495	0.005
Na : 2K	1.491	1.4965	0.0055
Na : 3K	1.493	1.498	0.005
Na : 4K	1.4935	1.500	0.0065

Table 7.19. Dependence of optic axial angle $2V$ (in deg) from composition of Tutton's salts $A_2B(SO_4)_2 \cdot 6H_2O$, acc. to [7.102]

A	B Mg	Zn	Mn	Fe	Co	Ni	Cu
K	47	68	–	67	68	75	46
NH_4	51	79	69	76	82	86	111
Rb	48	73	67	73	75	82	44
Cs	16	74	59	74	81	92	43
Tl	105	110	108	111	113	118	94

These findings confirm the assumption of subtractive or additive substitution to be expected if 1 Ca is replaced by 2 Na without any other charge compensation. Such a replacement must show up conspicuously in a changed interaction between lattice and light waves, whereas the unit cell dimension need not change appreciably. With that substitution mechanism assumed, the limited number of free lattice sites then accounts for the relatively narrow range of miscibility.

Figs. 6.27 and 7.13 (see Annex) exemplify the formation of mixed paraffin-olefin crystals and their characterization by optical data. The continuous change of the optic axial angle demonstrates that there is no miscibility gap. The second axial angle curve, recorded also during mixed crystal formation in the precipitate examined, indicates the simultaneous existence of a metastable phase.

When mixed crystals start to segregate, they become turbid by submicroscopic segregation products. As the process continues, the new formations adopt microscopic dimensions so that their spatial orientation in the crystal can be measured. Block and mosaic structures are typical segregation patterns, too.

Under the microscope, anomalous mixed crystals are distinguished from others by sectorial growth and stress birefringence. Areas of homogeneous extinction within a sector differ in birefringence because of the molecular alternation between host and guest components and their statistical distribution over the growth face. The stress birefringence may superimpose the inherent birefringence of the interstitial, optically anisotropic guest component.

Other optical parameters such as refractive index and optic axial angle, however, vary in proportion to the guest component same as in true mixed crystals.

7.2.3 Real structure

Information on real structure, i.e. on crystal lattice defects, can be obtained by direct microscopical imaging or by indirect methods such as etching (Table 7.20) [7.133 to 7.136].

Such examinations are highly important for assessing the real structure of semiconductor materials, such as silicon or germanium monocrystals. After etching, the polished surfaces show rows of etching pits that represent the sites where dislocations meet the surface. The number of pits per unit area is a direct

Table 7.20. Microscopical examination of real structures of crystal lattices

Lattice defect	Possible examination
Microscopic defects	
Cracks	Direct observation and measurement
Interstitial impurities >1 μm	
Twin boundaries, grain boundaries, homogeneous regions >1 μm	
Spiral growth	
Accessories growth >1 μm, terracing, etching pits	
Submicroscopic defects	
Edge and screw dislocation	Indirect examination by etching or anomalous extinction behaviour
Interstitial impurities	Turbidity and scattered-light measurements
Lattice stress in optically isotropic lattices	Birefringence measurement

Table 7.21. Microscopical measurement of etching pits by reflected light

Measured parameter	Equipment	Information obtained
Size, depth	Microscope with universal stage, goniomicroscope, interference microscope	Intensity of etching
Number/area	Microscope with integrator	Frequency of lattice defects
Distribution	Microscope, interference microscope	Distribution of lattice defects
Symmetry	Microscope, goniomicroscope	Symmetry and crystallographic orientation of the area examined
Dihedral angle	Microscope with universal stage, goniomicroscope, interference microscope	Miller indices

measure of the dislocation density. Comparative studies by X-ray topography have confirmed this [7.136]. By counting and measuring on the (111) face of germanium monocrystals, e.g., [2.6] one found

densities of 10^2 to 10^3 pits/cm^2 indicating low disorder of dislocation, and densities of 10^4 to 10^5 pits/cm^2 indicating high disorder of dislocation.

Silicon layers grown epitaxially have stacking faults, trigonal pyramids and dislocations, among other defects. Stacking faults show up in phase and interference microscopes without etching; trigonal pyramids are observed with plain bright-field illumination (Table 7.21, Fig. 7.14, see Annex).

7.2.4 Polymer structures

Organic polymer structures are analyzed primarily by measurements of refractive index and birefringence [7.137 to 7.159].

The refractive index of a polymer is mainly a function of its density and, thus, of the degree of order in its structure. Minor influences on light refraction are exerted by the degree of polymerization, the kind and number of foreign substances and the nature of the monomers and end groups.

The birefringence of a sample conveys a cumulative message about the orientation relationships of the structure, both within crystalline areas and within molecular groups, thus including the degree of order and that of orientation.

Several components contribute to the birefringence of polymers:
$\Delta n_{\text{total}} = \Delta n_{\text{int}} + \Delta n_{\text{def}} + \Delta n_{\text{form}}$.

The *intrinsic birefringence* Δn_{int} is the contribution of the molecular structure to the total birefringence. It is due to the bonding anisotropy of the molecules.
The *deformation or piezo-birefringence* Δn_{def} is the share of anisotropy that is due to mechanical load and stress, which changes the original bonding polarizabilities and the orientation of valence angles.
The *form birefringence* Δn_{form} expresses the different arrangements of crystalline and amorphous regions.
Intrinsic and form birefringence are frequently united under the term of *orientation birefringence.*

Calculations of refractive index and birefringence of synthetic fibres have proved the assumed models to be correct.

Examples for the relationship between high polymer structure and optical properties in material processing

Methods: Microtome section, transmitted light, orthoscopy, crossed polarizers, compensators for measuring low retardations (cf. 6.1.2.), transmitted-light interference microscopy (cf. 3.2), thickness measurement for calculating birefringence from retardations (cf. 6.2.2).

Injection-moulded polyamide and polyethylene parts show a marked increase of birefringence from the gripping head to the injection port (Fig. 7.15), which illustrates structural changes in the part.

Examinations of *ruptured tensile specimens of Miramid H* revealed maximum

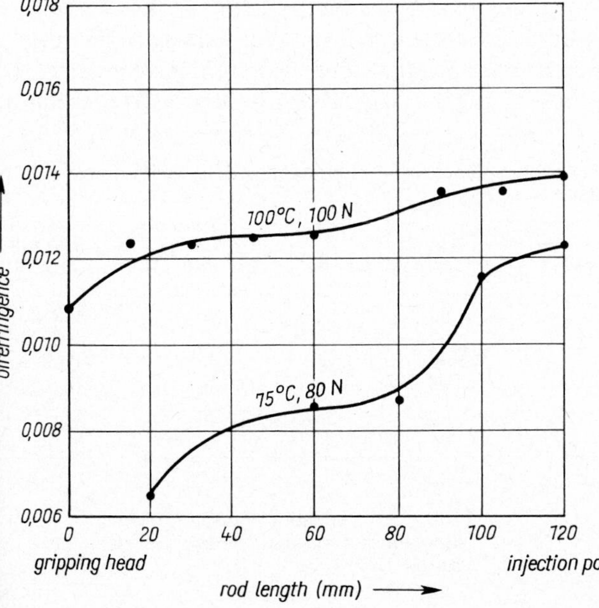

Figure 7.15. Variation of birefringence between gripping head and injection port in injection-moulded polyamide rods, at different annealing temperatures and different pressures (acc. to [7.139])

birefringence at the point of rupture. Here, the tensile load effected the greatest changes in the material structure.

Examinations of *weld seams and spot welds in optically anisotropic high polymers* prove that the tensile strength of the weld varies with the birefringence around it. The formation of many small amorphous regions within a homogeneously structured, low-stressed zone, identifiable by the lack of birefringence, goes along with highest tensile strengths.

In *materials and fibres* under strain, structure changes show up by a proportional rise of birefringence, strictly speaking, of deformation birefringence.

The structural change caused in a material by *annealing* can be read from the change in birefringence (Table 7.22).

The relationship between structure and birefringence is also present in *foils* of different thicknesses. Thin foils or fibres show substantially higher birefringence than thick ones of the same material. The differences are due to structural configurations varying with thickness.

Closely related to this is the relationship between *cooling time* and *birefringence*. Birefringence increases with crystallization time (Fig. 7.16, see pages 254 and 255). In polyamides, the degree of crystallinity depends on the symmetry of the molecular arrangement of the amide group; in polyethylene it is determined by the number of branch points along the chain.

The *temperature of polymerization*, too, affects birefringence. The lower the temperature of PVC, e.g., the higher will be its birefringence (Table 7.23).

Gloss effects in synthetic fibres are caused by structural changes in the surface layers, which can be revealed by measuring the change in refractive index of the copolymer particles and homogenizing the layer. The gloss of polystyrene copoly-

Table 7.22. Influence of manufacturing parameters on birefringence in injection-moulded plates of »Miramid H« polyamide

State	Injection temperature (°C)	Mould temperature (°C)	Dwell time (s)	Birefringence
Initial state	220	60	20	0.0197
	220	60	25	0.0200
	250	100	20	0.0211
After 17 h annealing at 140 °C	260	120	20	0.0242
	220	60	20	0.0259

Polymerization temperature (°C)	Birefringence
−50	0.00026
−30	0.00004
−10	0.000015
+20	–

Table 7.23. Dependence of birefringence on polymerization temperature in PVC, acc. to [7.139]

mers (styrene-butadiene copolymer particles in a polystyrene matrix) increases with the refractive indices of both the matrix and the copolymer particles.

7.2.5 Disperse systems of submicroscopic particles (imbibition method)

Dispersoid systems of submicroscopic particles may be regarded as mixed solids consisting of particles included in an interparticle matrix [7.13, 7.160 to 7.162]. If the two components have different refractive indices, the mixed solid is anisotropic. If the particles are small enough, their absolute size is of no importance to these considerations.

In the imbibition method, the interparticle matrix is replaced by liquids of different refractive indices. If the refractive index of the imbibition medium is equal to that of the particles, the retardation is zero or, in the presence of a residual birefringence component, a minimum (Fig. 7.17, see page 256).

The birefringence measured Δn_{meas} is composed of form and residual birefringence components. The amount of form birefringence directly corresponds to the difference in refractive index between particles and interparticle matter. The residual birefringence represents the particles' intrinsic birefringence.

With different shapes and orientations of the particles in the mixed solid, six types of birefringence can be distinguished (Fig. 7.17). The analysis of measured retardation curves can be correlated with these types to provide indications to the shape, orientation and intrinsic birefringence of the particles.

The imbibition liquids (Table 7.24) should cover the refractive index range of interest as precisely as possible. They must not attack or otherwise change the object under examination (say, by swelling, orientated absorption or chemical

Table 7.24. Some common imbibition liquids, acc. to the literature

Liquid	Refractive index n_D
Ethyl ether	1.357
Ethyl alcohol	1.361
Chloroform	1.449
Oil of turpentine	1.472
Paraffin oil	1.482
Terpineol	1.483
Toluene	1.496
m-Xylene	1.497
Benzene	1.501
Cedar wood oil	1.510
Methyl benzoate	1.517
Oil of cloves	1.533
Benzyl alcohol	1.541
Carbon disulphide	1.628
Mixtures of ethyl alcohol, benzyl alcohol, monobromonaphthalene	1.361 1.541 1.662

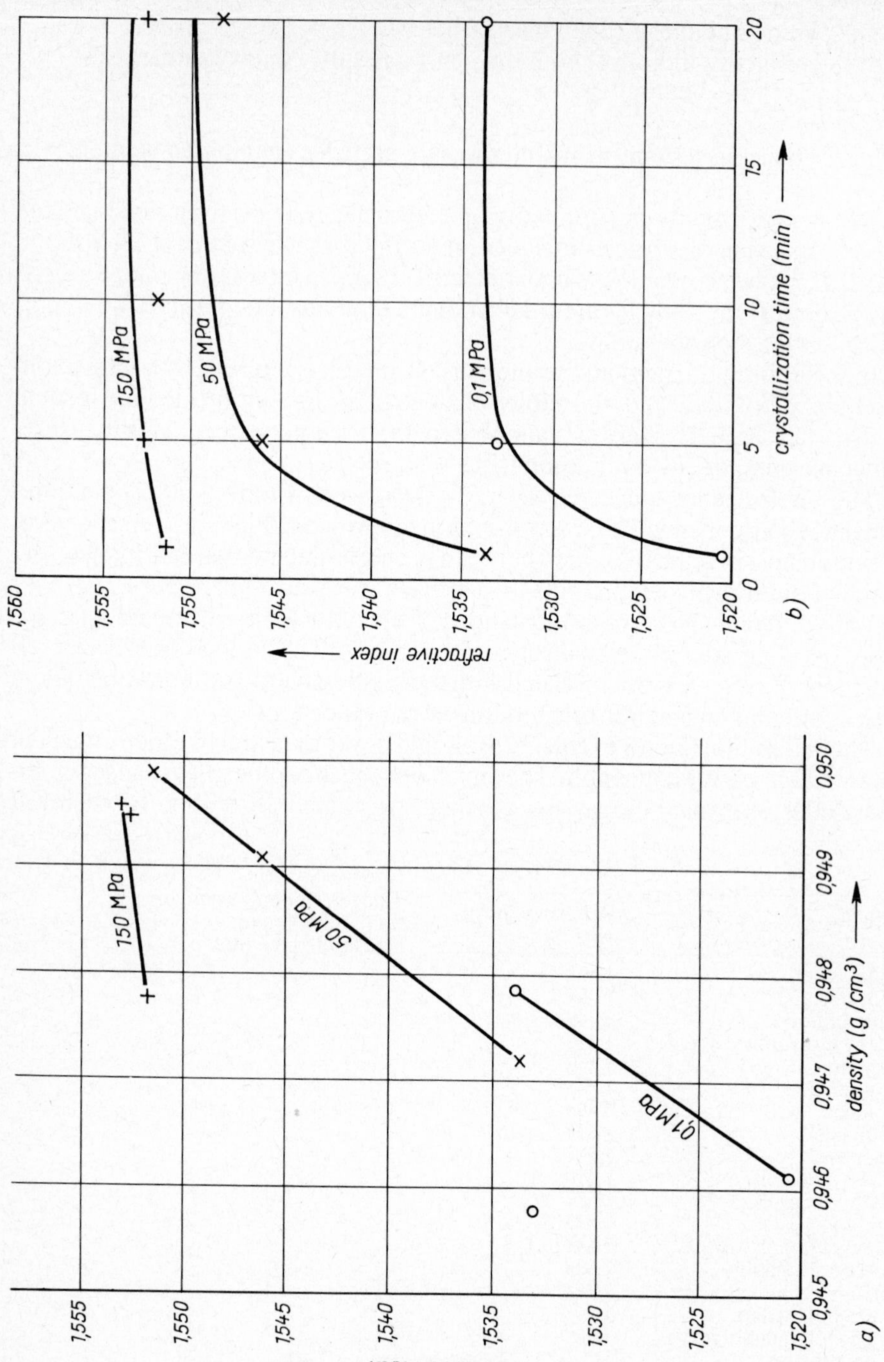

refractive index
1,520
1,525
1,530
1,535
1,540
1,545
1,550
1,555
0,945
0,946
0,947
0,948
0,949
0,950
density (g/cm³)
150 MPa
50 MPa
0,1 MPa
a)
refractive index
1,520
1,525
1,530
1,535
1,540
1,545
1,550
1,555
1,560
0
5
10
15
20
crystallization time (min)
150 MPa
50 MPa
0,1 MPa
b)

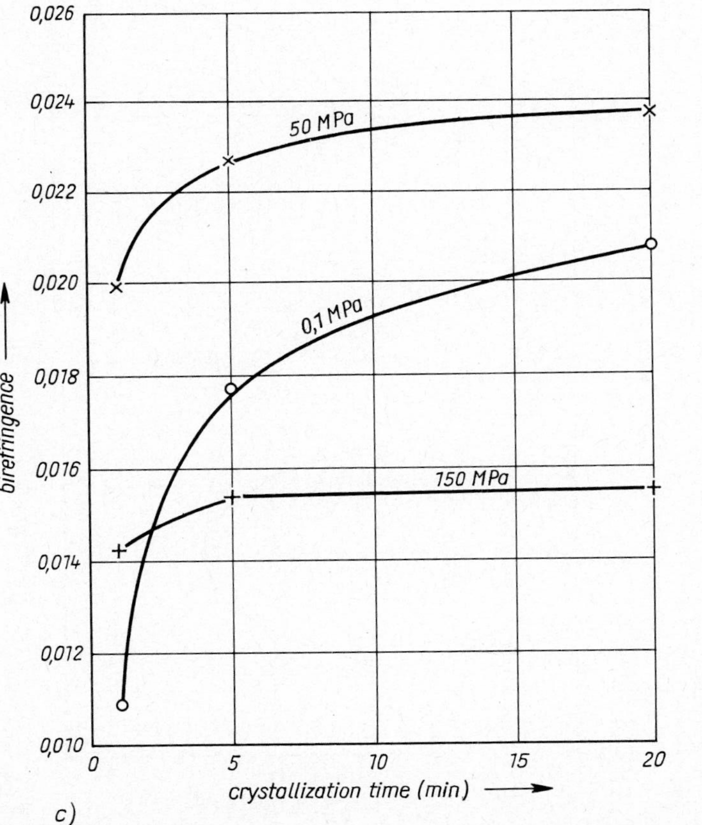

Figure 7.16. Relations between optical and material parameters in polyethylene at different pressures (acc. to [7.139])
a) refractive index as function of density
b) refractive index as function of crystallization time
c) birefringence as function of crystallization time

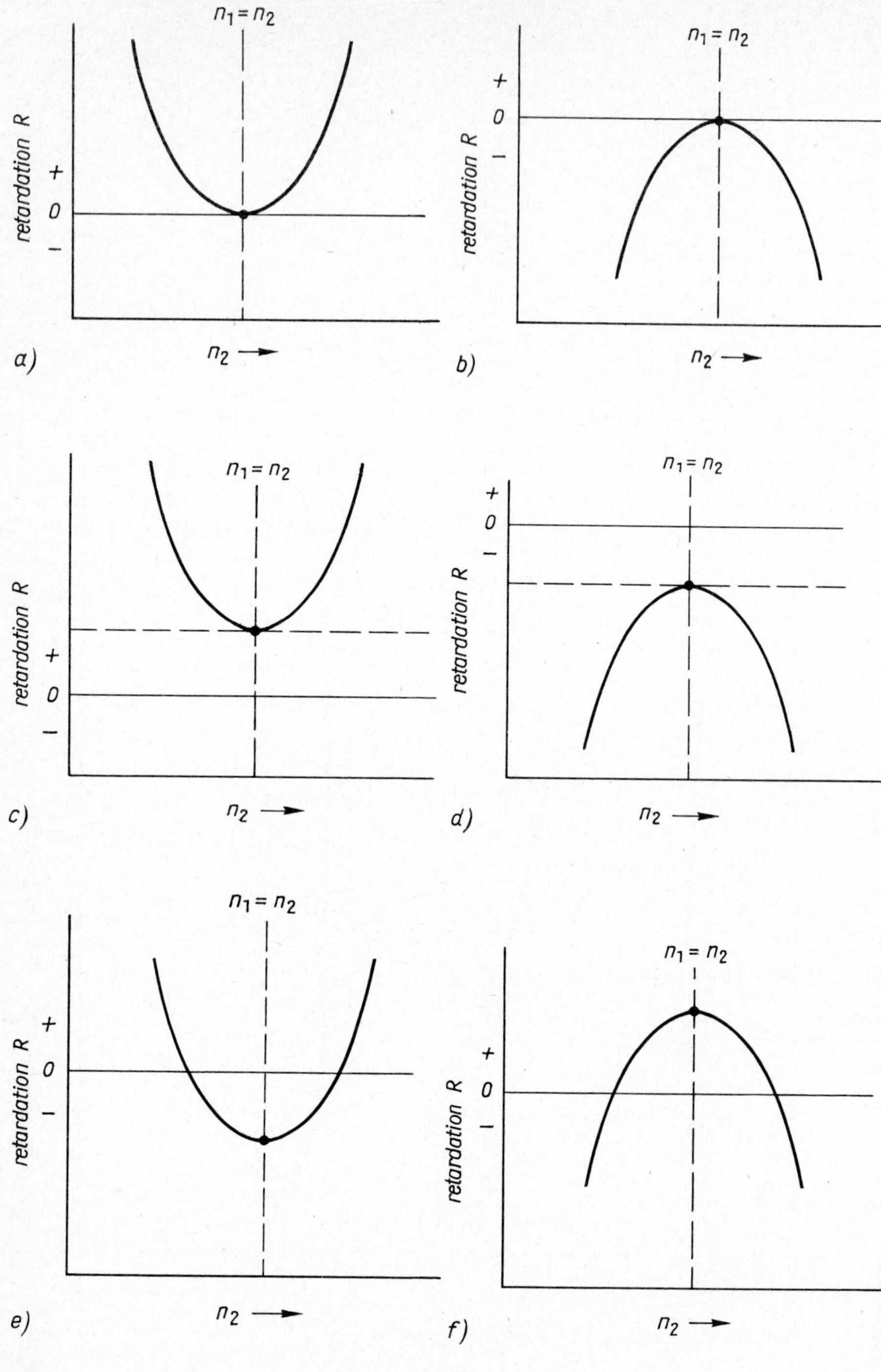

n1 = n2
retardation R
+
0
−
n2
a)
b)
c)
d)
e)
f)

change). To check this, one can perform comparative retardation measurements with various imbibition liquids having equal refractive indices. If retardations also are equal, it means that there are no disturbing effects.

The liquids should easily enter the specimen and easily be removable by washing with water. Liquids entering with difficulty are mixed with a low-viscosity carrier liquid of high vapour pressure, which is then allowed to evaporate (e.g. monobromonaphtaline with xylene). Imbibition examinations are possible on specimen slides or in microcells. A combination with orientation procedures enhances the scope of information provided by the method [7.13].

7.3 Phase transformations

7.3.1 Polymorphic transformation

Microscopical measurements can register very sensitively the structural changes involved in polymorphic transformations. As the temperature approaches the transformation point, the optical data indicate three ranges of lattice change (Tables 7.25, 7.26). Depending on the lattice examined, the boundaries of the three zones lie at a smaller or greater distance above or below the transformation point.

At the transformation point, the behaviour of the temperature curve of the optical parameters indicates whether the transformations are first or second order ones (Fig. 6.24).

Varying with thermodynamic and kinetic conditions, phase transition may be substantially delayed. In that case, modifications can be detected in temperature

Figure 7.17. Types of birefringence in examinations using imbibition method

a) positive form birefringence, no residual birefringence; **structural elements:** isotropic rods parallel to orientation axis, or isotropic tablets, normal perpendicular to orientation axis

b) negative form birefringence, no residual birefringence; **structural elements:** isotropic rods perpendicular to orientation axis, or isotropic tablets, normal parallel with orientation axis

c) positive form birefringence, positive residual birefringence; **structural elements:** optically positive rods parallel with orientation axis, or optically negative tablets, normal perpendicular to orientation axis

d) negative form birefringence, negative residual birefringence; **structural elements:** optically positive rods perpendicular to orientation axis, or optically negative tablets, normal parallel to orientation axis

e) positive form birefringence, negative residual birefringence; **structural elements:** optically negative rods parallel to orientation axis, or optically positive tablets, normal perpendicular to orientation axis

f) negative form birefringence, positive residual birefringence; **structural elements:** optically negative rods perpendicular to orientation axis, or optically positive tablets, normal parallel to orientation axis

n_1 refractive index of *structural elements*
n_2 refractive index of *imbibition liquid*

Table 7.25. Microscopical symptoms of solid phase transformations

Temperature range	Polymorphous transformation	Melt	Decomposition
Some deg. to abt. 40 deg. before phase transition	Birefringence change grows faster, no longer linear. The change is irreversible.		
Beginning of phase transition	Curve of optical data shows abrupt cut or knee, state of extinction changes, intensive development of microfissures	Transition to isotropic melt connected with change of transmittance, continuous process from edge rounding to homogeneously fused drop	Turbidity or intensive development of microfissures, separation of blisters in silicone oil specimen, state of extinction changes (frequently speckled or incomplete, or isotropy)
Range after phase transition	New phase has characteristic optical properties that change linearly with temperature.	Isotropic melt	New solid phase has fine crystals.

Table 7.26. Different optical constants of four modifications of *HMX* explosive (cyclotetramethylene tetranitramine), acc. to the literature

Modification	n_α	n_β	n_γ	Birefringence	Symmetry
I	1.589	1.594	1.73	0.141	Monoclinic
II	1.563	1.564	1.73	0.167	Orthorhombic
III	1.537	1.585	1.666	0.129	Monoclinic
IV	1.607 (o)	1.566 (e)		0.041	Hexagonal

ranges in which they are not stable. Transformation itself will essentially proceed very rapidly and spontaneously once it has started [7.166].

Irrespective of their diversity of appearance (Table 6.29), polymorphic transformations observed through the microscope may be divided into two types.

(1) Transformation is indicated by the sudden formation of fine cracks throughout the crystal ($K_2Cr_2O_7$, NH_4NO_3).

(2) Transformation is effected through fast-progressing transformation fronts (KNO_3, NH_4Cl).

7.3.2 Solid-state reactions

7.3.2.1 Solid-solid reactions

Reactions in the solid phase can be investigated microscopically if the grain size of the initial or reaction products is above 1 mm. To trace the reaction, one observes the proportional change of optical properties or directly measures the thickness of the growing or diminishing reactants. The study of the conversion

$$PbS + CdO \rightarrow PbO + CdS$$

in a temperature range of 463 to 690 °C by Leute [7.167] may serve as an example (Fig. 7.18). PbS was present as a monocrystal, and CdO as a polycrystalline aggregate. The thickness of the reaction layer was determined at different times during the reaction. Fig. 7.18 shows the curves of measurements of the volume and surface reactions. By evaluation one obtains the activation energy and the order of the reaction (Table 7.27).

A comparison between experimental reaction thicknesses and theoretical values reveals distinct deviations, allowing to conclude on the reaction mechanism. The two reaction products, PbO and CdS, deposit successively in separate sub-layers. The reaction layer is formed by mutual cationic diffusion; there is no substantial anionic diffusion. Diffusion into PbS proceeds four times as fast as that into CdO, i.e. the polycrystalline layer adjacent to the PbS crystal contains residual unconverted PbS.

At higher temperatures, the reaction mechanism changes, and the reaction product, analyzed microscopically, is found to contain $PbSO_4$.

7.3.2.2 Decomposition reactions

According to Heide [7.169], decomposition reactions in which a liquid or gaseous phase is released [7.168 to 7.172] may be divided into three types with clearly distinguished microscopical properties (Table 7.28). The variation of optical data as the decomposition point or range is being approached clearly indicates structural changes even before the separation of the new phase is noticeable (Table 7.25).

Depending on the bonding relationships of the separated phase in the lattice, the process takes place within a more or less narrow interval (Fig. 7.19). If the separation causes the old lattice to break down, the process starts at a characteristic temperature. In case of continuous supply from a widely-spaced framework lattice, the process will spread across a wider temperature span. With fast heating rates, one may be misled into noticing a decomposition interval in the first case, too, if the point of decomposition is passed without the process being completed.

Decomposition of crystal hydrates

(1) Decomposition of $2CaSO_4 \cdot 3Na_2SO_4 \cdot 2H_2O$

Table 7.27. Evaluation of kinetic microscopical measurements of the reaction PbS + CdO → PbO + CdS, acc. to Leute [7.167]

	Volume reaction	Surface reaction
Activation energy	40.3 ± 3.0 39.5 ± 3.0	31.0 ± 2.5
Time law	2nd order reaction	Zero order reaction

Table 7.28. Reaction types of dehydration of hydrates acc. to Heide [7.169] and their microscopical symptoms

Reaction type	Example	Microscopical symptoms
$E^+ = \Delta H$ The dehydrated fabric is unstable at first, turning stable at higher temperatures (exothermic). High E^+ values of the induction range, dependence of reaction process on external partial pressure	Gypsum, $CaSO_4 \cdot 2H_2O$	Surface decomposition by oriented crystallization, disordered new phase simultaneous with long-range ordering
$E^+ > \Delta H$ Very high E^+ values of the induction period (energy barrier) lead to high stability. Speed is determined by nucleation in the intermediate layer.	Kieserite, $MgSO_4 \cdot H_2O$	Reaction from margin inward (with distinct boundary surface)
$E^+ < \Delta H$ Evaporation predominant	Epsomite, $MgSO_4 \cdot 7H_2O$	Spontaneous reaction, lattice destruction or heavy disorder
$E^+ \ll \Delta H$ Evaporation dominating	Zeolites, addition compounds	Lattice preserved, solvate molecule removed

The course of the decomposition reaction corresponds to the following reaction equations:

$$2CaSO_4 \cdot 3Na_2SO_4 \cdot 2H_2O \xrightarrow{(150\,^\circ C)} 2Na_2SO_4 \cdot CaSO_4 + Na_2SO_4 + 2H_2O$$

$$Na_2SO_4 \cdot CaSO_4 \xrightarrow{(150\,^\circ C)} CaSO_4 + Na_2SO_4$$

Figure 7.18. Microscopical investigation into the conversion of PbS + CdO → PbO + CdS (acc. to [7.167])

a) schematic course of the reaction between A and B; growth of reaction layer AB after different times (I, II, III) b) measuring curves for volume reaction c) measuring curves for surface reaction

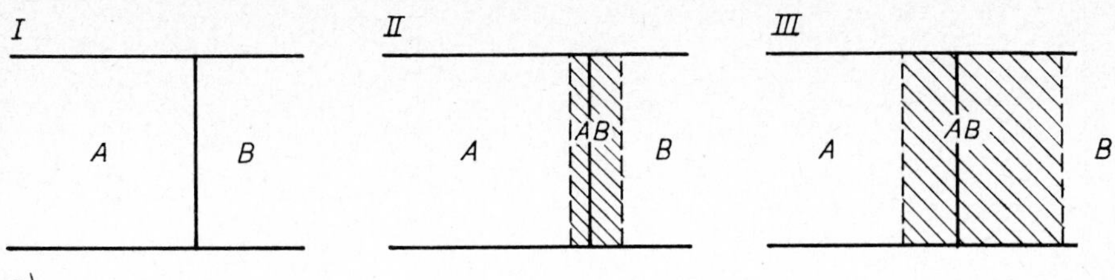

I
II
III
A
B
AB
a)

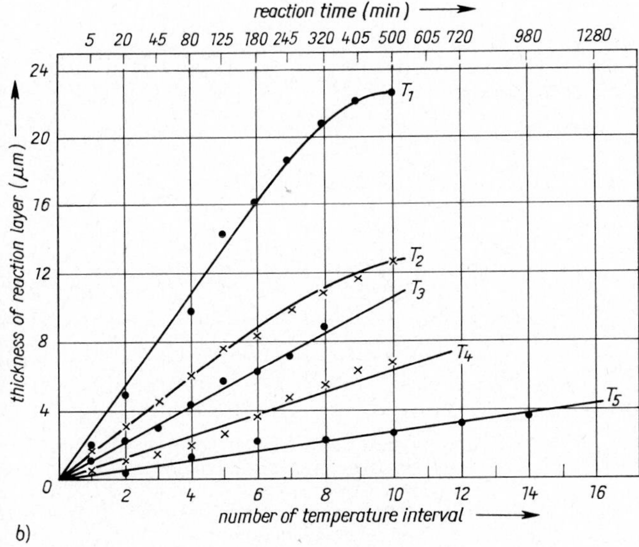

reaction time (min) →
5 20 45 80 125 180 245 320 405 500 605 720 980 1280
thickness of reaction layer (μm) →
24
20
16
12
8
4
0
T_1
T_2
T_3
T_4
T_5
2 4 6 8 10 12 14 16
number of temperature interval →
b)

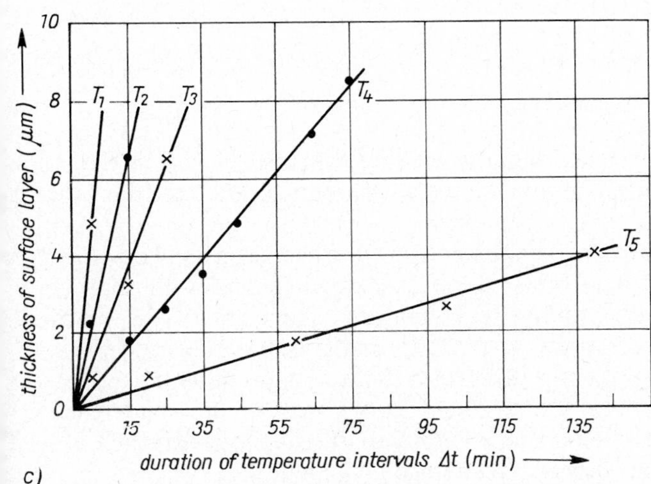

thickness of surface layer (μm) →
10
8
6
4
2
0
T_1
T_2
T_3
T_4
T_5
15 35 55 75 95 115 135
duration of temperature intervals Δt (min) →
c)

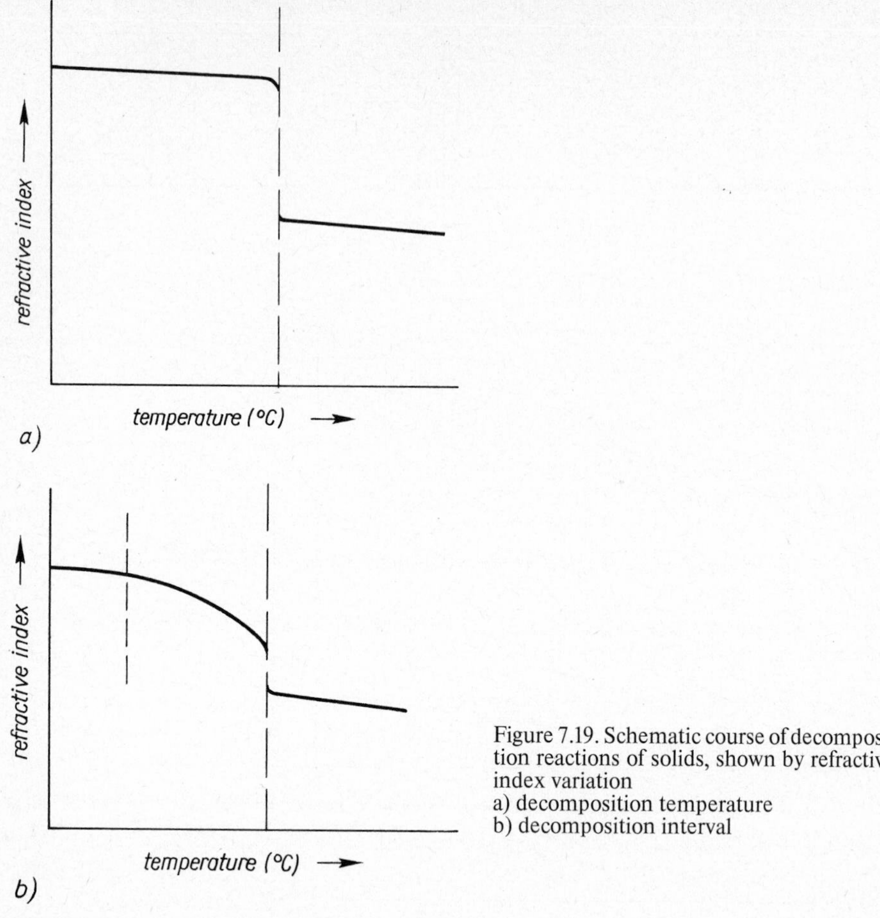

Figure 7.19. Schematic course of decomposition reactions of solids, shown by refractive index variation
a) decomposition temperature
b) decomposition interval

Observation of the substance with the hot-stage microscope will yield the decomposition temperatures, identify the reaction products and trace the various stages of the process.

At 150 °C, dehydration sets in spontaneously, together with a breakdown of the crystal lattice. The initial phase of acicular crystals characterized by homogeneous optical extinction is replaced by finely grained aggregates of the reaction products, forming a pseudomorph after the former crystal needles and showing oriented intergrowth mainly in parallel with them. Grain size in the aggregate increases with temperature. The oriented intergrowth of the solid reaction product suggests structural relationships with the lattice of the initial phase. Here, the nucleation of the new phases is encouraged energetically.

Above 500 °C, the glauberite portion of the reaction product is converted into a finely intergrown aggregate of Na_2SO_4 and $CaSO_4$. The pseudomorphic needles more or less disintegrate into elongated or irregularly shaped aggregates.

(2) Decomposition of $MgSO_4 \cdot 7H_2O$

The decomposition of hydrates containing much water is not always a single process but may take place in steps at different temperatures. Loose grains in silicone oil, placed on a hot stage and subjected to extremely slow heating, permit to observe these processes in vivid detail. Monocrystals of $MgSO_4 \cdot 7H_2O$ at first exhibit a distinct range of first dehydration. Further dehydration has rather blurred ranges (Table 7.29). Varying in parallel with this process is the optical behaviour of the crystal between crossed polarizers. The first dehydration phase comes together with a speckled brightening of the former homogeneous extinction. Individual lattice regions vary in extinction both between themselves and inside themselves. Marked cracks form in the crystal and, in case of a slightly faster heating rate, will lead to decrepitation. The combination of both observations suggests a structural interpretation by which the lattice bonding of the first-released water differs from that of the water released later. After the first dehydration stage, the lattice network is heavily disturbed so that no exact energetic definition of the activation energies for releasing the remaining water can be expected. Depending on the degree of disorder attained, dehydration adopts a statistical character.

These microscopical results are in accordance with those of structural investigations into $MgSO_4 \cdot 7H_2O$ dehydration, made by Heide [7.169].

(3) *Decomposition of* $MgSO_4 \cdot H_2O$ *acc. to Heide* [7.169]

For energetic reasons, the decomposition of kieserite ($MgSO_4 \cdot H_2O$) differs from the examples described above (Table 7.28).

About 40 deg. below the decomposition temperature, birefringence shows a marked, irreversible drop, the crystal still being pellucid.

Table 7.29. Microscopical measurements of thermal dehydration of $MgSO_4 \cdot 7H_2O$

Dehydration stages of $MgSO_4 \cdot 7H_2O$ → $MgSO_4$	Dehydration temperatures (°C)			
	Set I	Set II	Set III	Set IV
$7H_2O$ → $6H_2O$	68 to 73	69 to 76	65 to 76	68 to 75
$6H_2O$ → $2H_2O$	85 to 92.5 107 to 108 114 to 119 126 to 137	86 to 90 98 to 102 109 to 115 120 to 140	84 to 100 105 to 115 118 to 139	85 to 97 103 to 108 121 to 210
$2H_2O$ → $0H_2O$	 223 to 315	155 to 175 220 to 312	160 to 212	162 to 210

Decomposition starts at the periphery of the crystal, progressing towards the centre with a distinct boundary face. Contrary to $CaSO_4 \cdot 2H_2O$, no surface nucleation is observed. In some cases, decomposition proper is preceded by striation (twinning?), which also progresses from the margin to the centre.

Fig. 7.20 shows the change of the indicatrix due to decomposition. n_β undergoes the most significant change. In the lattice, it is coincident with the axis of the H_2O dipoles.

After conversion, $MgSO_4$ appears to be a homogeneous, optically negative biaxial crystal. The axial angle is 0° immediately after conversion and increases afterwards. Lattice vibration directions are rotated by 45° relative to the initial structure.

(4) Decomposition of $CaSO_4 \cdot 2H_2O$ *acc. to Heide [7.3]*

The decomposition of a gypsum monocrystal starts with nucleation of the newly forming γ-$CaSO_4$ (soluble anhydrite) at various energetically favourable surface centres. As crystallization proceeds, these centres expand and finally merge. The growth of the new phase is distinctly a function of the crystallographic orientation of the initial structure. Nucleation continues across the twinning plane without being disturbed. The nuclei in both elements of the twin have equal optical orientation.

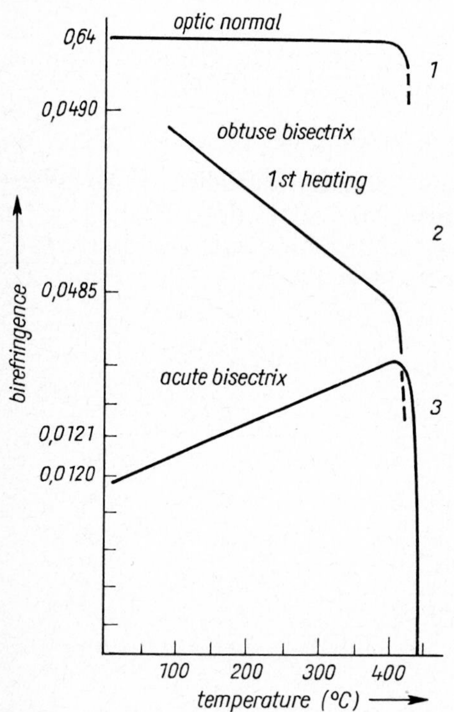

Figure 7.20. Variation of birefringence in kieserite with increasing temperature (acc. to [7.169])

1 section perpendicular to optical normal (no detectable change up to about 40 deg before proper turbidity sets in, $n_\gamma - n_\alpha = 0.064$)
2 section perpendicular to obtuse bisectrix (with increasing temperature, birefringence decreases linearly until 40 deg before decomposition proper, then steep drop)
3 section perpendicular to acute bisectrix (birefringence increases linearly with temperature and drops sharply short of the transformation. In that section, the new phase appears optically uniaxial immediately after the transformation.)

The formation of the new phase attenuates the intensity of light transmitted by the crystal. Fig. 5.6 is an automatic scan of the drop of intensity (cf. 5.2.2), showing continuous opacification as a function of time, which bears evidence to the gradual decomposition of $CaSO_4 \cdot 2H_2O$. The asymmetrical curve is typical for highly time-dependent reactions. Kinetic analysis is possible because turbidity increases in proportion to the expansion of the reaction product per unit of area. The final transmittance serves as a reference for complete conversion. The kinetic evaluation of intermediate values vs. time from the scan is reliable if heating is linear and if one uses the approximately linear voltage range of the photoelectric current.

Evaluation is by a modified Arrhenius equation:

$$\frac{m\omega}{T} = A_0 e^{-E+/RT}, \quad k = \frac{m\omega}{T}.$$

m heating rate
ω weight loss per unit area, here corresponding to increase of turbidity per unit area
A_0 constant
k rate constant
T temperature
E^+ activation energy

In a representation with log ($m\omega/T$) and $1/T$ as coordinates, the gradient of the straight line is a measure of the activation energy.

The boundary layer of the reaction has an interesting structure. According to Heide, it can be interpreted as a disordered transition range with a distinct long-range order relative to the initial structure. The [001]-direction of the γ-$CaSO_4$ has a preferred orientation parallel to the [001]-direction of the initial crystal.

The decisive factor in the process of reaction is the reaction heat or, in other words, the temperature coefficient of the vapour pressure. The rate of the reaction is determined by the dynamics of water separation at the boundary face of the reaction area. This accounts for the marked dependence of the course of reaction on the external partial vapour pressure.

Decomposition of addition compounds

A wide range of decomposition and continuous solvate separation are characteristic of a »zeolitic« bond of the solvate molecules. This is clearly obvious in a silicone-oil loose-grain specimen on the hot stage, contrary to the dehydration of hydrates having structurally incorporated H_2O molecules, for example.

The decomposition of $MgSO_4 \cdot 2CH_3OH$ may serve as an example for many similar compounds. Tables 7.30 and 7.31 and Fig. 7.21 demonstrate the variation of optical and structural properties during methanol separation. Measurement of the change of retardation is particularly suitable for kinetic studies. The thin-tabular habit of the crystals meets the requirement for a two-dimensional observation

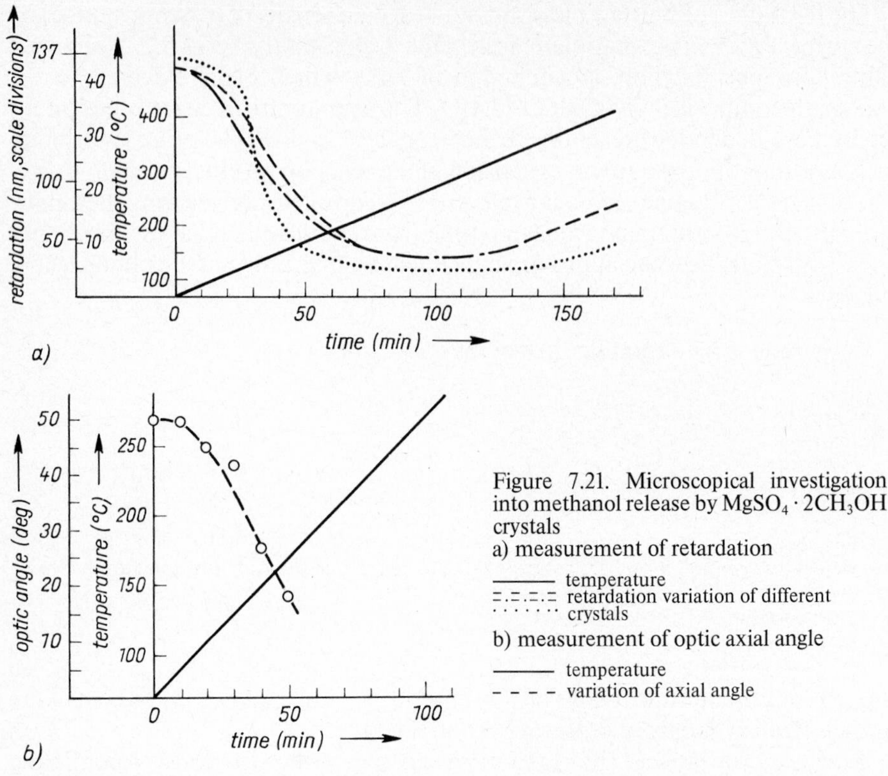

Figure 7.21. Microscopical investigation into methanol release by $MgSO_4 \cdot 2CH_3OH$ crystals

a) measurement of retardation

——— temperature

-·-·-·- ······· retardation variation of different crystals

b) measurement of optic axial angle

——— temperature

- - - - variation of axial angle

of the process. With a linear heating rate, activation energy may be calculated by a modified Arrhenius equation acc. to Flora [7.171]. In a graph with log h/log w_r as the ordinate and $1/T \cdot 10 \log w_r$ as the abscissa, the measured values form a straight line.

h rate of retardation change with time
w_r residual retardation
T temperature

Table 7.30. Change of optical constants of $MgSO_4 \cdot 2CH_3OH$ by decomposition

Parameters	Initial products (filled host lattice)	Decomposition products (host lattice)
Refractive index	1.47	1.44 (minimum)
Birefringence	0.002	0.000
Axial angle	$2V = 52°$	$2V = 0°$
Light intensity at 45° position	(rel. = 100%)	0 %

Table 7.31. Structural change of $MgSO_4 \cdot 2CH_3OH$ by decomposition

Initial structure	Decomposition structure	Temperature (310 to 338 °C)	Final structure
Cubic $MgSO_4$ framework lattice with cavities, large unit cell, orthorhombic-monoclinic downgrading of symmetry by irregular incorporation of polar guest molecules, bonded to host lattice by H bridges, bonding strength depending on position in the lattice, heterodesmic bonding of host lattice, with heteropolar and homopolar shares	**Type I** Guest components are removed without substantial destruction of host lattice. Cubic $MgSO_4$ lattice present in initial structure (solvate molecules slowly being removed). Optically isotropic pseudomorphs after the initial structure (continuous transition depending on decomposition rate). **Type II** After fast removal of guest molecules, host lattice is heavily deformed and disordered, partly roentgen-amorphous (ordered regions of several hundred Å). Residual birefringence still present, may differ between individual lattice regions of a grain. Weakly anisotropic pseudomorphs after the initial structure. **Type III** Under exposure to air, pseudomorphs after the initial structure are formed, consisting of $MgSO_4$ hydrates (?). Grain shapes filled by fine-grained, weak decomposition products.	Exothermic	Orthorhombic $MgSO_4$ lattice $a_o = 5.15$ Å $b_o = 7.89$ Å $c_o =$ Å *New formation:* starting at grain margins of types I and II. As temperature increases, orthorhombic $MgSO_4$ grains get coarser. Pseudomorphs form after the initial form.

The intersection of the straight line with the ordinate indicates the order of reaction, and the gradient represents the activation energy.

The values obtained slightly vary from grain to grain within a range of 6 to 9 kcal/mole. Sometimes even parts inside one grain differ in the rate of retardation change and, thus, in activation energy.

The fact that decomposition rates at a time t differ between individual grains in a batch of precipitates and even between various lattice regions in a grain attaches particular importance to the microscopical analysis of such compounds. Other than in X-ray, IR spectroscopical and derivatographical methods where results are integrated over all grains of a specimen, microscopy provides structural information *grain by grain* or *lattice region by lattice region.*

Photolysis

For studying the decomposition of photosensitive substances, microscopy provides two possibilities:

(1) Measurement of decomposition rate by the growth of stationary gas bubbles at the grain periphery (in favourable cases) (Fig. 7.22, see Annex). A study of deviations between series of measurements made on several grains will reveal how representative this auxiliary quantity is.
(2) Measurement of decomposition rate under monochromatic irradiation of different wavelengths.

7.3.3 Phase transformations via the liquid phase

The transformation of solid phases via an intermediate solution phase involves the disintegration of the initial lattice and the build-up of the new structure. Both

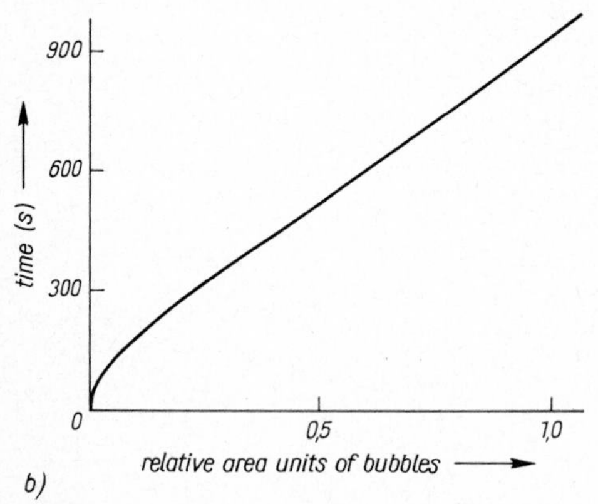

Figure 7.22. Measurement of decomposition rate by size increase of gas bubbles formed in monocrystal-silicon oil specimens (diazonium salts)
a) course of the process shown on micrographs 1 to 4 (see Annex)
b) curve of bubble growth

steps of the process, i.e. dissolution and crystallization, can be observed through the microscope.

Double conversion: $CaSO_4 + 2KCl \xrightarrow{(H_2O, NH_3)} K_2SO_4 + CaCl_2$

In parallel with dissolution, a K_2SO_4 reaction layer forms on the gypsum monocrystals. The increase in K_2SO_4 concentration is proportional to the increase in layer thickness with time (Fig. 7.23). According to [7.173], the microscopically determined curve reveals that layer thickness at first grows linearly; later the growth rate reduces constantly. The dependence of reaction rate on layer thickness indicates that diffusion of the solution through the layer is the rate-determining step. The calculated activation energy of 13.4 kcal/mole corresponds to that inhibited diffusion.

Industrial production of K_2SO_4

The mechanism of K_2SO_4 formation from mixed water-methanol solutions in the potash industry can be explored by examination with microcells. There are two different reaction mechanisms depending on the $MgCl_2$ content of the solution. If it is high, conversion is retarded, and the K_2SO_4 produced will be relatively coarse-grained. Absence or shortage of $MgCl_2$ leads to pseudomorphism after schoenite. New formation of K_2SO_4 starts at the grain boundary and progresses towards the centre. The pseudomorphs have low strength only; they easily disintegrate into fine K_2SO_4 particles, providing an undesirably high fine-grain fraction. Discontinuous microscopical studies performed by Scherzberg and Mohr have confirmed the results of continuous microcell experiments.

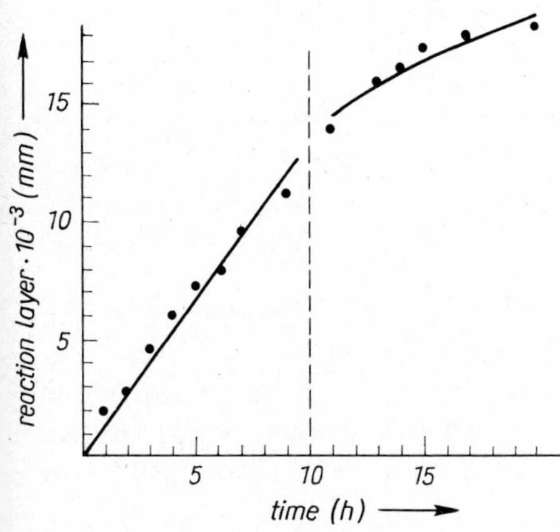

Figure 7.23. Investigation into the double conversion of $CaSO_4 + 2\,KCl \rightarrow K_2SO_4 + CaCl_2$, by microscopical measurement of reaction layer growth

7.3.4 Solid-liquid phase transitions

7.3.4.1 Melting and solidification processes

In approaching the melting point, a solid phase shows characteristic effects in the microscopical image (Table 7.25). The variation of optical data manifests the change of the lattice even before its breakdown.

Analogously to lattice changes below the melting point, the microscopical image of a melt in a cooling process indicates pre-ordered states in the liquid phase of some substances briefly before solidification. In paraffins and olefins, e.g., such pre-ordering is noticeable by the occurrence of weakly birefringent regions of uniformly cloudy extinction in the optically isotropic liquid phase, in which regions crystallization then starts.

Table 7.32 lists different types of crystallization occurring in the solidification of liquid phases [6.13, 7.174].

7.3.4.2 Dissolution processes

The three-dimensional process of dissolution of solids can be observed in the microscope by two-dimensional changes only. There are two possibilities (cf. Table 7.33) [7.175 to 7.177]:

(1) The measurement of the time-dependent reduction in size of the two-dimensional projected image of the grains supplies information on kinetics (Fig. 7.24).

(2) The measurement of refractive index gradients in the liquid phase reveals the course of substance transportation or the liquid-phase concentration states at the boundary face to the solid phase (Fig. 7.25, see Annex).

The kinetics of grain size reduction

Such measurements are possible under various experimental conditions (Table 7.33).

In an area-vs.-time graph, the measured values more or less follow a logarithmic curve. In a coordinate system with logarithmic scales, the measurements lie on a straight line if dissolution follows a first-order time law (Fig. 7.24). Towards the end of the process, the standard deviation between the measured values strongly increases. The influence of the Zapon lacquer or grease used to fix the grain prohibits reproducible measurements.

Various series of measurements were made starting from differing initial grain sizes, but the difference in size proved to be of no appreciable consequence to dissolution, so that it can be neglected. Therefore, all values are transformed to the initial value of the largest crystal. So far, the kinetic information obtainable from microscopical dissolution curves has not yet been studied theoretically. Two cases have to be distinguished:

Table 7.32. Crystallization symptoms appearing in a solidifying melt on the hot or cold stage, acc. to [6.13]

Type	Crystal configuration	Front configuration	Orientation	Frequency
I	Microcrystalline, no individual crystals visible	Smooth	Often large faces of uniform orientation	Relatively rare
II	Well-formed crystals	Ragged, crystals intergrown	Crystals mostly differ in orientation	Frequent in crystalline substances
III	Monocrystals	Straight	Uniform	Rare
IV	Isolated spherulites, further growth leads to interlocking	Round	Radial crystal orientation	Organic semi-crystalline high polymers or substances of similar structure

Table 7.33. Possibilities of microscopical measurement of dissolution and crystallization processes

Object	Liquid phase		
	stationary	flowing	agitated
Single grain	c	c	c
Grain mixture	c	c	c
Aggregate	c, d	c, d	c, d
Grains in aggregate	d	d	d

c continuous measurement
d discontinuous measurement

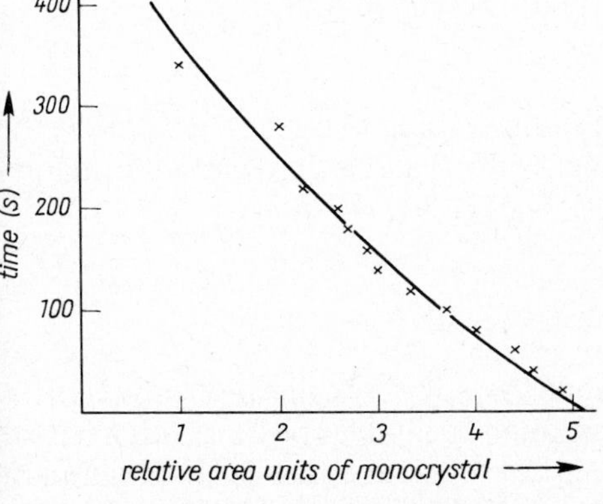

Figure 7.24. Microscopical measurement of dissolution rate of monocrystals, in two-dimensional image

(1) Integral dissolution of the total grain

The volume reduction of the grain as a whole has to be assessed especially where initial grain shapes are irregular. In earlier studies, the weight loss of very large individual grains served as an auxiliary quantity. In microscopical measurement, the weight change must be replaceable by a proportional area change.

The transformed equation then is

$$K = \left(\frac{3d^{2/3}}{f}\right)\left(\frac{A_1^{1/3} - A_2^{1/3}}{t}\right)$$

K specific dissolution rate
d density
f form factor
t time
A_1,
A_2 areas at times t_1, t_2

(2) Linear dissolution in case of congruent forms

The linear shift rate $(a_1 - a_2)$ of an area is described by the equation

$$K = \left(\frac{3d}{f}\right)\left(\frac{a_1 - a_2}{t}\right).$$

Concentration change in the liquid phase

To identify concentration states in the liquid phase, one has to resort to the phase or interference microscope. In the fringe method (cf. Table 3.3), the interference fringes mark lines of equal concentration (Fig. 7.25, see Annex). The concentration zone surrounding a dissolving grain is characterized by the distances measured between the fringes and by their variation in time.

7.3.4.3 Crystallization processes

Analogously as in dissolution, crystallization processes can be registered by the measurement of area and distance changes only. This also provides information on the crystallization mechanism (Table 7.34) [7.178 to 7.203].

Crystallization from solution

(1) Examination in flow-through cells

Suitable microcells (cf. 4.3.2) permit one to study a crystallization process in laboratory apparatus by pumping small amounts of liquid. The flow path from the crystallizer to the flow-through cell should be as short as possible. One example is the study of the course of precipitating crystallization of $NaHCO_3$ under continuous CO_2 supply in the ammonia soda process, by evaluating various series of

Table 7.34. Microscopical examination of crystallization and dissolution processes by transmitted light

Information desired	Parameters measured	1 polarizer	Crossed nicols
Kinetics			
Integral surface change	Areas, transmittances, intensities	×	×
Linear face or edge displacement	Lengths	×	×
Mechanism			
Aggregate size	Lengths	×	×
Aggregate form		×	–
Aggregate fabric		×	×
Crystal morphology			
Crystal habit		×	×
Crystal tracht		×	–
Face configuration		×	–

photomicrographs with varied technical parameters, i.e. intensity of gas injection, stirring speed, concentration and temperature.

Crystallization proceeds in three stages:

Stage I: Fine crystals and aggregates begin to appear after the start of gas supply.
Stage II: Crystals and aggregates grow, with a distinct shift towards aggregates.
Stage III: Many voluminous aggregates exist together with a few individual crystals.

Fig. 7.27 represents these relationships schematically. According to our findings, aggregates grow faster than single crystals as crystallization proceeds. We observed both the addition of substance to aggregate particles and a joining of aggregates into larger ones. Depending on stirring speed, a dynamical equilibrium between aggregation tendency, aggregate strength and turbulence will be established, determining a limit size of aggregates.

(2) Examination in stirred cells

In these studies, stirring speed can be varied in addition. If there is only one solid phase, the grain sizes measured at different times readily correlate with the phases of the system. These conditions prevail in cases of simple crystallization from solutions free from precipitates (analogous to the pure dissolution process of a solid phase).

If it is doubtful whether grains seen on a micrograph belong to the solid initial substances or to the crystallization products, it is essential to employ initial substances of known grain size fraction. The graphs of initial grain size reduction due to dissolution and of growing grain size of the crystallisate have an intermediate

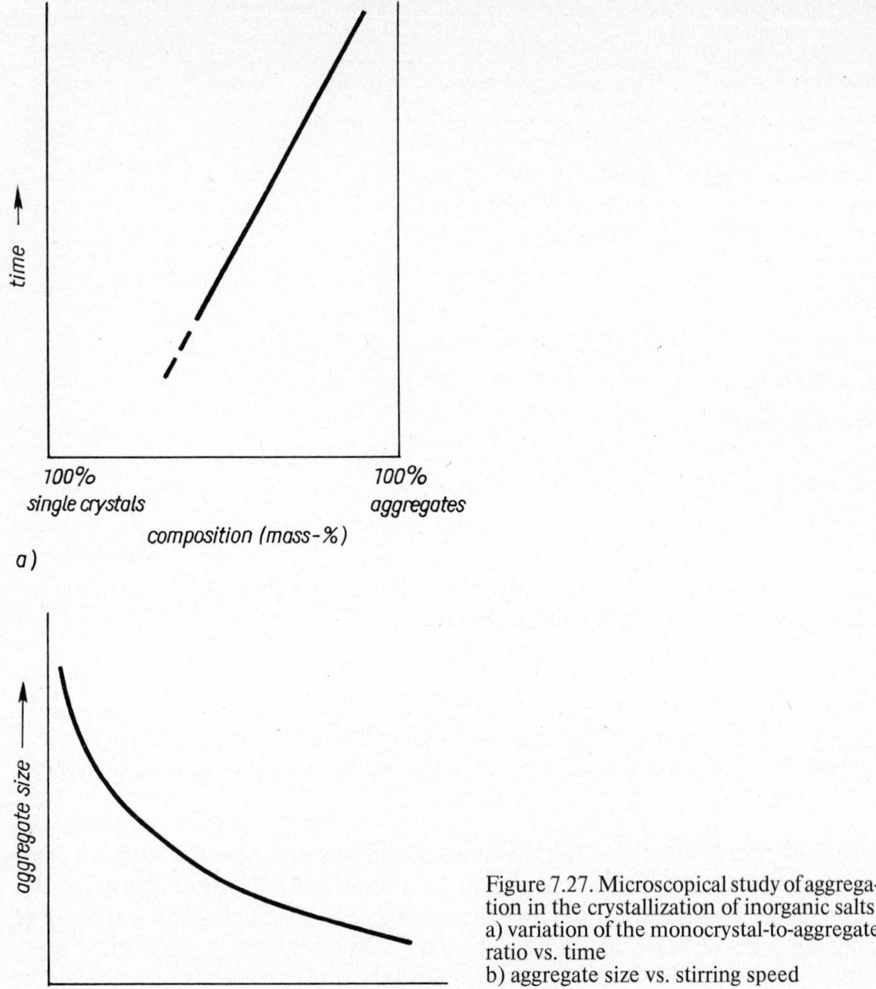

Figure 7.27. Microscopical study of aggregation in the crystallization of inorganic salts
a) variation of the monocrystal-to-aggregate ratio vs. time
b) aggregate size vs. stirring speed

range in which the experimentally established curves do not permit unambiguous evaluation.

Fig. 7.28 illustrates this state of affairs. The evaluation gap, which is not detrimental to the accuracy and significance of the method, can easily be bridged by interpolation in most cases.

Besides this general case, additional criteria may often be found by which to correlate grains shown on micrographs with the solid phases in question. For

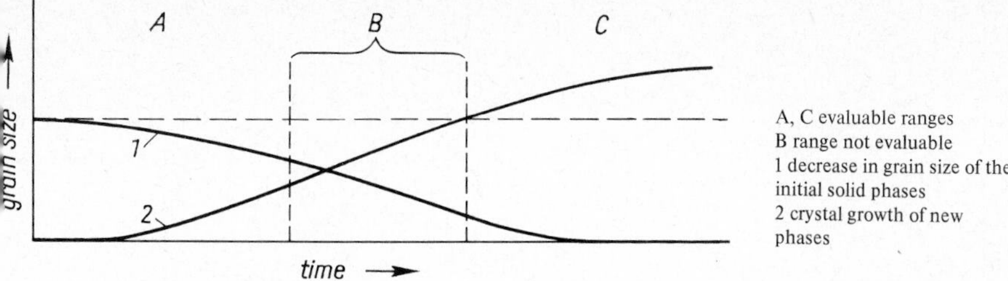

Figure 7.28. Evaluation of micrographs of simultaneous crystallization and dissolution processes in stirred cells

example, aggregates must be considered new formations whenever the microscopical examination of the initial solids registers single grains only. Between crossed polarizers, optically isotropic and anisotropic particles are readily distinguished. The characteristic grain shape of individual phases is another index for correlation.

We also studied the production of conversion salpetre in aqueous solution according to the equation

$$NaNO_3 + KCl \rightarrow NaCl + KNO_3$$

and the production of KCl from potash industry solutions. Here, too, small single crystals and aggregates existed side by side in the early stage, whereas further growth was biased in favour of aggregates. Part of the KCl crystals engaged in oriented intergrowth (Fig. 7.29, see Annex).

Crystallization from the melt

Crystallization from the melt can be examined on the hot or cold stage by varying the specimen temperature or creating a lateral temperature gradient within the visual field. Table 7.32 is a synopsis of various crystallization types.

(1) Linear crystallization measurements

Since the turn of the century, many model concepts of crystallization have been based on microscopical length measurements of growing crystals under isothermal or polythermal conditions.

The method may be demonstrated by the example of spherulite crystallization in organic polymers. The quantity measured is the spherulite radius at different intervals of time. The analysis of the curves of crystallization rate yields relationships with the degree of polymer cross-linkage, cooling rate and other factors of interest to process engineering (Fig. 7.30).

A plot of growth rate vs. crystallization temperature exhibits a distinct maximum, which is a function of polymer branching caused by additives. With strong cross-linkage, the crystallization rate diminishes progressively due to the decrease in viscosity.

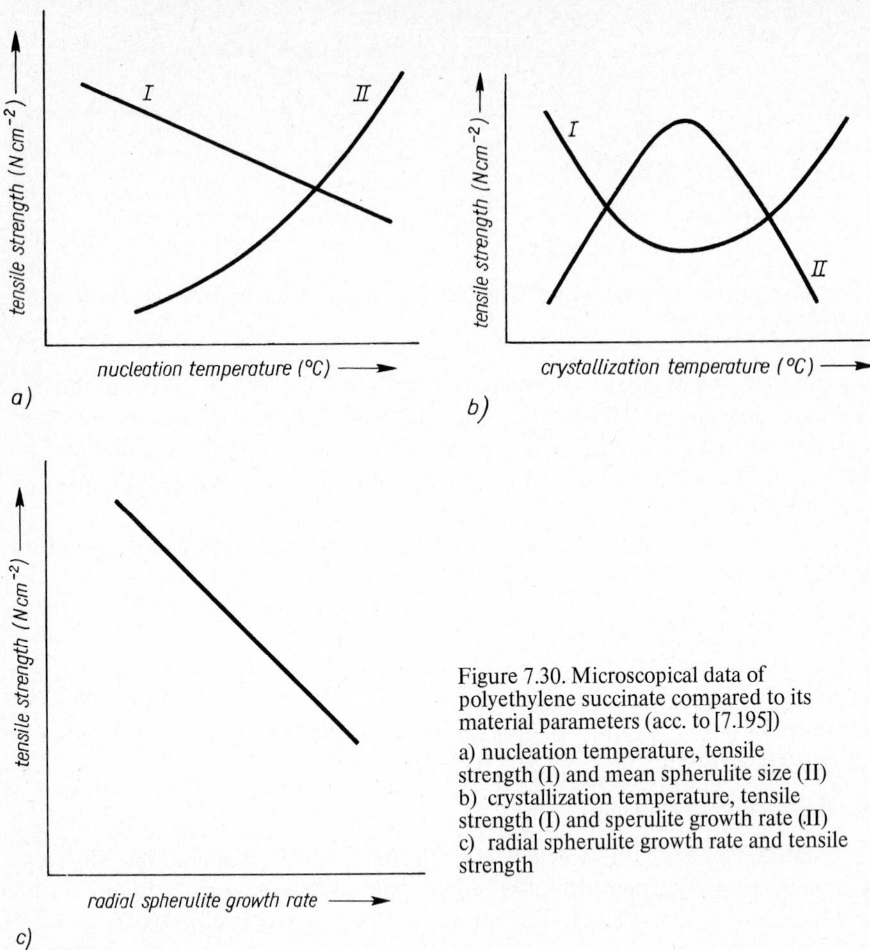

Figure 7.30. Microscopical data of polyethylene succinate compared to its material parameters (acc. to [7.195])
a) nucleation temperature, tensile strength (I) and mean spherulite size (II)
b) crystallization temperature, tensile strength (I) and sperulite growth rate (II)
c) radial spherulite growth rate and tensile strength

By a kinetic analysis of the curves one obtains the rate constant K, or the reciprocal half-time of the crystallization rate, $1/\vartheta$, for fast crystallization.

(2) Integral measurement

The growth of spherulites can be registered by integration over a measuring area (visual field) using a photometer to record the change of intensity of light in the orthoscopic ray path, and crossed polarizers. The curves obtained permit kinetic analysis.

A special case is the investigation of crystallization in melt spinning, say, of polyamide fibres.

The molten polymerisate is pressed through a nozzle, from which a thread is drawn and spooled on to a bobbin. A microscope installed horizontally in the spinning well permits retardation measurements at different distances from the nozzle, by which it is easy to register the orientation processes taking place in melt solidification within fractions of a second. The measurements allow conclusions on the change of birefringence as a function of time after the thread emerges from the nozzle. The increase in birefringence from zero immediately behind the nozzle to values around 0.006 indicates the increase of crystalline fractions in the structure during solidification.

7.4 Grain structure of aggregates

Solid aggregates are important chemical products, including granulates, sintering and solidification products, and manufacturing materials. Their grain structure can be characterized by three groups of parameters identifiable by microscopical methods (Table 7.35) [7.204 to 7.230].

The scope of information obtainable depends on the grain size of the aggregate constituents.

With *grain sizes* $>$ *1* µ*m*, each grain can be measured individually. Integral measurement over a larger aggregate region is possible, e.g. by microphotometric orientation measurement with crossed polarizers.

Grain sizes $<$ *1* µ*m* will yield integral information only on the orientation and phase distribution per unit area, provided that the properties of constituents

Table 7.35. Microscopically identifiable parameters characterizing the grain fabric of aggregate compounds

Parameter	Phase composition of aggregate	Fabric elements	Example
Orientation distribution (texture) of grains or crystals	Heterogeneous or homogeneous	Crystals, crystallites, grains, crystal faces, edges, longitudinal axes, crystallographic or morphological axes	Fibre array in high polymer composites
Phase distribution (special case: pore distribution)	Heterogeneous (homogeneous)	Crystals, crystallites, grains, pores	Phase distribution in multiphase granulate
Type of intergrowth	Heterogeneous or homogeneous	Faces, edges	Spherulitic texture in high polymers

differ measurably. Grain intergrowth will then defy investigation. With these minute grain sizes and with opaque (reflected-light) objects, microscopical methods of fabric analysis are less important than other techniques (e.g. X-ray diffraction). Microscopy has its main field of application in the study of coarser-grained fabrics, primarily of transparent objects.

Etching and selective staining enhance structural differences observed by reflected light.

Loose aggregates (such as filtration residues) can be embedded in hardening media to allow fixation and investigation of their grain distribution.

7.4.1 Orientation distribution

The distribution of orientation can be analyzed by measuring optical and morphological reference directions.

Optical reference directions:	optic axes, directions of principal refractive indices
Morphological reference directions:	face normal, edges (face intersections), longitudinal axes of needles and fibres, preferred linear directions, oriented etching pits

A mathematical description of the orientation distribution of the crystallites in an aggregate makes use of two orthogonal reference systems, with coordinates relating to the *specimen* and to the *crystallite*, respectively. The orientation of a crystallite (grain) within an aggregate is then described by the angle of rotation that transforms the coordinate system of the specimen into that of the crystallite (Fig. 7.31).

By combining the results of three planes of observation normal to each other (or their micrographs), one obtains a three-dimensional picture of the fabric. Orientation distribution can then be measured by integral or differential methods. A special case is presented by the spatial arrangement of ordered and orderless regions in partially crystalline organic high polymers. Here, microstructure and fabric analysis overlap, since the size of these uniform regions varies within wide limits.

Special application of fabric orientation measurements: Testing of air filter fibre mats in the chemical industry [7.218]

In ventilating installations, mechanical air filters are frequently used to separate low concentration impurities. These are webs of non-woven fabric consisting of fibres of suitable diameter and packing structure, arranged in air filter walls or as rolled strips. The point of interest is whether regeneration of such mats changes their structural orientation and thus affects their performance.

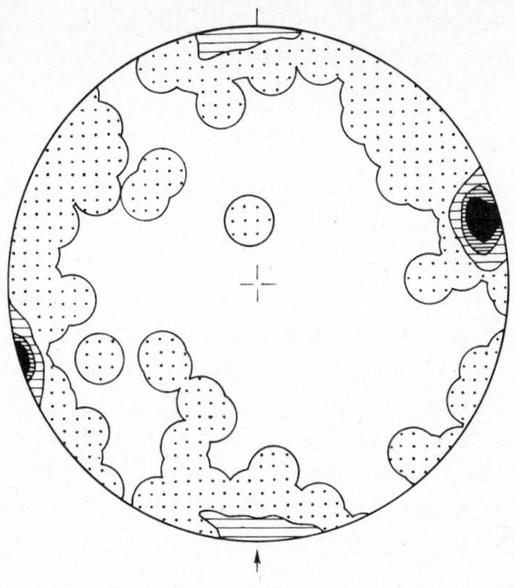

Figure 7.31. Diagram showing the orientation degree of aggregates in a Schmidt net (measured points were integrated into lines of equal »density«)

Alongside with the conventional, time-consuming inspection for the degree of separation and resistance as a function of service time (quantity of accumulated dust), microscopical examination of microtomed and thin sections can be applied to visualize the structure of the fabric. Specimens of new, washed and soiled fibre mats are consolidated by impregnation under vacuum with a thin embedding liquid. Some care is required in this preparation in order to avoid structural changes caused by displacement of loose fabric particles during impregnation. Specimen thickness has to be matched to the respective size of the structural units. As shown in Fig. 7.32 (see Annex), no changes in the three-dimensional irregular structure can be detected. The arrangement of fibres and the average distances between them are unchanged. Accordingly, the washing process has not affected the structural arrangement, so that the filter mat shown is still usable.

The microscopical result is confirmed by measurements of the degree of separation and of pressure loss as a function of service time of new and regenerated filter mats.

7.4.2 Phase distribution

Besides grain orientation, the type of phase distribution has an important influence on the properties of heterogeneous aggregate substances (containing phases A and B, for example) such as copolymers and fibre-reinforced materials. The distribution of phases is characterized by three parameters identifiable in the microscopical image.

Distribution density of A in B

The distribution density expresses the share of A within a unit volume of B, such as the share of crystallites (A) in a glass sample (B). It can be determined as an integral value by a photometric measurement of light transmittance depending on the number of crystallites, or differentially by measuring in many parallel sections through the aggregate.

Variation of distribution density of A in B

The variation of distribution density from measuring point to measuring point describes the spatial homogeneity or heterogeneity of the distribution of A in B.

Grain size distribution of A and B

The grain size ratio completes the characterization of phase distribution.

The determination of *porosity* is a special case of phase distribution analysis. Examples are foamed high polymers and macropores in catalytic compression granulates. Instead of phase A, one registers the distribution of pores in phase B. The specimens are vacuum-impregnated with coloured, self-curing embedding liquids to enhance contrast. In specimens whose pores are filled with embedding media of known refractive index, porosity can be determined by interference microscopy.

7.4.3 Type of intergrowth

The strength of an aggregate essentially depends on its type of intergrowth, an important parameter for the processing and use of aggregate materials. Intergrowths allow of direct microscopical observation. Staining or etching of individual phases improves contrast. The types of intergrowth observed microscopically can be described quantitatively by means of various parameters. One measure of intergrowth is provided by the surface-to-volume ratio of a grain. In the microscopical image, the surface is represented by the circumference *C*, and the volume by the area *A*.

The intergrowth is then defined by

$$I' = \frac{C_i}{a_i} .$$

Where intergrowth is of a high complexity,

$$\sum C_i = C \text{ and } \sum a_i = A .$$

$C_1, C_2 \ldots C_n$ are the circumferences, and $a_1, a_2 \ldots a_n$ the areas of a greater number of individual grains.

The sum of all circumferences of a phase (including cut-off protrusions of other grains) is measured by tracing the outlines on a micrograph, and the area of the grains is determined by planimetry.

In many applications, e.g. in *flotation*, it is imperative for all aggregates to be disintegrated into homogeneous single grains. Tables 7.36 and 7.37 show the *relationships in crude potash salts.*

The intergrowth parameter permits determining the *degree of disintegration in crushing* (percentage of the desired disintegrated phase in the total content). It has to be complemented by a characterization of grain size homogeneity. A favourable combination is that of grain size distribution curves and intergrowth parameter values.

Table 7.36. Relation between grain size class, degree of disintegration and intergrowth index in crude sylvinite salts, acc. to [7.39]

Intergrowth index 4 A/C (mm)	Grain size class (mm) at various degrees of disintegration (%)			
	25	50	75	90
0.79	1.0	0.66	0.48	0.26
1.08	1.3	0.84	0.50	0.33
1.16	1.2	0.76	0.58	0.44
1.43	1.47	0.87	0.51	0.38
1.54	1.6	1.0	0.54	0.32
2.05	1.65	1.12	0.61	0.45
2.75	1.74	1.45	0.75	0.50

Table 7.37. Intergrowth parameters of crude potash salts in the microscopical image, acc. to [7.39]

Intergrowth type with regard to a particular salt mineral	Intergrowth index 4 A/C (mm)	Grain size class of crushed salt rock giving a 75 % degree of disintegration of a particular salt mineral (mm)
Very coarse	3	1.1 - 1.2
Coarse	2 - 3	0.8 - 1.1
Medium	1 - 2	0.6 - 0.8
Fine	0.5 - 1	0.4 - 0.6
Very fine	0.3 - 0.5	0.3 - 0.4
Superfine	0.3	0.3

7.4.4 Textures

Textural patterns permit to conclude on the history, formation mechanism and strength properties of an aggregate. This concerns physical processes (e.g. the compression process of catalytic granulate, the baking of fine-grain fertilizers) as well as chemical ones (e.g. interphase reactions in sintering or melting processes).

The interpretation of textural patterns in a micrograph is made difficult by the two-dimensional section and the multitude of possible convergence forms. Different processes may result in similar textural patterns. In order to exclude incidental patterns occurring in a particular section, an analysis must always be based on a sufficiently large number of micrographs.

In the following, some frequent textural patterns are described together with the information they convey about their formation (Table 7.38, Fig. 7.33).

A-in-B textures, sieve textures

Besides genuine inclusions, similar patterns may occur depending on the section. Distinguishing features are the change in birefringence towards the inclusion and hose-type connections with the grain margin. True inclusions may be new formations or transformation products in the host grain, simultaneous formations, or unchanged earlier grains.

Reaction rims

Reaction rims around grains separate metastable intermediate phases (e.g. in peritectic transformations) from the further course of the reaction, thus preserv-

Table 7.38. Textures of aggregates (analogous to petrofabric nomenclature)

Descriptive nomenclature	Genetic nomenclature
Simple grain fabric	**Replacement (alteration) fabric**
All grains (grain pairs) intergrown in mosaic pattern - block texture - parallel texture - fibrous texture - sutural texture - amoeboid texture	*Grain A is decomposed from inside out:* B-in-A texture (beginning alteration texture) → sieve texture → A-in-B texture *Grain A is decomposed from outside:* 1. Grain is decomposed simultaneously on all sides (irregularly indented grain margins). 2. Grain is decomposed unilaterally (intermediate form: one grain half each A and B). *Grain B is newly formed* as uniform grain or fine-grained mixture
Composite grain fabric	**Intergrown fabric**
Penetration or implication texture (individual grains or pairs penetrate each other) - intragranular penetration texture - regularly oriented texture (e.g. eutectics) - cancellated texture	Grains A and B are stable and grow simultaneously (intergrowth forms depending on varying crystal formation trend) **Deformation fabric** Fabric is formed by mechanical stress on the aggregate.

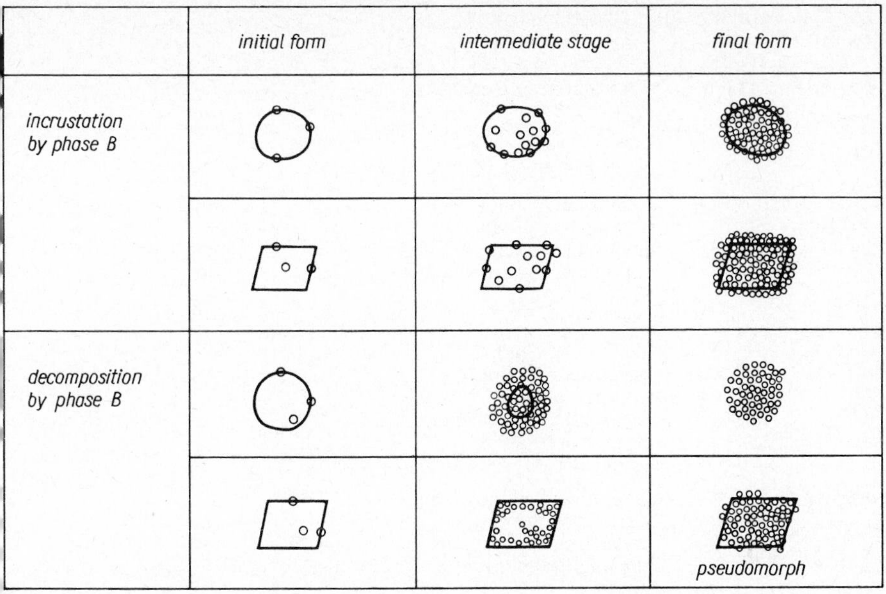

	initial form	intermediate stage	final form
incrustation by phase B			
decomposition by phase B			
			pseudomorph

Figure 7.33. Convergence forms in texture micrographs of individual grains

ing such phases. They provide valuable clues to the reaction mechanism. Patterns of similar shape are found in compression granulates, in encrustations of earliest separation products, or in new formations on intergranulars, with or without reaction with the grain observed.

Pseudomorphs (Table 7.39)

They supply essential information on the mechanism of reactions that took place in solids. Pseudomorphs are formed by substance replacement in which the former grain shape or characteristic inclusion arrangement is retained.

Intergranular fillings

They hint to substance changes that occurred on the grain boundaries of a solid aggregate.Elongated transformation residues may be mistaken for true intergranular fillings unless other structural criteria are taken into consideration.

»Reaction forms«

Reaction forms of phase B on A grains in solid aggregates, if observed in a single micrograph, may be confused with intergrowth types of phases showing different crystallization trends. Positively identified reaction forms indicate that reactions have not been completed.

Table 7.39. Possibilities of polycrystalline grain formation

Polycrystalline grain (secondary grain)	Information provided by microscopy
Pseudomorph	Initial substance and course of reaction
Host grain with inclusions	Age relationship (pre-, syn- or postgenetic inclusion relative to host), crystallization process or decomposition mechanism
Granulate	Texture of compressed granulate, durability, volume fractions and distribution of phases in the compressed aggregate
Agglomerate	Intergrowth of individual fine grains during crystallization or after separating the liquid phase

Recrystallization textures

They are distinguished by a gradual increase in grain size. Larger grains grow at the expense of smaller crystals. Microscopical examination shows two limit cases: the uniform increase of the mean grain diameter (*continuous grain growth*) and the preferred growth of certain grains while others retain their size (*discontinuous grain growth*). Fig. 7.34 shows the grain size curves of both types, established by microscopy [7.208]. Knowledge about both processes and their conditions is essential for deriving material properties from the various textural patterns.

Textures of organic high polymers [7.138 to 7.145, 7.223 to 7.229]

The textures of organic high polymers are manifold, depending on process parameters (Table 7.40). In many partially crystalline polymers, spherulitic textures are dominating.

Spherulites are composed of radially arranged, more or less helicoidal lamellae of folded molecular chains. The spaces between the lamellae and between the spherulites contain material of less or no order. Spherulites grow from a nucleus through sheaflike intermediate stages.

Spherical spherulites consist of optically uniaxial, mostly hexagonal regions being optically positive. Between crossed nicols, they exhibit a regular black Maltese cross figure.

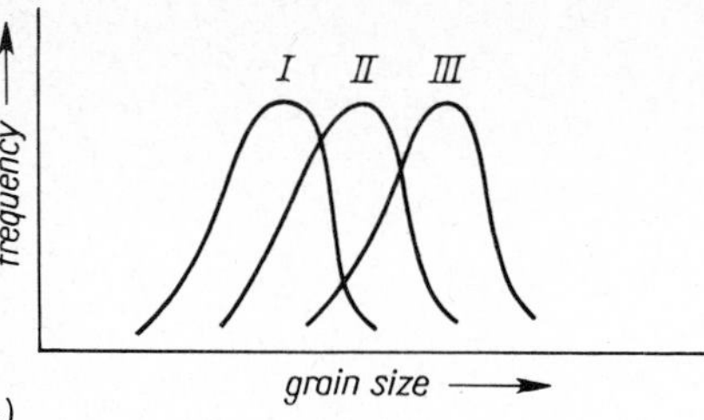

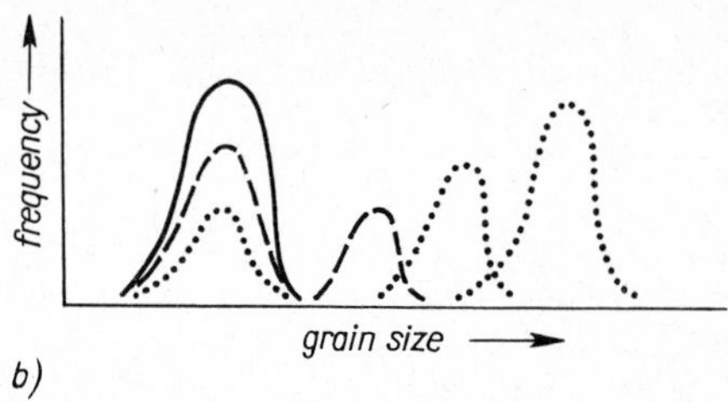

Figure 7.34. Microscopical registration of varying grain size distribution in recrystallization processes (acc. to [7.208])
a) continuous grain increase b) discontinuous grain increase
I initial state
II, III variation after different times

Deformed spherulites are shaped like a triaxial ellipsoid, the (mostly monoclinic) regions being optically biaxial and either positive or negative. The interference figure is that of a biaxial crystal.

In other textural patterns the spherulite shape is found to disintegrate. The interference figure is then no longer symmetric.

Hot stage examinations show that spherulite symmetry degrades as temperature is increased. Optically uniaxial specimens become biaxial. This symmetry will remain during cooling to room temperature.

Where several spherulites intergrow, the mutual growth rates determine the resulting boundary shapes in the microscopical image (equal growth rates produce plane boundary faces, different rates produce curved ones). The slower-growing spherulite will have convex forms, its faster-growing counterpart concave ones.

Spherulite growth rate decreases as the molar weight increases, with the number of spherulites increasing simultaneously. Spherulitic textures form primarily from slowly-cooling melts. The slower the cooling rate, the greater is the possibility of ordering, and the greater and more frequent are the spherulites. Higher pressure also favours spherulite formation. In a fast-cooling melt (e.g. in polyamide welding), optically isotropic (non-oriented) material only will form, in which a few sporadic, minute spherulites may be embedded at best.

In injection-moulded parts made of polyamide or polypropylene, e.g., the texture is a function of injection pressure, injection rate, mould temperature, annealing, and load. Near the injection port, the texture usually is vortical, whereas it gets calmer and more homogeneous at increasing distances from the injection port.

The faster-cooling outer zone is characterized by a fine-spherulitic texture, whereas spherulites are coarser internally. The sizes of spherulites in series of microscopical specimens yield lines of equal cooling rate. The texture suggests that this will not be changed basically by an annealing process.

Some specimens show alternating layers of coarser and finer spherulites. This indicates that the cooling rate of injection-moulded rod samples does not

Table 7.40. Textures of organic high polymers

Homogeneous textures (made up of constituents of a single phase)		Heterogeneous textures (made up of constituents of several phases)
Microscopically not resolvable into texture elements	Microscopically resolvable into texture elements	
Optically isotropic objects Optically anisotropic objects of uniform extinction	Optically anisotropic subdomains in various arrangements Regular spherulites (optically uniaxial subdomains, regular dark cross, with crossed nicols) Irregular spherulites (outline elliptic or sheaf-like, optically uniaxial or biaxial, subdomains seen with crossed nicols, oblique or displaced cross or 2 dark hyperbolae) Spherulitoid arrangement (ring texture of subdomains, extinction rings and spirals) Parallel or subparallel arrangement of anisotropic subdomains, parallel or oblique extinction of subdomains Submicroscopic crystalline subdomains of different arrangement or extinction Optically isotropic subdomains of different absorption and arrangement	Copolymer and graft polymer texture (rounded polymer particles incorporated in different matrix with or without orientation) Texture of fibre-reinforced high polymers (fibre incorporated in polymer matrix, with or without orientation)

diminish continuously from the margin to the centre. Accordingly, during injection the cooling and filling processes must be largely parallel in time.

The influence of microscopically discernible textural features on the property of a material was proved convincingly on polyethylene succinate [7.195].

Tensile strength increases with the size and growth rate of spherulites (Fig. 7.35, see Annex).

The interferometric measurement of the disintegration of the material texture revealed differences in strength. The texture is rather resistant to disintegration in the centre of the spherulites, whereas the texture between them is more loosely packed. The strength of the material is determined by that of the spherulites.

In the temperature range examined, the microscopical results proved that the tensile strength of the material is mainly a function of spherulite growth rate. This enabled the manufacturers to optimize their processes.

7.4.5 Special requirements for fabric analysis of decomposing substances; measurement at high and low temperatures

The fabric analysis of easily decomposing aggregates is of importance, e.g. in studying solidified carbide melts. The production of fresh surfaces presents the main difficulty. Same as the grinding of thin and polished sections, this has to be performed under an inert immersion liquid such as nujol. The examination of polished sections through the reflected-light microscope requires immersion objectives. Dry objectives may be used if the specimen is contained in a humidity chamber or if its surface is coated with a thin, colourless film of lacquer in the box.

The texture is highly sensitive to artefacts introduced by preparation. The specimens have to be observed carefully to avoid misinterpretation.

The study of aggregate textures at low temperatures has a practical importance in experiments on lowering the setting point of lubricating oils. The additives used change the shape and size of the paraffin crystals separated at low temperatures, from a bulky framework of tabular crystals to a suspension in oil of fine-grained, rounded paraffin particles, at the same time improving viscosity. Knowledge about the occurrence of bulky parrafin crystal frameworks in the liquid phase is of interest also with regard to optimizing separator performance [7.230].

Fabric analysis at high temperatures permits conclusions on material properties varying with heat treatment, such as of metals (Fig. 7.37, see Annex). In polished sections, textures will appear in particularly rich contrast by thermal etching at high temperatures. Nevertheless, the grain size measured on the surface should not without reservations be generalized for the specimen as a whole [6.150].

References

[7.1] Progress report »Chemical Microscopy« in Analytical Chemistry (biannual)

[7.2] *Kuhnert-Brandstätter, M.:* Gegenwärtiger Stand und Zukunft der Chemomikroskopie. Sci. pharmac. 33 (1965) 244–259

[7.3] *Brückner, H. D., and K. Heide:* Kristalloptische Analysenverfahren und ihre Bedeutung für die chemische Forschung und Ausbildung. Z. Chem. 10 (1970) 4, 125–132

[7.4] *Seyfarth, H.-H., Beyer, H., and H.-H. Emons:* Neue Zeiß-Mikroskope für die Anwendung in der chemischen Forschung und Industrie zur quantitativen Produktionskontrolle. Jenaer Rdsch. 15 (1970) 4, 211–214

[7.5] *El-Hinnawi, E.:* Methods in Chemical and Mineral Microscopy. Amsterdam/ London/ New York: Elsevier Scientific Publishing Company 1966

[7.6] *Chamot, E. M., and C. W. Mason:* Handbook of Chemical Microscopy. New York: John Wiley & Sons 1958

[7.7] *Schaeffer, H. F.:* Microscopy for Chemists. Dover: Dover Publications Inc. 1968

[7.8] *Reumuth, H.:* Die Mikroskopie des Chemikers. *Freund, H.:* Handbuch der Mikroskopie in der Technik. Frankfurt/Main: Umschau-Verlag 1957, 433–468

[7.9] *Benedetti-Pichler, A. A.:* Identification of Materials Via Physical Properties, Chemical Tests and Microscopy. New York: Academic Press 1964

[7.10] *Beljankin, D. S.:* Technische Petrographie. Berlin: VEB Verlag Technik 1960

[7.11] *Wallis, T. E.:* Analytical Microscopy. Boston: Little Brown and Company 1965

[7.12] *Thomas, S. L.:* The Use of the Microscope in Chemistry. Microscope 15 (1967) 4, 485

[7.13] *Ambronn, H., and A. Frey:* Das Polarisationsmikroskop. Seine Anwendung in der Kolloidforschung in in der Färberei. Leipzig: Akademische Verlagsgesellschaft Geest u. Portig KG 1926

[7.14] *Burrells, W.:* Industrial Microscopy in Practice. New York: Morgan and Morgan 1964

[7.15] *Benedict, H. C.:* The Polarizing Microscope in Organic Chemistry. Ind. Engng. Chem. Anal. Edit. 2 (1930) 91

[7.16] *Schumann, H.:* Metallographie. 5th. ed. Leipzig: VEB Fachbuchverlag 1964

[7.17] *Quitter, V., Seyfarth, H.-H., and H. Opitz:* Staubkartei. Magdeburg: VEB Kombinat Luft- u. Kältetechnik 1971

[7.18] *McCrone, W. C.:* The Particle Atlas. Michigan: Ann Arbor Science Publishers Inc. 1968

[7.19] *Waldmann, H.:* Qualitative Mikroanalyse und chemische Mikroskopie. Chimia 13 (1969) 224–230

[7.20] *Evans, R. C.:* Mikroskopie in der Elektrochemie. Microscope and Crystal Front 15 (1966) 43–56

[7.21] *Leigh-Dugmore, C. H.:* Microscopy of Rubber. Cambridge: W. Heffer and Sons 1961

[7.22] *Morehead, F. F.:* Modern Microscope of Films and Fibers. ASTM Bull. 163 (1950)

[7.23] *Loske, T.:* Methoden der Textilmikroskopie. Stuttgart: Kosmos-Verlag 1964

[7.24] *Isenberg, J. H.:* Pulp and Paper Microscopy. Appleton, Wisconsin: Institute of Paper Chemistry 1958

[7.25] *Stach, E.:* Lehrbuch der Kohlenmikroskopie. Kettwig: Verlag Glückauf GmbH 1949

[7.26] *Insley, H., and V. D. Frechette:* Microscopy of Ceramics and Cements, Including Glasses, Slags and Foundry Sands. New York: Academic Press Inc. 1955

[7.27] *Rösler, H. J., and K. Koch:* Salzmikroskopie. Freiberg: Internes Lehrmaterial der Bergakademie 1968

[7.28] *Crossmon, G.:* Chemical Microscopy in the Optical Industry. Philadelphia: ASTM Spec. Tech. Publ. (1959) 257

[7.29] *Keune, H.:* Bilderatlas zur qualitativen anorganischen Mikroanalyse. Leipzig: VEB Deutscher Verlag für Grundstoffindustrie 1967

[7.30] *Wolter, H.:*Das Phasenkontrastverfahren und seine Anwendbarkeit bei chemischen Untersuchungen. Fortschr. Chem. Forsch. 3 (1954) 1–40

[7.31] *Gumz, W., Kirsch, H., and M. T. Mackowsky:* Schlackenkunde. Wien: Springer-Verlag 1958
[7.32] *Neupert, M., Seyfarth, H.-H., Herrling, R., and B. Hahne:* Der Heiz- und Kühltisch −20 bis 80 °C und seine Anwendung. Jenaer Rdsch. (1972) 2, 75–83
[7.33] *Duncan, J. H.:* Technical Ceramics and the Optical Microscope. Trans. Brit. ceram. Soc. 67 (1968) 4, 147–169
[7.34] *Kuhnert-Brandstätter, M., and L. Müller:* Experience in Hot-Stage-Microscopy of Inorganic Compounds. Microscope 16 (1968) 3, 257
[7.35] *Grabar, D. G., and W. C. McCrone:* Application of Microscopic Fusion Methods to Inorganic Compounds. Chem. Educat. 27 (1950) 649
[7.36] *Ludwig, G.:* Mineralogische Methoden zur Bestimmung der Zusammensetzung von Aufbereitungsprodukten. Freiberger Forschungshefte A 269 (1963) 59–76
[7.37] *Jones, M.:* An Application of Microscopical Fusion Methods. Microscope 14 (1964) 7, 264
[7.38] *Rath, R.:* Fortschritte der Kristalldiagnose im Durchlicht. Fortschr. Mineralog. 46 (1969) 1, 73–86
[7.39] *Kühn, R.:* Die Mikroskopie der Kalisalze. Empelde-Hannover: 1950
[7.40] *Zberea, I.:* Möglichkeiten in der Strukturforschung durch Anwendung der Interferenzmikroskopie. Rev. roum. Sci. techn., Ser. Metallurgic 12 (1967) 1, 41–49
[7.41] *Doerbecker, K., and H. Joel:* Optische Eigenschaften keramischer Oberflächen (Streuindikatrix). Keram. Z. 19 (1967) 726
[7.42] *Brubaker, D. G.:* Light and Electron Microscopy of Pigments. Ind. Engng. Chem. 17 (1945) 184
[7.43] *Burri, C., Parker, R. L., and E. Wenk:* Die optische Orientierung der Plagioklase. Basel: Birkhäuser-Verlag 1967
[7.44] *Dreizler, L.:* Mikroskopie des Betons. Zement-Kalk-Gips (1966) 216–222
[7.45] *Frosch, C. J., and E. A. Hauser:* Einige mögliche Anwendungen der Fluoreszenzlichtmikroskopie zu industriellen Untersuchungen. Ind. Engng. Chem. Qualyt. Ed. 28 (1936) 423
[7.46] *Honigmann, B.:* Gleichgewichts- und Wachstumsformen von Kristallen. Darmstadt: Steinkopff-Verlag 1958
[7.47] *Crisler, J. P.:* Controlled Atmosphere Microscopy. Microscope 14 (1964) 4, 152
[7.48] *Bárta, R.:* Silikatmikroskopie. Jenaer Rdsch. 10 (1965) 307–310
[7.49] *Baumann, H. N.:* Microscopy of High-Temperature Phenomena. Amer. ceram. Soc. Bull. 27 (1948) 267–271
[7.50] *Reumuth, H.:* Methoden der Einzelfaser-Mikroskopie. Möglichkeiten und Grenzen ihrer Anwendung. Z. ges. Textil-Ind. 63 (1961) 3–9 and 82–89
[7.51] *Poling, F. D.:* Special Applications of Interference-Transmission Microscopy to Textile Fiber Problems. Microscope 16 (1968) 3, 201
[7.52] *Bobeth, W.:* Nutzungsmöglichkeiten der Textilmikroskopie. Dtsch. Textiltech. 15 (1965) 1, 39–48 and 2, 93–99
[7.53] *Felton, C. D., Thomas, E. J., and J. J. Clark:* Ultraviolet Microscopy of Fibers and Polymers. Textile Res. J. 32 (1962) 58–67
[7.54] *Schmidt, K. G.:* Phasenkontrastmikroskopie im Staublaboratorium. Staub-Reinhalt. Luft 22 (1962) 8, 307–312
[7.55] *Scheidl, H., and R. Mitsche:* Neue Beispiele aus der Mikroreflexionsmessung. Radex-Rdsch. (1967) 3/4, 596–607
[7.56] *Smithson, F.:* Some Aspects of Microscopic Study of Ceramic Glasses. Trans. Brit. ceram. Soc. 47 (1948) 191–207
[7.57] *Reumuth, H.:* Staubstrukturen durch neuzeitliche Mikroskopie. Staub-Reinh. Luft 22 (1962) 8, 301–306

[7.58] *Reumuth, H., and W. Buss:* Fortschritte der mikroskopischen Erforschung der Glasschmelzvorgänge bis 1600 °C. Glastechn. Ber. 32 (1959) 89–95

[7.59] *Rosch, S.:* Beobachtungen und Gedanken über das Sphärolithwachstum der Kristalle. Tschermaks mineralog. petrogr. Mitt. (3. F.) 12 (1968) 290–298

[7.60] *Rutter, F.:* Mikroskopische Charakteristik feuerfester Kohlekeramik. Trans. Brit. ceram. Soc. 66 (1967) 9, 423–441

[7.61] *Schreiter, P.:* Angewandte Mikroskopie in der Schlackenforschung. Jenaer Rdsch. 13 (1968) 3, 174

[7.62] *Trojer, F.:* Angewandte Mikroskopie in der Zement- und Betontechnologie. Leitz-Mitt. Wiss. u. Techn. 4 (1967) 1–7

[7.63] *Trojer, F.:* Mineralogische Untersuchungsverfahren auf dem Gebiet des Hüttenwesens mit Beispielen über Verschleißvorgänge an Gittersteinen. Ber. DKG 38 (1961) 12, 557–566

[7.64] *Seyfarth, H.-H., Wenke, L., and W. Schreiber:* Holographische Abbildung von Kristallwachstumsprozessen in bewegten Lösungen. Kristall u. Techn. 11 (1976) 4, 355–361

[7.65] *Steinbruch, U.:* Möglichkeiten der Untersuchung physiko-chemischer Vorgänge durch Küvettenmikroskopie. Diplomarbeit, Technische Hochschule für Chemie »Carl Schorlemmer« Leuna-Merseburg, Sektion Verfahrenschemie, 1972

[7.66] *Reumuth, H., and T. Loske:* Küvettenmikroskopie in Biologie und Technik. Mikroskopie 17 (1962) 149–216

[7.67] *Jarosch, R.:* Neue Erfahrungen zur Küvettenmikroskopie. Mikrokosmos 46 (1957) 250–255

[7.68] *Hauser, E. A.:* Über die Anwendung des mikrurgischen Verfahrens in der Kolloidchemie. Kolloid-Z. 38 (1926) 76–80

[7.69] *Mosebach, R.:* Zur mikroskopischen Bestimmung der Zusammensetzung heterodisperser Korngemische. Erdöl u. Kohle 7 (1954) 199–203

[7.70] *Spies, H. J.:* Einfluß der Prüfbedingungen auf das Ergebnis der mikroskopischen Bestimmung des Reinheitsgrades. Neue Hütte 11 (1966) 7, 420

[7.71] *Schmidt, G.:* Kristalloptische Nachweismethoden in der forensischen Toxikologie. Zeiß-Mitt. 1 (1957) 95–106

[7.72] *Wolff v., T.:* Methodisches zur quantitativen Gesteins- und Mineraluntersuchung mit Hilfe der Phasenanalyse. Z. Kristallogr. B, Mineralog. Petrogr. Mitt. 54 (1942) 1–121

[7.73] *Kofler, A., and J. Kolesek:* Beitrag zur mikroskopischen Identifizierung organischer Stoffe nach Kofler. Mikrochim. Acta (1969) 2, 408–436 and 5, 1038–1062

[7.74] *Hellmer, J. H., and H. Udluft:* Versuch einer quantitativen mineralogischen Staubanalyse. Zeiß-Nachr. 2 F. (1936) 1, 1–10

[7.75] *Mitchell, J.:* Microscopic Identification of Organic Compounds. Analyt. Chem. 21 (1949) 4, 448–461

[7.76] *Ernst, T.:* Spuren- und Phasenanalyse mit kristalloptischen Methoden. Z. analyt. Chem. 197 (1963) 119–133

[7.77] *Waldner, W. F.:* Die Genauigkeit von Mineralbestimmungen mit dem Phasenkontrastmikroskop. Bergbauwiss. 8 (1961) 79–84

[7.78] *Trojer, F.:* Die mikroskopische Bestimmung oxydischer Mineralphasen in anorganischen Industrieprodukten. Fortschr. Mineral. 33 (1955) 124–125

[7.79] *Kroll, J. M.:* Mikrophotometrische Bestimmung selektiv angefärbter Feldspate in keramischen Rohstoffen. Leitz-Mitt. Wiss. u. Techn. 4 (1967) 19–23

[7.80] *Forlini, L.:* Expanded and Revised Tables for the Determination of Unknowns by Dispersion Staining. Microscope 17 (1969) 1, 29

[7.81] *Wherry, E. T.:* Mineral Determination by Absorption Spectra. Amer. Mineralogist 14 (1929) 8, 299–303

[7.82] *Heide, K., and U. Brückner:* Grundlagen zur Phasenanalyse von Salzgesteinen. Chemie d. Erde 26 (1967) 4, 235–255

[7.83] *Böhm, G., and E. Bader:* Die Anwendung der Phasenkontrastmikroskopie bei der kristalloptischen Analyse von Kalisalzen. Bergakademie Freiberg 16 (1964) 11, 650–651

[7.84] *Vachrameeva, V. A.:* Kombinirovannyj metod opredellenija soljanych mineralov. Moskva: Trudy VNIIG (1954) 29

[7.85] *Grabar, D. G.:* Applications of Dispersion Staining to Microscopic Identification of Settled Dust. J. Air Pollution Control Assoc. 12 (1962) 560

[7.86] *Fulton, C. C.:* The Identification of Organic Compounds by Microcrystalloscopic Chemistry. Modern Microcrystal Tests for Drugs. New York: John Wiley Interscience 1970

[7.87] *Heinrich, E. W.:* Microscopic Identification of Minerals. New York: McGraw Hill 1965

[7.88] *Steyn, J. G. D.:* The Identification Mode of Occurance and Quantitative Determination of Crystalline Phases in Granulated Blast-Furnace Slag. Mineralog. Mag. 35 (1965) 269, 108–117

[7.89] *Katzschmann, R.:* Kristalline Einschlüsse in Gläsern. Silikattechnik 17 (1966) 5/6

[7.90] *Malyseva, T. J.,* et al.: Mineralogische Zusammensetzung der in der Industrie für die Ammoniaksynthese verwendeten Katalysatoren. Kinetika i Kataliz 9 (1968) 3, 700–702 (Russian)

[7.91] *Nassenstein, H.:* Über die Messung der optischen Konstanten von Farbpigmenten. Ber. Bunsenges. Phys. Chemie 71 (1967) 3, 303–313

[7.92] *Burakova, T. N.:* Kristalloptičeskie konstanty i ich ispolzovanič v mikrochimičeskom analize. Leningrad: Izd. Leningr. gos. univ. 1964

[7.93] *Schrader, H. J.:* Mikroskopische Identifizierung von nichtmetallischen und metallischen Einschlüssen in Nähten. Praktische Metallographie 4 (1967) 9, 457–469

[7.94] *Garrett, H. L., and P. A. Taylor:* Identification of Trace Contaminations in Polymers with the Aid of the Microscope. Microscope 16 (1968) 4, 295

[7.95] *Klein, A.:* Mikroskopische Identifizierungsmethoden von PVC. Kunststoffe 7 (1971) 432–496

[7.96] *Jörg, F.:* Die Schmelzpunktbestimmung von Chemiefaserstoffen – eine methodische Ergänzung zur Faseranalyse. Leitz-Mitt. Wiss. u. Techn. 4 (1967) 13–18

[7.97] *White, B. J.,* et al.: Optical Crystallographic Identification of Sulfanilamide. Analytic. Chem. 33 (1950) 950

[7.98] *Heyn, A. N. J.:* The Identification of Synthetic Fibers by Their Refractive Indices and Birefringence. Textile Research 23 (1953) 246

[7.99] *Nettelnstroth, K.:* Die Unterscheidung synthetischer Fasern mittels der spezifischen Doppelbrechung. Leitz-Mitt. Wiss. u. Techn. 4 (1967) 24–28

[7.100] *Kaufmann, G.:* Das Verhalten organischer Pigmente in Polymeren. Angew. makromolek. Chem. 10 (1970) 137, 83–96

[7.101] *Krause, J.:* Ruß-Analyse – Beschreibung einer Methode, die auf dem Vergleich mikroskopischer Präparate beruht. Kautschuk u. Gummi – Kunststoffe 20 (1961) 11, 630–652

[7.102] *Winchell, A. N., and H. Winchell:* Optical Properties of Artificial Minerals. New York–London: Academic Press 1964

[7.103] *Winchell, A. N.:* The Optical Properties of Organic Compounds. New York: Academic Press 1954

[7.104] *Tröger, W. E.:* Optische Bestimmung der gesteinsbildenden Mineralien. Stuttgart: Schweizerbartsche Verlagsbuchhandlung; Tabellenband 2nd ed. 1965, Textband 1st ed. 1967

[7.105] *Kordes, E.:* Optische Daten zur Bestimmung anorganischer Substanzen mit dem Polarisationsmikroskop. Weinheim: Verlag Chemie 1960
[7.106] *Utermark, W., and W. Schicke:* Schmelzpunkttabellen organischer Verbindungen. Berlin: Akademie-Verlag 1963
[7.107] *Trojer, F.:* Die oxydischen Kristallphasen der anorganischen Industrieprodukte. Stuttgart: Schweizerbartsche Verlagsbuchhandlung 1963
[7.108] ASTM-Index to the X-Ray-Powder Data File. ASTM Diffraction Data Sales Dept. Philadelphia, 19103 USA
[7.109] *Kraceck, F. G.:* Melting and Transformation Temperatures of Minerals and Allied Substances U. S. Geol. Surv. Bull. 1144-D 1963
[7.110] *Seyfarth, H.-H., Anger, G., and B. Hahne:* Mikroskopische Phasenanalyse von Nierensteinen. Münchener Mediz. Wochenschr. 114 (1972) 14, 670–677
[7.111] *Seyfarth, H.-H., Hahne, B., and J. Löscher:* Mikroskopische Identifizierung von Neben- und Spurenbestandteilen in Festkörpergemengen. Jenaer Rdsch. 4 (1971) 241–245
[7.112] *Seyfarth, H.–H., and B. Hahne:* Die mikroskopische Phasenanalyse von Festkörpergemengen durch Lichtbrechungsmessungen. Jenaer Rdsch. 4 (1971) 237–240
[7.113] *Seyfarth, H.-H.:* Zur Methodik der Phasenanalyse technischer Mischstäbe. Luft- und Kältetechnik (1969) 5, 259–263
[7.114] *Schareck, E., Hahne, B., and H. Holldorf:* Untersuchungen zu den Systemen $2KCl + MgSO_4 = K_2SO_4 + MgCl_2/CH_3OH - H_2O$ und Na^+, K^+, Mg^{++}/Cl^-, $SO_4^{2-}/CH_3OH–H_2O$. Dissertation, Technische Hochschule für Chemie »Carl Schorlemmer« Leuna-Merseburg, Fak. für Naturwiss., 1972
[7.115] *Salzmann, J. J., and C. K. Jorgensen:* Molrefraktion von Aquoionen metallischer Elemente und die Auswertung von Lichtbrechungsmessungen in der anorganischen Chemie. Helvet. chim. Acta 51 (1968) 6, 1276–1293
[7.116] *Ahrens, L. H.:* Variation of Refractive Index with Ionization Potential in some Isostructural Crystals. Mineralog. Mag. (1958) 242, 929
[7.117] *Bokij, G. B., and S. S. Batsanov:* Kristalloptische Methode für die Bestimmung der Struktur von komplexen Verbindungen. Izvest. Akad. Nauk. SSSR, Ber. chem. (1955) 173 (Russian)
[7.118] *Gerdzikov, S.:* Abhängigkeit der Ionenrefraktion von der Ionengitterenergie. Ž. strukturnoj Chim. 8 (1967) 3, 546–547
[7.119] *Baren v., F. A.:* Über den Einfluß verschiedener Flüssigkeiten auf den Brechungsindex von Tonmineralien. Z. Kristallogr. 95 A (1936) 464–469
[7.120] *Donnay, J. D. H., and G. Donney:* Optische Bestimmung des H_2O-Gehalts in sphärolithischem Vaterit. Acta Crystallogr. 22 (1967) 2, 312–314
[7.121] *Manning, F. G.:* Optische Absorptionsspektren der Granate und einige strukturelle Interpretationen von mineralogischer Bedeutung. Canad. Mineralogist 9 (1967) 2, 237–251
[7.122] *Adam, J.:* Absorptionsspektren natürlicher Granate im Bereich von 200 bis 1500 µm und ihre Deutung. Diplomarbeit, Humboldt-Universität Berlin, Mineralogisches Institut, 1968
[7.123] *Sackmann, H., and D. Demus:* Eigenschaften und Strukturen thermotroper kristallin-flüssiger Zustände. Fortschr. chem. Forschung 12 (1969) 2, 349–386
[7.124] *Gray, R. J., and J. v. Cathcart:* Mikroskopie von pyrolytischen Kohlenstoff-Abscheidungen im polarisierten Licht. J. nuclear Mater 19 (1966) 81–89
[7.125] *Engel, A.:* Die Bestimmung der Kristallinität von Schlacken. Kristall u. Technik 1 (1966) 3, 519–524
[7.126] *Kisfaludy, A., and P. Tardy:* Mikroskopische Untersuchung der ferromagnetischen Phasen von Legierungen. Acta techn. Acad. Sci. hung. 56 (1966) 131–137

[7.127] *Pepperhoff, W.:* Polarisationsmikroskopische Beobachtung ferromagnetischer Bereichsstrukturen mit dem 45°-Illuminator. Leitz-Mitt. Wiss. u. Techn. 3 (1966) 215–217

[7.128] *Kuhnert-Brandstätter, M.:* The Study of Isomorphism by Microthermal Methods. Microscope 14 (1964) 6, 223

[7.129] *Kofler, A.:* Über das Verhalten von Mischkristallen beim Schmelzen und Kristallisieren. Mikroskopie 11 (1956) 140

[7.130] *Emons, H.-H., Seyfarth, H.-H., and E. Stegmann:* Zur Beziehung zwischen Struktur und Optik bei den Sulfaten des Calciums und Natriums. Jenaer Rdsch. 4 (1971)235–236

[7.131] *Emons, H.-H., Seyfarth, H.-H., and E. Stegmann:* Zur Kristallographie von Doppelsalzen aus Calciumsulfat und Natriumsulfat. Kristall u. Techn. 6 (1971) 1, 85–95

[7.132] *Koritnig, S.:* Das Reflexionsvermögen opaker Mischkristallreihen. Neues Jb. Mineralog. Mh. (1964) 8, 225–232

[7.133] *Pelleg, J.:* Beobachtung von Ätzgruben und Subkorngrenzen bei Niob im optischen Mikroskop. J. less-common Metals 17 (1969) 1, 130–132

[7.134] *Fink, E.:* Der Nachweis von Versetzungen in Wolfram und Molybdän durch elektrolytisches Ätzen. Jenaer Rdsch. 12 (1967) 36–39

[7.135] *Fricke, P.:* Beitrag zum lichtmikroskopischen Nachweis von Versetzungsnetzwerken nach thermischen Behandlungsschritten im Silicium. Kristall u. Technik 4 (1969) 2, 253–264

[7.136] *Renniger, M., and W. Theis:* Quantitative Zuordnung von röntgenographischer und mittels chemischer Ätzung erfaßter Versetzungen bei einer Siliciumprobe. J. appl. Crystallogr. 2 (1969) 2, 48–52

[7.137] *Meyer, F.:* Polarisationsoptische Untersuchungen an transparenten hochpolymeren Werkstoffen. Allg. u. prakt. Chemie 20 (1969) 4, 121–124

[7.138] *Heinze, D.:* Charakterisierung von Hochpolymeren im festen Zustand. Z. analyt. Chem. 235 (1968) 1, 99–119

[7.139] *Lehmann, S.:* Mikroskopische Untersuchungen an Hochpolymeren. Diplomarbeit, Technische Hochschule für Chemie »Carl Schorlemmer« Leuna-Merseburg, Sektion Hochpolymere, 1969

[7.140] *Stuart, H. A.:* Optische Anisotropie – Orientierung kristalliner Anteile in hochpolymeren Körpern. Kolloid-Z. 120 (1951) 1/3 57–75

[7.141] *Ilberg. W.:* Die elektrische und optische Anisotropie von einachsig verstreckten Polymeren in einem weiten Temperaturbereich. Dissertation, Karl-Marx-Universität Leipzig, 1968

[7.142] *Hardy, G.:* Untersuchungen zur Polymerisation in fester Phase. Plaste u. Kautschuk 16 (1969) 10, 723–727

[7.143] *Vieweg, R., and J. Moll:* Fortschritte in der mikroskopischen Untersuchung von Kunststoffen. Kunststoffe 40 (1950) 317–321

[7.144] *Statton, W. O.:* Crystallographic Studies of Synthetic Fibers. Kristallogr. 127 (1968) 4, 229–260

[7.145] *Rüprich, G., and H.-H. Seyfarth:* Zur Anwendung mikroskopischer Methoden bei der Untersuchung teilkristalliner organischer Hochpolymerer. Jenaer Rdsch. 16 (1970) 4, 221–224

[7.146] *Kaufmann, S.:* Optische Bestimmung der übermolekularen Struktur von partiell kristallinen Faserstoffen, insbesondere von Polyamid-6. Dissertation, Pädagogische Hochschule Potsdam, 1967

[7.147] *Keedy, D. A., Powers, J., and R. S. Stein:* A Theoretical Calculation of the Birefringence of Polypropylene Crystals. J. appl. Physics 31 (1960) 11, 1911–1915

[7.148] *Géczy, I.:* Abhängigkeit des Brechungsindex vom mittleren Molekulargewicht in polymerhomologen Reihen einiger hochmolekularer Verbindungen. Vysokomol. Soed. 7 (1965) 4, 642–646 (Russian)

[7.149] *Gurnee, E. F.:* Theory of Orientation and Double Refraction in Polymers. J. appl. Physics 25 (1954) 10, 1232–1240

[7.150] *Thamm, F.:* Spannungsoptik in Kunststoffteilen. Allg. Prakt. Chem. 19 (1968) 4, 131; 5, 177 and 242; 8, 290

[7.151] *Eisenlohr, F.: and E. Wöhlisch:* Molekularer Brechungskoeffizient, sein additives Verhalten und seine Verwendbarkeit zur Konstitutionsermittlung. Ber. dtsch. chem. Ges. 53 (1920) 1746–1772 and 2053–2063, 54 (1921) 299–320

[7.152] *Nishioka, A.,* et al.: The Stress Birefringence of Vulcanizates of Polyisoprene – Polyethylene Blends. J. appl. Polymer Sci. 14 (1970) 700–806 and 1183–1187

[7.153] *Rosso, J. C., and B. Persoz:* Mechanische Doppelbrechung von Copolymeren aus ungesättigten Polyestern und Styrol. C R. heb. Séances. Acad. Sci. 261 (1965) 736–738 (French)

[7.154] *Andrews, R. D., and T. J. Hammack:* Temperature Dependence of Orientation Birefringence of Polymers in the Glassy and Rubbery States. J. Polymer Sci. 27 (1964) 101–112

[7.155] *Gurnee, E. F.:* Birefringence Dispersion in Oriented Polystyrene. J. Polymer Sci. A2, 5 (1967) 5, 817–828

[7.156] *Ilberg, W., and B. Mündörfer:* Vergleich der Doppelbrechung bei optischen und Mikrowellenfrequenzen an orientierten Hochpolymeren. Kolloid-Z. 198 (1964) 1/2, 23–27

[7.157] *Kimmel, R. M., and R. D. Andrews:* Doppelbrechungseffekte in Acrylonitrilpolymeren. J. appl. Physics 35 (1964) 11, 3194, 36 (1965) 3063–3071

[7.158] *Säuberlich, P.:* Die Messung des Einflusses der Lagerzeit auf die Strukturänderungen der Spinnfäden mit unterschiedlicher Elementarfadenzahl und Durchmesser mit Hilfe der Prüfmethode der Doppelbrechung. Ingenieurarbeit, VEB Chemiefaserwerk Schwarza, 1964

[7.159] *Möhring, A., and G. Duwe:* Veränderung der Doppelbrechung als Ausdruck der Strukturentwicklung während der Erhitzung und Wiederabkühlung von hochpolymeren Fasern. Faserforsch. u. Textiltechn. 11 (1960) 1, 7–15

[7.160] *Schmidt, W. J.:* Instrumente und Methoden zur mikroskopischen Untersuchung optisch anisotroper Materialien mit Ausschluß der Kristalle. Handbuch der Mikroskopie in der Technik, Bd. I, Teil I. Frankfurt/Main: Umschau-Verlag 1957, 147–315

[7.161] *Nahmacher, M.:* Analysis in Polarized Light of Submicroscopic Structures in Resins. ASTM Spec. Techn. Publ. 34E, (1964) 17–30

[7.162] *Janneschitz-Kriegl, H.:* Optik des mizellaren Mischkörpers. Kolloid-Z. 124 (1951) 1, 1–14

[7.163] *Teetsov, A. S., and W. C. McCrone:* The Microscopical Study of Polymorph Stability Diagram. Microscope 15 (1965) 1, 13

[7.164] *Kofler, A.:* Zur Polymorphie organischer Verbindungen. Mikrochim. Acta 34 (1949) 15

[7.165] *Rojtburd, A. L.:* Orientierungs- und Habitusbeziehungen zwischen kristallinen Phasen bei Umwandlungen im festen Zustand. Kristallografija 12 (1967) 4, 567–574 (Russian)

[7.166] *Kohlhaas, R., and K. H. Soremba:* Beiträge zur Struktur kristallisierter aliphatischer Verbindungen. Z. Kristallogr. 100 (1938) 47–57

[7.167] *Leute, V.:* Untersuchungen zur Kinetik der Festkörperreaktion zwischen einkristallinem PbS und polykristallinem CdO mit mikroskopischen Methoden. Z. physik. Chem. (N. F.) 59 (1968) 1/4, 76–108

[7.168] *Kofler, A.:* Mikroskopische Methode zur Untersuchung von Hydraten. Mikrochim. Acta 36/37 (1951) 302–306
[7.169] *Heide, K.:* Vortrag zur Tagung der Arbeitsgruppe Polarisationsmikroskopie der DVK am 18. 11. 1970 in Merseburg
[7.170] *Emons, H.-H., Seyfarh, H.-H., and F. Winkler:* Kristallographische Eigenschaften und Zersetzung der Methanolsolvate des $MgSO_4$. Kristall u. Technik 6 (1971) 4, 521–531
[7.171] *Flora, T.:* Über die kinetische Untersuchung der thermischen Zersetzung von Nickel(II)- und Kobalt(II)-amminhalogeniden auf Grund der derivatographischen Kurven. Acta chim. Acad. Sci. hung. 48 (1966) 225
[7.172] *Pearson, D.:* The Application of the Optical Microscopy in Studies of the Stabilisation of PVC. Microscope 16 (1968) 243–256
[7.173] *Emons, H.-H., and J. Näther:* Über die Reaktion $CaSO_4 + 2\,KCl = K_2SO_4 + CaCl_2$ und ihre Deutung in ammoniakalischer Lösung. Bergakademie [Freiberg] (1969) 310–313 and 355–359
[7.174] *Woodruff, D. P.:* Morphologie der Grenzfläche fest-flüssig während des Schmelzens. Philos. Mag. (8) 18 (1968) 151, 123–127
[7.175] *Heimann, R., and W. Franke:* Kinetik und Morphologie der Auflösung von Quarz I–V. Neues Jb. Mineralog. Mh. (1969) 145, 161, 305 and 413
[7.176] *Zatloukal, J., Nývlt, J., and L. Provaznik:* Mikroskopische Untersuchungen des Wachstums und der Auflösung von Kristallen. Chem. Listy 61 (1967) 11, 1521–1525
[7.177] *König, G.,* et al.: Vorgänge bei der Auflösung von Hart- und Weichbrandkalken in Schlacken aus dem Sauerstoffaufblas-Konverter. Stahl u. Eisen 87 (1967) 1071–1077
[7.178] *Felbinger, A., and H. Neels:* Die Natriumbikarbonat-Kristallisation im Ammoniak-Soda-Prozeß. Kristall u. Technik 1, (1966) 1, 137–146
[7.179] *Myl, J., and Z. Solc:* Beobachtung von Strömungen bei der Züchtung von Einkristallen aus Lösungen. Kristall u. Technik 2 (1967) 2, 217–219
[7.180] *Scherzberg, H.:* Bildungsverhältnisse der Alkalisulfate in organischen Selektivlösungsmitteln. Dissertation, Friedrich-Schiller-Universität Jena, 1964
[7.181] *Stag, A., and H. Reumuth:* Beobachtung über eine Kristallisationsbeeinflussung bei kristallinen chemischen Verbindungen durch Zugabe von nichtionischen Tensiden zur Mutterlauge, Mikroskopie 22 (1967) 242–252
[7.182] *Leray, J. L.:* Wachstumskinetik von Kupferchloridhydrat. J. Crystal Growth 3/4 344–349
[7.183] *Hide Toshi Miyazaki und Jun Mizuguchi:* Microscopical Observation of Crystallisation of $CaSO_4$-Hemihydrat. J. chem. Soc. Japan, ind. Chem. Sect. 70 (1967) 4, 423–425
[7.184] *Krings, A.:* Mikrokinematographische Aufnahmen an dynaktiven Flüssigkeitspaaren. Z. wiss. Mikroskop. mikroskop. Tech. 63 (1958) 391–396
[7.185] *Monro, P. A. G.:* Methods for Measuring the Velocity of Moving Particles under the Microscope, in Barer, R., and V. G. Cosslett: Advances in Optical and Electron Microscopy, Vol. 1. London/New York: Academic Press 1966
[7.186] *Hinze, E., and A. Neuhaus:* Auflichtmikroskopie und röntgenographische Untersuchungen an Cu_2S, Cu_2Se und Cu_2Te bei Drücken bis 80 kbar. Naturwiss. 56 (1969) 3, 136
[7.187] *Muzžerin, J. V.:* Mikroskop kak refraktometr. Zap. Vsesojuzn mineralog. Obšč. 89 (1960) 4, 473–483
[7.188] *Jones, F. T.:* Refractive Index Determination for Liquids by Crystal Rotation, Microscope 15 (1967) 8, 309–313

[7.189] *Conroy, J., Gottlieb, M., and M. Garbung:* Mikroskopie in supraflüssigem Helium (Auflicht). Cryogenics 5 (1965) 348–349

[7.190] *Shead, A. C.:* Mikrokristallisation durch spontane Verdampfung verdünnter gesättigter Lösungen schwerlöslicher Substanzen. Mikrochim. Acta (1967) 6, 1077–1079

[7.191] *Latiére, H. J.,* et al.: Beobachtung der Bildung und des Wachstums von Kupferwhiskern mit Hilfe des Mikroskops. Mém. Sci. rev. Metallurgie 64 (1967) 2, 169–176

[7.192] *Benedetti-Pichler, A. A., and J. Vikin:* Schmelzreaktionen unter dem Mikroskop. J. chem. Educat. 43 (1966) 421–422

[7.193] *Kofler, A.:* Die Kristallisationsvorgänge in unterkühlten Mischschmelzen organischer Stoffe. Mikroskopie 3 (1948) 193

[7.194] *Adamski, J. A.:* Polariskop zur Beobachtung des Kristallwachstums aus Schmelzen. Rev. sci. Instruments 39 (1968) 8, 1208–1209

[7.195] *Ueberreiter, K.,* et al.: Kristallisationskinetik von Polymeren. Kolloid-Z. u. Polymere 222 (1968) 97–102, 233 (1969) 849–856, 234 (1969) 1083–1092

[7.196] *Keller, A.,* Solution Grown Polymer Crystals. Kolloid-Z. u. Z. Polymere 231 (1969) 1/2, 386–418

[7.197] *DAS 1 097 167 of 12. 1. 1964*
Barett, P. T.: Polarisationsoptische Verfahren zur Regelung der Gleichmäßigkeit der Eigenschaften von schmelzgesponnenen Fäden während des Spinnens.

[7.198] *Fitchmun, D. R., and S. Nowman:* Surface Crystallisation of Polypropylene. J. Polymer Sci. A 2, 8 (1970) 1545–1564

[7.199] *Ishibashi, T.,* et al.: Studies on Melt Spinning of Nylon-6. J. appl. Polymer Sci. 14 (1970) 1597–1613

[7.200] *Harvey, E. D., and F. J. Hybart:* Rates of Crystallisation of Copolyamides. J. appl. Polymer Sci. 13 (1969) 2643

[7.201] *Kuczynski, G. C.,* et al.: Study of Sintering of Polymethylmetacrylate. J. appl. Polymer Sci. 14 (1970) 2069–2077

[7.202] *Kuhnert-Brandstätter, M.:* Spiralen- und Schichtenwachstum an Kristallen aus der Dampfphase. Ber. Bunsenges. physik. Chemie 56 (1952) 968

[7.203] *Glas, J. P., and J. W. Westwater:* Measurements of the Growth of Electrolyte Bubbles. Int. J. Heat Mass Transfer. 7 (1964) 1427–1443

[7.204] *Bunge, H. J.:* Zum augenblicklichen Stand und den Entwicklungstendenzen auf dem Texturgebiet. Kristall u. Technik 6 (1971) 3, 325–334

[7.205] *Exner, H. E., and G. Petzow:* Gefügeuntersuchungen mit neuen Methoden der stereometrischen Analyse. Bundesministerium für wiss. Forschung, Bericht K 67–92, 1967

[7.206] *Blaschke, H.:* Quantitative Gefügeanalyse feinkörniger Mineralparagenesen mit dem Polarisationsmikroskop. Z. wiss. Mikroskop. 67 (1965) 1–18

[7.207] *Schläfer, U.:* Fehlermöglichkeiten bei der metallographischen Texturbestimmung. Kristall u. Technik 3 (1968) 3, 467–471

[7.208] *Lücke, K., and R. Rixen:* Rekristallisation und Korngröße. Z. Metallkunde 59 (1968) 4, 321–333

[7.209] *Schimmel, F. U.:* Lichtmikroskopische Untersuchung von Faser- und Gewebeoberflächen. Melliand Textilber. 48 (1967) 3, 287–291

[7.210] *Mackowsky, M. T.:* Ergebnisse der Koksmikroskopie mit Hilfe der verschiedenen Untersuchungsmethoden. Brennstoff-Chemie 36 (1955) 304–314

[7.211] *Schüller, K.:* Gefügeuntersuchungen an Porzellan mit Auflichtmikroskopie und Elektronenmikroskop. Ber. dtsch. keram. Ges. 43 (1966) 649–653

[7.212] *Pavlova, N. Z.,* et al.: Mikrostruktur von Katalysatoren zur Synthese von Ammoniak in verschiedenen Stadien der Reduktion. Kinetika i Kataliz 9 (1968) 6, 1390–1391 (Russian)

[7.213] *Tardy, P.:* Die unmittelbare Beobachtung des Rekristallisationsvorgangs mit dem Jena-Elypovist-Gerät. Jenaer Rdsch. 14 (1969) 4, 242–244
[7.214] *Sander, B.:* Einführung in die Gefügekunde der geologischen Körper, Teil I und II. Wien: Springer-Verlag 1948/50
[7.215] *Fischer, G.:* Über die Auswertung von Gefügediagrammen. Abh. Dtsch. Ak. Wiss. Berlin, Kl. III (1960) 1, 283
[7.216] *Drescher-Kaden, F. K.:* Zur Darstellung des Regelungsgrades eines Gefüges. Tschermaks mineralog. petrogr. Mitt. 3 F, 4 (1954) 159–177
[7.217] *Durrance, E. M.:* Photometrische Methode zur Bestimmung der bevorzugten Orientierung von Kristallaggregaten im Dünnschliff. Geol. Mag. 104 (1967) 1, 18–28
[7.218] *Hahnheiser, H., and H.-H. Seyfarth:* Zur mikroskopischen Prüfung von Luftfilter-Fasermatten. Jenaer Rdsch. 15 (1970) 4, 225–226
[7.219] *Böhlen, B.:* Methoden zur Mikrostruktur- und Texturbestimmung an porösen Feststoffen. Keram. Z. 19 (1967) 726
[7.220] *Greif, H.:* Zur optischen Porositätsmessung an Koksen. Chem. Techn. 15 (1963) 490–491
[7.221] *Trojer, F.:* Verdrängungserscheinungen in Kristallgesellschaften von Industrieprodukten. Neues Jb. Mineralog. Mh. 94 (1960) 1425–1440
[7.222] *Petzow, G., and H. E. Exner:* Zur Kenntnis peritektischer Umwandlungen. Radex-Rdsch. (1967) 3/3, 534–539
[7.223] *Pelzbauer, Z.:* Morphologie der Polymeren in Pulverform. Staub-Reinhalt. Luft 27 (1967) 5, 233–237
[7.224] *Rüprich, G.:* Einige Fragen der Plastographie. Plaste u. Kautschuk 17 (1970) 6, 292–293
[7.225] *Brennschede, W.:* Spärolithische Struktur synthetischer Hochpolymerer. Kolloid-Z. 114 (1949) 1
[7.226] *Schönefeld, G., and S. Wintergerst:* Beeinflussung von Struktur und Festigkeit von Polypropylen durch Wärmebehandlung. Kunststoffe 60 (1970) 3, 177–184
[7.227] *Kowatschewa, R., and S. Somerdjiev:* Verfahren zur Entwicklung und lichtmikroskopischen Untersuchung der Überstruktur von Polyäthylen. Prakt. Metallographie 9 (1972) 3, 147–160
[7.228] *Wiegand, H., and H. Vetter:* Molekulare Orientierung in Spritzgußteilen als Folge der Verarbeitung. Kunststoffe 57 (1967) 14
[7.229] *Friess, S., and W. Gilde:* Mikroskopische Untersuchungen an PVC-Schweißungen. Plaste u. Kautschuk 17 (1970) 1, 29–36
[7.230] *Heymer, A., Schneider, W., and W. Helbig:* Die Kristallmikroskopie als Hilfsmittel bei der Optimierung der Paraffingewinnung. Chem. Techn. 19 (1967) 4, 253

8 *Future trends in chemical microscopy*

The development of chemical microscopy during the past decades and the present state of the art in methods and instrumentation permit to foresee a number of future trends, which are already beginning to be effective.

8.1 Standardization of measuring techniques, and automation of data acquisition

With its specific stamps in the chemical, biomedical and geo-sciences, microscopy as a method of measurement is now at the beginning of a new stage of development. Hesitantly, but unmistakably, it is taking the step from visual to automatic data acquisition – a step that other physical analysis methods, firmly established in chemistry, have taken before or even been able to skip. The lagging behind of microscopy is due to two reasons. Firstly, unlike »blind« methods of physical analysis, the presence of an optical image of the object minimized the necessity to automate data logging. And secondly, as measurement procedures with polarized light are relatively complicated, the type of data acquisition technology known from other measuring apparatus used in chemistry was not feasible to be manufactured economically and practically employed by microscopists before microelectronics achieved its present state. The first field in microscopy now being conquered by automation at a fast rate is geometrical image analysis. For chemical microscopy, though, it is of less interest than a breakthrough in the field of polarizing methods. We can expect easier-to-handle, largely automated instruments to appear on the market in the near future. The development of standardized measuring procedures that are widely applicable to problems in chemistry is still in its infancy, too. A broad variety of applications and measurements in chemical microscopy have been reported to date, but still there is a lack of published studies investigating the universal validity of such methods and their transferability to other chemical applications. Thus, future publications in chemical microscopy will have to deal no longer with favourable application examples only but, first of all, with widely applicable, standardized techniques. Apart from measurement itself, such standard procedures should more than before include sampling and specimen preparation.

Automation in microscopical data acquisition will first gain acceptance in serial tests performed in automatic continuous process analysis and process control systems, and in measurements of variations of optical data; high-speed data recording will open up new possibilities, especially in hot-stage microscopy.

8.2 Development in kinematic microscopy

Kinematic microscopy has so far lagged behind static methods, mainly because facilities for automatic data recording were lacking. The method holds untapped potential for a wide variety of interesting measurements in chemistry. Prospective applications are all physico-chemical processes that involve solid phases.

The development of methods and equipment tends towards the design of suitable microreaction vessels and standardized methods on the one hand, and towards continuous process analysis and control (combined with automatic data acquisition) on the other. Objectives are the measurement of kinetic quantities (rate and order of reaction, activation energy), the registration of the mechanism of complex processes, and the simulation of engineering processes in microreaction vessels.

8.3 Analysis of decomposing substances; measurement at high and low temperatures

To examine decomposing objects is the more difficult, the less stable the substances are. At high and low temperatures, measuring possibilities are restricted in most cases by unwieldy features of the equipment required to create the special conditions. Still, improvements of such techniques may be expected in the future, because microscopy alone offers the possibility to directly observe the stability or decomposition of the specimen substance during the measurement. Ways to improve things will be the use of new observation techniques (e.g., replacement of phase contrast by interference contrast) and modified equipment outfit (e.g., automatic compensators capable of high-speed registration).

The main emphasis will shift from the visual, qualitative observation of the objects in question towards fast quantitative measurement of a maximum number of optical data.

The information obtainable from such methods may be expected to expand further by new procedures of sampling and preparing specimens of decomposing substances. Box microscopy will, without a greater equipment outlay, permit examination of unstable specimens even under difficult conditions.

8.4 Higher accuracy and expanded possibilities of measurement

Among future developments, priority will be given to an expansion of microscopical methods to that large group of substances whose special properties still prevent the exhaustive use of measuring possibilities (lack of suitable mountants or immersion liquids; necessity to modify classical measuring techniques). This goes hand in hand with an increase in the accuracy of measuring the optical parameters of these substances, which has been lower compared to substances permitting the common precision measuring techniques to be readily applied. Methodical possibilities in combining polarizing with interference or phase microscopy may also be expected to expand.

With regard to methods, especially in chemistry, there is still much new ground to be broken. The implementation of fast, automatic data logging will open up new possibilities to kinematic microscopy, especially for quantitative hot-stage methods. The further development of combinations between microscopes and equipment for X-ray diffraction, electron microscopy, microprobing, laser microspectral analysis or micro-DTA largely depends upon the rate at which data acquisition in microscopy becomes easier.

Annex

Illustrations for main text

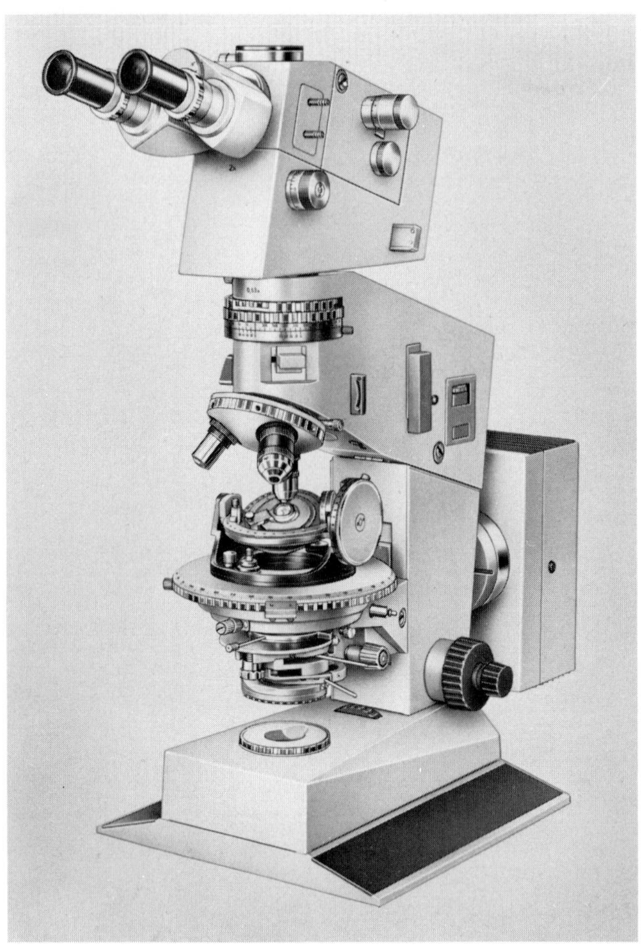

a)

Figure 3.1. Modern microscopes for transmitted and reflected light

a) Amplival-pol-interphako with universal stage (photo: courtesy JENOPTIK JENA GmbH)

b) Morphoquant (photo: courtesy JENOPTIK JENA GmbH)

c) liquid-cell microscope setup for interferometry on optical bench (joint laboratory development of »Carl Schorlemmer« Technical University, Leuna-Merseburg, and JENOPTIK JENA GmbH)

b)

c)

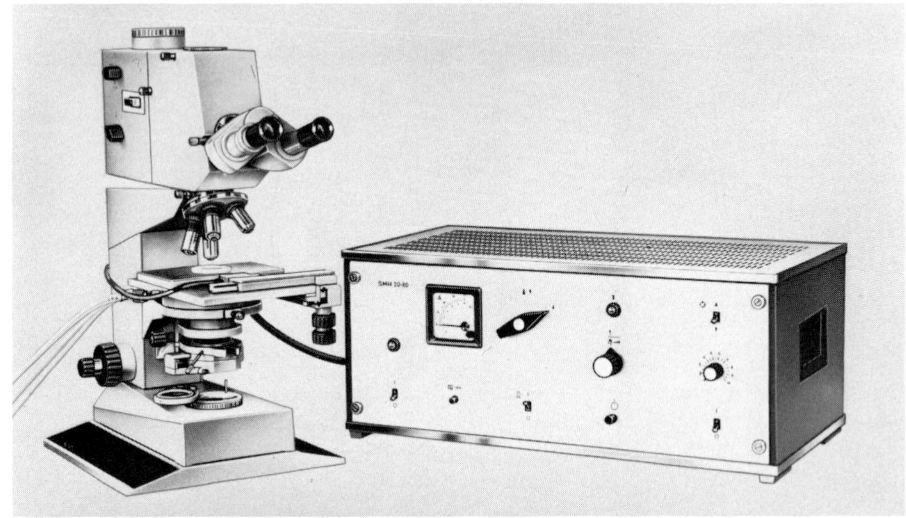

Figure 3.5. Peltier hot and cold stage, − 20 to + 80 °C (photo: courtesy JENOPTIK JENA GmbH)

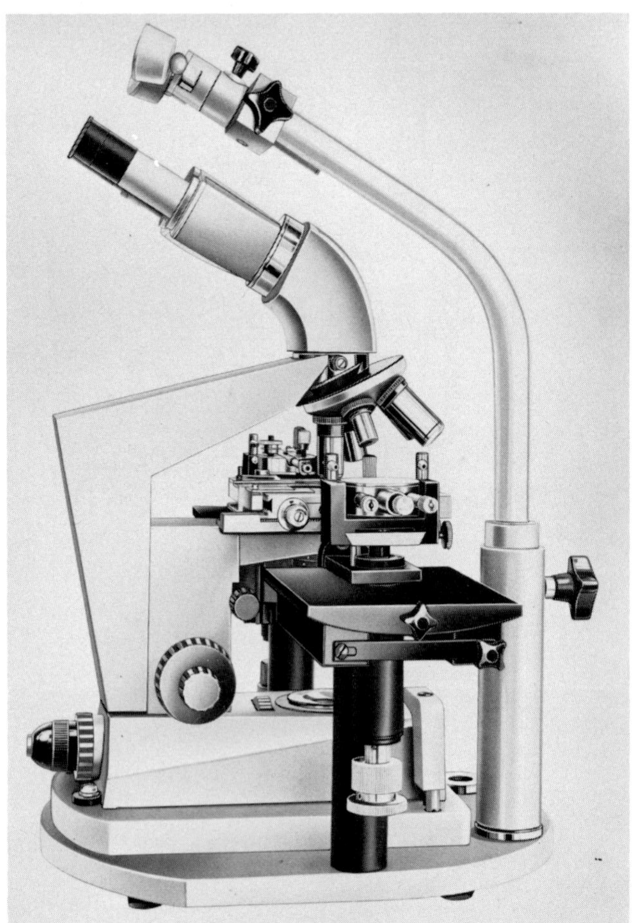

Figure 4.3. Microscopical selection of individual grains using the micromanipulator
a) micromanipulator (photo: courtesy JENOPTIK JENA GmbH)

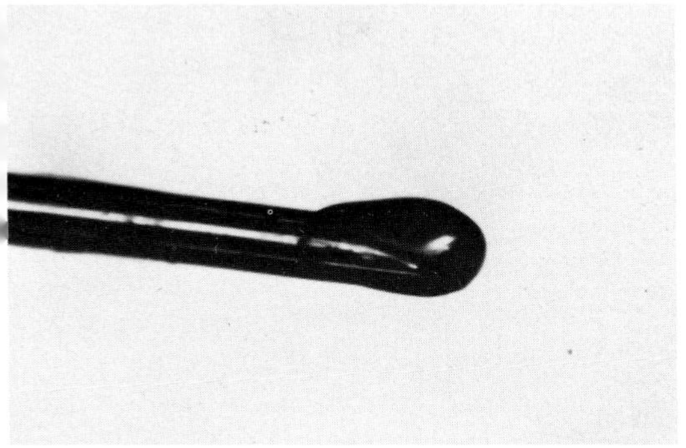

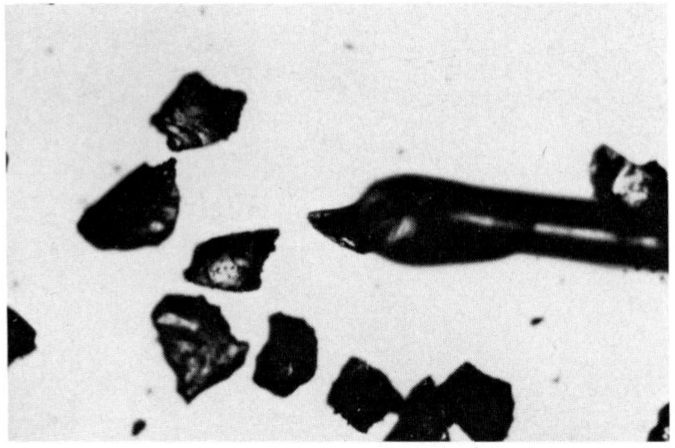

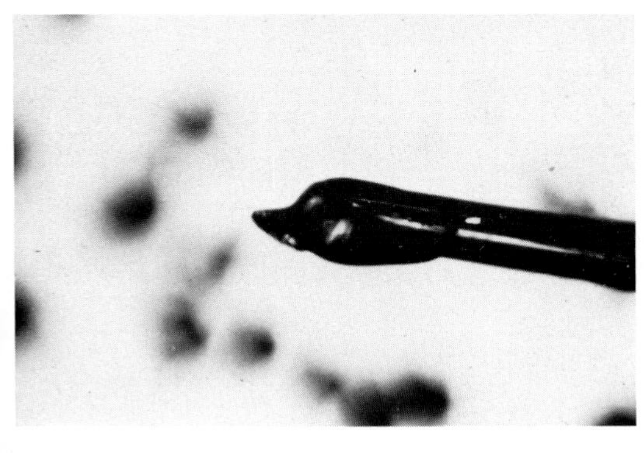

b) sticking a grain to a greased glass point (1 to 4)

Figure 4.4. Fluted sample divider

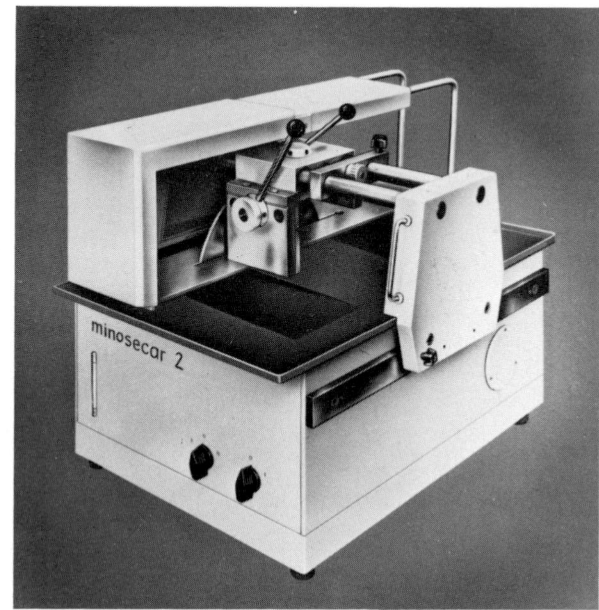

a)

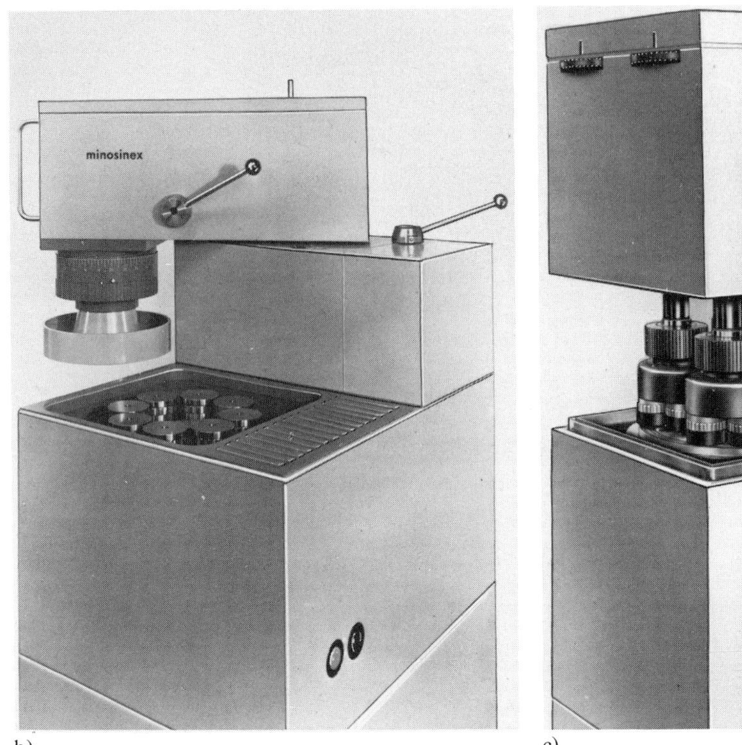

b)

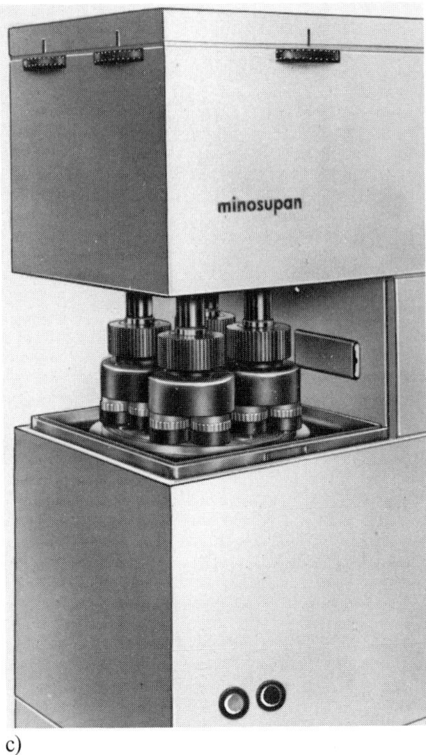

c)

d)

Figure 4.6. Machines for preparing thin sections and polished sections, made by VEB Rathenower Optische Werke RATHENOW (photos: courtesy manufacturers)

a) abrasive cutter
b) thin section grinder
c) polished section grinder
d) polished section (top view)

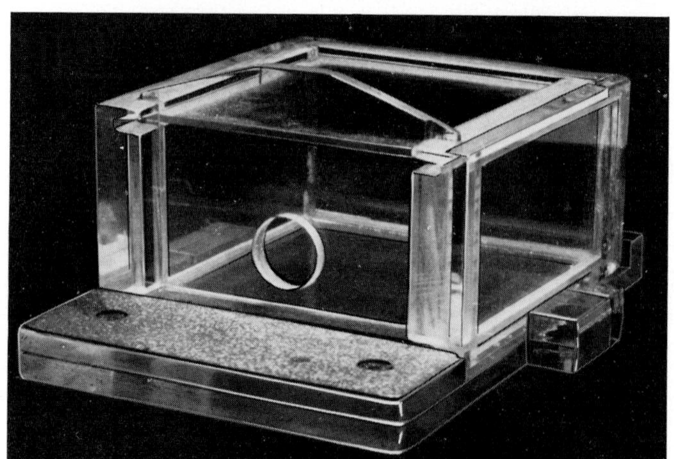

Figure 4.8.
k) humidity chamber

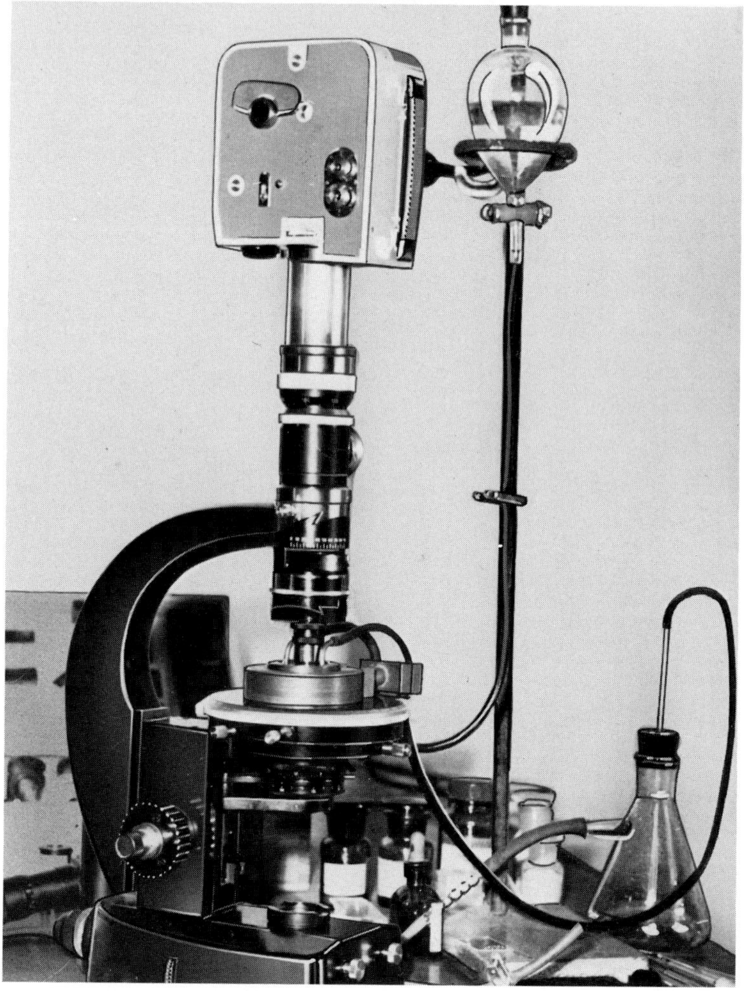

a)

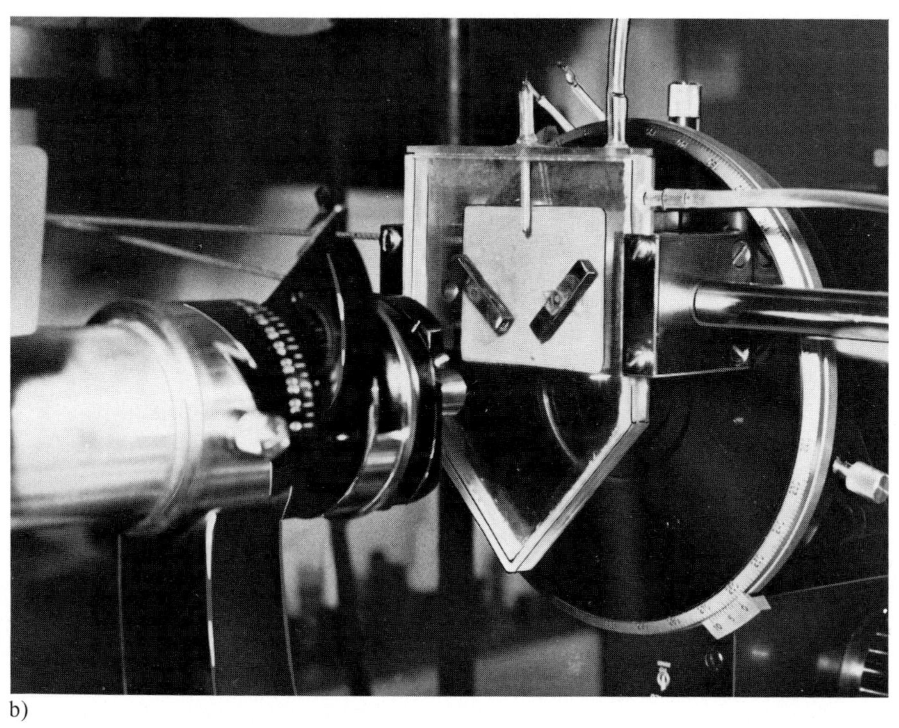

b)

Figure 4.9. Microcells for transmitted-light microscopy
a) flow-through equipment, vertical beam arrangement
b) cell with stirrers, horizontal beam arrangement

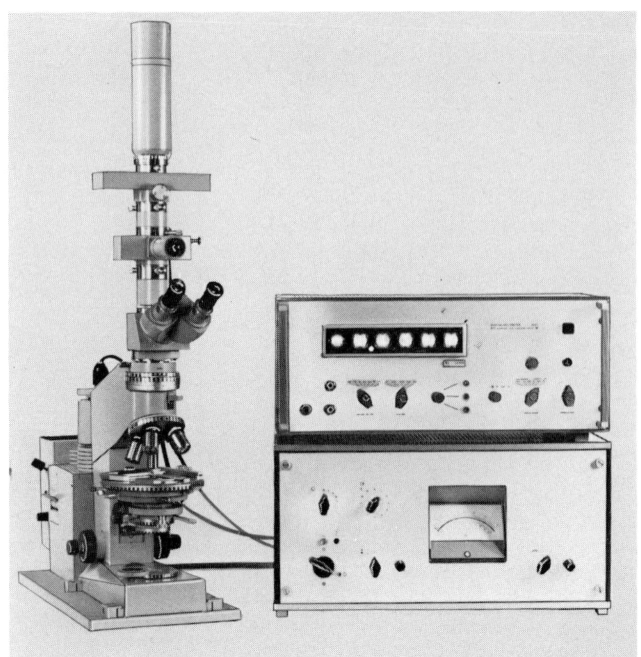

Figure 5.1. Amplival-pol-photometrie, a polarizing microscope with photometer (photo: courtesy JENOPTIK JENA GmbH)

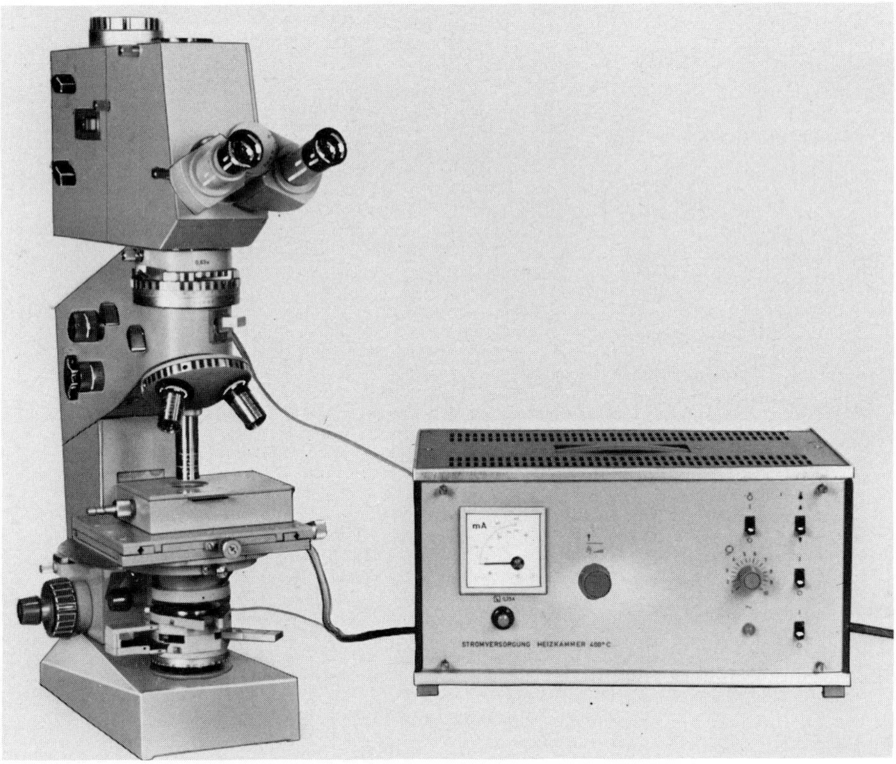

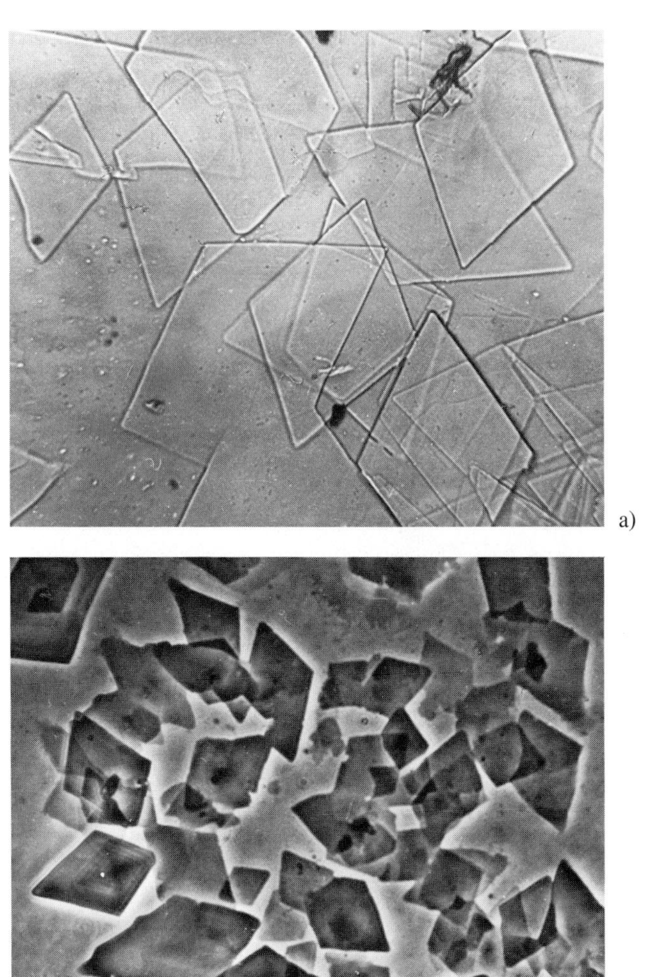

a)

c)

Figure 6.2.
a) Becke line along the edges of paraffin crystals
c) halo around paraffin crystals seen through phase microscope, magnified 220 ×
(photos: courtesy VEB Hydrierwerk Zeitz)

◄
Figure 5.2. Amplival-pol-u, a polarizing microscope for transmitted and reflected light, shown with 400 °C-pol hot chamber (photo: courtesy JENOPTIK JENA GmbH)

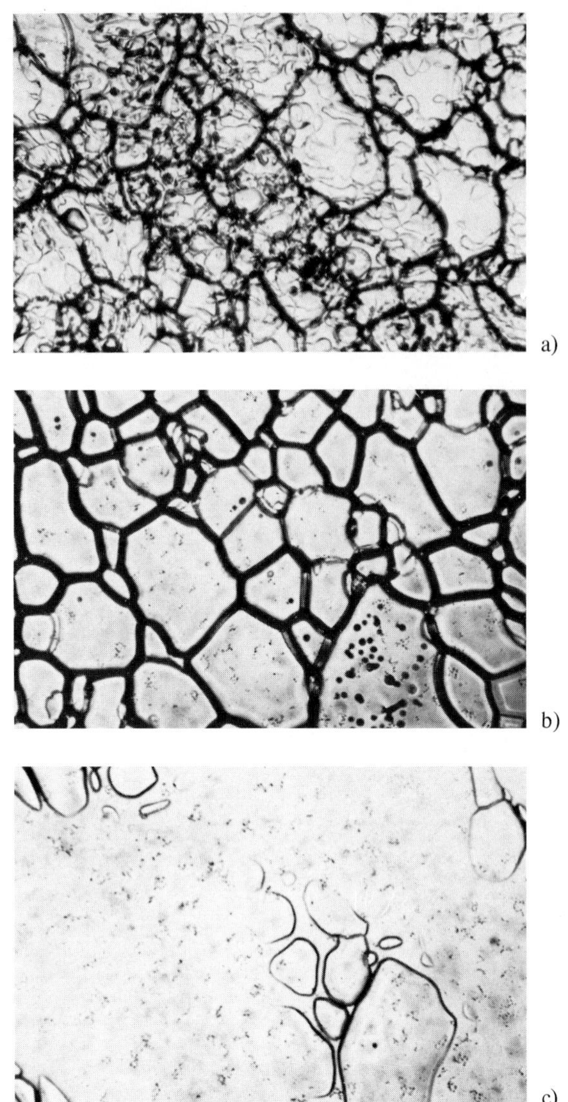
a)
b)
c)

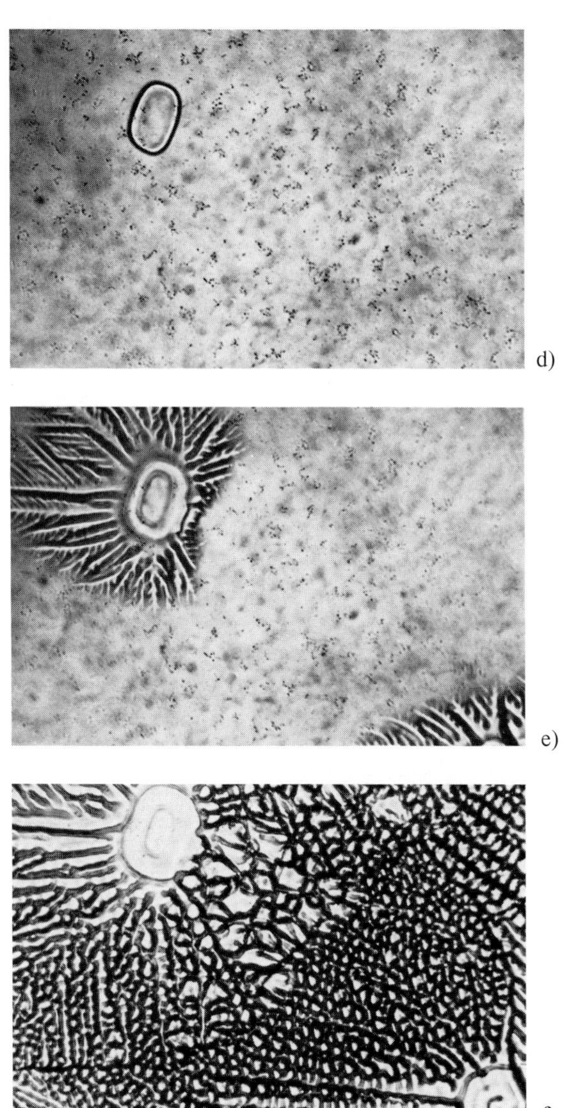

Figure 6.24. Melting and freezing of 1,2-dibromoethane, magnified 220 ×
(photos: courtesy JENOPTIK JENA GmbH)

a) specimen wholly frozen (8.2 °C)
b) grain boundaries melt (9.2 °C)
c) advanced state of melting (9.3 °C)
d) small unmolten remnants only (9.4 °C)
e) crystallization from the melt (9.3 °C)
f) specimen wholly frozen again (8.9 °C)

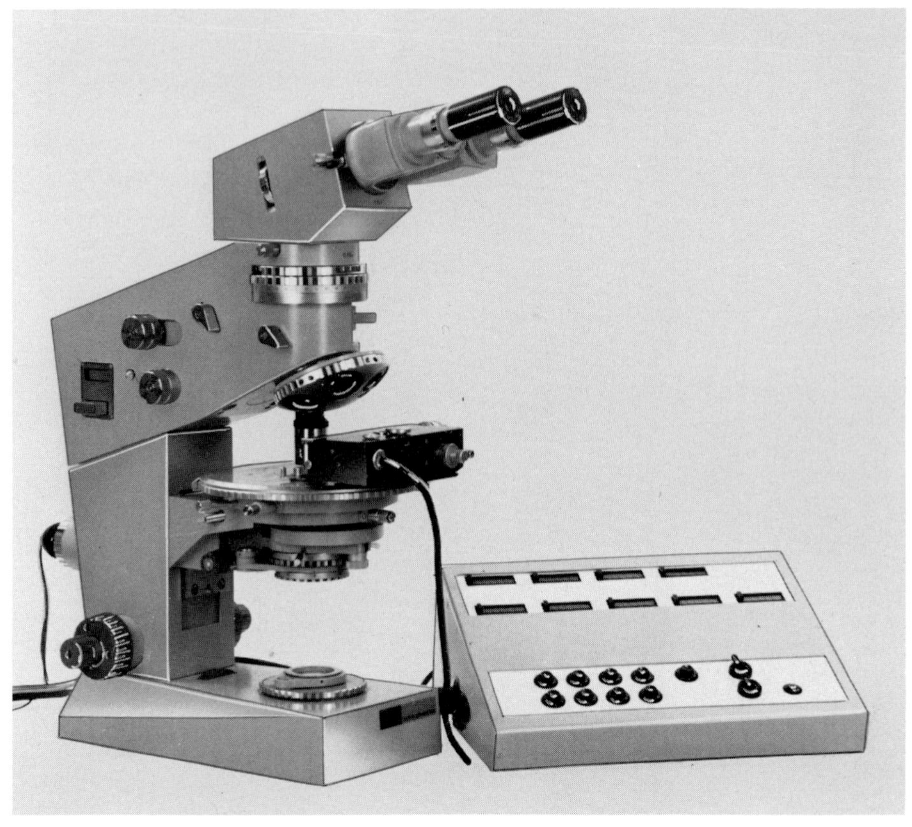

Figure 7.1. Semi-automatic integrating stage (photo: courtesy VEB Rathenower Optische Werke RATHENOW)

Figure 7.5. Preparation box-microscope unit with cold stage for examining decomposing substances and low-temperature work (laboratory setup developed jointly by the »Carl Schorlemmer« Technical University, Leuna-Merseburg, and JENOPTIK JENA GmbH)

Figure 7.13 see page 316

Figure 7.14. Texture of a compressed granulate, 70 ×
(photo: courtesy VEB Kombinat Kali, Sondershausen, DDR, Mineralogical Laboratory)

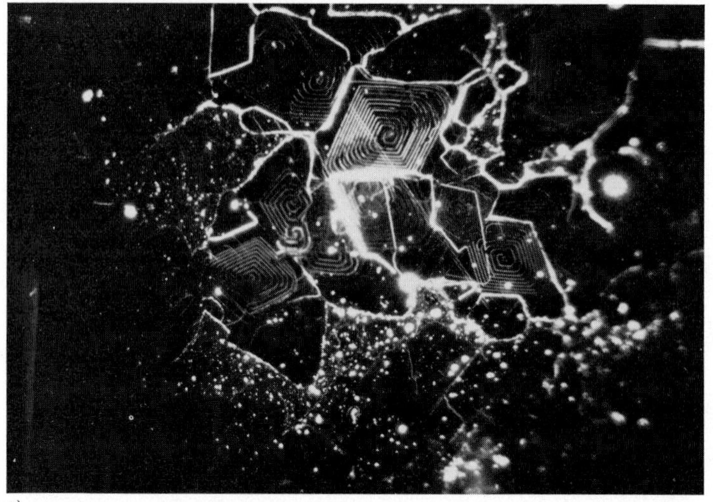

a)

b)

Figure 7.13. Growth forms of paraffin crystals, 220 × (photo: courtesy VEB Hydrierwerk Zeitz)

a) dark field
b) phase contrast

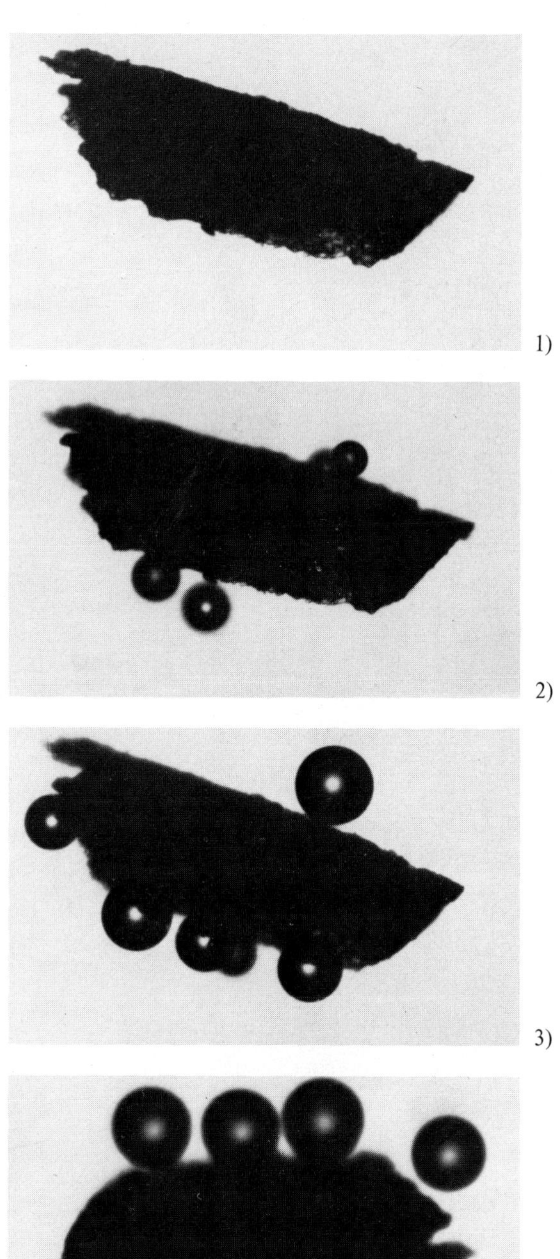

Figure 7.22. Measurement of decomposition rate by size increase of gas bubbles formed in monocrystal-silicon oil specimens (diazonium salts) Course of the process shown on micrographs 1 to 4, 200 ×

1)

2)

3)

4)

Figure 7.25. Interference-microscopical examination of concentration in the liquid phase around a dissolving NaCl crystal in methanol-water mixture (1 to 4, 150 ×)

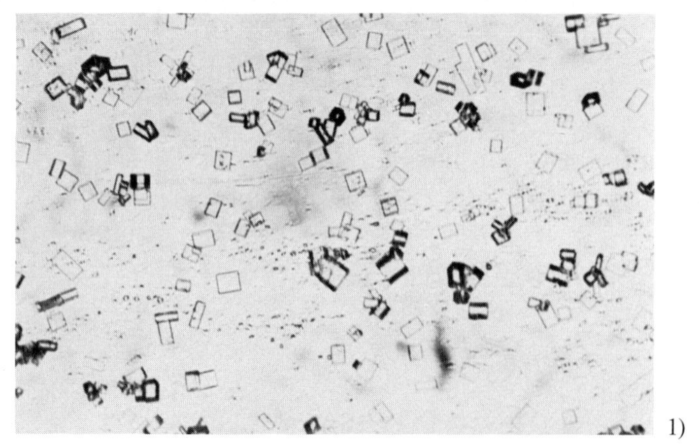

1)

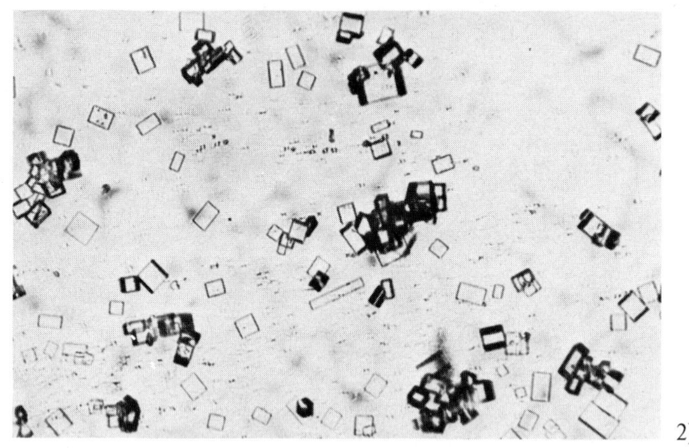

2)

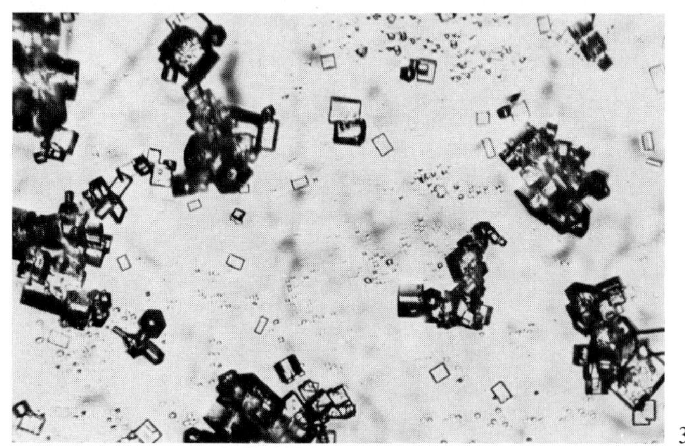

3)

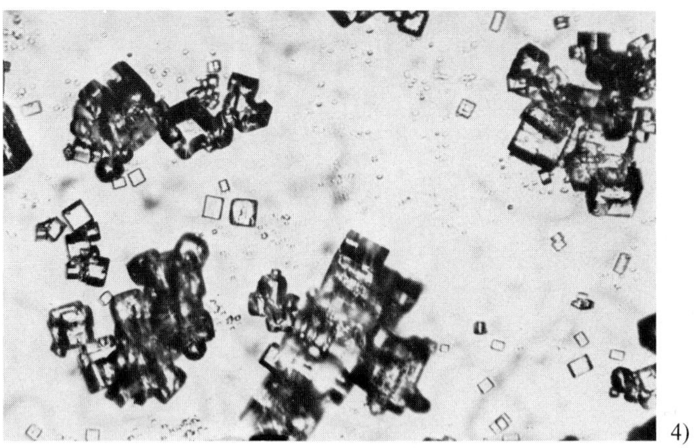

4)

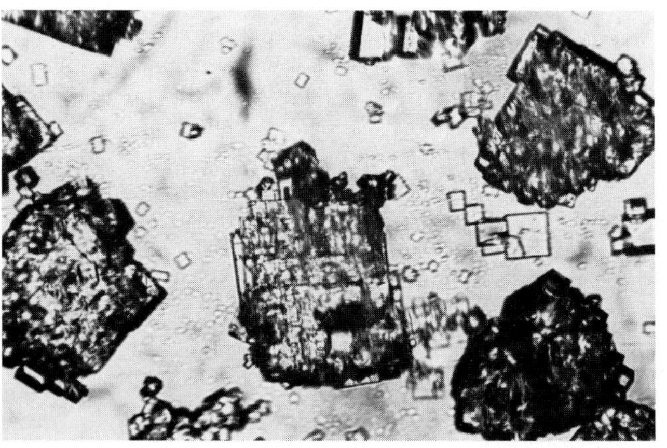

5)

Figure 7.26. Investigation of progressive agglomeration in KCl crystallization by cooling method, using microscopy through sample cell (1 to 5,50 ×)

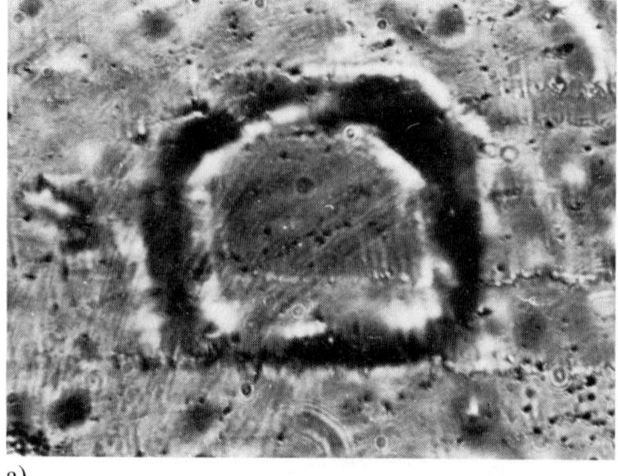

a)

Figure 7.29. Microautoradiographs of KCl crystallisates
a) autoradiograph of a crystal obtained by cooling method from solutions containing ^{42}K
b) autoradiograph of an agglomerate obtained by cooling method from solutions containing ^{42}K

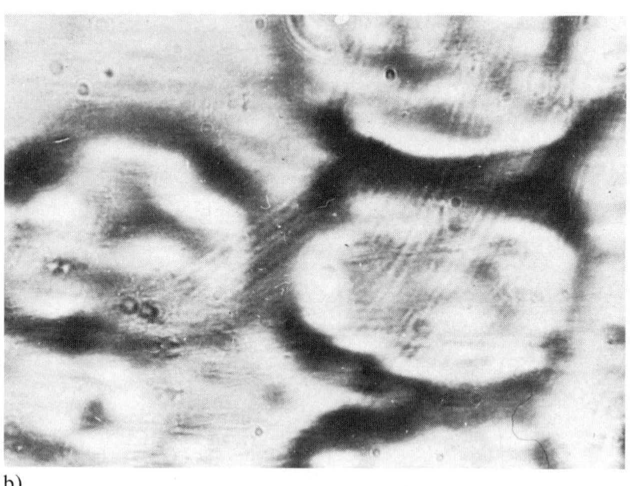

b)

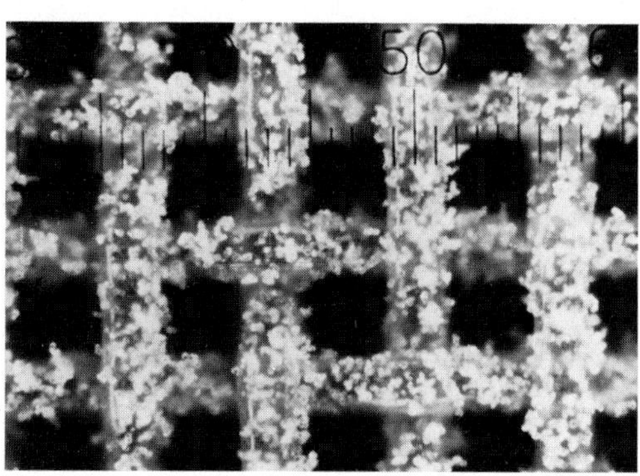

Figure 7.32. Texture of fibre fabric filter (300 ×)

Figure 7.35 see pages 324/325

a)

b)

Figure 7.36. Various forms of paraffin crystals, formed by low-temperature crystallization in lubricating oils (photo: courtesy VEB Hydrierwerk Zeitz)

a) platy crystals (200 ×)
b) acicular crystals (220 ×)

a)

b)

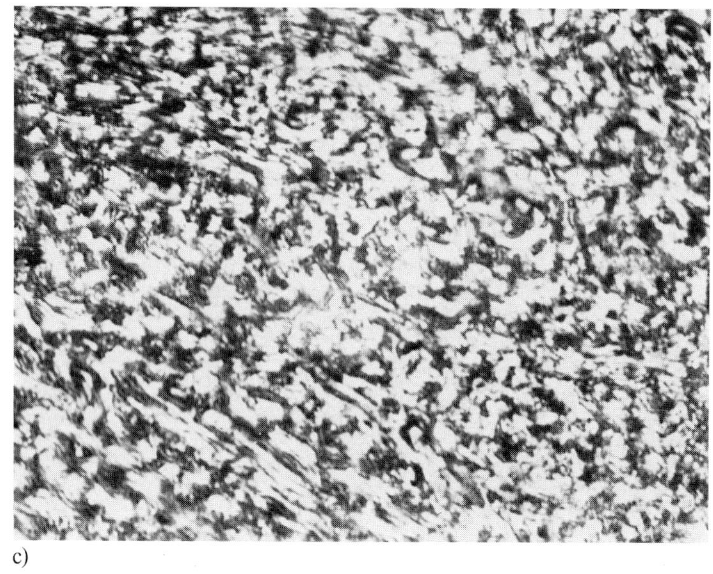

c)

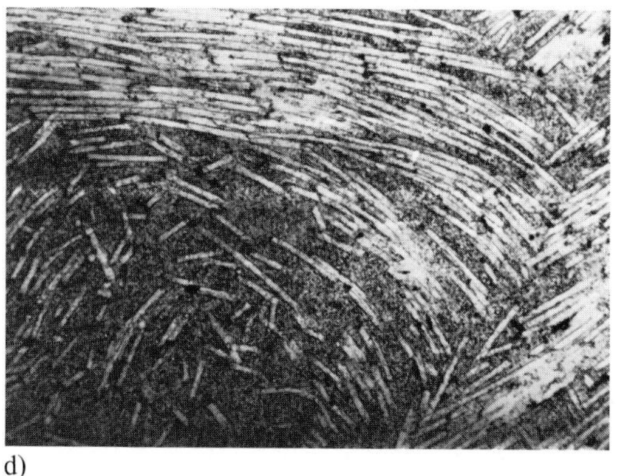

d)

Figure 7.35. Micrographs of semicrystalline organic high polymers

a) individual isolated spherulites (270 ×)

b) closely interlocked spherulite texture (270 ×)

c) blended texture (270 ×)

d) phenolic resin with long glass fibres, oriented, partly broken; thin section, bright field (110 ×)

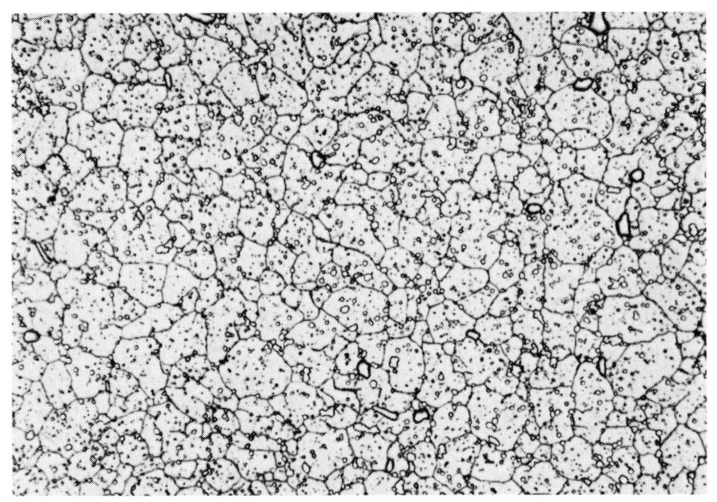

a) 1180 °C

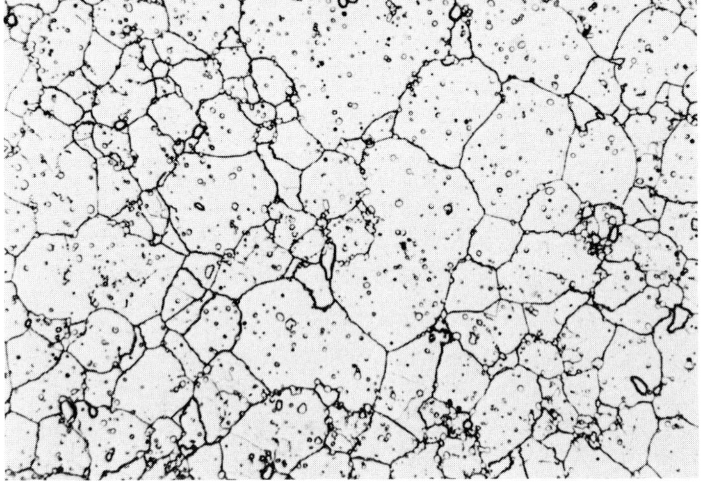

b) 1220 °C

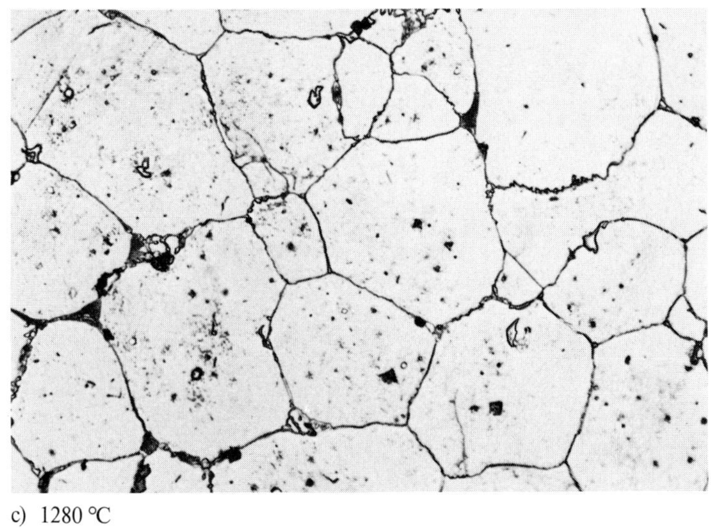

c) 1280 °C

Figure 7.37. Influence of hardening temperature on the texture of highspeed steel, 500 × (photo: courtesy JENOPTIK JENA GmbH)

Thermomicroscopy of Organic Compounds

by Maria Kuhnert-Brandstätter

in association with P.D. Lark
University of New South Wales, Australia

1. Introduction *

More than a century ago, Otto Lehmann was the first to observe microscopically the changes brought about in crystals by varying the temperature. In the course of his life, he gathered a wealth of knowledge by means of hot-stage microscopes, much of which has unfortunately fallen into oblivion. Lehmann developed several hot stages, the last of them, even at this early period, being heated electrically. Although his main interest lay in the fields of physics and crystallography, Lehmann also developed a "chemical" microscope intended for use in the chemical laboratory but even though others, crystallographers such as Groth and botanists like Klein, immediately recognised the value of the microscope equipped with a hot stage, its use in chemistry grew very slowly. It was still not very popular in 1931 when Kofler and Hilbck, after building a series of experimental models, produced a successful hot stage. Compared with Lehmann's last microscope, the numerous subsequent models were simpler in design and construction and in 1934 Kofler even replaced the expensive thermoelectric temperature measuring system with a mercury-in-glass thermometer.

It is certain that Kofler's hot stage would have remained just one among many had he been satisfied to use it only to solve special problems. Instead of this, he devoted all his energies to the development and application of thermomicroscopic methods. As student and collaborator, I observed the work of this indefatigable scientist during the last decade of his life and he inspired in me, at an early point in my scientific career, an interest in and enthusiasm for this field of research. The greatest of Kofler's many achievements was his demonstration that organic compounds could be analyzed qualitatively and quantitatively by purely physical means. This, today, is accepted as a matter of course; he had to prove it in the face of opposition from the chemists of his time.

Thus, Kofler not only advanced the use of the microscope in chemistry; he also advocated the substitution of physical constants for chemical reactions because the former seemed to him to provide a surer and more clear-cut basis for identification. His aim was to use the combination of three physical constants, melting point, eutectic temperature and refractive index of the melt, to make a sure identifi-

* A list of the symbols and abbreviations used in this chapter is given on p. 498.

cation regardless of interference or decomposition of reagent solutions used in colour reactions. Besides this, however, many applications have been found for the hot-stage microscope by other workers. First and foremost among these was Kofler's wife, Adelheid Kofler, who must be remembered for her pioneering work in the fields of polymorphism and isomorphism. Her papers on crystallization in supercooled melts, especially on the isothermal crystallization of supercooled binary mixtures, go far beyond organic chemistry in their importance and serve, among other things, as models in metallography. These special investigations and her methods of analysis of three-component systems will not be described here. The latter depend, in principle, on methods of analysis for binary systems which are described in detail in a later section.

Of Kofler's pupils, R. Fischer deserves special mention for his independent and important contribution to the development of methods for the determination of critical solution temperatures and numerous adaptations of micro methods for use with the hot stage. In the United States, W.C. McCrone has contributed to the development of thermomicroscopy and has helped to popularize it. He is a committed thermomicroscopist even today, although all the newer alternative methods are practiced and taught in his laboratories.

Since it is not our intention to provide a history of thermomicroscopy, we forbear to mention others who have either constructed hot or cold stages or, alternatively, have applied the techniques in fields besides those dealt with in this contribution.

2. Equipment

Since Lehmann's [1,2] invention of the hot-stage microscope in the latter part of the last century, so many hot and cold stages have been described that it is not possible to detail all their modifications, simplifications or special features. Much information on this subject is to be found in books by Emons et al. [3] and Chamot and Mason [4].

In principle, thermomicroscopes or their stages may be classified into three types according to their temperature of operation: viz.

(i) cold stages which normally operate from room temperature down to about −50°C but, in some cases to as low as −265°C with liquid helium as coolant;

(ii) ordinary hot stages which operate from room temperature up to about 300—350° C, a suitable range for the investigation of organic solids; and

(iii) high-temperature devices for the investigation of metals, ceramics and other materials.

Relatively accurate temperature measurement up to about 1800° C is possible with thermal probes but, beyond this, measurement is difficult and unreliable. Of the three types, the last is of little interest in connection with organic substances and therefore only ordinary hot stages and cold stages need be dealt with here.

Commercial hot stages have as their primary requirement long working distance objectives with which overall magnifications of about ×100 can be achieved. This is adequate for most work but if, for some special investigations, higher magnifications are desired, stages which can be operated with shorter working distance objectives are more suitable. Some such are to be found in the literature but they are not available commercially. Hartshorne [5] describes two models which are suitable for workshop construction. The procedures discussed below, with few exceptions (e.g. the observation of spiral growth), call for low magnifications and long working distances.

(A) THE KOFLER HOT STAGE

This hot stage may be mounted on the stage of any commercially available microscope which offers adequate objective working distance but the makers, C. Reichert Optische Werke AG, Vienna, provide a specially adapted microscope fitted with polarizer, analyzer and suitable filter holders to which the hot stage is attached (Fig. 1).

The hot stage itself (Fig. 2) consists of a round metal body which is divided by a horizontal plate into a closed heating chamber below and a specimen chamber above. In the lower chamber is the heating coil, which is fed by an adjustable transformer, the connection to which is on the right-hand side (looking over the limb), a hole through the middle for the passage of light and a thermometer well on the left-hand side (1) which lies close to the underside of the separating plate and ends near the central hole. In the wall of the specimen chamber opposite the thermometer well is a space to take the pin of the object guide. This consists of pin, frame (2) and spindle (3). By means of it the slide may be moved, laterally by the

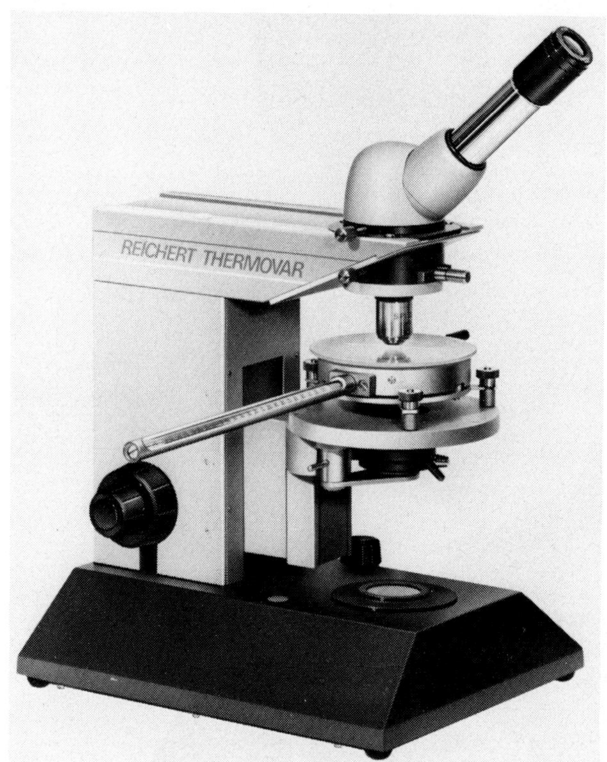

Fig. 1. Reichert Thermovar microscope with Kofler hot stage.

spindle and transversely by swivelling on the pin, when the specimen chamber is closed. A glass bridge (4) is placed over the slide; this helps to maintain a uniform temperature within the preparation and to prevent, especially in the case of highly volatile substances, the formation of sublimates in the field of view. A circular glass plate with a ground edge (5) closes the specimen chamber. The stage can be cooled quickly between individual determinations by placing on it a 2 cm thick aluminium cooling block after removal of the cover plate, glass bridge, slide and guide frame.

The heating rate and temperature are controlled by a low voltage adjustable transformer (Fig. 3) whose control knob (1) is surmounted by a temperature scale (2). When the knob is turned so that a temperature graduation on the scale is set against the moulded pointer on the case (3), the temperature rises but the heating rate decreases

Fig. 2. Kofler hot stage.

until it is about 2°C min^{-1} at the set point. The hatched segments of the scale (4) mark ranges over which setting of the knob does not change the rate of heating. Since the line voltage is subject to variation and the room temperature also has some influence, it is best to check the rate of heating at various settings of the transformer scale from time to time.

If a hot stage is being newly commissioned or if it is about to be put back into use after lying idle for a long time, it is desirable to raise it to a high temperature without the glass cover plate in order to dry it. In addition, it is advisable to check its proper operation by

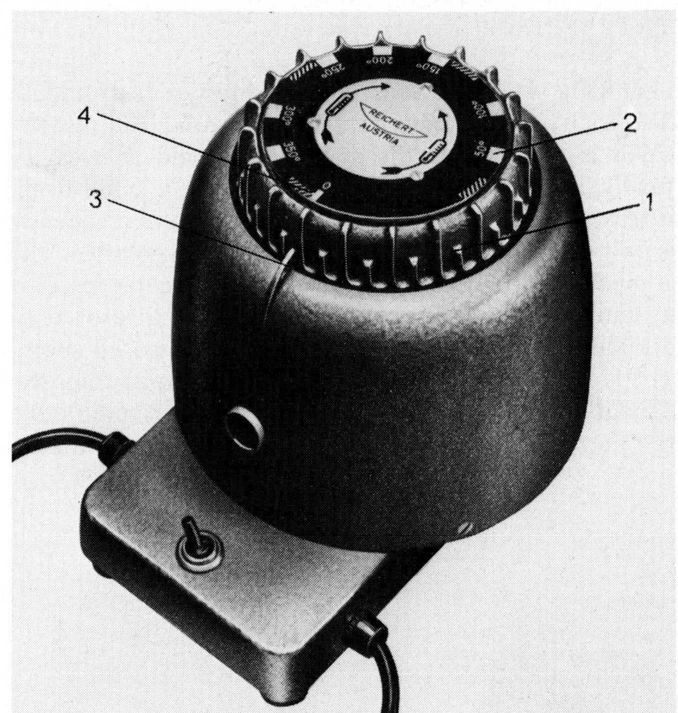

Fig. 3. Transformer for the Kofler hot stage.

determining the melting point of a reference substance, for example phenacetin (m.p. 134.5°C). Errors may arise because the thermometer is not properly pushed home into its well or because the slide does not lie flat on the stage. If the latter occurs, the air gap between the stage and slide has an insulating effect which results in a high temperature reading. Consequently, the slides, whose dimensions are 26 × 38 mm, i.e. the international standard sized slide cut in half, must be correctly registered and slightly loose within the frame of the guide. The Kofler hot stage is provided with two thermometers which range up to 230 and 350°C, respectively. They can be interchanged during an investigation. The manufacturers also advertise a digital thermometer.

Descriptions of Kofler-type cold and high-temperature stages are to be found in the earlier literature [6—8], but they have not attained great significance in thermomicroscopic work.

(B) THE METTLER FP 52 STAGE

This instrument (Fig. 4) is manufactured by Mettler Instrumente AG, Greifensee-Zürich, Switzerland. It is rectangular (unlike the Kofler stage which is round) and the specimen is held in a cavity which is electrically heated top and bottom (1) like a small oven. This allows the temperature to be maintained with greater constancy and to be measured more accurately than is otherwise possible. The measurement is made by a platinum resistance thermometer inserted below the oven and near the observation port. The indicated temperature is the same as the temperature of the sample so no subsequent adjustment is required. The slides used are narrower than the international standard. They are mounted at one end on a small table (2) which offers a narrow range of lateral and transverse movement.

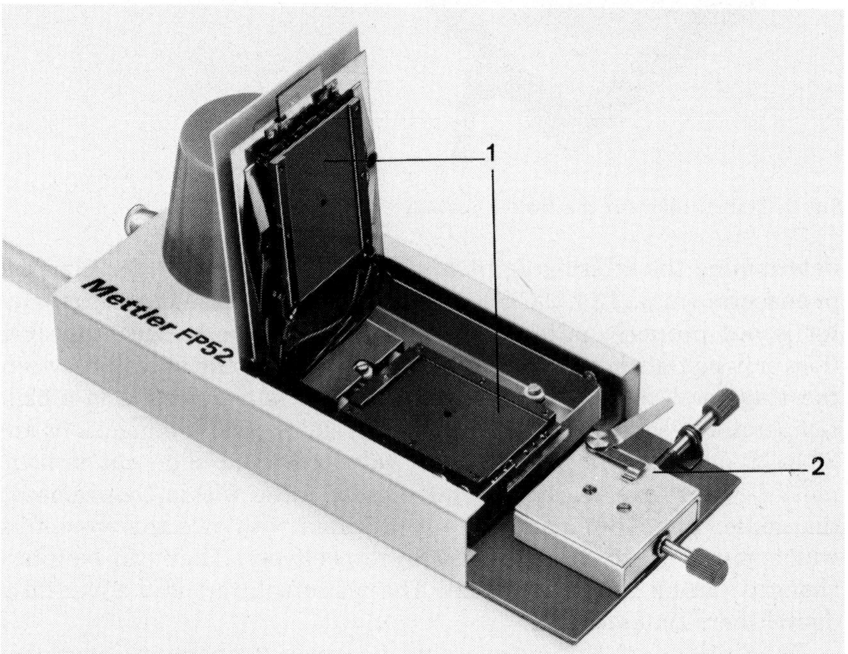

Fig. 4. Mettler FP 52 hot stage (opened).

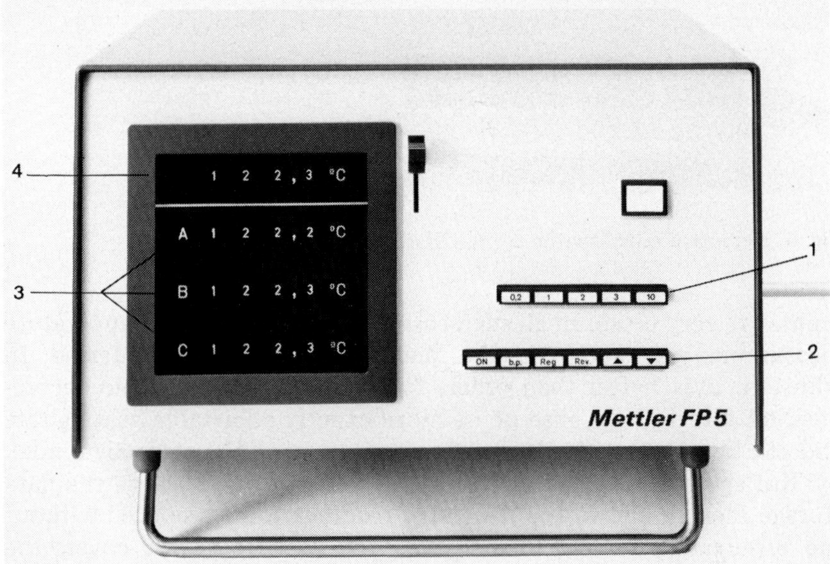

Fig. 5. Control unit for the Mettler FP 52 stage.

An electronic unit (Fig. 5) is used to control the temperature and to display the reading. Very precise heating and cooling rates, ranging from 0.2 to 10°C min^{-1}, are available by push-button selection (1) as are also rapid pre-heating and rapid cooling (2), so that the apparatus can be brought to the desired starting temperature in a short time. The temperature read-out is digital and the associated data storage (3) permits uninterrupted observation of the preparation during critical phases of the investigation. The top display (4) gives the stage temperature continuously but by operating buttons A, B and C on the remote control switch (Fig. 6), three temperature readings of interest can be stored and displayed in the lower windows to one tenth of a degree.The control of the heating rate is exceedingly precise; at a rate of 0.2°C min^{-1}, the variability is only ±0.1°, a precision never before attained with any other stage. Such precision is useful in the thermomicroscopic determination of the melting point if it is to be used for purity testing or in any application in which slight temperature differences are important, for example in differentiating between crystalline modifications of a substance whose melting

Fig. 6. Remote control switch for the Mettler FP 52 stage.

points are very close. In all such cases where very precise temperature measurement is called for, the Mettler stage is to be preferred. In addition, it is better than others for obtaining reproducible decomposition intervals because of its more exactly adjustable heating rate and starting temperature. The determination of the refractive index by Kofler's method (p. 366) is easier to carry out because the data storage facility allows the necessary readings to be recorded without the observations being interrupted. The Mettler FP 52 covers the range from room temperature to 300°C. It may be extended downward to −20°C by passing a cold gas stream directly into the gas inlet port.

With the addition of a lightmeter and potentiometric recorder, the Mettler hot stage can also be used to plot the variation of intensity of transmitted polarized light with temperature [9].

(C) THE LEITZ 350 HOT STAGE

The Leitz 350 hot stage, like the Mettler FP 52, is also a cold stage since it operates from −20 to 350°C, four thermometers being required to cover the range. It works as well with incident as with transmitted light.

The stage (Fig. 7) consists of an outer ring to which the inner working part is attached. The outer ring contains the connections for the power supply and cooling conduits, the object guide support and provision for attachment to any microscope stage. Within this and attached in such a way as to minimize thermal contact are the resistance heating coil, cooling chamber and, in the centre, an adjustable plate with the aid of which the thermometers can be calibrated. The object guide, which enables the slide to be moved 8 mm in two directions at right angles, contains a novel pair of tongs for holding the

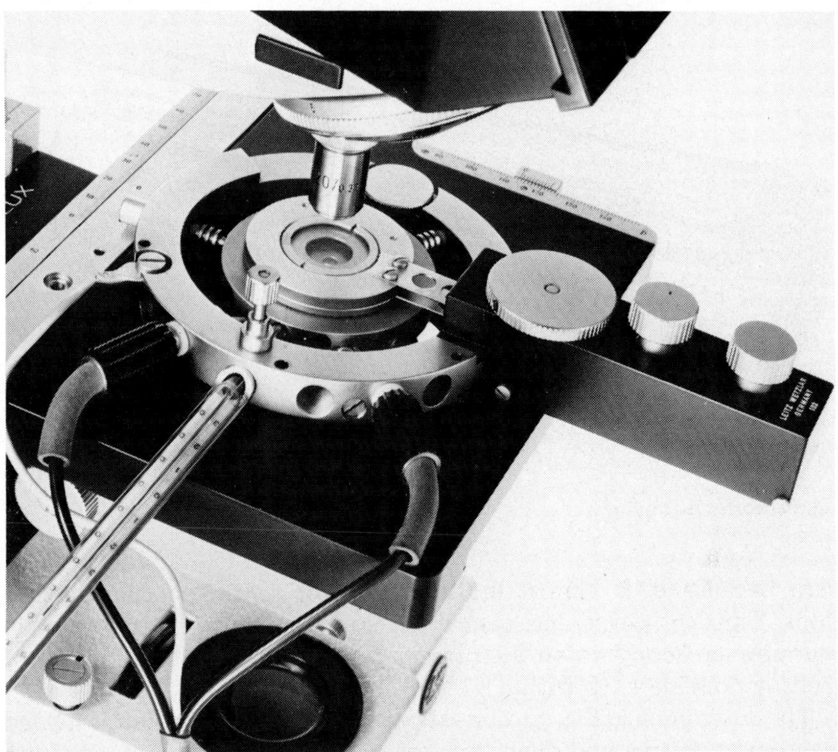

Fig. 7. Leitz 350 hot stage.

square (26 × 26 mm) slides used. The heating is controlled by a low-voltage transformer. A feature worthy of special mention is the thermometer ocular (×10) developed by Leitz. This makes it possible to see part of the thermometer scale together with the specimen in the field of view. The part of the thermometer shown can be changed by moving a prism. In this way, the temperature can be read in comfort without losing sight of the specimen. This ocular can be used only with monocular tubes.

(D) THE KOFLER HOT BENCH

Although this instrument is a piece of apparatus in its own right and not just an adjunct to the hot-stage microscope, it is described

Fig. 8. Kofler hot bench.

here because it is almost indispensable for some special investigations. Time, however, has seen it develop from an auxiliary device in thermomicroscopy to an instrument for the quick determination of melting points etc. to within ±1° without difficulty.

The instrument (Fig. 8) consists of a metal bar (1) which is heated electrically at one end. The temperature gradient so produced (from about 265 to 50°C) is uniform along the bench. The temperature is read from a scale (2) along which moves a sliding block we shall call the runner (3). This has attached to it a pointer (4) which can be lowered on to the middle of the bench. Since the temperature on the surface of the bench depends to some extent on the room temperature, the scale consists of a series of slanting lines, the upper ends of which apply when the room temperature is about 28°C and the lower when it is about 14°C. To take care of variations within this range, another slide with a pointer attached, which we shall call the rider (5), is made to move vertically on the horizontally moving runner. If the instrument is to be used for melting point determination, the reading device must be set by determining the melting point of a reference substance (p. 377) in the same temperature region. For this purpose, some crystals of the reference substance are strewn along the hot bench and the pointer (4) is set to the boundary between crystals and melt (Fig. 9). The rider (5) is then moved so

Fig. 9. Melting boundary and reading device on the Kofler hot bench.

that its pointer indicates the melting point on the temperature scale. When further determinations are made in the same region and under the same ambient conditions, only the runner is moved to the melting boundary (the rider being left in its position) and the correct melting point is read from the position of the pointer on the scale.

This calibration procedure is unnecessary when the hot bench is used simply as an adjunct to the hot stage for making contact preparations or crystal films, ageing or tempering specimens, subliming, etc. Details about the determination of melting points, mixed melting points, eutectic temperatures etc. together with the necessary tables may be found elsewhere [10—12].

(E) EQUIPMENT FOR PHOTOMICROGRAPHY AND CINE PHOTOMICROGRAPHY

All the larger microscope manufacturers offer fully automatic equipment for 35 mm photomicrography which can be used with hot-stage microscopes. With such equipment, focusing and exposure timing are very simple and no preliminary knowledge is required to get good pictures. As a rule, however, such apparatus is mounted on large microscopes with strong light sources and fitted with a separate tube to carry the camera so that it is permanently mounted and ready for use while observation can continue unhindered. Some firms also supply heads containing eyepiece and photographic tubes for their simpler laboratory microscopes. Alternatively, many investigators use an ordinary 35 mm camera without its lens and fix it to the eyepiece tube with a microscope adaptor. The camera manufacturers facilitate this by supplying adaptors with different bayonet fittings to take the better known cameras. Exposure timing is not a problem with through-the-lens metering cameras. Simple microscope illuminators are generally sufficiently powerful for photomicrography by ordinary transmitted light but stronger light sources are often necessary when crossed polars are used.

Commercial equipment is also available for cine photomicrography. As in the case of ordinary photomicrography, there is, on the one hand, elaborate equipment on the market which is fully automatic and offers every convenience and, on the other, we have stock cameras which can be adapted to existing microscopes. Since most cine cameras today have built-in exposure meters, a microscope can be adapted for filming at relatively small expense. Various firms supply a suitable camera adaptor and a strong stand to which it can be attached. It is recommended that a 16 mm camera be used since too much detail is lost with the 8 mm format. Of the various cameras available, the Bolex EL made by Paillard, St. Croix, Switzerland, is especially recommended.

In future, however, it may well be that the video recorder will supersede the camera. Already equipment has appeared which is not unduly expensive for its purpose.

3. Qualitative and quantitative analysis

(A) GENERAL METHODS FOR THE IDENTIFICATION OF FUSIBLE SUBSTANCES

Although it is nearly half a century since L. and A. Kofler introduced their thermomicroscopic method for the identification of organic compounds, scarcely any new developments have occurred in this field. There have, of course, been innovations and improvements in apparatus (see Sect. 2), but the methods for identification have not been improved upon in any essential aspect. It must be said that the Koflers did their work thoroughly.

(1) The determination of melting point and other properties

(a) Sample preparation. For the determination of the melting point, a few crystals (say about 0.1 mg) of the substance are sufficient. The crystals are placed on a suitable slide and covered. (The best size of cover slip has been found to be 15 × 15 mm.) By lightly pressing on the cover slip and rubbing, if necessary, over-large crystals are comminuted and distributed. When doing this, it is important to avoid touching the cover slip with the bare finger because the droplets of fat deposited interfere with vision; the finger tip should be covered with a piece of cloth or, better still, a blunt object (e.g. a pencil eraser) should be used. Extremely large or soft crystals should be ground in a small mortar or between two slides. If this is not done, the cover slip will be lifted too much by the larger crystal fragments and an erroneously high melting point reading may result. Preparations containing too much material (i.e. ones that are too dense) may also result in falsely high values so no more than the above-mentioned quantity should be used, regardless of the amount available. The crystal distribution must be sparse enough for there to be sufficient gaps for the transmission of light and for the observation of sublimation. However, if a substance is highly volatile, a greater amount must be taken so that some crystals still remain under the cover slip at the melting point.

Because the crystals lie loosely scattered between slide and cover slip, water present (either bound as water of crystallization or as free moisture) can escape during heating so that previous drying in a dessicator is seldom necessary. There are exceptions to this rule; in particular, hygroscopic substances or those which have a very low melting point.

The microscopic preparation is placed on the hot stage and, with the aid of the object guide, a suitable part is brought into the field of view. It is almost always best to select the middle of the sample because sublimation or decomposition first involve the edges. Before heating, the preparation is covered with the glass bridge and then the Kofler hot stage is covered with the round glass plate referred to in Sect. 2. The Mettler stage is totally enclosed in any case.

The melting point of a substance may also be determined in a crystal film or solidified drop. This is often necessary, particularly in the investigation of unstable modifications. For the preparation of a crystal film, the substance is melted completely between slide and cover slip and solidified by transfer to a cold surface. If crystallization does not occur immediately, seed formation may be induced by scratching along the edge of the cover slip with a needle. Often, crystallization only occurs on reheating the cold melt. Melting points determined on crystal films are often found to be about one degree lower than on powder preparations. This is to be expected from the fact that hot stages are calibrated with powder preparations whose thermal conductivity is lower than that of a crystal film. This difference does not arise with the Mettler apparatus which provides heating both above and below the preparation. We are assuming, here, that the substance melts without decomposition. The advantage of the crystal film is that it is more uniform in thickness than the powder preparation and gives better reproducibility in the tenth-degree region. Thus this type of preparation will be more efficacious for purity testing or for determining the melting points of polymorphic modifications which differ very slightly. In addition to this, crystal film preparations are used routinely in the thermal analysis of two-component systems (p. 442).

(b) The melting range. In determining the melting range, the temperature of the hot stage is allowed to rise without interruption until the substance has completely melted. The rate of heating is chosen so that the rise in the vicinity of the melting point is about 2° C min^{-1}, so long as the substance is one which melts without decomposition. The smallest crystal fragments melt first, i.e. at the beginning of the range, then the larger crystals follow but the rounding of their corners and edges precedes complete fusion (Fig. 10). The interval

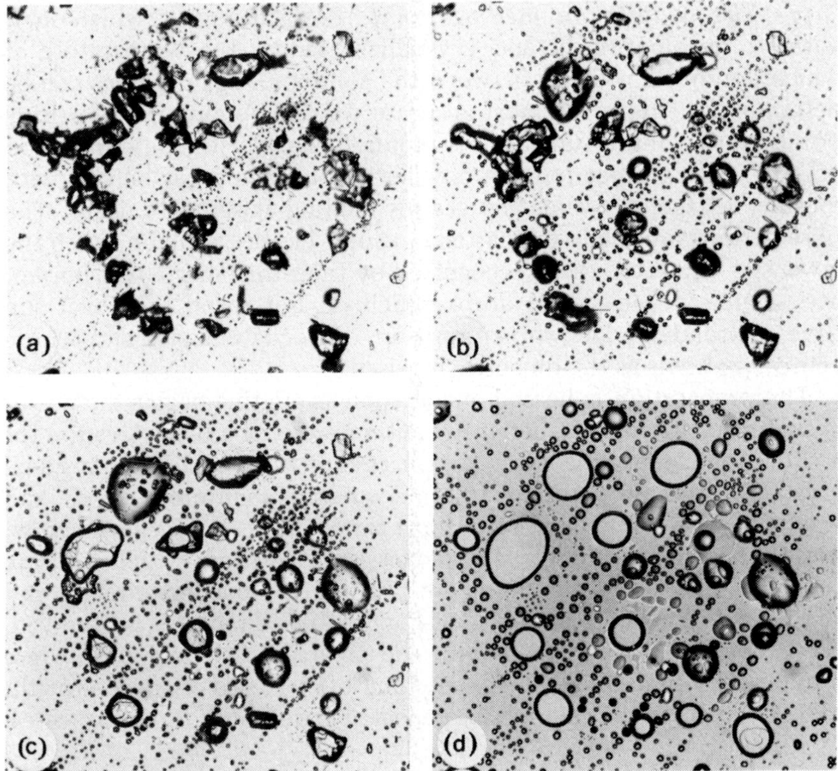

Fig. 10. Melting range of benzil.

between the appearance of the first drops of melt (b) and the disappearance of the last crystal residues (d) is taken as the melting range. It may be as much as one or two degrees for pure substances. Greater ranges are observed for impure substances or if some decomposition takes place. In some cases, polymorphism may also be the cause of a longer melting interval. For example, phenobarbital occurs as modification II, which melts at 172—174°C, but inhomogeneously with the formation of crystals of modification I whose melting point is 176°C. Consequently, the melting process extends over the range from 172 to 176°C.

(c) The equilibrium melting point. For substances which melt without decomposition and crystallize readily, the temperature at which equilibrium exists between the solid and liquid phases can be determined. This is the most accurate way of obtaining the melting point. Rather more substance is required than for the determination of the melting range but the powder preparation is, as before, with loosely scattered crystals and groups melting as separate drops. The observations are made on flattened drops, i.e. those which touch the cover slip. These are easily recognized by their thin sharp dark border; the smaller dome-shaped drops which do not reach the cover slip have a broader edge whose position of focus changes and whose centre brightens as the objective is raised.

The preparation is heated and as soon as melting begins the power is switched off; the temperature continues to rise a little, steadies for a short time, then falls again. In the first of these stages, the residual crystals in the drops of melt shrink quickly, then more slowly and finally remain unchanged for a short time before they begin to grow, more and more rapidly. The edges and corners, which became rounded during the melting phase [Fig. 11(a)] become straight and sharp as soon as growth restarts [Fig. 11(b)]. When this is observed, the preparation is reheated until the crystals lose their sharp edges and corners once more; then switching off leads to renewed growth, and so on. The process can be repeated indefinitely in many cases.

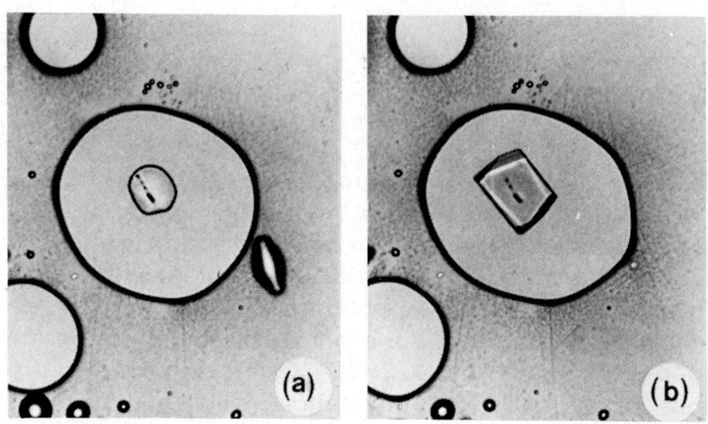

Fig. 11. Equilibrium melting point of pentetrazole.

The temperature should be read on the rise when only a few residual crystals remain in the drop of melt.

Unpracticed observers find difficulty at first in holding the preparation at equilibrium. The commonest mistake is switching off the power too late with the result that the material melts completely. Since organic substances supercool very readily as a rule, there is little point in cooling the melt until it recrystallizes spontaneously; it is better to start again with a new preparation. The opposite mistake is to switch off the power too soon so that only a small amount of melt forms beside the largely intact crystals. It is necessary that the "final equilibrium" be achieved; i.e. that between small residual crystals and relatively large amounts of melt.

The operation of adjusting the temperature to restore equilibrium is best carried out with the specimen between crossed polars. The thin residual crystals are often not easy to distinguish from the melt because of similarity in refractive index. However, one should avoid the use of crossed polars from the very beginning of heating when determining the melting point because, first, it makes the onset of melting (important as a guide to purity) hard to detect and, secondly, it is possible to make a mistake if a transformation from an optically anisotropic to the isotropic cubic crystal form takes place; melting of the crystals may be simulated by the loss of birefringence so that the somewhat higher real melting point is missed. If either the polarizer or the analyzer of the microscope is rotatable, it is best, in case of doubt, to work with the polars partially uncrossed so that the crystals still show polarization colour but the isotropic surroundings do not appear too dark.

As mentioned above, the determination of the equilibrium melting point in this way presupposes the ability to crystallize so close to the melting point that a fraction of a degree of supercooling is enough to cause visible growth. However, it is often the case that supercooling of even a degree or more causes only slow growth. This leads to a "pseudo-equilibrium" since there is too large a temperature interval between growing and melting. In the equilibrium method, pseudo-equilibrium is not a source of error since the melting point is read as the temperature is rising.

The results of the melting range and the equilibrium melting point procedures are mostly in conformity, but occasionally the end of the melting range is as much as one degree higher than the equilibrium melting point. Only sluggishly melting substances are so affected.

In addition to the exact melting point, the equilibrium point method provides another criterion for the identification of substances, viz. the shape of the crystals which grow from the smallest grains at equilibrium. Such data are reported in the "special remarks" columns of the identification tables [7,11], notations taking the form "Equilib.: rectangles, rhombs, etc.", or, in the case of pseudo-equilibrium: "residual crystals grow into needles, stems, etc."

(d) Sublimation. On heating, many substances are observed to suffer the changes caused by sublimation. The original crystals on the slide evaporate with the preliminary rounding of edges and corners and the vapour condenses again on the underside of the cover slip to form sublimate crystals or condensation droplets. The latter are more likely to be produced close to the melting point (for lack of seed formation) and may flow together on contact to produce larger drops. Sometimes, condensation drops crystallize due to contact with sublimate crystals produced earlier during heating. Figure 12 shows

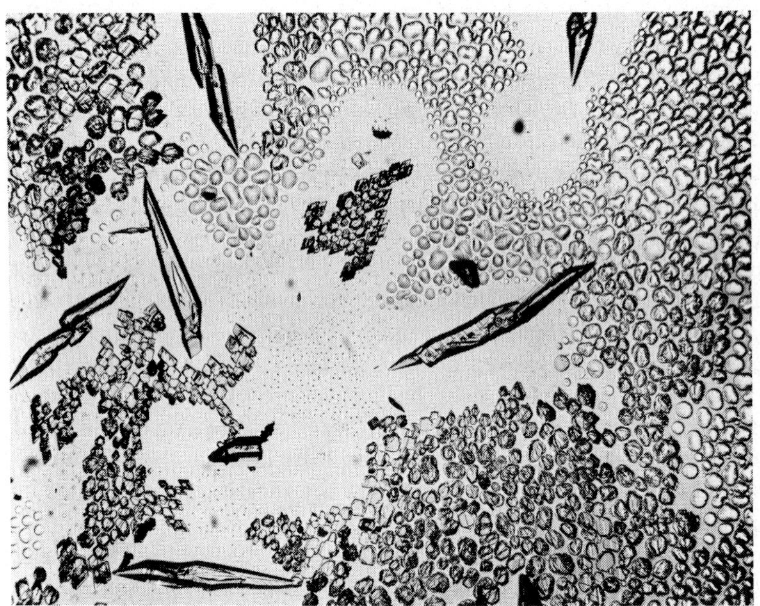

Fig. 12. Sublimate crystals and condensation drops of allobarbital.

sublimate crystals and condensation drops, some of which have been solidified by seeding from contact with the crystals. To the unpracticed observer, the large condensation drops sometimes produced might simulate prematurely the beginning of the melting range of the substance.

Sublimate crystals assume various shapes and sometimes their habit is characteristic of the substance. If the rate of sublimation is very high, the original sample may vaporize completely and the cover slip be completely covered by aggregates of polygonal crystal plates, the whole having a tesselated effect. Camphor and caffeine behave in this way. Such sublimate films melt with unusual rapidity because there is a slight temperature difference between slide and cover slip. If some crystals remain when melting begins and the larger crystals supporting the cover slip collapse, it sinks onto the slide some tenths of a degree warmer and the sublimate film melts immediately.

If the rate of evaporation is very high, it may be difficult to alter the position of focus of the microscope from the crystals on the slide to the sublimate on the underside of the cover slip sufficiently quickly to follow all the events during the heating process. On the other hand, if the rate of evaporation is low one can, during heating, focus alternately on the slide and the underside of the cover slip so as to observe the onset of sublimation and not overlook changes in the original crystals.

Substances which are extremely volatile (e.g. camphor, borneol) not only form sublimate crystals but the sublimate itself may evaporate from the cover slip and it is possible for this to be complete before the melting point is reached. In this event, the unpracticed observer could mistake the temperature at which the sublimate disappears for the melting point. One must make sure, therefore, not only that crystalline material (original or sublimate) has disappeared, but also that the melt is in evidence.

If, during a determination, all the original sample has been observed to have evaporated, it is expedient to continue with a new preparation containing a large amount of substance starting from the hot-stage temperature already reached. In this way, it is often possible to determine the melting point or range without further trouble. If the melting point of a very volatile substance under test is already known, it is advisable to preheat the stage to within 5 or 10 degrees of this temperature before inserting the preparation.

In the case of very highly volatile substances, the use of polyethylene sealing can be recommended [13]. For this purpose, a square leaflet, big enough to overlap the edges of the cover slip by about 3 mm all round, is cut from polyethylene film of about 0.3 mm thickness. The sample crystals are covered by the cover slip and leaflet and the slide is put on the hot stage at about 120° C. At this temperature, the plastic melts to a viscous transparent film and its borders are pressed against the slide with a glass rod or spade forceps. If the melting point of the substance is lower than the melting point of the plastic, a cut can be made in the leaflet. This prevents the leakage of vapour which might otherwise disturb vision by producing a sublimate or by escaping violently. [This method of enclosure has also proved useful in the determination of refractive index (see p. 366).]

Some substances, hexachloroethane is an example, escape by vaporization even through polyethylene film and their melting points cannot be determined by ordinary hot-stage methods. In such cases, even though the temperature of complete volatilization is not reproducible, identification is still possible by means of eutectic temperatures (p. 376). The melting point of such substances can be determined only in closed capillary tubes.

The temperature at which sublimate crystals are first observed depends mainly on the variation of vapour pressure and consequent rate of vaporization with temperature, the difference in temperature between cover slip and slide, and the ease of nucleation of the solid on the cover slip surface from the vapour. Since the vapour pressure and rate of vaporization of any substance both increase as the temperature increases, the chance of sublimate crystal formation and the amount and rate of growth, once formed, increase correspondingly. Consequently, if the properties of the substance favour sublimation and the temperature is high enough, the formation of sublimate will be noticed sooner the greater the separation of the cover slip and the cooler it is relative to the slide, and vice versa. This means that reproducible values for the onset of sublimate formation can be obtained only if carefully standardized experimental conditions are observed. In the course of melting point determinations using the Kofler stage, in which the sample is heated from room temperature at a steadily decreasing rate, fairly reproducible values are obtained. These are quoted for sublimable substances in the "special remarks" column of the thermomicroscopical identification tables. They represent, to within ±10°, the temperature at which sublimate crystals can be expected to be first observed on the underside of the cover slip when the Kofler stage is used. The different design of the Mettler stage precludes observations of the kind described here.

Only rarely is the outward shape of the sublimate crystals sufficiently characteristic to serve as a useful criterion for identification and little importance should be attached to it. Considering the great number of organic compounds and the relatively limited possibility of differences in crystal habit, there must naturally be numerous substances whose sublimates are similar to one another. Furthermore, many substances produce during sublimation, simultaneously or successively, crystals of different habit and this quite apart from any additional complication due to polymorphism. Nevertheless, provided the same experimental procedure is followed as in the determination of the melting point, sublimate crystal shape may be helpful as a subsidiary indicator of identity. In principle, one must distinguish between the acicular, bladed, platy, tabular, prismatic and grainy habits. Some of these, leaflets and plates especially, are easy to describe in geometrical terms, but with others it is difficult. In the case of prisms and, above all, grains, the ascription of differentiating features requires a knowledge of systematic crystallography and the investigatory techniques of crystal optics. A good account of optical crystallography is given by Hartshorne and Stuart [14,15] for those not content with the simple habit description usual in thermomicroscopy. A very elaborate, but still non-technical, scheme for the description of habit is given by Fulton [16].

The ability of a substance to sublime is also of practical importance since it is the basis of a useful method of purification or separation of mixtures for which thermomicroscopic equipment can be used when required (p. 385).

(e) Decomposition. Gross decomposition can be recognized by discoloration of the crystals or melt or by the formation of gas bubbles. As a rule, it gives rise to a long melting interval. Slight decomposition, unaccompanied by the manifestations just mentioned, is revealed by a lowering of the equilibrium melting point on repetition with the same preparation and also by the impossibility of getting a reproducible value for the temperature of a particular refractive index match (p. 368).

Since the temperature at which a substance begins to decompose depends on various factors such as particle size, amount of sample, heating rate and initial temperature, these must be taken into account in determining the decomposition interval. In the case of Kofler's method, one starts at room temperature, routinely, and heats so that the rate in the vicinity of the melting point is 4°C min^{-1}. This cannot be done in the case of substances (e.g. morphine hydrochloride) whose crystals carbonize at temperatures below the beginning of the melting range. In the case of the Mettler stage, because of the linear rate of heating, the temperature from which heating begins must always be stated because it may greatly affect the melting range of easily decomposable substances. In general, the higher the starting point, the higher the end of the decomposition range.

In determining the melting points of very unstable substances, it is best to take a larger sample than usual and to observe the middle of the preparation since decomposition generally begins earlier at the edges. The onset of melting is often poorly defined and one must disregard the melting of small isolated particles and take the initial reading only when the bulk of the sample begins to melt (often preceeded by jerky movements) so as to avoid reporting unrealistically long melting intervals. With high melting substances, especially, the reproducibility of the decomposition range is much worse than that of the melting point so one must make more allowances than in the case of stable substances.

For compounds which are decomposed by both heating and reaction with atmospheric oxygen, protective measures must be taken in order to achieve higher melting temperatures. In some cases, embedding in a silicone gel is sufficient. The recommended silicone gel "Sil Gel" is a silicone polymer of viscosity 500—1000 cP and supplied by Wacker-Chemie, Munich. For example, histidine hydrochloride melts in Sil Gel at 265—270°C but, if unprotected, decomposes in the range 245—255°C [17]. However, only rarely may decomposition be prevented completely; in most cases, it is only possible to raise the temperature interval.

It is unfortunate that the proportion of known substances which melt with decomposition is constantly increasing. The newer pharmaceuticals in particular, as a result of the great size and complexity of their molecules, have high melting points. Decomposition interfers badly not only with the determination of the melting point but also with investigations into polymorphism and it renders the determination of refractive index impossible. It follows that decomposition is the main reason why fewer polymorphic forms are known for high melting substances than for those melting in the middle to low regions. This is because the most fruitful method of finding polymorphic forms is by observing crystallization from the supercooled melt (p. 423). Sometimes the melting of a substance is accompanied by the separation of a crystalline decomposition product in such a way that the inhomogeneous melting of a polymorphic modification is simulated. For example, cinnamic acid *p*-ureidophenyl ester (Elbon ®) melts in the range 190—200°C with the slow growth of new crystals which, in turn, dissolve at 260—275°C with slight decomposition. Formerly, these observations were explained by polymorphism, but differential scanning calorimetric and infrared investigations have shown that the crystals growing in the original melt are already a decomposition product [18]. Molecular addition compounds in which one component is highly volatile are especially liable to behave in this way since the volatile component evaporates and the melting point registered is that of the remaining component. Examples are theophylline diethanolamine [11] and quinhydrone [19]. In the first case, decomposition leaves modification II of theophylline and, in the second, hydroquinone.

(f) Crystal solvates. Many substances crystallize from particular solvents in the form of molecular complexes with the solvent. Such solvates vary greatly in stability. Some lose the solvent molecules bound in the crystal lattice as soon as they are removed from the solvent and brought into air. In other cases, it has not yet proved possible to prepare the desolvated crystal form. Solvates range in their behaviour between these two extremes.

When determining the melting point microscopically, crystals which have already lost their solvent on drying are often recognizable by being turbid or punctate or, if they are thicker, by appearing black in transmitted light. However, this is not an entirely reliable indication because crystals which have undergone a polymorphic transformation without change of shape (paramorphosis) may have a similar appearance. Nevertheless, pseudomorphosis (loss of solvent without change of shape) occurs much more frequently than true paramorphosis.

While solvates involving organic solvent molecules only rarely survive the usual drying after preparation, solvates with water are very often more stable so that, in practice, it is mainly hydrates with which one has to deal. Accordingly, we shall confine our attention to hydrates but note that other solvates behave similarly.

Hydrates give off water on heating in different ways.

(i) With pseudomorphosis. The hydrate crystals change to a polycrystalline aggregate but retain their outward shape. Thus, the originally clear and transparent crystals become turbid or opaque and, depending on the thickness, brown or black in colour within a wide temperature range (Fig. 13). When observed between crossed polars, their polarization colour changes or, in the case of small crystals, disappears altogether. Small needle-like crystals are sometimes observed to jump about and burst. Very small crystals whose birefringence is not recognizable may desolvate without visible effect.

(ii) With the formation of new crystal aggregates. The crystal loses its original shape and a new aggregate of anhydrous crystals is formed

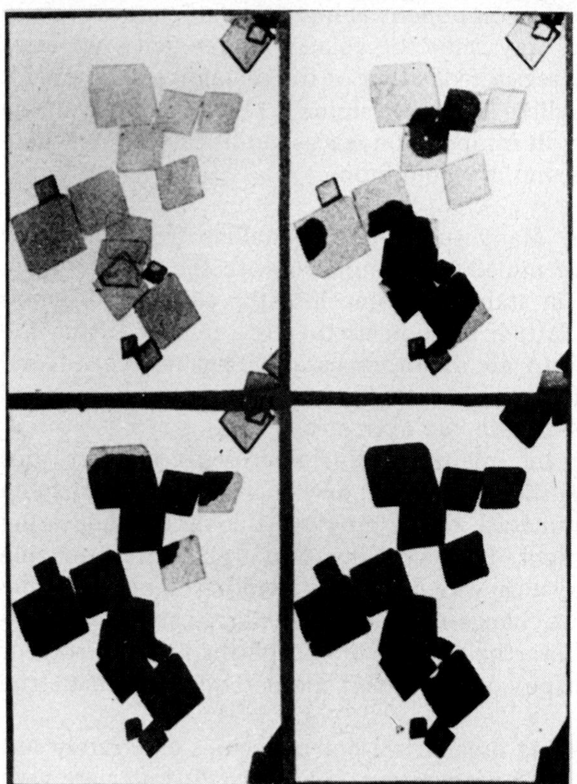

Fig. 13. Pseudomorphosis of phloroglucinol hydrate.

Fig. 14. Loss of water with formation of crystal aggregates (methandriol).

(Fig. 14). This process, in its essentials, bears a certain similarity to that of sublimation.

(iii) With incongruent melting of the hydrate. Part of all of the hydrate crystals melt with the simultaneous separation of the anhydrous form. Crystals which have already lost their water do not take part in this melting process. After the incongruent melting point of the hydrate has been passed, the melting point of the anhydrous form may be determined. Codeine hydrochloride [7,11] exemplifies this process.

(iv) With congruent melting of the hydrate. In most cases, the congruent melting point of the hydrate cannot be attained since molecular compounds with water belong, as a rule, to the incongruently melting group above. Congruent melting can occur only if no anhydrous crystals are formed before the congruent melting point is reached (e.g. piperazine hexahydrate [11]). In some cases after congruent melting, the anhydrous form crystallizes from the melt only after further heating. For instance, papaverine sulphate melts at 95—100°C as the hydrate, new anhydrous crystals being formed only above 110°C [11], due to the gradual loss of water. (Also, in the case of incongruent melting dicussed above, it often happens that some small crystals of the hydrate melt to clear drops and partly crystallize again only on further heating; the reason for this is that water escapes from the melt only gradually, but seed formation of the anhydrous form is often delayed or does not occur at all.)

For the reproducibility of the observations attendant upon desolvation, standardization of starting temperature and heating rate is necessary as in the case of observations of the processes of decomposition. Statements in the Kofler identification tables refer, for instance, to starting at room temperature and adjusting the temperature regulator to the expected melting point of the anhydrous form. This will generally mean a very quick rise of temperature through the desolvation region. It is also possible that the amount of sample plays some part; water often escapes early from isolated crystals, forming the anhydrate, while the dense mass of a preparation melts incongruently as the hydrate.

Substances which have been stored for a long time not infrequently suffer a surface dehydration (partial efflorescence) while the interior crystals are still hydrated. In such cases, the behaviour on heating may differ from that of the freshly prepared hydrate.

As a rule, the melting of hydrated crystals does not occur sharply but over an interval of some degrees, the length of which depends on how far the congruent melting point is above the peritectic point (p. 446). It follows from the phase diagram for a hydrous system that the incongruent transition of the hydrate to the anhydrous form ought to occur at the temperature at which all three phases are in equilibrium: the hydrous phase, the anhydrous phase and the melt (which is, of course, a solution). This process is, as a rule, very sluggish, resulting in a wide melting range.

In general, the melting point of the hydrate is lower, often very much lower, than that of the anhydrous form, but in exceptional cases the reverse obtains. On normal heating, terpine hydrate melts homogeneously as the anhydrous form at 105°C following loss of water accompanied by the appearance of turbidity, but in a sealed capillary tube, the substance melts as the hydrate at 117°C.

The loss of water may also occur in a stepwise fashion, either before reaching the melting point of the hydrate (accompanied by pseudomorphosis or transformation) or on melting incongruently. Higher hydrates are converted to lower and only the latter transform to the anhydrous form. However, not every polyhydrate gives off its water in stages although this is more likely than in the monohydrates.

The behaviour of sparteine sulphate, marketed as the pentahydrate, is quite exceptional. During heating under the microscope, the original crystals are first seen to transform and become turbid and then from about 100°C they melt very sluggishly to a viscous melt with the incongruent separation of small grains of yet another crystal form; these dissolve in the melt up to about 130°C. Thus, we have a melting range of about 30°, which is exceptionally wide in this temperature region. Analysis of the evolved gas helps to explain this curious behaviour [20]. There are three different hydrates (even the last observed crystal form melts with the release of water) and the loss of water peaks about 75, 110 and finally at about 145°C for the form with the least water of crystallization. If the preparation is protected by Sil Gel, the top of the melting range is also 145°C (see below). In this unusual and fortunately rare example, the results of the gas analysis show that the anhydrous form does not occur and microscopic examination does not reveal it. Sparteine sulphate is the most complicated case met with in the investigation of many hundreds of organic hydrates.

Although mentioned only briefly above, polymorphic transformations and incongruent melting are sometimes not easily distinguish-

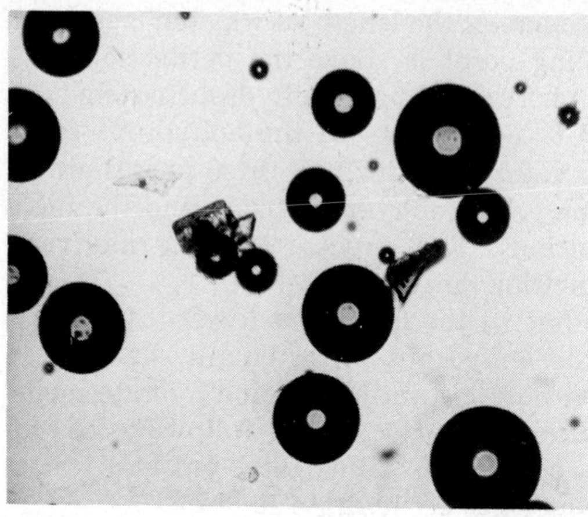

Fig. 15. Evolution of gas bubbles from hydrate crystals in Sil Gel.

able from desolvation because the escaping vapour cannot be seen. In order to detect it microscopically, crystals can be heated in a liquid in which they are not soluble; in this way, the emergence of gas bubbles from the crystals can be observed directly (Fig. 15). It is interesting that embedding in Sil Gel can also produce the opposite effect. While the prevention of desolvation and consequent congruent melting of the hydrate is normally expected, there are some cases where the anhydrous form appears specifically in Sil Gel after stepwise loss of water (e.g. sulphatriazine [21]).

The Sil Gel test is very sensitive and it is possible to demonstrate the evolution of quantities of water vapour which formerly, in classical analysis, escaped detection. This was the case with pyrogallol whose quarter mole of water of crystallization can be detected very easily with the hot stage [22].

In order to distinguish water from other liquids of crystallization, Fischer [23] has combined the paraffin oil method first used by us with potassium lead iodide as an indicator for traces of water (due to Biltz [24]). In this procedure, the reagent is triturated with the oil used to embed the sample. The water of crystallization which escapes on heating reacts with the potassium lead iodide with the separation of pure lead iodide and this can be seen by the yellow coloration or by the formation of yellow crystals close to the crystals of the hydrate. Today, in cases of doubt, the solvate vapour is identified chromatographically.

Since the production of gas is also the result of ordinary chemical decomposition, unstable substances develop gas bubbles in Sil Gel during heating and sometimes long before melting starts. As opposed to desolvation, in which the production of gas ceases after transformation to the desolvated form, accompanied by the formation of new crystals, has occurred, in chemical decomposition the formation of gas bubbles continues increasingly up to the end of melting and beyond. It is possible for a substance, originally hydrated, to produce two crops of gas bubbles on heating; the first of water vapour during dehydration and the second of some decomposition product during chemical decomposition.

In general, the heating of a sample in Sil Gel will be stopped after the melting of the hydrate or crystallization of the anhydrous form since the bubbles of water vapour seriously interfere with further observations. However, in special cases it may be of interest to continue heating the sample up to the melting point of the anhydrous form. If so, the gas bubbles can be removed by pressure on the cover slip.

(g) Polymorphic modifications. It has already been said in the discussion of solvates (p. 353) that the appearance of turbidity in crystals during heating or their congruent melting does not always mean that desolvation is occurring; these phenomena can be the result of polymorphism. Perhaps as many as one organic substance in ten, when recrystallized from solvents, does not occur in the highest melting crystal form, but either in a different form stable at room temperature or in a metastable form. The behaviour on heating is analogous to that of the solvates, i.e. the crystals may transform or melt inhomogeneously if seeds of the stable modification can form or homogeneously if they can not. Paramorphosis, whereby the transformation to an aggregate of the new modification takes place within the framework of the original crystal shape, is commonly encountered. The phenomenon may occur over a narrow temperature range or take place very sluggishly; some enantiotropic transformations take place especially quickly, sometimes with explosive force causing the crystals to jump about.

For identification, however, the most relevant data are the melting points, inhomogeneous or homogeneous, of polymorphic modifications which melt below the final most stable modification. Phenobarbital is a good example. Commercial preparations are mostly in

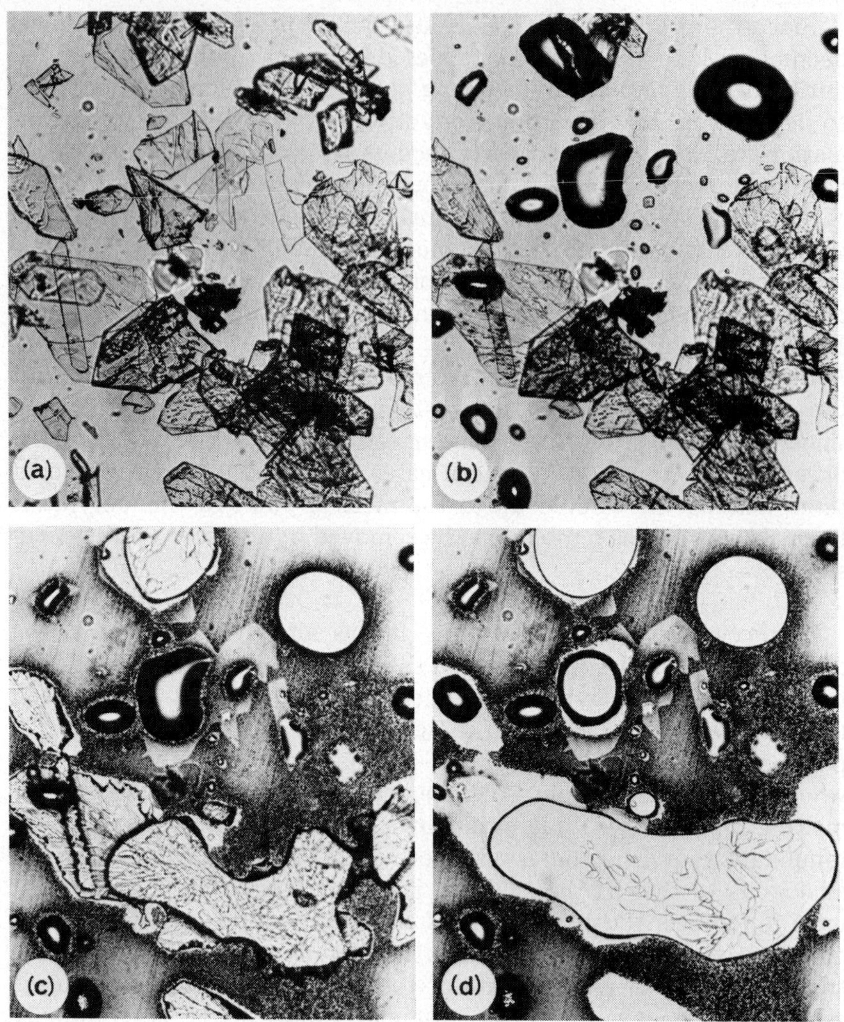

Fig. 16. Melting of two modifications of sulphathiazole.

modification II (very occasionally as the hydrate) which appears to be the commonest form of this substance and is enantiotropic with respect to modification I. Modification II melts at 174°C, inhomogeneously with the precipitation of crystals of modification I, which, in turn, melt at 176°C. Since the melting points are so close, the actual relationship between the two modifications escaped notice for a long time.

Sulphathiazole, which may be compared with phenobarbital, is a classic example of a pure substance showing two melting points. As in the former case, modification II of sulphathiazole (Fig. 16) is enantiotropic with respect to modification I and is partly transformed to it during heating (a) but what remains melts homogeneously at 175°C (b), while the newly formed crystals of modification I melt only at 202°C (c, d). If the observer were not to know that polymorphism existed, he could be misled and believe the partial melting so far below the final melting point to be the result of impurity. However, exact observation distinguishes between the effects of impurities and polymorphism. A very impure substance begins to melt at the eutectic temperature (p. 376) and the remaining crystals dissolve as the temperature rises, becoming more and more rounded in the process. On the other hand, if we have two modifications, say I and II, crystals of modification I are not involved in the homogeneous melting of modification II so their edges and corners remain sharp as the temperature rises towards their own melting point. They may even grow or multiply if they come into contact with and are able to seed drops of the melt from the modification II crystals.

It should be noted that the behaviour of many substances depends to some extent on the experimental conditions, such as particle size, amount or condition of sample, initial temperature and heating rate, and on the provenance of the sample. The latter, if it causes different or different mixtures of modifications to be present, may give rise to quite distinct phenomena on heating; e.g. a transformation followed by homogeneous melting of a stable modification as opposed to the inhomogeneous melting of a lower melting modification. This happens quite frequently in the case of different commercial products. For instance, the investigation of over one hundred steroid hormones has shown that many exist in various commercial products in the form of different modifications or as mixtures of them.

One of the best known examples of differing commercial products is provided by progesterone. The highest melting modification, modification I, is known as α-progesterone and a lower melting modification as β-progesterone. The latter is, in fact, relatively stable and is transformed to the α form only via the melt or solution so it behaves as a separate substance, having a melting point 8 degrees lower. The physical constants are given separately in the literature.

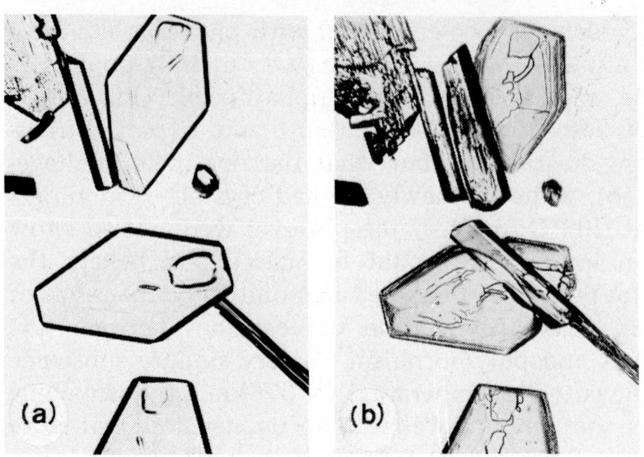

Fig. 17. Transformation of sublimate crystals of hydroquinone.

This is a rare case, relatively speaking; the more usual situation is that, if two different forms are available on the market, heating will reveal a transformation or an inhomogeneous melting point in one case. L-Hyoscyamine is an example of this. For a long time, only the highest melting modification (m.p. 107–109°C) was known but some years ago we were able to demonstrate that the then new commercial product of Merck, Darmstadt, which melted at 104°C with the separation of a few thin needles of modification I, consisted of a previously unknown metastable modification [25]. Another example is dihydrocodeinone enol acetate hydrochloride (Acedicon), which is available on the market in two different modifications; modification I melts homogeneously at 230–250°C, whereas modification II melts inhomogeneously at 210–230°C, thus giving the appearance of a wide melting interval from 210 to 250°C [26].

Transformation phenomena and melting points of metastable modifications can often be observed in sublimates which have formed in the course of melting point determinations. Hydroquinone is an example. In Fig. 17(a) are to be seen platy crystals of modification II alongside prisms of modification I. Because of their higher vapour pressure, the plates of modification II diminish in size while the prisms of modification I grow or, where the crystals touch one another, a transformation takes place [Fig. 17(b)] [27]. Barbital is another substance whose sublimates may contain several modifications and show transformation phenomena and, occasionally, metastable melting points.

(h) The solidification of the melt. Characteristic behaviour, which provides additional criteria for identification, is shown not only when crystals are heated but also when the melt is cooled. It must be said first of all that the melts of most organic substances supercool readily. Exceptions are, for example, camphor derivatives which belong to the cubic crystal system and which begin to crystallize spontaneously only slightly below their melting points. Further, substances like anthracene and phenanthrene, which adopt a platy habit, are generally harder to supercool than those which adopt prismatic habits, but the rule is far from invariable.

Solidification of the melt can be investigated in three ways.

(i) Slow solidification. The power to the hot stage is switched off as soon as all the crystals have melted or, rather, a few degrees above the melting point so that the observer can be certain that there are no unmelted crystals beyond the field of view, and the apparatus is allowed to cool down slowly. When this is done, crystals or crystal aggregates frequently appear but not before the preparation has cooled to 10 or 20 degrees below the melting point. If crystallization fails to take place, it can sometimes be induced by touching the edge of the cover slip with a needle or in some other way. The temperature of spontaneous crystallization is not a constant but depends on such factors as surface imperfections on the glass of the slide or cover slip or the presence of dust particles and so on which may act as nuclei for seed formation. It should also be noted that repeated melting and solidification or holding the melt at an elevated temperature, especially if prolonged, may influence the spontaneity of crystallization.

(ii) Quick solidification. In this procedure, the completely molten preparation is cooled rapidly on a cold surface. In most cases, a metal plate is most suitable but a wooden surface is to be preferred for substances which produce too many seeds and badly developed crystals if cooling is too rapid. Substances with a low melting point should be cooled on an ice-cold metal block especially if polymorphism is to be investigated (p. 410).

As a general rule, the more the melt is supercooled during the crystallization process, the more fine-grained will the crystal film be. The rule may apply, however, only for certain degrees of supercooling

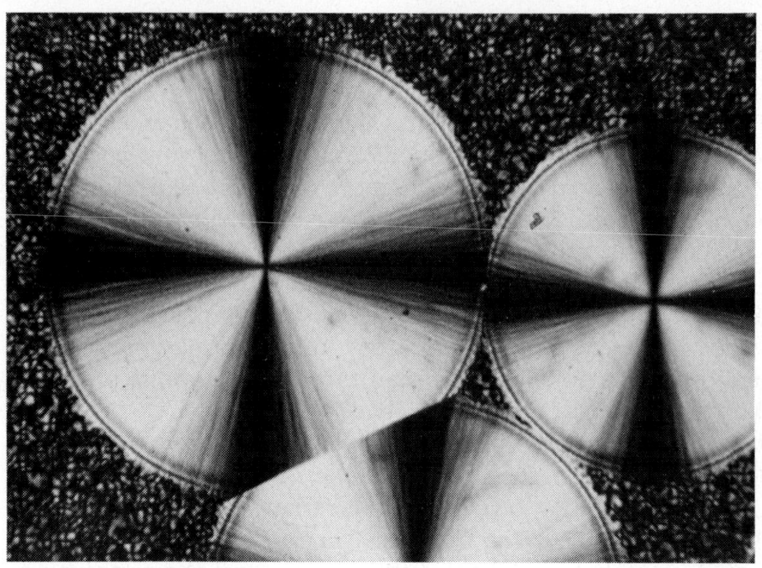

Fig. 18. Spherulites of chloramphenicol palmitate.

and its operation depends on the specific properties of the substance. For instance, if, when the melt starts to crystallize, it is already in the region in which the rate of crystallization is much retarded, the crystal structure will be coarse rather than fine-grained.

(iii) Reheating of the quickly solidified melt. Many organic melts cannot be induced to crystallize by cooling either slowly or quickly but remain as a glass or highly viscous liquid. In such cases, it is often sufficient to reheat the slide or to put it on a preheated hot-stage.

Small droplets often crystallize to beam-like aggregates; larger drops or continuous films often produce spherulites. Although normally spherical, when confined between slide and cover slip these take the form of disc-shaped crystal aggregates in which needle-like or prismatic crystals are distributed regularly about a centre. The single crystal needle may be so fine that the radial structure is revealed only between crossed polars. When so viewed, spherulites show a black cross whose arms correspond to the vibration directions of the polarizer and the analyzer (Fig. 18). In addition to spherulites, which may be from fine to coarse-rayed in appearance, there are

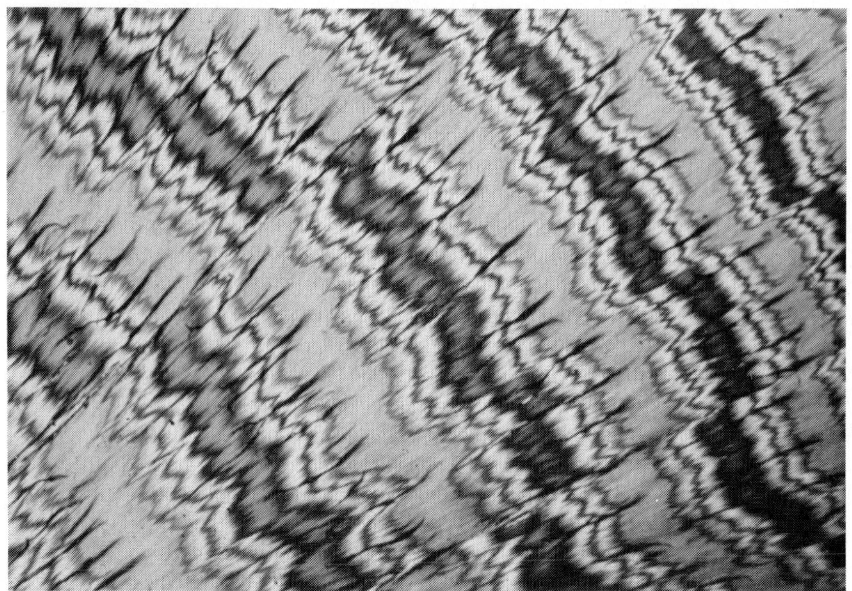

Fig. 19. Fibre twisting.

other less characteristic crystal aggregates made up of plates, grains, matted needles and so on.

Related to spherulite formation and sometimes associated with it is the phenomenon and structure of "fibre twisting", investigated principally by Bernauer [28]. In this, fibrous crystals are twisted round their longitudinal axis at regular intervals during growth. In consequence, there is repetitive rise and fall of polarization colour which causes characteristic stripes when viewed between crossed polars (Fig. 19). The phenomenon may be confined to crystallization within a particular temperature range. It is favoured by impurities.

Finally, two other phenomena which may accompany crystallization should be noted. The first is the contraction fissures caused by volume changes in the cooling crystal aggregates as they grow in the film of melt and the second bubbles of air or other gases which were dissolved in the melt but which are rejected on crystallization. Both contraction fissures and gas bubbles may give rise to characteristic patterns which may be of assistance in identification.

(2) The refractive index of the melt

In addition to the melting and eutectic points (p. 371), there is a third basic element of the Kofler identification scheme which depends on refractive index. Since the determination of the refractive indices of birefringent crystals, because of their dependence on crystal orientation, is not as simple as for optically isotropic crystals or liquids, Kofler "escaped" to the liquid phase [29] when devising a confirmatory identification test. In other words, the refractive index of the melt is determined, not those of the crystal. This has the additional advantage of involving temperature as a variable. Since the refractive index of a liquid decreases as the temperature rises, we can determine the point at which the melt of the substance has the same refractive index as that of a reference solid added to the preparation before melting.

The principle of this determination is that of the well-known immersion method. It is based on the fact that a colourless transparent object can be seen against the surrounding medium only when their refractive indices differ and, conversely, when it has the same index as the medium, it is invisible.

In the microscopic preparation, a particle with a distinctly different refractive index from that of the melt in which it is immersed seems to be delimited by a dark outline when in sharp focus. If the objective is now raised or the stage lowered by a small amount, a bright line, called the Becke line, appears alongside the dark border

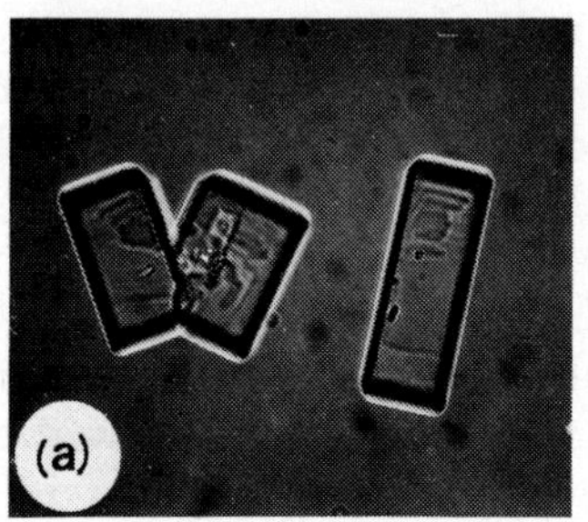

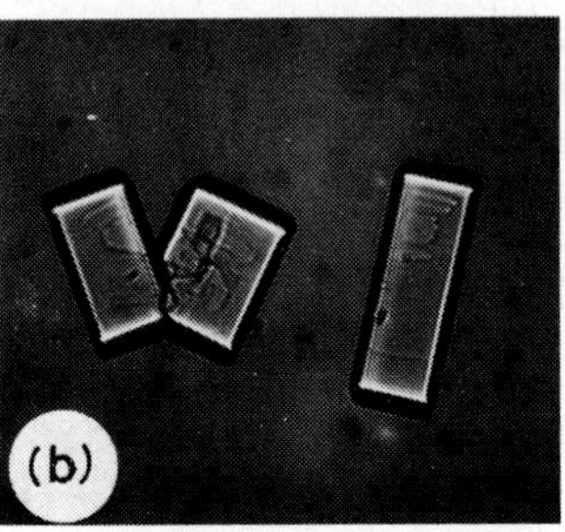

Fig. 20. Crystals embedded in a medium of higher refractive index (a) with objective raised above position of best focus, (b) with objective below position of best focus.

on the side of the more highly refracting medium [Fig. 20(a)]. If the objective is defocused in the opposite direction, i.e. downwards below the plane of exact focus, the Becke line crosses into the medium of lower refractive index [Fig. 20(b)]. The explanation for this phenomenon is given by Hartshorne and Stuart [15] and many other writers.

In order to be able to observe the behaviour of the Becke line properly, it is necessary to narrow the cone of illumination by stopping down the condenser iris diaphragm or aperture stop. In addition, monochromatic light should be used because, in white light, the interfaces between the two media disappear completely only when their dispersions as well as refractive indices are identical. If there are differences in dispersion, coloured borders appear on reaching equality of refraction for a particular wavelength and, instead of the single bright Becke line, a double coloured band of greenish-blue and orange-red can be seen. If the objective is now defocused, one colored band moves outwards from the particle border and the other inwards, thus confusing the determination.

Kofler devised a set of powders having refractive indices between 1.34 and 1.69 as reference standards and specified a filter to mitigate the disturbing effects of dispersion differences between organic melts and the higher refracting glass powders. The filter he introduced had a definite red light transmittance and was supplied with the reference powder set by Optische Werke C. Reichert, Vienna. For the last twenty years, an interference filter having λ_{max} 589 nm corresponding to the wavelength of sodium light has been used in Innsbruck. In the identification tables [11], the matching temperatures for particular glass powders are given for both red and sodium light.

Because organic substances in the molten form can be neither greatly overheated nor supercooled at will, it is necessary to have, in the standard set, powders whose refractive indices are not too far apart. Kofler adapted his set from that of Linck and Köhler [30] which he extended. The precise sodium D-line refractive indices of his 24 powders (2 minerals, 22 glasses) are given in Table 1. This set is defective in the lower region in that there is so great a gap between n_D for cryolite (1.3400) and for fluorspar (1.4339). On the other hand, differences in the D-line refractive indices of the glass powders above fluorspar average a little over 0.01. There is no need for further extension of the scale since few organic melts have an index

TABLE 1

Refractive indices, n_D, of the Kofler glass reference powders

1.3400	1.4842	1.5204	1.5611	1.6011	1.6483
1.4339	1.4936	1.5299	1.5700	1.6126	1.6598
1.4584	1.5000	1.5403	1.5795	1.6231	1.6715
1.4683	1.5101	1.5502	1.5897	1.6353	1.6877

higher than 1.6877. Chamot and Mason [4] give a list of 41 isotropic crystalline substances, some of which have indices in the gap in the Kofler set and others which are above its upper limit. However, although a few of these might be as stable as the glass powders and be able to withstand temperatures up to about 300° C, others, particularly in the lower range, are hydrates or have low melting points or suffer transitions.

To determine the temperature of match, an amount of substance (a little more than for the determination of a melting point) is taken, some particles of a suitable glass powder (as supplied; there is no need to grind further) are added and the preparation is heated until molten. Then the substage iris of the microscope is closed and the filter inserted in the beam. With the aid of the object guide, a search is made for a part of the preparation where some glass fragments are to be found in one of the larger drops of melt. If the temperature of equality of refraction is close to the melting point, further heating to 5 or 10 degrees above it is recommended so that the glass fragments are clearly visible, or, in the case of sensitive compounds, cooling to the same amount below. By raising the objective and observing the position of the resulting Becke line, it may be seen whether the melt or the glass fragments are the more highly refracting. Depending on which is the case, the preparation must be heated or cooled further. Powders for identification testing have frequently been chosen to give a temperature of match below the melting point so as to avoid having to overheat the specimen, thus reducing the danger of decomposition. The temperature of the hot stage is allowed to fall some degrees below the match point and the determination is made on reheating. The temperatures at which the glass fragments become invisible and at which they reappear are read. This interval is generally one or occasionally two degrees. If, however, the degree of supercooling is too great in that it leads to spontaneous crystallization of the melt, it is easier to carry out the determination as the temperature falls. The values read on rising and falling temperature usually differ by about two degrees.

If there is a very great difference in dispersion between melt and glass powder, it may happen that, even when a filter is used, the outlines of the powder particles do not disappear at any temperature. In such cases, the Becke line crosses the border at the matching temperature in such a way that, with the objective raised above the position of best focus, it lies within the particles at first then suddenly crosses into the melt as the temperature falls, and vice versa if the temperature is raised. This change may take place very sharply, even within one degree, but its determination requires greater watchfulness than the observation of the disappearance of the fragments.

The method described above assumes that the refractive index of the glass powder which is to be matched to the melt is known. The identification tables [7,11] are based on this assumption. If the appropriate powder for confirmation of identification has yet to be determined, the first step is to try one approximately in the middle of the set. The Becke line test quickly reveals whether the powder or the melt just above the melting point is the more highly refracting. If the latter, a second trial is made with a glass somewhat above that already tried. If it is now possible to match the refractive indices at a practicable temperature, there is no need to proceed further; otherwise another trial must be made, and so on. A practiced observer can often estimate how far a trial powder is from the "right" one from the strength or weakness of the appearance of the outline of the fragments in the melt. For such an investigator, three trials are generally sufficient.

In order to avoid the disturbance caused by heavy sublimation, the use of a polyethylene seal is recommended when the refractive index match of a highly volatile substance is to be found. If sublimation experienced during the determination of the melting point is sufficient to interfere with vision, such a seal should be used from the outset when determining refractive index, even if it is not actually required during the melting point determination.

The temperature of refractive index match cannot be determined if the substance decomposes on melting, even if only to the extent that the melt has a brown colouration. If only slight decomposition takes place, the procedure still gives values higher or lower by several degrees when the determination is repeated.

The Mettler hot stage simplifies the carrying out of the determination. The fact that, with this instrument, the observer can continue to watch the preparation at the critical stage without the interruption of reading the thermometer is a great advantage. However, temperatures determined with the Mettler stage are about one degree lower than those found with the Kofler stage. This is due to the fact that the glass fragments added to the preparation cause an abnormal separation between slide and cover slip and this has more effect when the Kofler stage, which is heated only from below, is used than in the case of the Mettler stage in which the preparation is heated on both sides. The identification tables [7,11] give matching temperatures determined with the Kofler stage and are designed for use with it. Consequently, matching temperature data must be corrected to allow for the slightly lower Mettler stage values, unlike melting point data which are valid for either stage.

Since untrained observers encounter difficulties with the refractive index determinations, phase contrast illumination has been tried [31]. Its use facilitates the determination to some extent since with it the glass chips are still clearly visible when the difference in refraction is very small and, even at the point of equality, they remain faintly visible. This means that a more precise determination is possible with phase contrast than with the Becke line method.

When using phase contrast illumination, it should be noted that the objective is not defocused. When sharply focused, optically denser detail viewed in positive phase contrast appears darker than the background and, at the same time, is surrounded by a brighter

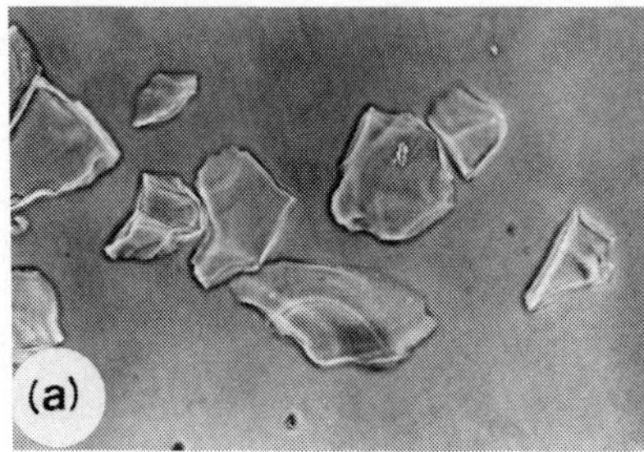

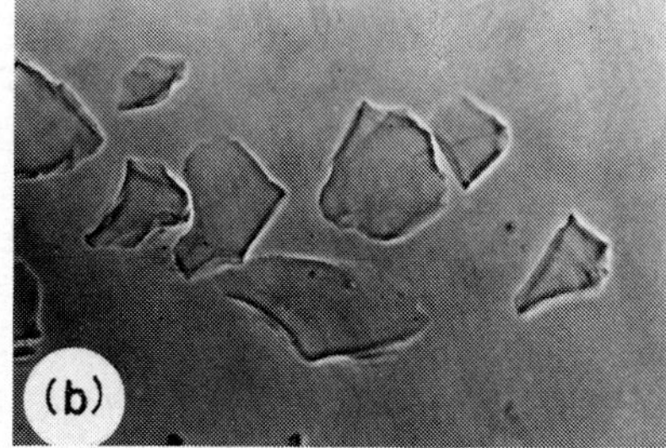

Fig. 21. Glass chips; phase contrast illumination. (a) The refractive index of the melt is the higher; (b) the refractive index of the chips is the higher.

border on the outer side. Less dense optical detail appears brighter than the background and is surrounded by a dark border.

This is illustrated in Fig. 21(a) which shows chips whose refractive index is lower than that of the surrounding melt and are therefore surrounded by a bright then by a dark border. If, on heating, the chips come to be more highly refracting than the melt, the bright border moves to the outside and the dark to the inside as in Fig. 21(b).

A specially constructed long working distance condenser is needed for the phase contrast illumination with the hot stage because the distance of the object on the stage from the condenser is substantially greater than in the situation for which commercial phase contrast devices are usually constructed. Since the accuracy and reproducibility of the method are quite satisfactory without the aid of phase contrast, the advantages of the latter do not seem to compensate for the cost and complication of altering the illumination system.

(3) Mixtures

(a) The simple eutectic phase diagram. For the better understanding of what follows, it seems advisable to consider first the simplest possible phase diagram for a two-component system, i.e. a simple eutectic system in which the liquid phases are miscible with each other in all proportions but the components form neither compounds nor mixed crystals. In Fig. 22, the composition is plotted on the x axis and temperature on the y axis. The equilibrium freezing point curves meet at the eutectic point, E, at which the melt of substance A is saturated with B and vice versa. Below the eutectic temperature (ET), the liquid state is unstable and both components crystallize out side by side.

A finely powdered mixture of the eutectic composition melts completely at the eutectic temperature. Mixtures of any other composition will have a melting range which will be greater the further the composition from the eutectic point. Melting will begin at the eutectic temperature and end on one or other of the freezing point curves, according to composition. (This does not apply, of course, to the pure substances. In practice, there is always a slight mutual solubility and the melting range diminishes towards zero as the composition approaches 100% purity.)

If (see Fig. 22) a mixture of composition X, in which A is in excess, is heated, substance B melts completely at the eutectic tem-

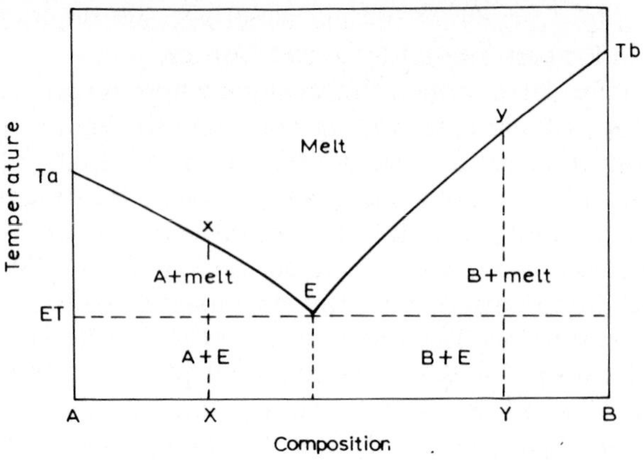

Fig. 22. Simple eutectic phase diagram.

perature together with as much of A as is required (in accordance with the eutectic composition ratio) to dissolve it. As the temperature increases, more and more of A melts until finally, at point x, the last crystals of A dissolve. A mixture of composition Y, in which component B is in excess, behaves similarly on heating but, in this case, the residual crystals are of component B.

The plotting of phase composition diagrams will be described elsewhere (p. 457).

(b) Mixed melting points. The determination of the mixed melting point is the quickest and most convenient method to decide the question of the identity of two samples (one generally being of a known reference substance). For the purpose of this test, crystals of both samples are ground together and a preparation made from the mixture and heated. The beginning and the end of the melting range are compared with the already determined melting points of the individual substances. In general, if the two samples are not identical, the mixture will begin to melt at the eutectic temperature and melting will be complete somewhere between that and the melting points of the samples; just where depends on the accidental variation of the mixture composition. Non-identity can be seen much more easily under the microscope than by the old capillary tube method. This is particularly the case with unstable substances since the

beginning of melting at the eutectic temperature is, on average 20–30 degrees below the melting points of the pure substances.

In the case of unstable substances, it is necessary for the initial temperature, heating rate and amount of sample to be the same for the pure substances and the mixture. Attention should also be paid to the particle size. Often, samples of pure substances are not comminuted while material for the mixed melting point determination is ground during preparation. Even if the samples are identical, the smaller crystals in the "mixture" may dissolve quicker than the larger ones in the original samples so that the end of the melting range appears to have been lowered. To overcome the difficulty, all that is required is to grind the individual samples to the same extent as in the preparation of the mixture so that an accurate comparison may be made.

In the case of volatile samples, it is best to preheat the hot stage to about 10 degrees below the expected melting point to avoid the possibility of one of the substances evaporating completely during a prolonged period of heating, identity of the two samples being thereby simulated.

If only a very small amount of volatile substance is available, the procedure just described may be modified as follows [32]. The known reference substance is sublimed on to a cover slip, a crystal or fragment of the volatile unknown is put on the slide and the sublimate pressed gently on to it. In this way, the small crystal is crushed and its irregular fragments can be easily distinguished. On heating, the sample fragments sublime to the cover slip so that the crystal film there is thickened and becomes an intimate mixture of the two substances. Now, if the two substances are in fact identical, the melting behaviour will be that of a pure material. If they are not, droplets appear round the fragments at the eutectic temperature and, as the temperature rises further, these coalesce into larger drops containing the particles of the second component. Figure 23 illustrates the effect.

The roles are reversed if it is the reference substance which is available in only small amount. A sublimate (p. 385) is produced of the unknown substance and this is pressed on to a crystal or fragment of the known reference substance.

Drugs, when sublimed, occasionally produce unstable modifications and it may happen that, on this account alone, the sublimate and the reference substance differ in melting point. As a rule, how-

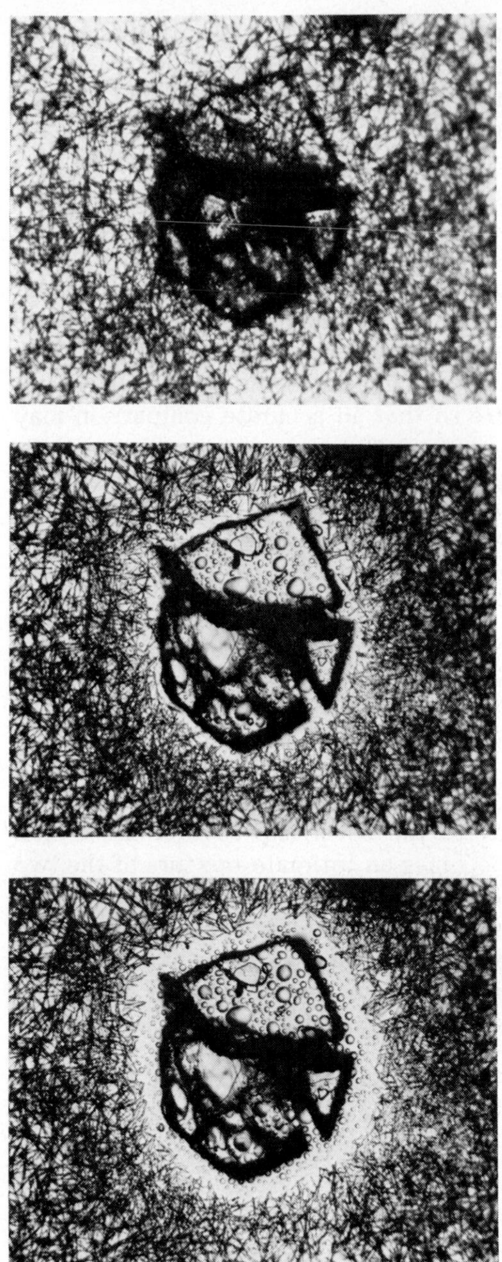

Fig. 23. Identification test on sublimate crystals.

ever, in this situation the sublimate will be transformed during heating by contact with the stable modification.

It is well known that the mixed melting point test fails if the two substances are isomorphous and form mixed crystals or if there is a miscibility gap between the liquid phases. The first difficulty appears in the case of substances which form continuous series of mixed crystals of Bakhuis-Roozeboom Types I, II or III (p. 450). Camphor provides a classic example; *d*-camphor and *l*-camphor form a series of mixed crystals of Type I and the melting point of *dl*-camphor is the same as that of the pure substances [33].

Mixtures, other than those of enantiomers, rarely behave in this way. Investigations of more than 40 barbiturates [34—36], a class of substance noted for polymorphism, have produced only one case of relevance here. Butethal and idobutal melt at 127 and 129°C, respectively, and the mixed melting point is 124°C [37]. We have here Bakhuis-Roozeboom Type III behaviour, but the minimum mixed crystal melting point of 124°C is only 3 degrees below the melting point of butethal and this is not enough for there to be no possibility of misinterpretation. An even greater number of steroid hormone pairs have been investigated [38], but, again, only one pair has been found whose mixed melting point might falsely indicate identity. This is 17-α-methylandrostane-3β,17β-diol (m.p. 213°C) and methandriol (m.p. 208°C) which, in a mixture of equal parts, melt at 209°C.

Undoubtedly, the second situation mentioned above, viz. the existence of a miscibility gap between the liquid phases (p. 447), is to be found more often than mixed crystal formation. It arises when the two substances are totally different from a chemical point of view and which, being insoluble in each other's melt, fail to depress the freezing point significantly. The investigation of a great number of drugs shows that this is a situation whose possibility must be borne in mind [39]. As examples, levonordefrin hydrochloride gives no depression when mixed with diethylstilbestrol, nor does cinchocaine hydrochloride with noradrenaline bitartrate. The miscibility gap between liquid phases can be recognised by careful observation because the drops of melt are not homogeneous, drops of one of the substances swim within drops of the other. Mistakes can be avoided by determining the eutectic temperature with a suitable reference substance (see below) because there are great differences in such cases.

(c) Eutectic temperatures. The determination of the mixed melting point allows us not only to answer the question "Are these two substances identical?", but also, if they are not the same, it permits systematic identification via the eutectic temperature [40]. For this purpose, we do not test with the supposed identical substance but with a certain reference substance (chosen by rule according to the melting point of the substance under test) whose melting point it depresses. It should be noted that here it is only the beginning of melting which is of interest and not the final melting point nor the melting range. The eutectic temperature is a constant for each pair of substances and, as we have seen above, is unaffected by the relative amounts of each present in the mixture provided the simple eutectic phase diagram of Fig. 22 applies. To determine the temperature at the end of melting would be valueless because it depends on the unknown mixing ratio.

For the determination of the eutectic temperature, as for the mixed melting point, approximately equal amounts of the sample to be identified and of the reference substance are taken, ground between two slides, and a suitable amount of the powder gathered and covered with a slip. The preparation is then heated and attention directed to the temperature at which melting is just beginning, this being taken as the eutectic temperature. At this point, part of the crystals melt together to form drops within which are contained crystals of the component in excess. After this reading, it is advisable to heat the preparation a further two or three degrees in order to make sure that melting is continuing. Isolated particles of an impurity, accidentally present in the field of view, could possibly simulate the eutectic temperature by melting prematurely and, for this reason, it is recommended that heating be continued for a few degrees to avoid any misinterpretation.

Reference substances should have the following desiderata [41].

(1) Be obtainable in high degree of purity.

(2) Have a sharp, easily reproducible melting point determinable, as far as possible, by the equilibrium method and not too far from that of the substance under test.

(3) Be available as the highest melting modification (polymorphism being undesirable although unfortunately unavoidable).

(4) Have not too high a vapour pressure in the temperature range of use (since high vapour pressures presage drop condensation on the underside of the cover slip and this complicates the observation of the eutectic temperature).

(5) Have not too small a cryoscopic constant so that acceptable eutectic temperatures may be expected.

(6) Be non-toxic.

(7) Be readily available at reasonable cost.

(8) Be easily purified from inferior grades in the event of failure to meet requirements (1) and (7). In this connection, it is to be noted that, of the reference substances in the original set chosen by Kofler (see below), salophen is no longer obtainable commercially and azobenzene is no longer available in the required degree of purity. Experience has shown that it is not possible to select a set of reference substances which fulfil all these requirements, particularly in the high-temperature region.

The reference substances introduced by Kofler are listed in Table 2 together with the melting points of the unidentified substances with which they are to be used. Apart from salophen being now unavailable, there is a big gap between dicyandiamide (m.p. 210°C) and phenolphthalein (m.p. 263°C) which, although the last of the series, has a melting point rather lower than is desirable. Table 3 gives a modified list of substances in which these defects have been overcome by replacing salophen with racephedrine hydrochloride, inserting saccharin between dicyandiamide and phenolphthalein and adding *p*-sulphamoylbenzoic acid.

Substances with the same or similar melting point are generally easily distinguishable by determination of the eutectic temperature.

TABLE 2

Series of reference substances (Kofler)

Reference substance	M.p. (°C)	M.p. limits of substance under test (°C)	Reference substances to be used for ET determination
Azobenzene	68	20—100	Azobenzene, benzil
Benzil	95	100—120	Benzil, acetanilide
Acetanilide	114.5	120—140	Acetanilide, phenacetin
Phenacetin	134.5	140—170	Phenacetin, benzanilide
Benzanilide	163	170—190	Benzanilide, salophen
Salophen *	191	190—350	Salophen, dicyandiamide
Dicyandiamide	210		
Phenolphthalein	263		

* Acetyl *p*-aminophenylsalicylate.

TABLE 3

Wider series of reference substances

Reference substances	M.p. (°C)	M.p. limits of substance under test (°C)	Reference substances to be used for ET determination
Azobenzene	68	20—100	Azobenzene, benzil
Benzil	95	100—120	Benzil, acetanilide
Acetanilide	114.5	120—140	Acetanilide, phenacetin
Phenacetin	134.5	140—170	Phenacetin, benzanilide
Benzanilide	163	170—190	Benzanilide, racephedrine hydrochloride
Racephedrine hydrochloride	190	190—210	Racephedrine hydrochloride, dicyandiamide
Dicyandiamide	210	210—240	Dicyandiamide, saccharin
Saccharin	228	240—280	Saccharin, phenolphthalein
Phenolphthalein	263	280—350	Phenolphthalein, *p*-sulfamoylbenzoic acid
p-Sulfamoylbenzoic acid	288—295 decomp.		

This may be seen from the examples in Table 4, but the table also shows that completely reliable identification cannot be achieved by the melting point and eutectic temperatures alone.

TABLE 4

Eutectic temperatures of substances with similar melting points

M.p. (°C)	Compound under test	ET (°C) with	
		Acetanilide	Phenacetin
119—121	Etilefrin	86	105
121.5	Clofedanol	95	108
120—122	Chloral formamide	79	93
120—122	Estradiol methyl ether	86	101
122	Picric acid	71	89
122	Testosterone propionate	74	93
122.5	Benzoic acid	76	89
122.5	β-Naphthol	60	69

The identification must be confirmed by the refractive index determination and the necessity for this is greater the more numerous the possibilities which have to be taken into consideration. In spite of this, the method is valuable for characterizing substances which do not have sharp melting points as a result of decomposition but which, when mixed with suitable reference substances, give sharp eutectic temperatures. For example, acetylsalicylic acid has a wide melting range (130–136°C), but the eutectic temperatures with acetanilide and phenacetin are sharp at 81 and 97°C, respectively.

The reproducibility of eutectic temperatures is generally very good for pure substances whose melts are not very viscous and deviations of more than ±1° are not to be expected in such cases. However, drugs with high melting points often melt sluggishly, even at a lower eutectic temperature, and this then is poorly defined because the onset of melting at the eutectic temperature, instead of being sudden, is rather a gradual process which is difficult to fix on the temperature scale. The reason is that such substances tend to form viscous melts and the flow and coalescence of droplets is retarded. Water of crystallization and impurities are sometimes the cause of unsharp eutectic melting.

Polymorphism causes even greater deviations than the characteristics alluded to above. This is especially true if the modifications are enantiotropic (p. 411) because the commercial product can then occur in the form stable at room temperature and not in the highest melting form. The eutectic temperature is lower or higher according to whether or not the crystals transform before reaching the eutectic temperature. For example, sulphaguanidine has two enantiotropic forms, I (m.p. 187–191°C) and II (m.p. 174–176°C), whose eutectic temperatures are 166 and 161°C with salophen and 158 and 151°C with dicyandiamide, respectively. The examination of numerous cases of enantiotropic and metastable commercial products has shown that the crystals are mostly transformed during heating, even before the eutectic temperature is reached. This is all the more likely because grinding with the reference substance favours the formation of seeds of the higher melting modification so that transformation occurs at lower temperatures when the mixture is heated than in ordinary melting point determinations.

In most cases, hydrates give the same eutectic temperatures as the corresponding anhydrous compounds. This is because they lose their water of crystallization at comparatively low temperatures. However, if the eutectic temperature is below 100°C, the possibility of two values must be considered. For example, pyrogallol occurs in some commercial products as the hydrate, while it is anhydrous in others [22]. With acetanilide, the hydrate gives a eutectic temperature of 49°C compared with 59°C for the anhydrous form. Consequently, the observer will find one or other eutectic temperature depending on the source of the product.

If the substance under test contains solvent, it is sometimes not possible to obtain reproducible eutectic temperatures unless the solvent is removed. To achieve this it is generally sufficient to heat the sample, prior to mixing with the reference substance, beyond the temperature at which, according to the identification tables [7,11], loss of solvent takes place. If the sample is held for some minutes above this temperature, it may be cooled and tested directly for the eutectic temperature. Drying at a suitable temperature on the Kofler hot bench (p. 339) is even more time saving.

Occasionally, a mixture of hydrate and reference substance initially produces only a small amount of eutectic melt, whereupon the anhydrous form crystallizes out and its eutectic melting with the reference substance is only observed later. For example, when a mixture of oxalic acid (m.p. of hydrate 101°C) with dicyandiamide (m.p. 210°C) is heated, signs of melting are observed at 85°C but the real eutectic temperature with the anhydrous oxalic acid (m.p. 190–192°C) does not occur until 177°C.

Very occasionally different eutectic temperatures may be the result of a molecular addition compound being formed between sample and reference substance. Since such compounds are formed during heating at places of contact between two components, it depends on the accidental mixing ratio whether the eutectic whose melting is observed is that formed between the compound and the substance under test or between the compound and the reference substance. In addition, there is the possibility that if, for example, heating commences with the stage at an already elevated temperature, the molecular addition compound does not form at all so the observed eutectic is the lower melting one between the reference substance and that under test.

Another possible occurrence is the apparent absence of any depression in the melting point of the lower melting component. This means that the eutectic temperature falls within the melting range of that component, be it the test or reference substance. The cause is a miscibility gap between the two liquid phases. For example, the eutectic temperature of mannitol (m.p. 167°C) with phenacetin (m.p. 133—134.5°C) is 133°C and the eutectic temperature with benzanilide (m.p. 161—163°C) is 161°C. A survey of the identification tables (e.g. ref. 11) shows that highly polar compounds like sugars, sugar alcohols and tartrates do not, generally, give significant depressions with the usual reference substances; dicyandiamide is an exception. A search for satisfactory reference substances reveals that phenazone is suitable for testing substances melting in the range 30—170°C and DL-malic acid for substances in the range 70—220°C, but dicyandiamide remains the best for the wide region from 130 to 300°C [42]. It is another indication of the specially fortunate choice Kofler made of dicyandiamide as a reference substance.

Substances of extreme volatility, such as camphor, borneol and so on, may evaporate before the eutectic temperature is reached and the onset of melting of the reference substance may be taken as the eutectic by mistake. This may be avoided by resorting to lower melting reference substances as recommended by the Austrian Pharmacopoeia [43] or by the special sublimate film procedure described above (p. 374; Fig. 23).

(d) Testing for purity. The determination of the melting point with the hot-stage microscope automatically includes a test for purity. Often, levels of impurity can be recognized which, in the ordinary course of events, would not be detected at all. Premature melting will reveal as little as 0.25—0.5% of an impurity [44]. It is fair to state, as a general principle, that the degree of purity can to be judged better by microscopic methods of melting point determination than by macroscopic methods.

It must be recognized that organic substances, as a class, cannot be expected to have absolutely sharp melting points since the crystal lattices are held together only by van der Waals forces and these tend to give rise to crystal defects. Due to dislocations in the crystal lattice, crystal growth often takes place via spiral growths (p. 472), so it may be assumed that the breakdown of the lattice accompanying heating occurs first at the sites of dislocations or crystal defects and gradually spreads to undisturbed lattice regions. However, a melting interval rather than a sharp melting point is to be expected not only on this account; differences in thermal conductivity also affect the apparent length of the melting interval. It follows that, in order to judge the purity of a given sample, comparison must be made of the melting point or interval with that of the purest preparations. Experience shows that even the purest substances (melting point standards or analytical reagent grades) may have intervals of up to one or two degrees.

Provided there is only a single impurity present and the simple eutectic type of phase diagram is applicable, melting should begin at the eutectic temperature. As a rule, only about 1% of the impurity is required for this to happen but it is impossible to say definitely how much is necessary since the amount of melt formed depends largely on the position of the eutectic point (see Fig. 22) and this can be very far towards one end of the concentration axis. In addition, the degree of dispersion of the impurity also exerts an influence. Melting will not begin sharply at the eutectic temperature but at a rather higher temperature if there is only a small amount of impurity and the position of the eutectic point is not favourable. This is because so small an amount of eutectic melt is formed that it adheres to the surfaces of crystals and does not appear in the form of separate drops. Only when more liquid is formed by the crystals dissolving does the onset of melting become clearly evident by droplet formation.

If several impurities are present in small amounts, the beginning of the melting process will appear to be gradual and unsharp. As a rule, impurities are unevenly distributed and sometimes a particular microscopic field of view might be completely free of foreign particles. In the search for impurities, therefore, the vigilant inspection of more than one field of view by moving the slide during heating is recommended. If the melting interval, and especially its lower limit, could be fixed exactly, it should be possible to specify the desired degree of purity of a substance. Of course, the use of the melting interval for purity testing or specification presupposes that the substance melts without decomposition.

An especially sensitive test for the purity of organic substances can be carried out with crystal films [44]. If a substance is allowed to solidify slowly close to its melting point, the conditions for separation of dissolved impurities are at their optimum. The last of the melt to solidify in this way, therefore, starts to melt at much lower temperatures than the bulk of the crystal film and the onset of melting is clearly signalled by the movement of vacuoles. These are small bubbles trapped in the last of the melt which become distorted into irregular cavities when it freezes or which are the result of contraction on crystallization. When, on reheating, the vacuoles change shape to spheres or discs and begin to move around, the first appearance of the liquid phase is indicated.

If an impurity consists of a particular substance, e.g. an isomer of the principal constituent, it may sometimes be seen directly in slowly solidified preparations. In this case, the last of the melt to solidify is close to the eutectic composition and is recognizable as a fine crystalline layer on the larger crystals of the substance itself. On heating, this layer melts at the eutectic temperature. For example, preparations of Analytical Reagent grade *m*-dinitrobenzene show the effect because of the presence of *o*-dinitrobenzene [44]. Minute amounts of impurities that have no effect whatever on the melting point or the end of melting can still be recognized in this way.

Inorganic impurities in organic substances are perceptible because, on heating beyond the melting point, they remain as solid particles in the drops of melt. This also happens if an organic impurity is present which has a much higher melting point than the substance itself or if, due to a miscibility gap between the liquid phases or an extreme position of the eutectic point, there is no distinct depression of the melting point. Undissolved residues are very frequently found on examination in the molten drops of pharmaceuticals. This proves that specified methods of examination fail to reveal some impurities which do, however, show up microscopically.

In rare cases, impurities are mimicked by polymorphic modifications. For example, an investigator examining a sample of modification II of L-hyoscyamine (p. 362) might suppose he had an impure sample of the stable form which melts 3—5 degrees higher. Phenobarbital provides another example. In most commercial production samples, it occurs as modification II [45] or, rarely, as the hydrate. Now the overall melting range is four degrees since modification II melts inhomogeneously at 172—174°C with the precipitation of modification I which starts to melt immediately and has itself an interval of two degrees, so that melting is not complete below 176°C. If the polymorphism of the system were unknown, it might be supposed that the sample was impure because of the unusually long melting interval for a substance not obviously decomposing. Finally, the existence of substances with widely separated polymorphic melting points like sulphathiazole (202 and 175°C) should be noted (see p. 360). The possibility of distinguishing between polymorphism and an impure sample or a mixture has already been referred to (p. 361).

(4) Methods of purification and separation

Since pure substances are required for the determination and use of thermomicroscopic data, simple methods have been developed in order to obtain the requisite degree of purity without having to resort to recrystallization.

(a) Drawing off the eutectic melt. With this method, advantage is taken of the fact that organic impurities melt at some eutectic temperature so that, by drawing off the melt at that point, it is often possible to purify a sufficient quantity of a substance for identification [46]. Unstable substances cannot, of course, be purified in this way.

Hardened filter paper, cut into 2 cm squares, may be used to absorb the melt. The procedure is to place such a square on a slide, spread a thin layer of the sample crystals over it and lay a second slide across the first. The preparation is heated to above the eutectic temperature on the open hot stage while the upper slide is pressed down firmly with spade forceps. The eutectic melt will be absorbed into the filter paper and form translucent spots. At this stage the slide to which the crystals are adhering is lifted off, the filter square changed and the process repeated. Changes of filter paper are made until no more melt is absorbed. At this stage it will be found best to continue to heat cautiously until signs of renewed melting appear but, as this is an indication that the melting point of the pure substance has been reached, the preparation should now be removed quickly. The sample being now sufficiently pure for the melting point determination, the crystals are scraped off the slide and the necessary determinations carried out. The filter paper squares contain not only the impurity, but also a certain amount of the substance under test, at least as much as was required to dissolve the impurity. Some loss of sample is unavoidable when using this method. It is best for substances which do not sublime readily and which exhibit drop condensation instead of forming crystalline sublimates, as is likely with compounds melting below about 80° C.

For working at higher temperatures, above 300°C, clay wafers about 1.5 mm thick may be used. The idea of purifying substances by heating on small clay plates is due to von Braun et al. [47]. The method was successfully used by Lindner [48] to separate quinoline and quinaldine derivatives and it is possible that his work inspired

Kofler to see if filter paper would work.

A modified procedure based on the same principle of drawing off the eutectic melt was devised by Fischer [49]. Instead of filter paper squares or clay wafers, he used a specially constructed glass capillary of about 2 mm diameter containing a sintered glass disc. The capillary was attached to the table of the hot stage and was manipulated so as to draw off the melt. It is questionable if the advantages over Kofler's method claimed by the author outweigh the extra trouble and greater cost.

Martinek [50] has described a method for the purification and identification of very small amounts of contaminated organic crystals such as many result from thin layer chromatographic procedures. In this method, the eutectic melt is squeezed away from the crystals by pressing the cover slip with a specially mounted pin while the preparation is under observation on the hot stage.

(b) Sublimation. Sublimation is a very old method of purification as is shown by the production of camphor, for example. Many volatile substances which are insufficiently pure or which have decomposed on long storage may be adequately prepared for melting point determination in this way. Sublimable substances from the material of biological investigations may also be similarly isolated and the method is suitable for isolating drugs from tablets and the like, thus saving extraction.

It is not our purpose to describe the numerous pieces of apparatus for microsublimation whose descriptions have been published over many years, especially as, with few exceptions, they are not designed for direct use with the hot stage microscope. Detailed descriptions may be found in ref. 51. In special cases of insufficient volatility or decomposition, vacuum sublimation may be advantageous [52]. The simple procedure we recommend here is based on one proposed by Molisch [53] for subliming constituents from plants.

The more volatile a substance, the greater is the difference in temperature between the onset of sublimation and melting. It follows that the applicability of sublimation as a method of purification can be judged from such information and that data relating to the onset of sublimation in the identification tables are valuable not only as aids to identification, but also as a guide to the applicability of this method.

To obtain useful amounts of material in a short time, it is best to sublime at 10—20 degrees below the melting point. On the other hand, if the object is to produce especially well-formed crystals, it is better to let the temperature rise gradually from the point at which sublimation begins to the optimum temperature. A glass vessel may be used or, more simply, a slide with a glass or metal ring attached. A cover slip serves as receiver. Since numerous crystal seeds are formed on a cold receiver, it is best to preheat the cover slip so as to reduce the number of seeds and increase the size of the sublimate crystals, especially if they are to be used in the determination of optical properties. If condensation droplets appear instead of crystals, the cover slip should be lifted and the droplets scratched with a needle to stimulate crystal seed formation. Alternatively, the receiver can be inoculated by sprinkling some crystals on it, rubbing them over it and then wiping them off again.

It is best to prepare as many sublimate deposits as are needed for the planned determinations by changing the cover slip as many times as required. The cover slips are placed sublimate side down on slides and the preparations used as soon as possible in case the crystals evaporate, even at room temperature. In the event that no more than these very small amounts of the sample have been purified, the eutectic temperature and mixed melting point are determined as described on p. 373.

It should be noted in this connection that not too much importance is to be attached to the shape of crystals. In an investigation of the influence of temperature and amount of sample taken, both factors have been shown to exert a considerable influence on the shape of sublimate crystals [54]. Figure 24 shows four sublimates of mephobarbital prepared under different conditions. These are all of the same polymorphic modification and the crystals differ only in habit due to their different combinations of faces.

The importance of not relying only on crystal shape and melting point can be illustrated by a case in which the author was personally involved. In the investigation of a case of death from an overdose of sleeping pills, the problem was to discover the cause from the biological material available. As Luminal ® had been found in the possession of the dead person, phenobarbital poisoning seemed obvious. Microsublimation of the extract produced rhombic leaflets as is usual for this substance. The melting point was found to be 176° C, which is also the melting point of phenobarbital. I was convinced that I had verified phenobarbital, but when I carried out a mixed melting point

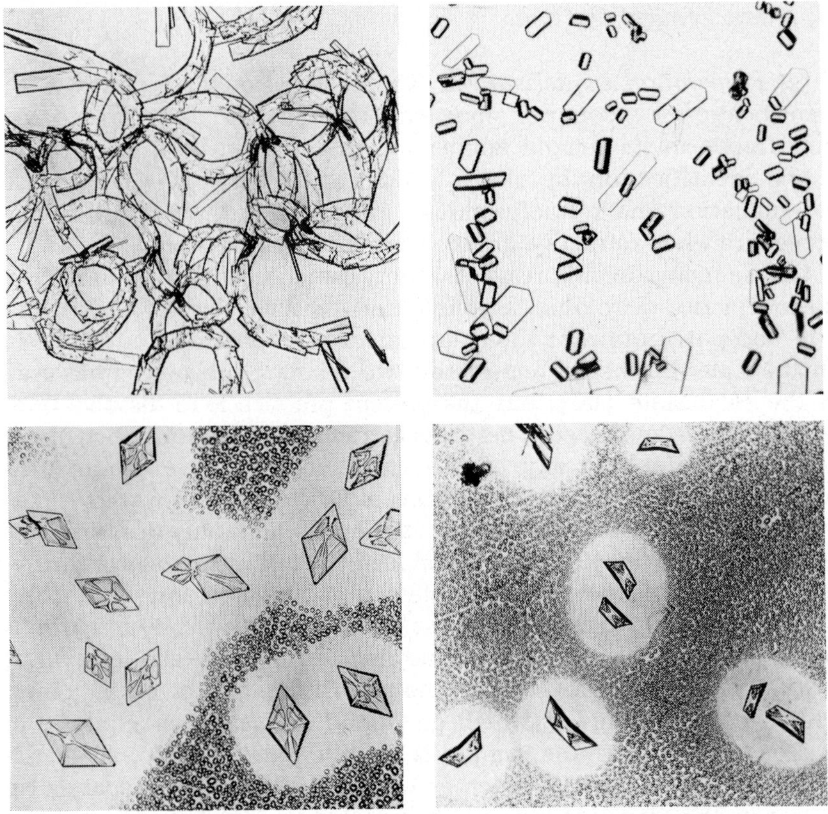

Fig. 24. Sublimates of mephobarbital prepared under different conditions.

with an authentic sample, I was astonished to observe a melting point depression. I then remembered that barbital has a modification which melts at 176° C, although I encountered it previously only in the form of rectangular leaflets [55]. The mixed melting point determination with barbital gave a definite result because the original crystals of the sample underwent a transformation and at 190° C, the melting point of the stable form of barbital, the sample and the authentic substance melted together. On this account the suicide theory appeared improbable.

(5) Microchemical reactions

(a) Preparation of derivatives. The preparation of derivatives for identification is a common chemical procedure. It has been rendered superfluous in its general application in the field of thermomicroscopic identification by the introduction of eutectic temperature determination and refractive index matching, but we return to the procedure when refractive index matching is not possible [11].

Of the many organic reagents, those which have proved most useful are picric, picrolonic, styphnic and dilituric acids. The preparation and purification of their derivatives can be carried out directly on the slide. Formerly, one of the two reacting substances was used in the solid state [56], but the present practice is to dissolve both test substance and reagent before reaction unless the substance under test is not soluble in any of the usual solvents. The reagents mentioned are used as saturated solutions in 20% ethanol in water.

Since picric acid is undoubtedly the most commonly used of these reagents, we shall concentrate upon it in the following description.

Picric acid forms water-insoluble compounds not only with many organic bases but also with phenols and a variety of hydrocarbons which contain at least one benzene ring [57]. Above all, however, it is most usefully employed with the salts of organic bases which melt with decomposition but which give good crystalline precipitates in aqueous solution. If the sample is readily water soluble, all that is required is to dissolve as much as will be held by a lancet spatula tip in one or two drops of water on a slide and to add about three drops of picric acid solution. The onset of precipitation should be marked by turbidity immediately or shortly after mixing.

If the test substance is only slightly soluble, it can often be brought into solution by weak acidification or by heating briefly. If necessary, the substance can be treated in its solid form with the dissolved reagent. In every case, care must be taken that the picric acid is in excess, as indicated by the yellow colour of the surplus clear liquid. This is particularly important when dealing with diacidic bases from which may be formed both mono- and dipicrates. Since the pure monopicrate is not obtainable without weighing the reagents and a mixture of the two picrates is the likely result in the ordinary course of events, it is best to aim at preparing the pure dipicrate by using an excess of the picric acid.

A derivative may be seen under the microscope in the form of a crystalline precipitate or as droplets. If the latter, crystallization can often be stimulated by rubbing with a glass rod or by quickly heating to about 80°C, but the reaction mixture may not, of course, be heated to dryness since pure picric acid would crystallize out. Amorphous products in droplet form may also be induced to crystallize on occasion by washing and drying and then dissolving in one or two drops of an organic solvent (e.g. acetone, benzene, pure ethanol). The solvent is then allowed to evaporate slowly under a watch glass.

The simplest way of isolating the picrate precipitate is by drawing off the excess liquid with a piece of filter paper with its edges cut. The liquid is taken from both sides alternately, then a drop of the washing liquid (20% ethanol solution or pure water) is added and the process repeated. The precipitate must be washed more than once.

Another method has been described by Fischer [58]. In this, the precipitate crystals are covered and a 2 cm × 10 cm strip of filter paper is placed so as to touch the edge of the cover slip. Washing liquid is dropped on to the slide at the opposite edge of the cover slip for as long as it takes for the filter paper strip to become moist. Alternatively, the wash liquid can be removed by pipette with a flat ground tip through a triangle of filter paper [59].

If the precipitate is practically insoluble, washing should be continued until the washing liquid is no longer yellow in colour from excess picric acid. In some cases, proper washing is prevented by relatively high solubility and, in this case, the process should be followed under the microscope and stopped if the crystals dissolve noticeably. Picrate derivatives are themselves yellow or orange in colour and if they are soluble, the wash will never become colourless as is the case with insoluble precipitates. Water should be used, not 20% ethanol solution, when washing soluble picrates. After washing, the precipitate should be pressed with filter paper and dried at about 80°C on the hot bench or hot stage.

Very fine crystalline precipitates can be recrystallized from organic solvents in order to obtain larger crystals. After adding one or two drops of solvent, the slide is covered with a watch glass so as to slow down the rate of evaporation, thus favouring well-shaped crystals.

It should be noted that picrate derivatives may occur in unstable modifications and this can happen not only in the case of recrystallized material, but also in the primary precipitate or when a hydrate is dehydrated. Benzocaine picrate may be cited as an example. The metastable modification II (m.p. 129° C) is the one generally formed while the stable highest melting form I (m.p. 162° C) seldom appears [60]. The melting point—composition diagram (p. 460) shows that the equimolar 1 : 1 benzocaine picrate has four polymorphic forms and that there is, in addition, a dimorphic inhomogeneously melting picrate formed from two moles of benzocaine and one of picric acid [61]. Crystallization from an aqueous medium often yields a hydrate which changes into modification I or II on drying.

Because picrates often show symptoms of decomposition on heating, the melting interval is generally determined rather than an equilibrium melting point. In the case of soluble picrates which are difficult to purify properly, the melting interval may extend over several degrees. In order to avoid reporting wide intervals whose length depends on the degree of purity, the beginning of melting is read only when droplets have begun to form over the whole of the field of view.

(b) Salts of organic acids and bases. In the analysis of drugs especially, compounds occur which are the salts of organic acids and inorganic bases or of organic bases with inorganic acids. The salts of organic bases are generally directly determinable and the isolation of the base is necessary only in rare instances. Salts of organic acids, on the other hand, decompose at high temperatures as a rule and are not accessible to thermomicroscopic identification. Before this can be done, the acid must be liberated and purified.

To liberate pure acids and bases from their salts, Mayrhofer [62] has devised an extremely efficient method based on that of Molisch (p. 385). It is applicable only to sublimable substances. To liberate an organic acid from its salt, one or two drops of 6 M hydrochloric acid are added to 2—5 mg of the sample on a slide and the mixture stirred with a glass rod and then dried at about 50° C. (Since hydrochloric acid will corrode the hot stage, another heating source must be used for this purpose.) After the excess of hydrochloric acid has been completely volatilized, a 2—3 mm high glass ring and receiver are put on and the residue sublimed (p. 386). A Kofler hot bench is the best source of heat but a hot stage, adjusted to remain steady at the desired temperature, can be used. If the optimum temperature of sublimation is unknown, a hot stage should be used so that the receiver can be inspected. If polymorphism is to be expected, it has been found best to sublime only a few degrees below the melting

TABLE 5

Examples for the identification of salts

Compound	Sublimation temp. of acid liberated (°C)	Melting range (°C)	Eutectic temperature (°C)	
Barbital sodium	130—150	188—191 (182—184)	Salophen	164
Bismuth tribromophenate	80— 90 (dry at 50°C)	90— 94	Azobenzene	48
Calcium (or sodium) *p*-aminosalicylate	200—220 * (*m*-aminophenol)	120—122	Phenacetin	94
Cerous oxalate	160—180	188—191 **	Salophen	132
Cyclobarbital calcium	130—150	166—173	Salophen	151
Phenobarbital sodium	130—150	174—176	Salophen	154
Saccharin sodium	190—210	226—228	Salophen	165
Sodium benzoate	100—110	121—123	Phenacetin	89
Sodium cinnamate	120—130	132—134	Phenacetin	99
Sodium-4-methoxy benzoate	150—170	182—185	Salophen	159
Sodium salicylate	120—130	157—159	Phenacetin	93

* Heated without acid.
** Vigorous decomposition.

point so as to lessen the tendency for unstable modifications to form. The residue on the slide is heated until it begins to melt and this is chosen as the optimum temperature for sublimation. (With regard to further processing, see p. 386.)

For melting point determinations, it is advisable to put sublimed preparations on the hot stage not more than five to ten degrees below the expected melting point since otherwise the generally thin sublimate may evaporate completely even before the melting point is reached. If the temperature of refractive index match is required, it is better to use a slide rather than a cover slip as receiver and correspondingly more substance is required in the first instance.

In Table 5, some salts are listed whose acids can be isolated and identified by the methods just described.

Organic bases can be isolated from their salts by reaction and sublimation in the same way as acids. Cold saturated sodium bicarbonate solution is generally the best reagent, but for almost insoluble salts 15% sodium hydroxide solution or, sometimes, a 10% ammonia solution is better. The reagent is added to a small amount of sample dissolved or suspended in one or two drops of water, warmed if necessary. After stirring with a glass rod, the base generally precipitates immediately. The base will also precipitate slowly from suspensions of almost insoluble salts, the salt dissolving in the reagent meanwhile. After drying the reaction mixture, sublimation can be carried out as for liberated acids.

If the liberated acid or base cannot be sublimed, conventional methods must be used for its extraction but microchemical procedures should be adapted as far as possible so as to save material.

(B) THE ANALYSIS OF MIXTURES

The introduction of chromatographic methods has greatly simplified the separation of mixtures, so much so that thermomicroscopic methods are no longer of great importance in this field. Details are included here because of their historical interest, for their possible usefulness to those not possessing chromatographic equipment, and for completeness.

(1) Identification of components

By a combination of the method of purification (drawing off the liquid from the partially melted mixture) (p. 384) with the general Kofler method of identification, one component of a two-component mixture can be found relatively quickly [63]. The component obtained on drawing off the liquid depends on which branch of the two-component diagram (see Fig. 22) the composition of the mixture lies. To isolate the second component in the same way as the first, the relative proportions of the components in the mixture must be changed by treatment with solvents, by sublimation or by some other suitable procedure, so as to shift the composition of the mixture to the other side of the eutectic point. Naturally, it is much easier to do this than to purify the second substance completely because of the difficulties involved in the final stages of purification.

Having found one component, the search for the second generally proceeds as follows. The literature is consulted for the solubility of the component already found so as to be able to choose a very good and a very poor solvent. First, one endeavours to remove the identified component from the mixture, wholly or partially, with the aid of the good solvent. It is sufficient to leach a small heap of the mixture on a piece of filter paper; the undissolved part is treated by the drawing-off method and the melting point of the residue determined. If the melting point differs from that of the identified component, the residue can be subjected to further tests to identify it; if the melting point is unchanged, the mixture under investigation is treated with the poor solvent, the liquor evaporated and the residue further treated by drawing off the remaining impurities and testing by melting point determination, etc. A solution can be concentrated directly on a slide if it can be prevented from spreading. The Behrens—Kley form of ring oven is useful for this purpose [64]. A slide is placed on a warm metal plate perforated by a hole of suitable size and the solution is dropped slowly on to the cooler part of the slide which lies above the hole. The residue can be drawn off on the slide and tested immediately. In some cases, separation or enrichment of the second component can be achieved by microsublimation. If this is to succeed, the two components of the mixture must differ adequately in volatility. In general, the wider apart are the melting points, the greater the difference in volatility.

The possibility of fractional sublimation in combination with microchemical reactions deserves special reference. Mixtures, like those of sodium theobromine with sodium salicylate (Diuretin), for example, are first treated with acid (p. 390) and the dry residue sublimed at 100—120°C to remove the pure salicylic acid. After this first separation, the temperature is raised to 230—250°C to sublime the pure theobromine. Identification is made as described on p. 373. In the case of mixtures of caffeine with salts of benzoic acid or salicylic acid, the caffeine is first sublimed off quantitatively by heating to 200—220°C and, subsequently, the organic acid is liberated by treatment with hydrochloric acid, purified by sublimation at the appropriate temperature (see Table 5) and identified.

The isolation of *one* component from three- and four-component mixtures is generally no more difficult than its isolation from a two-component mixture, but it is harder to separate further components from multi-component mixtures. Again, in such situations the problem is primarily one of altering the ratio of components so as to gain a different branch of the melting point—composition phase diagram. If this can be done, the drawing-off method can be used to separate components.

The procedure of separation is complicated if the components of a mixture form a molecular compound. Here, depending on the mixing ratio and the stability of the compound, drawing off the eutectic melt leaves one or other of the pure components or the addition compound. In the latter event, it is necessary to alter the ratio (and it is generally possible to do so) so that the drawing-off process leaves one or other of the pure components. It is necessary to have a simple method, such as the contact method described by Kofler [65] (p. 442), for recognizing addition compounds. The same applies to mixed crystals; if these are present, the drawing-off method is generally not applicable.

With respect to a mixture which is supposed to consist of only two substances, the question often arises as to whether these alone are present or if other substances are also present. Here again, the mixed melting point test may be applied provided there is a wide enough gap between the eutectic temperature of the binary system which is supposed to obtain and that of the ternary or multi-component system which is actually under test; very often this condition is, in fact, met. A control mixture having the same composition as that of the supposed mixture is prepared and its eutectic temperature determined as a check on its composition. Then this and the sample under test are ground together and the eutectic temperature again determined. If the sample has, in fact, the supposed composition, there should be no depression of the eutectic. If only one of the components is common to the control and the sample, a ternary eutectic point, lower than the binary, would result. If there are no common components, an even lower quaternary eutectic point would be observed.

Four two-component mixtures are given in Table 6, one component of which, phenacetin, is common to all and which all have the same eutectic temperature. They have been chosen specially so as to show that the eutectic temperature alone is insufficient as a proof of identity and a control mixture must be used. Of the sixteen possible combinations, only with that for which the control mixture and sample are identical does the onset of melting occur at the same temperature as for the original mixture. Where different mixtures have been combined, a very much lower (ternary) eutectic temperature has been observed, the depression of the eutectic point being at least 16 degrees.

The identity of three-component mixtures may be established in

TABLE 6

The use of the eutectic temperatures for the identification of mixtures

Mixture under test	Eutectic temp. of mixture (°C)	Eutectic temperature (°C) of mixture with reference mixture			
		Carbromal phenacetin	Allylpropylbarbituric acid—phenacetin	Salicylsalicylic acid—phenacetin	Salicylamide—phenacetin
Carbromal—phenacetin	102	<u>102</u>	68	75	81
Allylpropylbarbituric acid—phenacetin	102	68	<u>102</u>	86	79
Salicylsalicylic acid—phenacetin	102	75	86	<u>102</u>	82
Salicylamide—phenacetin	102	81	79	82	<u>102</u>

the same way. Every different component in a control mixture with which the original sample is mixed will cause a depression of the eutectic point.

The method is subject to the same restrictions as the mixed melting point test (p. 372) for single substances; no depression is to be expected and the test will therefore fail if the liquid phases of the components are of limited miscibility or if mixed crystals are formed.

In the case of mixtures, a further difficulty arises if a very high melting substance is combined with one or more low melting ones. The eutectic point for a mixture of a high and a low melting substance is often only a few degrees below the melting point of the lower melting component and the addition of another high melting component will give a ternary eutectic point which is only slightly lower, if it is lower at all. For example, consider mixtures of caffeine (m.p. 236° C), phenacetin (m.p. 134.5° C) and aminophenazone (m.p. 108° C). For the various possible two- and three-component mixtures, we have the following eutectic temperatures.

Aminophenazone + phenacetin	82° C
Aminophenazone + caffeine	103.5° C
Phenacetin + caffeine	125° C
Aminophenazone + phenacetin + caffeine	81° C

This mixture may be tested for the presence of aminophenazone and phenacetin by a mixed melting point with a reference mixture of those two substances. If the original mixture contains an impurity from the same melting point region, a quaternary eutetic would result with a distinctly depressed melting point. The reference mixture would not, however, reveal the presence or absence of caffeine. Since the melting point of the binary eutectic of aminophenazone and phenacetin is 82° C and that of the ternary eutectic 81° C, the absence of caffeine means a rise of only one degree, which is within the limits of error. It also follows that the presence of, say, theobromine (m.p. 350° C) or codeine phosphate (m.p. 240° C) instead of caffeine would have no significant influence on the ternary eutectic point.

Although it is possible, in particular cases of organic mixtures, to recognize as little as 0.5% of a component by the eutectic point (p. 381), this method is recommended for general use only if all components are present to the extent of at least 5% so that the development of a eutectic melt is assured.

(2) *Quantitative analysis*

(a) Refractive index of the melt. Refractive index measurements are not only of use in the identification of substances; they can also be applied to the analysis of two-component [66—69] and even (under restricted conditions) three-component mixtures [70]. This is made possible by the fact that temperatures at which molten mixtures have particular refractive indices with few exceptions vary linearly between the lower and higher component values. This is also true of most liquid mixtures in the region of room temperature.

The data are best presented by plotting the matching temperature on the y axis against composition on the x axis. If the two components do not differ greatly in refractive index, it may be possible to obtain all the necessary data with one reference powder as in Fig. 25. If they differ greatly, the matching temperature—composition curves are steeper and two or more reference powders must be used since the melt cannot be overheated or supercooled at will (Fig. 26). The steeper the calibration curve, the better it is for the purposes of quantitative determinations because a change in concentration of, say, 1% may result in a matching temperature change of several degrees.

When dealing with pairs of substances which differ greatly in refraction, care must be taken to grind the components well and to mix them intimately, otherwise values may vary even within a single field of view. With careful grinding and mixing, the accuracy and

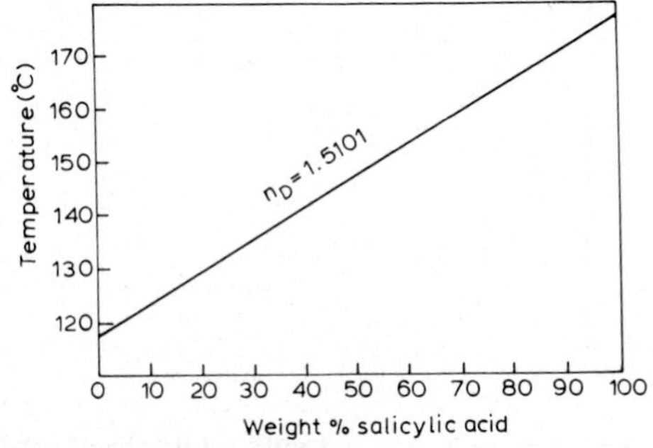

Fig. 25. Matching temperature for reference powder 1.5101 versus composition of ethyl *p*-hydroxybenzoate and salicylic acid mixtures.

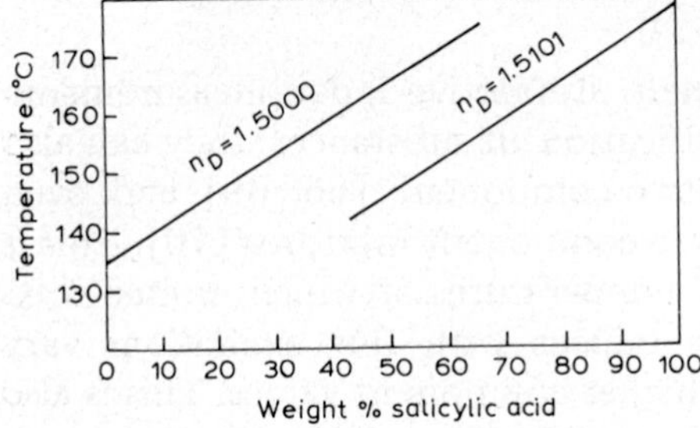

Fig. 26. Matching temperature for reference powders 1.5000 and 1.5101 versus composition for benzoic acid and salicylic acid mixtures.

precision of determination of the composition of such mixtures can be very good, the range of experimental error being generally within 1%.

Deviations from linearity are to be expected if the two components form a molecular addition compound. Aminophenazone forms such a compound with phenobarbital. Both substances can be matched with the glass reference powder of n_D = 1.5299 (amino-

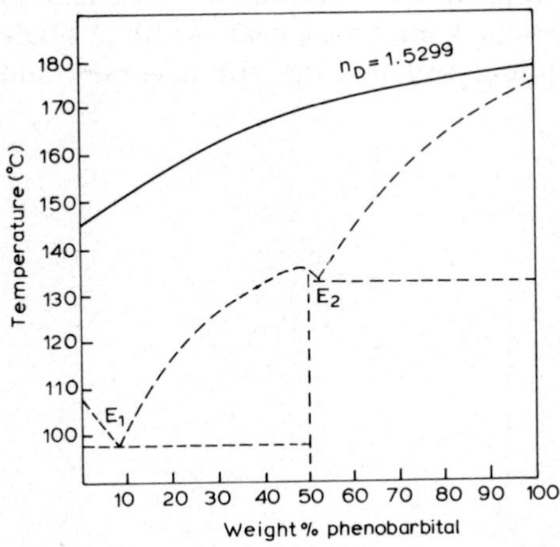

Fig. 27. Matching temperature for reference powder 1.5299 (solid line) and melting point (broken line) versus composition of aminophenazone and phenobarbital mixtures.

phenazone at 146°C, 34 degrees above its melting point, and phenobarbital at 179°C, three degrees above the melting point of its stable modification) and the matching temperature versus composition dependence is curvilinear. Figure 27 illustrates this situation; the melting point versus composition curve given below the matching temperature curve shows that an addition compound is formed [70].

Provided deviations from linearity are not too great, the method of Lennartz [71] may be used to calculate diagrams approximately, given the matching temperatures for pairs of reference glasses for each component. The temperature coefficient of refractive index, or rather its inverse, is calculated from the differences between the refractive indices of two glass powders and between their matching temperatures. This value is then used to calculate the matching points for all the reference powders, whether realizable or not, for the substance in question. A comparison of calculated with experimentally determined diagrams shows complete agreement in some cases, while in others, discrepancies are as great as 5%, regardless of whether the red filter (p. 367) or sodium light is used for refractive index matching. In the light of the possibility of errors as great as this, the precalculated diagram cannot be relied upon for exact determinations of composition, but it has the advantage of facilitating the choice of calibrating mixtures for substances of widely different refractivity.

Decomposition of the components in the vicinity of the melting point, wide separation of the melting points (over 80—100 degrees), high volatility, and immiscibility of the liquid phases all preclude the use of this method. It may be possible to overcome these disabilities by the addition of a third substance. The literature [72] should be consulted on this point.

Choice of the method is favoured if mixtures of suitable components have to be controlled continuously for composition and if simpler methods cannot be used.

(b) Depression of the melting point. The first attempts to use the melting point—composition diagram in quantitative investigations of mixtures were made a long time ago [73,74]. Kofler [75] recommended the method in cases where the refractive index method failed and this alone indicates that, even in his time, it was a method of second choice because of its low precision and the many restric-

tions which apply to it. The method is generally not practicable if there is a miscibility gap in the liquid phase region, if addition compounds are formed, or if mixed crystal formation occurs. In other words, it is really only suitable if the melting point—composition diagram is of the simple single eutectic form. In addition, melting must occur without decomposition and neither component should be highly volatile.

For these reasons, it is necessary to know not only the thermoanalytical behaviour of the individual components, but also the type of melting point—composition diagram for their mixtures. The latter is quickly revealed by a contact preparation (p. 442) and, if it does in fact show a single eutectic, the eutectic temperature can also be determined. With knowledge of this and the two melting points, three or four mixtures are generally all that are needed to establish the melting point diagram (p. 457). Once the diagram has been plotted, the sample whose composition is to be determined can be treated like a mixture as described on p. 458, viz. a crystal film is prepared, the temperature corresponding to the end of the melting found, and the composition read from the graph.

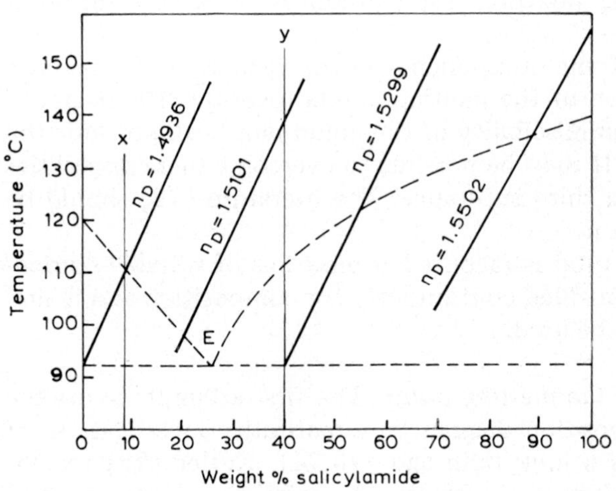

Fig. 28. Matching temperatures for various reference powders (solid lines) and melting points (broken lines) versus composition of carbromal (bromodiethylacetylurea) and salicylamide mixtures.

Suppose, for example, we are testing a mixture known to contain only carbromal and salicylamide and find that melting is complete by 111°C. Figure 28 shows the melting point and refractive index—composition diagrams for mixtures of these substances and we see that the sample might have either 8% (x) or 40% (y) of salicylamide. Since both substances have a needle-like crystal form which does not indicate composition, this must be found from a contact preparation between carbromal and the sample. If a eutectic strip shows up in this when heated, we have the 40% salicylamide mixture, but if not, it contains only 8% salicylamide. The range of error, using the melting point diagram is ±2%. By comparison, the matching temperature determination is both unambiguous and more precise. The mixture x has a refractive index of 1.4936 at 111°C and the mixture y matches 1.5101 at 140°C (red filter) and the temperature of match can be found more reliably than that corresponding to the end of melting.

(C) CHARACTERIZATION OF LIQUIDS

Although substances in the liquid state are less amenable to thermomicroscopic investigation than those in the solid state, nevertheless techniques have been developed which allow the hot-stage microscope to be used for their identification.

(1) Determination of the boiling point

Several micro-methods for the determination of boiling points are described in the literature and, of these, the methods of Emich [76] and Siwoloboff [77] can be adapted to the Kofler hot stage. Fischer [78] described a modification of Emich's method but it is somewhat difficult to handle for those untrained in general microchemical methods. On the other hand, Siwoloboff's method, described here as adapted by Brandstätter and Thaler [79], is more suitable for general use.

A glass tube, about 40 mm long and of 2 mm internal diameter, is filled to a height of about 15 mm with the aid of a drop pipette whose end has been drawn out to a fine tip. The tube is inserted in the hole in the metal block with the viewing aperture whose use in molecular weight determination is described elsewhere (p. 481). The block is inserted into the frame of the object guide and the microscope inclined about 40° by tilting the limb (if this is possible) or by supporting the foot. A capillary (see below) is now placed in the tube with its open end about level with the middle of the window and clear of the bottom of the tube and the hot stage is covered with the round glass cover plate, this having two small glass strips fixed to its underside to prevent it from slipping off. The control unit is set to the temperature at which boiling is expected and heating is commenced. As the temperature rises, the short column of liquid in the capillary moves out slowly until finally a small bubble escapes; this appears to move downwards after leaving the tip due to the inversion of the microscopic image. The stream of bubbles increases as the temperature rises further but when the rate of escape reaches three to four per second, the heating is switched off. Generally the temperature will rise a further one or two degrees to the real boiling point. This is taken as the temperature at which single bubbles can no longer be distinguished but only a continuous chain stretching downward as a dark line from the tip of the capillary. The heating having been switched off, the stream of bubbles starts to slow down and finally ceases completely and the liquid moves jerkily back up the capillary. The thermometer is read again at this point, which is generally about a degree or so below that at which boiling begins.

The setting of the control transformer depends on the boiling point itself; the rate of heating in its vicinity should be about 4°C min^{-1}. If the real boiling point has not been reached because the heating has been switched off too soon, the heating is switched on again as soon as the development of bubbles has been observed to slacken.

The capillary used here should not be quite as fine as those which are customarily used to facilitate boiling in vacuum distillations. To prepare it, a 5 mm piece is broken off the stock capillary and is held with tweezers with about 3 mm jutting out and this free end is melted to a small ball in a micro-flame. A 35—40 mm long glass rod having the same external diameter is fused to the ball on the end of the capillary, care being taken not to make this too thick.

The values given by this method are too high because the Kofler hot stage is calibrated for use with ordinary slides and not for the metal block used in this procedure. The correction to be made is about one degree in the 50—100°C region and rises to three degrees or so in the vicinity of 200°C. The method is not suitable if the amount of liquid available is small, say only a few microlitres. In this case, Fischer's method [78] is recommended; it enables the boiling point of as little as two microlitres to be determined. It should also be mentioned that an alternative method has been described by Wiberley et al. [80].

(2) Determination of refractive index matching temperatures

Fischer [78] has also described a method which enables the refractive index of a small quantity of liquid to be matched with the hot-stage microscope. It resembles Kofler's procedure (p. 366), the determination being carried out in a micro-cell [81] in the form of a flattened glass tube 1—3 mm in width and 0.1—1 mm in depth.

A small amount of a glass or crystal reference powder of known refractive index is introduced into the cell which is then heated at the point where the flattened part joins the original cylindrical tube and about 1 cm is drawn out to form an even narrower capillary. The cylindrical part of the cell is now cut off leaving a length of about 1 cm which thus forms a short funnel. The liquid under test is put in this and the entire sample is pulled down into the cell by spinning for about 30 sec in a hand centrifuge. The cell should be only one third to one half full and no liquid should remain in the narrowed capillary, any surplus being expelled by gentle heating after which the capillary is sealed quickly in a flame. With highly volatile liquids, the cell can be held with moistened filter paper or by tweezers cooled in iced water and sealed by touching with the flame from one side. An alternative way of filling the cell after inserting the reference powder is to draw out its end into a capillary, break in the middle, warm slightly and then insert the broken end into a drop of the liquid under test, a small amount of the latter being sucked in as the cell cools. The cell is then centrifuged and sealed as already described. If the reference powder has been chosen correctly, a temperature can be found for the refractive index match.

Before sealing particles of the reference powder in with the liquid in the cell, it is necessary to find the best powder for the determination and a search must be made in the same way as for a suitable reference powder for the identification of an organic melt (p. 369). Liquids, however, tend to have refractive indices below the coverage of Kofler's set (p. 368) and additional reference powders are required. Fischer added sodium fluoride and lithium fluoride (n_D = 1.3255 and 1.3919, respectively, at room temperature). Undoubtedly, substances in Chamot and Mason's [4] set of isotropic crystals would also be of interest for such determinations.

(3) Critical solution temperatures

Another property of use for the identification and purity testing of liquids and liquid mixtures is the upper critical solution temperature. Fischer et al. [82—84] introduced its determination to the range of thermomicroscopic techniques by adapting Harend's earlier method [85] for its determination based on the observation of schlieren.

In principle, it is related to the contact preparation method (p. 442) but suitably modified for liquids, reference compounds whose liquid phases are immiscible with the liquids under test at room temperature being chosen. A list of suitable reference liquids is given in Table 7.

A small amount of the substance to be tested is run into a capillary and an approximately equal amount of the reference liquid is added (the required quantities being in the range 0.1—0.3 μl, depending on the bore; no weighing or measuring being necessary) and the

TABLE 7
Reference liquids for critical solution temperature determinations

Class of substance under test	Reference liquids
Aliphatic hydrocarbons	Aniline, benzyl alcohol, tricresyl phosphate
Halogen derivatives of low aliphatic hydrocarbons	Formic acid
Naphthenes	Methanol, aniline
Aliphatic alcohols	Liquid paraffin (lower alcohols); water (middle range alcohols)
Aliphatic aldehydes and ketones	Glycerine
Fatty acids	Liquid paraffin
Benzene and its homologues	Glycol, ethylene cyanide
Benzene derivatives	Glycol
Halogen derivatives	Ethylene cyanide, glycol
Phenols	Water
Essential oils	Glycol, butandiol, ethylene glycol diethylene glycol, triethylene glycol [86]
Fatty oils	Ethanol, ethylene chlorhydrin [87], nitromethane, acetonitrile [88]

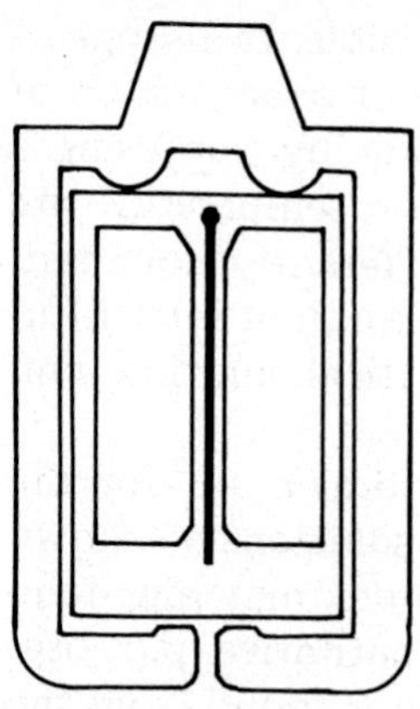

Fig. 29. Half-slide capillary mount for critical solution temperature determination in Kofler hot-stage object guide.

capillary is sealed. The capillary is held in the gap between two glass strips, 0.7—0.8 mm thick and cemented with a suitable silicone resin to a half slide held in the Kofler hot-stage object guide as shown in Fig. 29. The meniscus between the two liquids can be easily seen and, with the aid of the object guide, can be positioned exactly above the aperture of the hot stage. The temperature at which the meniscus disappears on heating is the upper critical solution temperature. On cooling, the meniscus reappears at practically the same temperature.

As is well known, the precondition for the visibility of the meniscus is a sufficient difference between the refractive indices of the two liquids. It may happen that the indices match below the critical solution temperature and thus the meniscus appears to vanish prematurely. For example, in the system *m*-xylol—glycol, this phenomenon is observed at 215°C, 22 degrees below the critical solution temperature. Fischer demonstrated the correctness of this explanation by determining the temperature—refractive index curves using the reference powder method and showing that they cross at 215°C because of the difference in temperature coefficients. If this really is the critical solution temperature, the meniscus would not have reappeared on further heating; since it is not, the meniscus reappears because the refractive index—temperature curves begin to diverge again. The Becke line phenomenon (which we use to detect small differences, and their sign, between crystals and mountants; p. 366) is in evidence in this case, even when the meniscus is nearly invisible. It disappears within less than half a degree of the temperature of exact match of refractive indices. Mistakes which might be made by confusing a refractive index match with a real critical solution temperature are obviated if the test and reference liquids are differently coloured. Such a difference enables the investigator to observe their interdiffusion and this can only occur readily above the critical solution temperature.

A temperature—composition diagram can be drawn up for the quantitative analysis of binary solutions and salt solutions of known components. This is analogous to the refractive index matching temperature—composition diagram as used for quantitative purposes (p. 397) except that, in this case, the composition is found from the critical solution temperature value [89—91].

(4) Preparation of fusible derivatives

The preparation of crystalline derivatives for the identification of organic liquid substances is an old-established method which, although less important today than it once was, offers some advantages when carried out in micro-test form. As a rule, only one derivative need be prepared if the Kofler eutectic and refractive index confirmatory tests are carried out in addition to the ordinary melting point determination. It has been shown that the crystal habit of a derivative is insufficient for identification [64,92] and so the Kofler physical constants are still needed [93].

Since liquid organic substances belong to a wide variety of chemical classes, a considerable number of reagents may be used. The first microchemical tests of this sort were primarily concerned with the volatile aldehydes and ketones [92,94]. For their detection, semicarbazide, *o*-, *m*- and *p*-nitrophenylhydrazine, and *m*- and *p*-nitrobenzhydrazide have proved especially useful in conjunction with the micro-beaker method introduced by Griebel [95]. The procedure is of the nature of a micro-distillation and is suitable only for volatile substances. Volatile aldehydes or ketones from solid or liquid material of biological origin may be identified very elegantly by this method. Volatile alkaloid bases from plant organs, such as coniine and nicotine, can also be isolated and identified in this way [96].

The material under investigation is placed in a micro-beaker and covered with a slide or cover slip on the underside of which hangs a drop containing the reagent either in solution or suspension. The volatile aldehyde or ketone is released by gentle heating and reacts with the reagent to form a crystalline precipitate in the hanging drop. Originally, identification was made from a study of the morphological aspects of the precipitate and, for this purpose, a cover slip was suitable for closing the micro-beaker but, following Fischer [93], a melting point determination has been added and the usual Kofler hot stage half-slide is to be preferred. Before the melting point is determined, the precipitate is washed as described above (p. 389) with distilled water or a special solvent.

Opfer-Schaum [97] recommends purification of the derivative crystals by Kofler and Wannenmacher's eutectic melt drawing-off method (p. 384) in preference to washing and also, as a further simplification to Griebel's micro-beaker method, the immediate determination of the eutectic temperature between reagent and derivative, thus obviating the necessity for purifying the latter at all [98].

Brandstätter and Thaler [79] have derived esters of 3,5-dinitro-

TABLE 8

Kofler identification data for some esters of 3,5-dinitrobenzoic acid

Ester of	M.p. (°C)	Eutectic temp. (°C) with	Reference glass n_D	Match. temp. (red light) (°C)
Methanol	107	Benzil 66, acetanilide 81	1.5203	127–128
Ethanol	93	Azobenzene 44, benzil 58	1.5203	93– 94
n-Propanol	74	Azobenzene 37, benzil 50	1.5101	97– 98
Isopropanol	122	Acetanilide 92, phenacetin 105	1.4937	125–126
Isoamyl alcohol	60	Azobenzene 31, benzil 42	1.5101	69– 70
Benzyl alcohol *	105 (113)	Benzil 70, acetanilide 85	1.5609	107–108

* The 3,5-dinitrobenzoic acid benzyl ester occurs in two modifications with the melting points shown. When the derivative is heated, the lower melting modification is seen to melt but this mostly recrystallizes to the stable modification.

benzoic acid, *p*-nitrobenzoic acid and benzoic acid with various alcohols by conventional microchemical methods and have determined the Kofler identification data. These are given in Table 8 for the esters of some alcohols and 3,5-dinitrobenzoic acid.

They have also shown that aldehydes and ketones give readily identifiable derivatives with dimedone, *p*-nitrophenylhydrazine and 2,4-dinitrophenylhydrazine, aromatics with tribromoaniline, pyridine and quinoline with picric acid, and that aliphatic acids can be converted to their *S*-benzylthiuronium salts.

(5) Cold stages

Although some cold stages are commercially available today, their manipulation is more difficult and usefulness limited by comparison with hot stages. The need to freeze a liquid and to identify the solid form as described earlier can seldom, if ever, arise although, in principle, all the necessary data (melting points and eutectic and refractive index matching temperatures) could be determined using a cold stage. Similarly, polymorphic transition phenomena and the behaviour of two-component systems can be studied. However, it must be stressed again that such investigations cannot be carried out as easily with the cold as with the hot stage. In the low region, even the determination of the melting point may be difficult because of the great tendency towards supercooling due to the tardiness of seed formation. Generally, the mode of operation now to be described is recommended [6].

A drop of the liquid under investigation is mounted between slide and cover slip as usual. If it is easily supercooled (water, for example, rarely freezes spontaneously above -20°C), this should be minimized as far as possible by pressing on the cover slip, rubbing its edge or inoculating with seed crystals to induce crystallization. Seeds for inoculation can be prepared by putting a droplet of the substance on a small aluminium plate alongside the slide carried on the cold stage. During cooling, this is stirred steadily with a dissecting needle until crystallization sets in and the edge of the cover slip is immediately touched with the dissecting needle to initiate crystallization. Once crystals have been formed, their melting point can be found by allowing the temperature of the stage to rise and this can be accelerated if need be by turning on the electric heating element. The melting point should be determined at equilibrium if possible.

When investigating polymorphism, it is best to cool the stage as far as possible before mounting the preparation and so encourage the spontaneous crystallization of an unstable form. If this does not occur, crystallization may be induced by pressing on the cover slip. This is not necessary if there is a special enantiotropic relationship between the polymorphic modifications. For example, crystals of dioxane transform spontaneously on cooling below −1° C, the transition point, to the low temperature form and transform again on heating to the high temperature form.

Contact preparations (p. 442) are almost as easy to make with liquid organic substances as with those which are solid at room temperature. Only in the case of light, mobile substances of low viscosity does one get films so thin that it is difficult to introduce a second component under the cover slip. The difficulty may be overcome by scattering a few fragments of a glass powder over the slide before applying the drop of the first component or, even better, by using a fine hair to divide the preparation space and govern its thickness [99]. If a Kofler cold stage is used, after the first substance has been solidified the second is introduced by a long thin micropipette through a small hole in the side wall of the stage. The tip of the micropipette is touched to the side of the cover slip so that the liquid runs in and forms a contact zone with the first substance. The preparation is now cooled so as to solidify the contact zone as quickly as possible; it is then reheated and the contact zone inspected in the usual way. Another way of preparing a contact zone is by placing two drops side by side on the slide and covering them with a cover slip before cooling but this method gives broad and somewhat ill-defined contact zones.

4. Phase investigations of one-component systems

Although thermomicroscopy finds its greatest application in the identification of organic compounds, it cannot be held to be indispensible in this regard. The situation is different in the investigation of polymorphism and the analysis of two- or multi-component systems. In spite of differential calorimetry, X-ray analysis and infrared spectroscopy, the hot-stage microscope is an essential tool in such investigations, although it should be pointed out that the best results come from a combination of thermomicroscopy with the other methods. In Kofler's time, the aim was to do everything with the hot-stage microscope and this had been achieved in the identification method, but it is now our view that it is best operated in conjunction with other instruments in the fields under discussion here. In this way, thermomicroscopy serves as the basic tool and the results obtained by it are confirmed or supplemented by the other methods.

In research into polymorphism, many more methods can be added to the list of those already mentioned. Haleblian and McCrone [100] suggest, in addition, optical crystallography, dilatometry, proton magnetic resonance spectroscopy, NMR spectroscopy and electron microscopy, although these, with the exception of the first two, are somewhat limited in scope.

(A) POLYMORPHISM

(1) General aspects

Many substances, both elemental and compound, can crystallize with their atoms, ions or molecules arranged in two or more different ways. This property, although observed early in minerals, was first scientifically demonstrated in 1821 by Mitscherlich in sodium phosphate [101] and, later, sulphur [102]. It was he who invented the term "polymorphism".

In polymorphism, the various crystal forms, known as modifications, have the same chemical composition but differ in respect of their physical properties such as density, conductivity, refractive indices, solubility, melting point, latent heat of fusion, infrared spectrum, X-ray scattering and so on. The outward shape of the crystals is different in many cases but this is not a sure sign of polymorphism since, even with a single crystal lattice, there can be great morphological differences.

Polymorphic modifications represent separate crystalline phases in the sense of the definition formulated by Gibbs [103]. It is a general requirement of thermodynamics that of all possible crystalline phases of the same composition, the stable phase under any given pressure is that which has the lowest Gibbs free energy. The transition from one phase to another can be brought about (at least in principle) by changing the temperature because different phases have the lower free energy in different temperature regions, the transition point itself (at which they are in equilibrium) occurring when the free energies are equal. The fact that the stable modification has lower free energy than the unstable means, among other things, that the former has the lower sublimation pressure and, also, solubility in a contiguous phase. Sublimation pressure arises from the ability of molecules to escape from the crystal surface and the extent to which they can do so is related to the free energy. Thus, the sublimation pressure is just as good an indicator of the relative stability of a crystalline form as the Gibbs free energy. Lehmann himself [104] was able to lay down the stability relationship of two polymorphic modifications by measuring their sublimation pressures at different temperatures.

(a) The kinds of relationship between phases.. Enantiotropy and monotropy. The term "enantiotropy" was chosen by Lehmann because it refers to a reversible process while the converse ("monotropy") implies that the transformation can occur in one direction only. In the enantiotropic system represented in Fig. 30, the sublimation pressure curves of modification II and modification I intersect at the transition point (t.p.) before the latter intersects the vapour pressure curve of the liquid. Since, as stated above, the stable form is that which has the lowest sublimation pressure (and the lowest free energy), modification II is stable up to the transition point and modification I from the transition point to its melting point. At the transition point, the two can co-exist in equilibrium and they have the same sublimation pressure, solubility and free

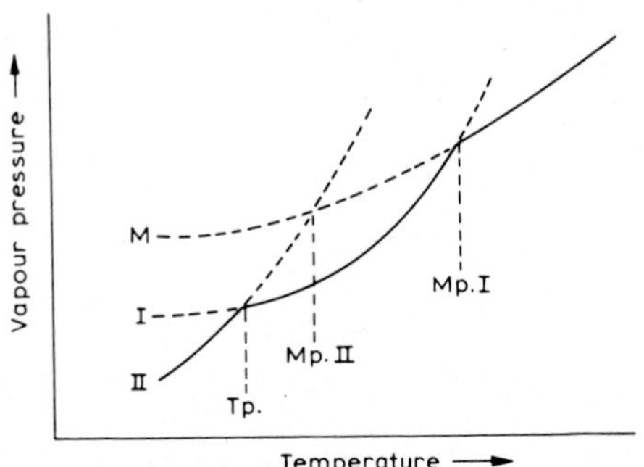

Fig. 30. Sublimation pressure curves of an enantiotropic system.

energy. Below the transition point, modification I is transformed into modification II with the reverse occurring above the transition point. The melting points of the two forms are m.p. I and m.p. II, but m.p. II may be difficult to attain because transformation of modification II to modification I will frequently be complete before this temperature is reached. Since, in enantiotropic systems, the modifications have their own ranges of stability, they are sometimes referred to as low and high temperature forms (or even as α and β forms) instead of modifications II and I as denoted here. The numerical system is more convenient, especially when there are many modifications.

Instead of bearing an enantiotropic relationship to each other, polymorphic modifications may form a monotropic system. In this case, modification I has a lower sublimation pressure than modification II over the whole of its range of existence and is therefore the more stable and transformations can only occur from II to I; transformations in the other direction, as observed in enantiotropic systems, cannot occur. The sublimation pressure curves for both modifications intersect the liquid vapour pressure curve below the transition point and this means that the latter is unattainable (Fig. 31).

Lehmann's classification of polymorphic systems as enantiotropic or monotropic is based on their behaviour under atmospheric pressure and is therefore not complete. As Tammann [105] has pointed out, the external pressure affects the relationship between the phases. Since the melting point can be altered considerably by a substantial

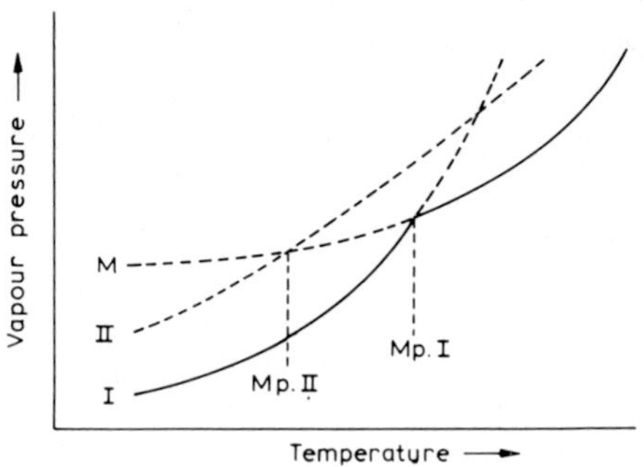

Fig. 31. Sublimation pressure curves of a monotropic system.

change in pressure, there are cases where a monotropic system is changed into an enantiotropic one and vice versa. Bridgman [106], Deffet [107], Tammann himself [108,109] and others have investigated polymorphic relationships under higher pressures. In spite of the qualification, however, the enantiotropic—monotropic classification is still retained for convenience in describing systems under atmospheric pressure.

(b) The kinds of transformations between phases. In recent years, Buerger and his co-workers have given a fresh impetus to the investigation of polymorphism. Not only have they developed its interpretation in terms of crystal structure and so clarified our views on the nature of polymorphic transformations but, in addition, they have provided a suitable working classification [110—112].

Buerger and Bloom [110] proposed that there are unlimited possibilities, from the loosest arrangement to the densest close-packing, for the ordering of the basic building blocks in the crystal structure of a chemical substance. All these possible arrangements represent conceivable polymorphic modifications each with its own lattice energy, but only a few forms actually exist (in some cases, perhaps, only one) and the others cannot be realized.

From structure analysis it is known that crystals are built up out of repeated groups of atoms. These groups, which Buerger called "atomic clusters", are generally ordinary molecules in crystals of organic substances and, in such structures, the forces acting in the so-called secondary co-ordination sphere region (i.e. between the molecules) are relatively weak. In the case of inorganic substances, the atomic clusters generally take the form of co-ordinated groups and here the bonds holding an atom to its neighbours and them, in turn, to their neighbours are the same so the primary and secondary co-ordination forces are equally strong. In either case, the clusters are coupled to each other. They cannot move in isolation and are influenced by their neighbours. Any crystal is, therefore, to be considered as a system of coupled oscillators and the transmission of thermally induced motion through the crystal is to be regarded as a wave phenomenon.

The picture we have, then, is as follows: the atoms vibrate within their clusters and the clusters themselves vibrate within the general framework. When the temperature is raised, the amplitude of the cluster vibration increases and this leads to a weakening of the bonds between them with consequent increase in the sublimation pressure. Now, before the bonds are disrupted sufficiently for melting to occur, a new situation may arise. Another structure may be realizable which allows for greater amplitude of vibration or, perhaps, for rotation which could not occur in the more constrained structure without actual disruption of the bonds between the clusters. If this is so, the new structure will have the lower sublimation pressure and the old structure will tend to transform into the new.

The old and the new structures may differ in various ways according to whether the changes, in going from the one to the other, are within clusters or between them (changes in primary or secondary co-ordination) and according to whether they involve only the position of the atoms or clusters (displacive or dilatational changes) or the actual bonds between them (reconstructive changes). In addition, there may be changes in the degree of disorder of the constituent bodies. For example, free rotation (the acquisition of spherical symmetry) may suddenly become a possibility or free translation of a constituent or freedom of arrangement of two constituents. Lastly, if the constituent bodies can be linked by different types of bond, the bond type itself may change. This phenomenon is most noteworthy in the case of the Group IV elements. For example, in tin the covalent bond between the atoms changes to a metallic bond at about 18° under atmospheric pressure while the covalent and van der Waals layered structure of graphite can be converted to the purely covalent diamond if temperature and pressure are increased sufficiently.

On the basis of the considerations we have briefly outlined, Buerger distinguished four main kinds of transitions, each with subclasses. These are given here together with an indication of the speeds with which they are likely to occur, given that crystalline systems can adjust rapidly to altered external conditions if only a slight change in the positions of constituent bodies is involved, but rather sluggishly if bonds must be broken and reformed.

(1) Transformations involving the primary co-ordination (i.e. the number of nearest neighbours):

(a) dilatational (such processes occur rapidly)

(b) reconstructive (sluggish)

(2) Transformations involving the secondary co-ordination (i.e. the next-nearest neighbours):

(a) displacive (rapid)

(b) reconstructive (sluggish)

(3) Transformations involving a change in the degree of order or disorder:

(a) rotational (rapid)

(b) substitutional (sluggish)

(4) Transformations involving changes in bond type (generally sluggish):

(a) covalent and metallic

(b) etc.

Although Buerger has further refined his classification and subdivided the four types listed above, Winkler [113] has criticized it as being too formal since, in his opinion, many transformations must be regarded as complex processes. His classification contains twenty types but, as in the case of Buerger's, inorganic substances which show a much greater variety of structure than organic substances provide most of the examples. A rather different phenomenological classification, based on the temperature variation of heat capacity, has been put forward by Westrum and McCullough [114].

The reason that polymorphism is of such great importance in inorganic chemistry is that there are so many cases of thermodynamically unstable modifications which can exist metastably over wide ranges without showing any tendency to transform. The example of carbon in the diamond form comes immediately to mind, but many other minerals exist in metastable states. Their structures can be determined without difficulty by X-ray analysis which is unaffected by considerations of stability.

The situation is different with respect to organic compounds. For a long time, polymorphism appeared to be of merely theoretical interest but with the increasing utilization of physical methods for identification, the problem has become of greater importance. To take but one example from pharmaceutical science; it is now recognized that polymorphism plays a role in the stability of solid drugs and in the bioavailability of their active agents so that studies in organic polymorphism take on a more practical meaning. X-Ray structure analysis of polymorphic organic substances has certainly increased in the last ten years but it is often difficult to prepare crystals of a size suitable for single-crystal photographs and, besides, only stable and metastable but not unstable modifications can be investigated in this way. As a result, only two or three structure investigations may be possible even for an organic substance which can exist in numerous forms.

As a rule, there is a great difference between inorganic substances whose crystals are composed of atoms or ions and held together by very strong covalent, metallic or ionic bonds and organic substances whose crystals are molecular. The latter are bonded mainly by London dispersion forces enhanced, in some cases, by the interactions between permanent dipole moments or by hydrogen bonding in the numerous compounds containing electronegative oxygen, nitrogen or fluorine atoms [115]. Basically, however, these sorts of bonding forces (which come under the general heading of van der Waals bonds) are weak by nature and this manifests itself in the low melting points of organic compounds compared with those of inorganic compounds.

Even though it was designed with organic substances in view, Buerger's classification embraces the vast majority of transformations exhibited by organic substances [116]. Even early attempts at classification like Schaum's can be incorporated within it. Schaum [117] distinguished between associative and configurational polymorphism and subdivided it into (a) orientational, (b) twisting, (c) distortional, (d) kinetic, and (e) bonding polymorphism. The first categories fall within types 1 and 2 of Buerger's classification and the last two correspond to types 3a and 4, respectively.

Commonly, organic molecules themselves are unaffected in the course of polymorphic changes and the structures differ only by reason of their different packing under the influence of van der Waals forces. In other words, we have displacive transformations of type 2a. If, however, directional hydrogen bonding is present and the bonds are broken or rearranged in the transition, it will belong to type 2b. For example, resorcinol exhibits a sluggish transition involving the disposition of the hydrogen bonds at above 71°C, the higher temperature form being the denser.

Undoubtedly, transformations involving a change in the degree of order or disorder (type 3) play an important role in the polymorphism of organic substances as is to be expected if the asymmetry of the molecules is not too great [114]. A good example is provided by benzothiophene; at the transition of modification II to modification I of this substance, the entropy gain indicates that the higher and more disordered modification involves all four of the possible configurations [118]. This is an example of substitutional change of Buerger's type 3b but substances with very symmetrical molecules like borneol, carbon tetrachloride, adamantane and others exhibit ordinary rotational transformations. In the transition to the higher modification, individual groups (e.g. hydroxyl groups) or entire molecules undergo random jumps in orientation which appear as rotation on a time average basis [114].

It should be mentioned here that Bernstein and Hagler [119—121] have recently studied the effect of lattice forces on molecular conformations in terms of the phenomenon of "conformational polymorphism." The term implies that two or more modifications of a substance are due to the molecules being able to exist in significantly different conformations.

Rotational transformations occur spontaneously on heating and proceed rapidly so that the melting point of the low temperature form can seldom be realized. The systems undergoing such transformations are generally enantiotropic.

While classical rotational transformations may be relatively easily classified in the case of organic substances, it is often more difficult to differentiate between ordinary structural transformations and order—disorder transformations. It is not to be expected that all the transformations observed microscopically or calorimetrically can be identified by type and classified as for inorganic substances, if only for the reason that in many cases it is not possible to prepare the unstable modifications for X-ray examination.

The idea that, in principle, the same transformation mechanisms apply to organic as to inorganic substances has not been unopposed. Mnyukh et al. [122—124] have concluded, after extensive study, that only one general mechanism exists for polymorphic transitions in the class of molecular crystals as a whole and that there is no evidence that such transformations occur by means of shifts, displacements, deformation etc. This implies that the problems of polymorphic transformations are much simpler than has hitherto been assumed and are, in fact, reduced to two only: the problem of nucleation and that of crystal growth. The authors are of the opinion that all rearrangements within a crystal reduce to the phenomenon of the growth of one crystal within another. They call their model process a "contact mechanism" and distinguish two possibilities. The first, which should be the commoner, is that the growth of the secondary crystals is unoriented because it is not influenced by the structure of the primary crystalline material. The second and rarer situation is that the secondary crystals grow in an oriented or epitaxial fashion. This implies, however, that both modifications have a layer structure or good cleavage and that the structures of the molecular layers, which lie parallel to the cleavage planes, are nearly identical for both phases.

Mnyukh and his collaborators propose that it will be possible in future to explain all the many mechanisms of polymorphic transformation in terms of these two variants, the non-epitaxial and the epitaxial, of the one mechanism.

(c) Chemical properties. It was originally supposed that polymorphic modifications, being of the same chemical composition, have the same chemical properties but this cannot now be maintained. The modern view is that they possess similar chemical properties [125] but they may differ in reactivity and even give rise to different reaction products. Differences in reactivity at the surfaces of solids occur because of the decisive importance of crystal geometry, the reacting molecules adhering to the surface in a certain configuration so as to minimize atomic or molecular movement. A much quoted example is the photodimerization of *trans*-cinnamic acid [126,127]. *Trans*-cinnamic acid is trimorphic and different products result from the differing orientations of the molecules in their lattices. In the stable modification I, the molecules are arranged in head-to-tail fashion and the distance between their double bonds is 0.36—0.41 nm; irradiation of this modification produces α-truxillic acid. The intermolecular separation is similar in the metastable modification II at 0.39—0.41 nm but the molecules are positioned in a head-to-head arrangement and dimerization leads, in this case, to β-truxinic acid. Modification III is photostable because the intermolecular separation of the double bonds is 0.47—0.5 nm and this is too great for cyclobutane formation.

In the field of biopharmacy, the example of chloramphenicol palmitate has aroused much interest. It was noticed long ago that suspensions in which the metastable modification II originally present had transformed to modification I lost much of their efficacy. It was thought at first that the great difference between the solubilities of the two modifications provided a sufficient explanation [128], but Andersgaard et al. [129] and also Burger [130] came to the conclusion, after their investigations, that the bioavailability depends on the rate of splitting of the ester and this is fifty times faster in the case of modification II than in that of modification I. The obvious implication is that the crystal lattice of modification II presents a stereochemical arrangement which is favourable to the attack of the esterase concerned. Although this hypothesis has not yet been confirmed by X-ray structure analysis, the evidence is clear that the polymorphic modifications are not chemically identical as was once supposed.

(d) Nucleation. Although we may now suppose that the majority of chemical substances are polymorphic, we do not know why one modification will crystallize from one solvent and another from a different solvent. We observe experimentally, for example, that a certain metastable modification will crystallize from one solvent but we cannot predict from which other solvents we will get other modifications. We have no better explanation for this than for the fact that two different modifications crystallize side by side from the melt under the same conditions. We may anticipate that the explanation for these phenomena will eventually be found in the theory of nucleation but as yet they are far from being completely clarified.

Nucleation plays a part in all crystallization processes, whether from solution, the gas phase, the melt or by transformation from another crystalline phase. Nucleation in melts has probably been best investigated and this is fortunate for the study of polymorphism because more polymorphic modifications arise by crystallization from the melt than in any other way. The rate of nucleus formation depends on the degree of supercooling. Tammann [105] found that, as the temperature falls below the melting point, the rate first rises, then passes through a maximum and finally falls off again. It is a familiar observation, to be seen also in metals, that the rate of nucleation is decreased if the melt has previously been heated well above the melting point.

Crystallization from the melt when greatly supercooled tends to promote the formation of unstable modifications. Weygand [131] observed that the repeated appearance of an unstable modification at a particular position in a film of melt could be attributed to heterogeneous nucleation. It is indeed striking that an unstable form, after transformation to a stable one and subsequent melting, should reappear on cooling at exactly the same place as before. If homogeneous nuclei had remained after melting, they would of necessity be those of the stable modification so we may infer that a rough spot on the slide or the cover slip or an impurity must act as nucleus for the unstable modification.

Fundamentally, all solid surfaces provide a stimulus to crystallization without their having of necessity any structural relationship to the new phase. Generally, to touch the supercooled melt with a needle point is sufficient to induce crystallization. In addition, pressure, shock, and the presence of electric or magnetic fields favour nucleus formation. The distinction between crystallization so induced and spontaneous crystallization is sometimes blurred [131]. In preparations for the microscope, spontaneous crystallization may occur because of quick cooling or while holding the preparation at some particular temperature; on the other hand, it can be induced or forced by pressure on the cover slip, scratching the edge of the cover slip, or by seeding with crystals of similar or dissimilar substances. The production of unstable modifications in contact preparations (p. 442) is always, in my opinion, crystallization of the induced or enforced category. When growth occurs through the contact zone from an isomorphous solid into the melt of another, modifications (the nuclei of which never form in the ordinary course of events) arise because of the specific stimulation of crystallization (i.e. seeding by the isomorphous surface) persists over an uninterrupted layer of mixed crystals in the contact zone. For example, in the contact preparation of 5-allyl-5-phenyl barbituric acid and barbital, modification II of the latter (that with the second highest melting point) is formed. Hitherto, its production has never been observed in any other way [35].

It has often been observed that, when the temperature changes, one modification can grow out of another already present. Clearly, the crystallization of the second modification is specifically stimulated by the presence of the first and this suggests a morphological relationship between the two which enables the one to seed the other [131]. As a rule, this occurs when the velocity of crystallization of the first form is falling off rapidly but that of the second is still relatively high.

The number of modifications which can be prepared by crystallization from solvents is less than from the melt, although they are of more practical interest. It can be laid down as a basic principle that the tendency for unstable or metastable forms to occur is increased by the rapid cooling or rapid evaporation of hot saturated solutions. Also, the modification which generally appears is that which is stable at the temperature at which crystallization is occurring. In consequence of this, if the system is enantiotropic, the form which separates at room temperature is the ordinary stable form. Very many examples of this could be given, but a few will suffice, e.g. acetamide, resorcinol, succinic acid, sulphanilamide, sulphathiazole. Certainly, there are exceptional cases in which unstable forms crystallize out and these may either prove to be metastable on drying (they may have a storage life of years) or they transform to the stable form as soon as the solvent is removed, this being shown usually by the phenomenon of paramorphosis associated with it.

It may also occur that a particular unstable or metastable modification always crystallizes from solution in a particular solvent. Weygand [131] suggests two reasons for this behaviour: (a) traces of accompanying impurities which may be incorporated in one modification but not in an other, and (b) the first material to separate from solution is in the form of droplets of the melt which subsequently solidify. If the second of these suggestions were to be accepted, the conditions of crystallization from solution would be the same as those for crystallization from the melt and this could explain why the former process sometimes produces a quite unexpected modification.

Unstable modifications of volatile substances are likely to be produced as the sublimate, and the probability is the greater the greater the difference between the temperature of sublimation and the melting point of the stable modification. The advantages of sublimation are that crystals formed are well separated so that the melting points of unstable forms can be determined more easily and, secondly, they are generally well formed and better suited to the determination of optical properties.

Finally, we must consider the nucleation of crystallization of one form in another crystalline form, either unstable or metastable but not including mesophases. Like nucleation in the melt, nucleation in a crystalline modification is primarily a characteristic property of the substance and is influenced by various factors [131—134]. Dislocations certainly play an important part in the process but so also do fissures and cracks of greater dimensions. Inclusions of foreign particles may favour nucleation and impurities in general certainly do so. Spontaneous nucleation of a more stable in a less stable modification is also to be expected if there is a similarity between the two structures. Indeed, this can lead to so rapid a transformation from the unstable to the stable form that further investigation of the former is impossible. Again, as in the case of nucleation in the melt, nucleation in the crystal can be induced by the unspecific processes such as rubbing, scratching, application of pressure or change of temperature. Often, crushing a crystal is enough to activate latent seeds of a more stable form.

(2) Experimental investigation of polymorphism

Undoubtedly it is mainly due to the hot-stage microscope that so many organic substances are known today to be polymorphic. As early as 1942, Deffet [135] was able to list more than twelve hundred compounds whose polymorphism had been described up to that time. The greatest contribution to the discovery of so many cases of polymorphism has been made by workers in the school founded by L. and A. Kofler in the Institute for Pharmacognosy of the University of Innsbruck. Many of these examples were found while determining data for Kofler's thermomicroscopic identification method.

At the present time, polymorphic modifications are investigated not only with the hot-stage microscope but also with the infrared spectroscope and the differential scanning calorimeter. As it is often difficult to distinguish between enantiotropy and monotropy by thermomicroscopic methods, investigators in the Institute for Pharmacognosy at Innsbruck have attempted for many years to make the differentiation from regularities in the infrared spectral shifts of the O—H stretching frequencies [136,137]. Modifications stable at room temperature should absorb at lower frequencies than high temperature forms because of their stronger hydrogen bonds. In cases of monotropy, we would expect, as a matter of principle, that the stable modification would show lower O—H or N—H stretching frequencies than the unstable one. The rule is followed in very many cases and it has proved useful, especially in the investigation of the very many steroid hormones [138,139].

Burger and Ramberger [140] have recently reassessed the infrared rule and at the same time investigated the possibility of making the distinction between monotropy and enantiotropy with the aid of the latent heat of the transition, the latent heat of fusion and the difference in density. They came to the following conclusions.

(1) The enthalpy of transition between two enantiotropic forms is always positive above their transition point and negative below. It is always negative between two monotropic forms (the "heat of transition" rule).

(2) The difference between the latent heats of fusion (the "heat of fusion" rule) can generally be used instead of the latent heat of transition.

(3) The less dense modification is the less stable at absolute zero.

(4) In the case of hydrogen-bonded crystals, the modification whose first absorption band in the infrared spectrum is the higher in frequency has the greater entropy.

Among 228 substances examined, only a few exceptions were found to these rules and most of these few were in conflict with the infrared and density rules. The heat of transition and heat of fusion rules are obeyed in over 99% of cases. By means of these rules, the stability relationships in polymorphic systems can easily be clarified.

It must, nevertheless, be pointed out that, in some cases of enantiotropy recognized by these rules, the transition point is so far below room temperature that the behaviour of the modifications with respect to solvents, heating etc. resembles that of a monotropic pair. In my opinion, the practical implications of these rules are that when monotropy is indicated, no further examination (for enantiotropy) is necessary, but if enantiotropy is indicated (by the heat of fusion rule, for example) further investigation is required to see if the transition point is above room temperature or if it is too low to be recognized by the usual methods.

(a) The preparation of polymorphic forms in crystal films. We have already stated that crystallization from a film of melt is, generally speaking, the most productive way of obtaining different forms of a substance. To prepare the film, enough substance is heated between slide and 15 × 15 mm cover slip so that the whole area is filled with the melt, about 8 mg being required for the purpose (p. 466). If insufficient material is available, a smaller cover slip may be used or not all the area under the cover slip need be filled with melt. The preparation should not be too thick since this makes it difficult to distinguish the modifications. As a rule, the molten film is cooled to room temperature on a metal cooling block and this is generally sufficient to produce one or more unstable modification. If the melt fails to crystallize at first, it can often be induced to do so by heating slowly or by holding for some time at different temperatures above ambient. As an example, we have phenobarbital. When the molten film, after cooling to room temperature, is reheated to about 100°C, finely fibrous spherulites of modification II (m.p.

174°C), narrow rays of modification IV (m.p. 163°C) and broad rays of modification VI (m.p. 157°C) all form. Modifications II and IV are shown in the crystal film in Fig. 32. Modification III (m.p. 167°C) can be produced by holding the melt at about 90°C and scratching the edge of the cover slip. Modification I (m.p. 176°C) is formed when modification II melts inhomogeneously. The latter is the form which generally recrystallizes from solvents, except when a hydrate is produced. Modifications I and II are enantiotropic [140]. The preparation of other forms of the substance is described on p. 456.

The way in which one modification can nucleate another has already been described above in the general introduction. This is illustrated, in the case of nicotinamide, in Fig. 33. The blue spherulite of modification V formed at first has grown somewhat irregularly owing to the fact that crystals of modification III have been nucleated at several places and have developed to fan-like aggregates, thus hindering the further growth of modification V. The red-brown spherulite without fissures belongs to modification IV.

In the case of substances with relatively low melting points, cooling below 0°C often leads to the development of unstable forms. (Occasionally, if the melt has solidified to a glass, touching with a drop of a solvent will induce crystallization. In this event, care must be taken, of course, to distinguish between an ordinary crystalline form and a solvate.) On the other hand, there are also cases in which sudden cooling on the cooling block always gives rise to the stable modification I, but cooling at a much slower rate on the hot stage results in the formation of one or more unstable modifications. Crystal films may also serve for the preparation of specimens for the infrared spectrometer when recrystallization from solution fails. The same applies to investigations by differential scanning calorimetry; modifications are formed in the crystal film, the cover slip removed, and the material scraped off carefully into the aluminium pan of the calorimeter. In this way, one may obtain more melting and transformation peaks in one thermogram than if the substance is crystallized directly in the pan.

Care is necessary in the interpretation of morphological aspects and the appearance of specimens between crossed polars on the polarizing microscope; these are not sufficient to prove the existence of different modifications. In some cases, crystal aggregates of the same modification lying side by side in the same crystal film differ so greatly in appearance that the temptation to conclude that they are polymorphs is overwhelming. In fact, the only proofs of polymorphism by microscopic observation are the observation of different melting points, or quite definite transition processes or, alternatively, exact determinations of the crystallographic properties.

(b) Transformation processes in crystal films. Since transformations from one modification to another are generally associated with a change in polarization colour, they can be followed very easily in the hot-stage microscope. Such processes show differences which are partly specific to the substance and partly the result of extraneous circumstances (temperature conditions, etc.). In spite of their extraordinary multiplicity, these processes may be divided into two types; diffuse transformations and transformations by a continuous crystal front. The former derive from many scattered nuclei and, since the arrangement of the new crystals is generally random, the crystal film will often appear opaque or, macroscopically, to have a milky turbidity. Because of this, regions containing other structures can often be seen with the naked eye. A particularly good example of this type of transformation is afforded by the formation of modification I of β-naphthol in a crystal film of modification II (Fig. 34).

The progressive transformation front is the more frequent phenomenon. It may be accompanied by an increase or a decrease in grain size or there may be no great change. The transition may release gas bubbles from the original crystal film and often these are recognizable macroscopically as turbidity in the new crystalline aggregate. Figure 35 illustrates a crystal front transformation accompanied by coarsening of the crystal grains. The fine-rayed irregular spherulites of modification III (m.p. 153°C) of estradiol monopropionate on the left of the figure are transformed to large coloured crystals of modification I (m.p. 200°C), the process starting from a small fine-rayed spherulite near the lower right-hand corner. The difference in appearance between crystals of modification I formed spontaneously in the melt and those formed in the transformation is striking and demonstrates that morphological differences by themselves are an unreliable guide to phase differences.

This kind of transformation, described also as "gradual" or "reconstructive", can be either monotropic or enantiotropic [141]. It is connected with the availability of nuclei of the other modification so it is often possible, by reducing the chance of nucleation, to determine the melting points of unstable forms. These transforma-

Fig. 32

Fig. 33

Fig. 34

Fig. 35

Fig. 36a

Fig. 36b

Fig. 36c

Fig.37

Fig. 32. Two modifications of phenobarbital.

Fig. 33. Three modifications of nicotinamide.

Fig. 34. Diffuse transformation of β-naphthol.

Fig. 35. Transformation with grain coarsening of estradiol monopropionate.

Fig. 36.(a), (b) and (c). Transformation without change of grain boundary.

Fig. 37. Recrystallization in a crystal film of DDT.

tions take place with a measurable velocity which depends on the temperature and which diminishes to zero, in the case of enantiotropy, at the transition point (p. 411). On the other hand, there are cases of unstable modifications in which the velocity of transformation is so great that no sooner has the modification been formed than it is destroyed and, in fact, it is never seen at all. That it did exist and that a transformation has already taken place is revealed by rows of bubbles and fissures in the new structure.

There is, in addition to the above, a rather slower and differently nucleated type of enantiotropic transformation which was at one time supposed to be analogous to the α-β transformation observed in metals [142,143]. The first organic substances found to exhibit the phenomenon were suberic acid [144] and methyl butyl ketone-2,4-dinitrophenylhydrazone [145]. The transformation of the low temperature phase sets in spontaneously and proceeds by jumps in such a way that crystals of the new modification are arranged in twinned lamellae. It cannot be prevented, is not associated with visible seeds and the reverse process proceeds in a similarly jerky fashion. Whether it is really a unique "concertina" mechanism [8] or only a rare morphological variant of the uniform homogeneous mechanism postulated by Mnyukh et al. [124] cannot be decided at present.

Another mode of transition, commoner but possibly related to the preceeding, is that which manifests itself by abrupt changes in polarization colour. The same effect is observed on both cooling and heating. The grain boundaries are completely unaffected. Generally, the change is very rapid but in the case of 4,4-diaminodiphenylsulphone (Dapsone) it is slow enough for the intermediate phase to be recorded photographically. Figure 36(a) shows modification III, which is stable at room temperature, in the form of a spherulite with numerous concentric fissures. On heating, the individual crystal rays acquire a herring-bone pattern [Fig. 36(b)] due to simultaneous epitaxial growths which eventually cover the original crystalline substrate, the polarization colour meanwhile remaining unchanged. On further heating, the pattern disappears and the crystals regain their earlier smooth appearance but now with a different polarization colour [Fig. 36(c)]. If the preparation is now allowed to cool, the sequence is reversed; the herring-bone pattern reappears, then disappears and finally the original polarization colours return.

If the transformation from one modification to another is rapid, the melting point of the low temperature form cannot be determined and, associated with this, the high temperature form is difficult to supercool. This kind of behaviour is most frequently found when the high temperature form belongs to the cubic system (although, of course, other transformations proceed rapidly as shown by the examples given above). On the other hand, when the rate of transformation is slow, it is difficult to distinguish between enantiotropy and monotropy because, even though there may be no difficulty in observing the transformation or inhomogeneous melting during heating, it is often hard to obtain useful evidence from observations made while the crystal film is cooling. Even if the two modifications are present side by side, often nothing can be discovered because the velocity of transformation is too low.

In cases like the last, a solvent may sometimes initiate the transition. The test is made by carefully loosening the cover slip over a crystal film of the high temperature modification and touching a drop of solvent to its edge so that the solvent flows in under the cover slip. The solvent must not be so good that the entire film is dissolved away. Because of the differences in solubility, crystals of the modification stable at room temperature form directly in the crystal film. [If there is any doubt whether crystals of the modification itself or of a solvate have been produced, a test must be made after evaporating the solvent and completely drying at low temperature then heating the crystals in Sil Gel (p. 358). If no crystallization occurs then a place where the modifications are in contact must be observed; the less soluble and more stable form will grow at the expense of the more soluble and unstable form under the influence of the solvent.

By using this test, we have succeeded in elucidating the complicated conditions of the polymorphism of the other modifications of Dapsone [146]. Modification II, which is formed by spontaneous transformation from modification III (the form stable at room temperature), is for its part enantiotropic with respect to modification I. As a matter of fact, this transformation does not occur readily and the enantiotropic relationship can only be demonstrated by the solubility test just described, which in this case had to be carried out above the modification III to modification II transition at 80°C. In addition, modification I is enantiotropic with respect to modification III, the relationship being established in the same way but in this case at room temperature. Thus, in the Dapsone system, there are three transition points of which only one could actually be deter-

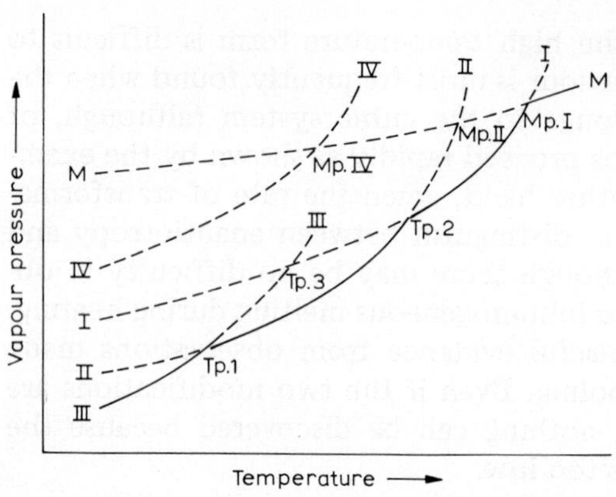

Fig. 38. Sublimation pressure curves of dapsone (schematic).

mined. The schematic vapour pressure phase diagram is given in Fig. 38.

It must be pointed out in this connection that the term "transition point" is often applied incorrectly to the temperature at which a transformation is observed by optical or calorimetric means. At the (enantiotropic) transition point itself there is neither growth nor decay of the crystals of the modifications in equilibrium and the transition velocity is zero. Only in a few favourable cases is the velocity close to the transition point fast enough for the reversal of growth to be observed by heating or cooling within a few tenths of a degree. Only when we have a very rapid kind of transition can the displacement of the equilibrium between two crystalline forms be observed like that between crystal and melt (p. 346).

In obstinate cases, an enantiotropic transition may be attained by suspending the metastable crystals for several hours in a solvent and gently grinding them to produce nuclei and stimulate the transformation.

(c) Determination of the melting point of unstable modifications. Sometimes, the melting point of an unstable modification can be determined by heating the crystal film at a normal rate without taking any special precautions. This fortunate situation may occur if a more stable modification is not present nor forms during heating or, alternatively, the rate of transition is extremely low. For exam-

ple, the melting point of modification II of *N*-ethylthiourea (88°C) can be determined by the equilibrium method in a preparation which actually contains some modification I in spite of the fact that the latter does not melt below 107—110°C.

Often, however, the rate of transition increases so much with increasing temperature that all the crystals of the lower melting modification are transformed before its melting point can be reached. To overcome this difficulty, it is best not to allow the preparation to solidify completely but to put it back on the hot stage (preheated but some degrees below the expected melting point) as soon as some reasonably large aggregates of crystals of the modification of interest have formed. If crystallization of higher modifications is very rapid, the melting point of several modifications may be determined in the same way, even on complete crystal films.

Sometimes, it is only possible to determine the melting point of a modification when it is in the form of a small solidified drop [8]. When preparing a slide, by scattering a number of crystals over it, care must be taken to keep them separate so that after melting there are many small drops as well as larger ones. If the substance crystallizes sufficiently readily, not only will the larger drops solidify but also the smaller ones with the formation of unstable modifications. Generally, it is not hard to find a small solidified drop of an unstable modification and the melting point can be determined without it being seeded by a more stable form during heating.

If the procedure just described should fail because, for example, the small drops of melt do not solidify, the following variation may be tried. A thin crystal film of the unstable modification is prepared and the cover slip lifted off, many parts of the preparation being torn away with it. The slide is now covered with a new cover slip, inspected for a region containing some isolated fragments and then returned to the pre-heated hot stage. If the volatility of the substance is high, it may be better to use the cover slip if this has more of the original preparation adhering to it; it should be placed on a clean slide and this put on a preheated hot stage. Whether slide or cover slip is used, the main part of the unstable crystal film will probably transform during heating but there should be enough small isolated fragments to enable the melting point to be observed. The melting points of, for example, the unstable second modifications of chloro- and bromoacetanilide, which are only found in company with the stable forms, have been determined in this way. It is important that the preparation be cooled only so far as is necessary for crystallization to occur because if it is cooled further, contraction fissures may form and these may act as centres from which the stable modification may be seeded.

The melting points of unstable modifications of substances which sublime readily may be found by observing the sublimate. Here again, the preparation (which in this case is the sublimate) should not be allowed to cool before the determination of the melting point is attempted. If cooling is permitted, reheating over wide range, say from room temperature, involves the dangers of loss by volatilization and transition to more stable forms.

In addition to direct determination, indirect determinations of the melting points of unstable modifications may be based on the fact that they can be stabilized to some extent by the addition of other substances, particularly if isomorphous. By extrapolation of the data for a series of mixed crystals or of the freezing point curve in a two-component eutectic system (p. 443), the melting points of very unstable modifications can often be found. Similarly, if the eutectic point between the unstable modification and some reference substance can be found, its melting point can be determined approximately by a graphical method based on the two-component phase diagram for the stable modification with the reference substance [147]. Unfortunately, by comparison with the direct methods, the indirect methods are very costly.

(d) Non-polymorphic recrystallization. It must be appreciated that the term "recrystallization" itself does not necessarily imply the involvement of polymorphism in the process but the qualification "non-polymorphic" has been applied and this section is added here because of the danger of confusion. Today, recrystallization is often referred to as if related only to solidification of the melt. This, however, is incorrect since the term has a very definite meaning in metallography and mineralogy, i.e. the alteration or re-formation of the crystalline structure of plastically deformed solids. The deformed material has greater energy than the un-deformed and therefore tends to revert to the latter state, this often being brought about by a slight thermal activation. The nucleation of the new crystals always occurs at places where the density of dislocations is very high. In the case of metals, the nucleation process or primary recrystallization is followed, after prolonged heating, by secondary recrystallization in

which the crystal structure coarsens. This is a process in which the larger single crystals grow at the expense of the many smaller ones which are consumed until finally only relatively few very large crystals remain with consequent increase in the average grain size. In general, it is typical of recrystallization that a crystal growing at one end at the expense of another crystal should itself be attacked simultaneously at the other end by a third.

McCrone [148] has carried out most extensive studies on recrystallization and he distinguishes two types. The first, in which the new crystals are orientated in growth, is exemplified by *p,p'*-DDT, and the second, in which grains grow without any particular orientation, is exemplified by octachloropropane. The octachloropropane type, which is shown particularly well by compounds like camphor and borneol and is also observed when sublimate films of highly volatile substances like saccharin and caffeine (Fig. 39) are heated, more obviously typifies the grain enlargement process described in the preceding paragraph.

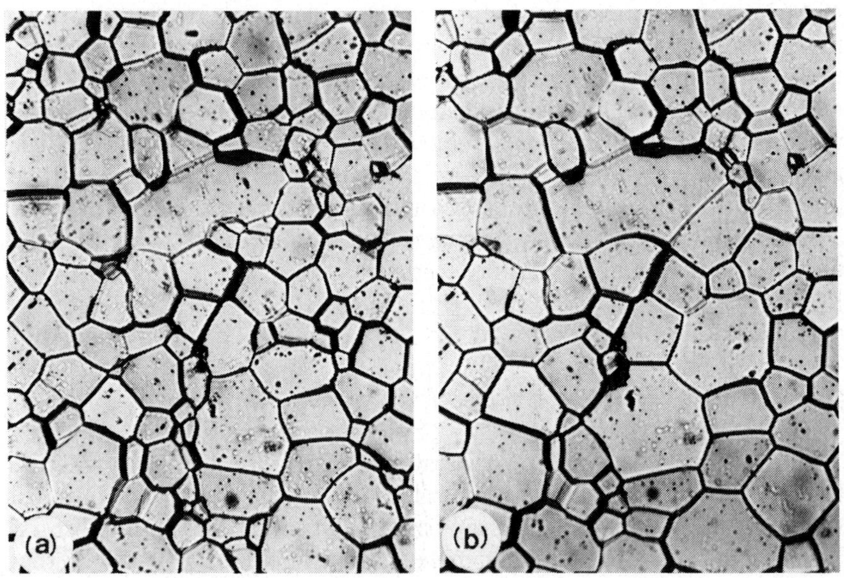

Fig. 39. Grain enlargement of caffeine (high temperature form).

The situation is different with McCrone's DDT-type of recrystallization. Here when the crystal film is heated, most of the newly enlarged crystals reveal a particular orientation. When heated, they may cease to grow at a particular temperature or growth may continue until the melting point is approached, as in monotropic transformations.

In the first case, it is relatively easy to distinguish recrystallization from polymorphic transformation although it becomes harder if the whole of the original crystal film is affected.

Recrystallization can sometimes occur when preparations are allowed to stand for a long time (an ageing process analogous to the ageing or "Ostwald ripening" of precipitates). Figure 37 shows the beginning of recrystallization of fine radiated spherulites of DDT, the original crystal film having been prepared by quick cooling of the melt. The recrystallization is not complete and when the preparation is heated, the old and the new crystals can be seen to melt at the same temperature.

Undoubtedly, recrystallization is a phenomenon which can cause the less practised observers much trouble. Obviously, here, as in the case of transformations without change of polarization colour or grain boundaries, we are at the frontiers of investigations into polymorphism but we are supported in our contention that one method by itself is not enough to deal with such special situations. It is fortunate that DDT-type recrystallization, which is the more likely to cause confusion, occurs rarely. By comparison, ordinary classical grain growth, exemplified by the octachloropropane type, is not so likely to mislead; even the occasional microscopist should not make a mistake if he has once observed the behaviour of, say, caffeine during a normal melting point determination.

(B) LIQUID CRYSTALS

(1) Definitions and descriptions of the mesomorphic phases

The earliest discovery of a substance of the kind loosely referred to as "liquid crystals" can be traced to the Austrian botanist Reinitzer [149]. In 1888, he reported his observations on cholesteryl benzoate which he found to melt quite sharply at 145°C but to a cloudy liquid which became completely transparent, again quite sharply, at 178°C. Other esters of cholesteryl were also found to behave in the same way. He described the state as "trübe Schmelze" or "cloudy melt". Lehmann [150,151] reported similar observations made on ammonium oleate and *p*-azoxyphenetole and interpreted the cloudy liquid as consisting of flowing crystals. Today, the term "liquid crystals" is in common use but since the state to which they refer lies between the ordinary solid and liquid phases, the terms "mesophase" or "mesomorphic state" are to be preferred.

Among the other early investigators of mesophases, mention must be made of Friedel [152]. On the basis of his optical studies, he distinguished three different types of mesophase: the smectic, the nematic and the cholesteric. The characteristic structures of the smectic phase were first observed using the polarizing microscope on soaps and from this they derive their name (the Greek "smectos" meaning "soapy"). The nematic phase owes its name to the thread-like (the Greek "nematos") appearance of some specimens. The term "cholesteric" derives from cholesterol whose esters revealed the first observed mesophases, they being of this type.

The mesophases produced by heating certain pure organic solids are said to be thermotropic. Under appropriate conditions, some solutions are also intermediate in character between the true solid and true liquid states and as such are said to be lyotropic mesophases. Their structures are analogous to the thermotropic smectic and nematic phases and they are converted to true solutions by the addition of excess solvent. They are believed to be an important factor in the metabolism of living cells. We shall deal only with thermotropic mesophases here.

Various combinations of methods are appropriate for the study of the liquid crystalline states: (a) crystal optical and thermomicroscopic methods, described in extensive detail by Hartshorne [153]; (b) differential thermal analysis of the individual substances and investigations of the miscibility relationships in binary systems; and (c) spectroscopic methods in the X-ray, ultraviolet and infrared ranges and proton magnetic resonance studies [154].

The ordered three-dimensional lattice structure of ordinary crystalline solids confers on them, unless they belong to the cubic system, anisotropic properties. At the melting point, the thermal energy possessed by the molecules is sufficient to overcome the potential energy of attraction by which they are held in the crystal lattice so that the solid long-range order decays into the short-range order characteristic of the liquid state and this, since the molecules are now random, is isotropic. If, however, the cohesive energy of the crystal lattice varies greatly in the different directions, then the thermal energy of the molecules at a given temperature may be sufficient to overcome the weaker bonds while the stronger still remain effective. Thus, the three-dimensional order of the solid lattice may be broken down to a two- or one-dimensional order and the phase produced possesses both fluidity and optical anisotropy. The general conclusions as to structure are given below.

(1) Smectic states. Their characteristic is that the cohesive forces of the solid lattice are overcome only in one direction, say that of the x axis, and as the forces remain operative in the other directions, the molecules in this state are ordered in layers. Within the layers they are strongly held and cannot move in the direction of the longitudinal axis but the attractive forces between the layers themselves are slight and they can slide over one another easily. There is still a high degree of order in the resulting fluid and this explains the high viscosity and surface tension of smectic phases. There are at least five smectic states (designated by the letters symbols A to E) which correspond to different arrangements of the molecules and a further three (designated F, G and H) are considered possibly distinct. The principal exponents of this scheme for the identification and assignment of smectic mesophases are Sackmann and his collaborators [155,156]. Some substances, on heating or cooling, pass through several smectic mesophases with ordinary first-degree transition points between them.

(2) The nematic state. This state arises because the thermal energy of the molecules overcomes two weaker and approximately equal solid-state bonds in, say, the x and y directions but not the stronger attraction between the molecules in the z direction. Consequently, the molecules preserve a parallel arrangement in the melt and lack the layered structure preserved in smectic phases. The parallel arrangement of the molecules in the longitudinal direction is localized in both time and space so the nematic phase is much less viscous than any smectic phase. Although there is some degree of lateral cohesion between the molecules, it is insufficient for them to form a layered structure but they are able to rotate about their longitudinal axes and to move freely in their direction.

(3) The cholesteric state. This is closely related to the nematic in structure, so much so that the cholesteric state is generally regarded as a special case of the nematic state. The structure is characterized

by parallelism of the longitudinal axes of the molecules but with the parallelism confined to planes each of which is turned by a certain angle with respect to its neighbouring planes. The combination of superimposed planes gives rise to a helical arrangement of the molecular axes, the structure and left- or right-handedness being due to the chiral nature of the molecules.

Compounds exhibiting mesomorphic states fall into one of the following groups according to the number of mesophases between the highest melting solid state and the isotropic liquid:

(a) those having one or more smectic phases,

(b) those having a nematic or a cholesteric phase (no substance can have both a nematic and a cholesteric phase),

(c) one or more smectic phases and a nematic *or* a cholesteric phase.

This is in consequence of the difference between the more highly ordered smectic state and the more disordered nematic or cholesteric state. The sequence of transitions crystalline solid $\rightleftharpoons$ smectic phase(s) $\rightleftharpoons$ nematic *or* cholesteric phase $\rightleftharpoons$ isotropic liquid is immutable.

A substance which exhibits mesophases may, of course, also have several solid states; in fact, mesomorphism is very often accompanied by polymorphism in the solid. On account of this, it is also possible for an unstable polymorph, as a result of its low melting point, to transform to an unstable mesophase.

(2) The chemical constitution of liquid crystalline substances

(a) The molecular structure of nematogenic and smectogenic substances. The molecules of substances which give rise to nematic and smectic phases have certain essential structural features. They have a long, rigid and more or less flat skeleton (composed of aromatic rings disposed about a centre (see Fig. 40, X) with substituents such as *n*-paraffin, alkoxy or carboxylic acid ester chains (Fig. 40, Y) carried in the *para* positions. The polarity and polarizability of the molecules also play important parts. The liquid crystal interval is likely to be long if strongly polar or easily polarizable groups are attached to the centre (X) or the ends (Y) but short if they are aligned along the stiff skeleton in such a way as to increase the breadth of the molecule significantly. The *trans*-stilbenes, tolanes, azobenzenes, nitrones, phenylbenzoates and other compounds with different end groups conform to the model which is represented in a simplified fashion in

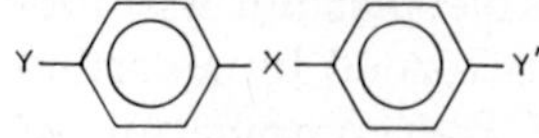

Fig. 40. Model of nematic and smectic mesophases.

Fig. 40. The *cis*-stilbenes, whose molecules are also flat, do not form liquid crystals because of the non-linearity of their molecular structure. The long skeleton can either be made up of two rings or arise by the fusion of one- or two-ring systems. Gray [154] gives many examples of molecular structures which sometimes give mesophases and sometimes not, in spite of their conformity to the model.

(b) The molecular structure of cholesteric phase forming substances. Cholesteric phases have been observed mainly in the derivatives of cholesterol and other related steroids which possess asymmetric centres. In most of these substances, the hydroxyl group in the C_3 position is esterified with an aliphatic or an aromatic acid. Wiegand [157] has been the principal investigator of the effect of alterations to the steroid skeleton on trends in the stability and range of the cholesteric phase, while Gray [158] has investigated the effect of the length of the substituent chains in the case of the cholesterol esters.

A few compounds, apart from the steroids, form cholesteric phases. The pre-condition is that the molecules have the structural requirements to form a nematic phase but at the same time carry optically active substituents, obviously only one chiral centre being required to set up the helical arrangement which distinguishes the cholesteric state from the nematic. As an example, the optically active amyl esters of various *p*-substituted cinnamic acids form a cholesteric phase but their inactive isomers form a nematic phase.

(3) The applications of liquid crystals

The modern revival of interest in liquid crystals can be traced to the fact that various fields of application have now been opened up for them [159]. At least this is true of nematic and cholesteric materials; the smectic state is still of very little practical importance.

The importance of nematic liquid crystals is due to their property of being influenced by an electric field. Because the molecules are polar, a nematic film is electrically anisotropic and when placed in an electric field a turning moment is imposed upon the molecules which aligns them in the field direction if the dipole moment is parallel to the molecular axis (positive anisotropy) or normal to the field of direction if the moment is normal to the molecular axis (negative anisotropy). The application of an electric field to a thin already ordered nematic film will, however, destroy its ordered condition by virtue of the phenomenon known as "dynamic scattering" with consequent changes in its optical properties like birefringence, colour, transparency and so on. The electro-optical effects can be utilized in the liquid crystal displays which have become a part of everyday life in the form of digital wrist watches, pocket calculators and so on. They have also found use in alphanumeric and analogue displays, image converters and matrix-type picture screens.

Cholesteric liquid crystals prepared by mixing cholesterol esters are used in thermotopography. As mentioned earlier, the molecules are ordered in a helical fashion and selectively reflect transmitted light. By varying the composition, the pitch of the helix may be so adjusted that visible light is reflected. The effect of temperature is such that a rise lowers the pitch of the helix and shifts the reflected light to shorter wavelengths. This effect is used for the measurement of surface temperatures, a technique which has found wide application in technology and medicine. The technological applications include the examination of sub-surface defects in tool and other materials since these cause a discontinuity in thermal conductivity which is revealed by a change of colour. Similarly, defects in electronic devices are recognizable. There are also thermometers which work on the same principle and which indicate the ambient temperature by colour or digital display. In medical diagnosis, infections and tumors which are located close to the skin and whose temperature is higher than surrounding regions can be localized without interference with the patient. Thus, it is possible to achieve an early diagnosis of mammary cancer or to investigate pathological circulatory conditions of the extremities.

In science, the use of liquid crystals is increasing. Two fields of application stand out at the present time; first, the use of liquid crystals as anisotropic solvents for the determination of molecular properties in spectroscopy and, second, for separations in analytical chemistry, especially in gas chromatography [159].

(4) The thermomicroscopic investigation of liquid crystals

Although there are some substances which exist in a liquid crystalline state at room temperature (e.g. ammonium oleate), most exhibit their mesophases above room temperature. Such states can be recognized by the appearance of a turbid melt in ordinary capillary melting point tubes but investigations on films with the hot-stage microscope are much more informative because the mesomorphic states, being birefringent, reveal their different textures between crossed polars [153,154].

In place of the single melting point marking the transition from solid to liquid of an ordinary substance, liquid-crystalline substances pass through a number of steps. The transition between the solid state with its three-dimensional order to the first two-dimensionally ordered smectic, or to the nematic or cholesteric phases ordered in one direction only, is regarded as the transformation corresponding to melting and the transition point is called the melting point. The transition from the last mesophase to the disordered isotropic melt is called the "clearing point".

The observable optical characteristics of smectic phases are manifold [154] and cannot be discussed in greater detail here. They show different textures between crossed polars but it is not always easy to distinguish them (especially the fan-like A and B type phases) from the regular solid state. To be sure that it is a mesophase with which one is dealing, the cover slip can be subjected to slight pressure to displace it with respect to the slide. As a rule, bright or coloured streaks appear which reveal the liquid character of the film by their movement. Of course, after the pressure has been removed the smectic layers may reorder themselves parallel to the glass surfaces and with their optical axes parallel to the axis of the microscope so that the interference colours disappear either straight away or in the course of time. The process is not to be confused with the final transformation to the isotropic liquid at the clearing point. Some smectic phases are so viscous that a considerable effort is required to move the cover slip.

The nematic phase is less viscous than the smectic phases and its fluidity is easier to recognize. It is most easily demonstrated by the observation of dust particles which move by convection through the anisotropic liquid without altering the texture. The texture is characterized by threadlike mobile lines which mark discontinuities in the medium. When the nematic phase is produced by heating a crystal film, the transformation often takes place with the retention of the original crystal boundaries. When the nematic phase is allowed to form by cooling the melt, many small birefringent "droplets" are formed first and then these unite to form larger regions and finally a complete mobile nematic film. The texture sometimes gives rise to moiré patterns. On further cooling, crystals are nucleated just as in true melts. Spherulites are often formed after supercooling.

Cholesteric phases are easily distinguished from the nematic by their appearance when formed from the melt. As a rule, they first form diffuse clouds which merge into one another and, on further cooling, a random confused fine-grained texture.

Generally speaking, the mesomorphic phases do not supercool readily, probably because tiny residual seeds remaining after transformation to a higher mesophase seed the lower mesophase on cooling. Solids, however, often separate only after considerable supercooling and this provides the opportunity for the formation of a variety of polymorphic modifications. Consider, for example, 4′-pentyl-4-biphenyl-4-butyl benzoate (S 1012 Licristal ®, Merck, Darmstadt). When the isotropic melt of this substance is allowed to cool slowly on the hot stage, a nematic phase forms spontaneously at 171°C without any supercooling, but this phase may itself be supercooled to below 80°C with the formation of the solid modification III which is subsequently transformed to modification I. If the preparation is allowed to stand for some time at room temperature, modification I is transformed into the enantiotropic modification II. (Modification II is also present in the commercial product.) The melting points of the three modifications are: III, 84°C; II, 95°C; I, 97°C; the clearing point is 171°C. If the crystalline modifications are made to melt separately, all form the nematic phase, which is unstable between 84° and 97°C and is partly transformed to crystals of modification I in this region.

As an example of the mesophase sequence: smectic ⇌ nematic/cholesteric, we have cholesteryl myristate. The substance melts at 71°C to its smectic phase, the transition to the cholesteric phase takes place at 79°C and the clearing point is 83°C. In this case, all the relationships are enantiotropic, but this is not so with 4-pentylphenyl-4′-pentyl-4-biphenylcarboxylate (S 1011 Licristal ®, Merck, Darmstadt), one mesophase of which exists below the melting point of the solid phase and is formed only on cooling the melt. On cooling, the nematic phase forms at 176°C and this transforms at 151°C to the smectic A phase which can be supercooled with respect to the solid until transformed to the smectic B phase at 85°C. The solid forms only on further cooling to between 70 and 60°C. When the preparation is reheated, the solid melts at 94°C to the smectic A phase, the transformation to the nematic form takes place at 151°C and the clearing point is at 176°C. The smectic B phase is clearly monotropic with respect to the solid which is stable over the entire range of that mesophase's existence but all the other relationships are enantiotropic.

By way of conclusion, we may say that, quite apart from their increasing practical importance, the study of liquid crystals is a very attractive and rewarding field for the application of thermomicroscopic techniques.

5. Phase investigations of two-component systems

(A) THE CONTACT PREPARATION METHOD

The contact preparation method of investigating binary systems verifies the proverb that all good things are discovered at least twice. When Adelheid Kofler hit upon the idea of bringing the melts of two different components into contact beneath a cover slip and then allowing them to freeze so that the contact zone could be remelted as a crystal film, she was unaware that Otto Lehmann had carried out similar experiments fifty years before. Obviously, the time had not then been ripe for the development of hot-stage microscopy and Lehmann's work was forgotten. Only after the publication of her first papers [160] did Lehmann's book *Molekularphysik*, published in 1888, come into Adelheid Kofler's possession. She was not discouraged by the knowledge of prior discovery but went on to develop the method for the systematic investigation of the phase diagrams of two-component systems. So important has the method become that one cannot imagine carrying out such thermal investigations without its aid.

A contact preparation is made as follows. Some crystals of the higher melting of the two substances are placed at the edge of a cover slip on a slide. The slide is then heated until the crystals melt and the liquid drawn by capillarity into the space between slide and cover slip, the amount of substance taken being enough to fill about half of it. The slide is now cooled and the material allowed to solidify. The second, lower melting component is treated in the same way as the first. Some crystals are melted alongside the cover slip so that the liquid fills the remaining space under it. In the zone of contact, the melt of the second component dissolves a little of the now solid first component so that a "mixing zone" is formed. The second component is now allowed to solidify. If this does not crystallize spontaneously, it must be encouraged to do so by one of the methods described on p. 363. It is sometimes desirable to hold the preparation above the melting points of both components for some time so as to broaden the mixing zone, especially if molecular addition compounds are likely to be formed. When the contact preparation is heated, the main features of the binary phase diagram can be recognized by observation (with crossed polars) of the processes occurring within the contact zone. More detailed explanations of phase diagrams for two-component systems will be found in monographs on phase relationships, e.g. that by Findlay [103].

(1) Systems without solid solutions

(a) Simple eutectic systems. If two components neither react with one another nor form mixed crystals (i.e. if they are completely mutually insoluble in the solid), the melting point diagram simply consists of two curves which meet at the eutectic point and a horizontal line which passes through it [Fig. 41(a)]. The mixing zone in the contact preparation contains all the mixtures represented in the diagram, including the eutectic mixture. Accordingly, on heating the preparation, melting first takes place at the eutectic temperature in accordance with the phase diagram. The eutectic temperature is read as soon as a continuous dark strip appears in the field when observed between crossed polars [see Fig. 41(b)]. An example of this type of system is azobenzene, m.p. 68°C, with benzil, m.p. 95°C. The eutectic point is 52°C.

(b) Systems involving compound formation. If two substances react with one another to form an addition compound which itself shows no tendency to mixed crystal formation with either pure component, then a third (one might say a third and a fourth) curve appears in the phase diagram. There are two cases between which it is

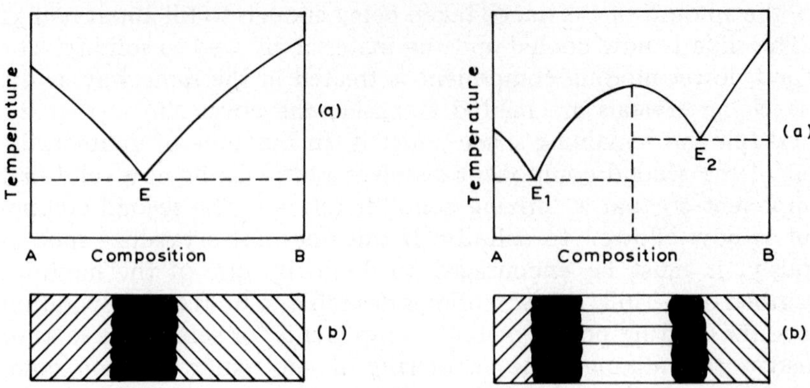

Fig. 41. Phase diagram (a) and contact preparation (b) of a simple eutectic system.

Fig. 42. Phase diagram (a) and contact preparation (b) of a system with a congruently melting addition compound.

necessary to distinguish; these are characterized by congruent and incongruent melting of the compound, respectively.

(1) Congruent melting of the compound. If the compound has a definite congruent melting point which is lowered by dissolving either of the pure components in its melt, the melting point of the compound corresponds to a maximum on the equilibrium curve [Fig. 42(a)]; When the contact preparation is heated, the first strip of melt to appear is the lower melting eutectic [Fig. 42(a), E_1] which is followed by the higher melting eutectic, E_2, and at this stage the molecular addition compound survives as a birefringent strip between the two dark molten eutectic strips [Fig. 42(b) and Fig. 43]. Finally, the melting point of the compound is reached. If the compound does not crystallize spontaneously on cooling because of a paucity of seeds, the contact zone remains as a broad dark strip. Rewarming and scratching the edge of the cover slip are generally sufficient to initiate crystallization, if desired. The system *p*-nitrophenol, m.p. 113°C, with benzamide, m.p. 128°C, is an example. The first eutectic temperature, E_1, is 84°C, the second, E_2, is 91°C, and the melting point of the compound is 98°C.

If two or more compounds are formed, then the contact zone will contain three or more eutectic strips with the strips of the molecular addition compound lying between them.

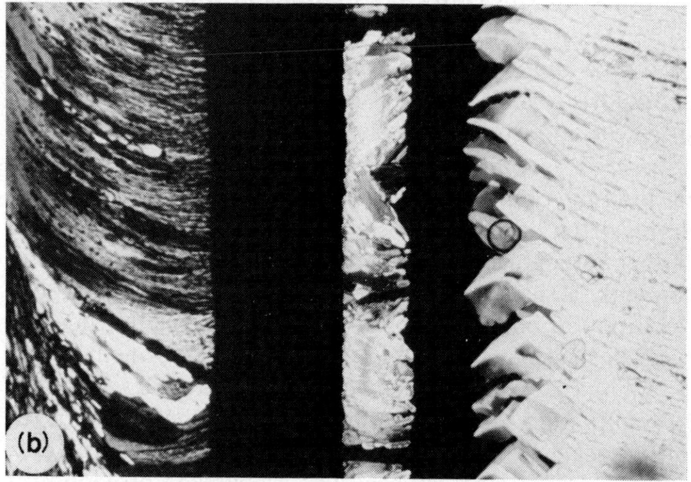

Fig. 43. Contact preparation of benzamide and resorcinol showing an addition compound.

(2) Incongruent melting of the compound. If the compound formed decomposes below what would be its congruent melting point, there is only one eutectic mixture (Fig. 44). In place of the second eutectic we have a peritectic point, P, at the intersection of

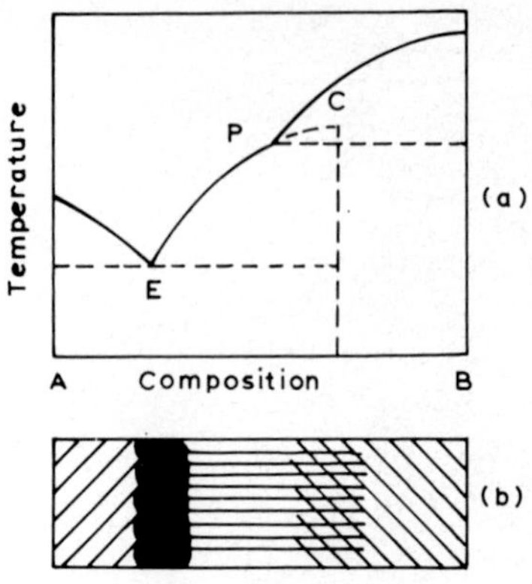

Fig. 44. Phase diagram (a) and contact preparation (b) of a system with an incongruently melting addition compound.

the liquidus of the compound with the liquidus of the higher melting component. The temperature corresponding to this point is the incongruent melting point of the compound at which it should "melt" or decompose with the separation of crystals of pure component B. If, however, no seeds of this component are formed on heating, the expected decomposition will not occur immediately and the congruent melting point of the compound (Fig. 44, C) may be reached. Hydrates often behave in this way.

In the contact preparation, the compound and component B generally meet with a sharp straight boundary but the mixing zone between the compound and component A is more irregular because of the eutectic. On heating, the eutectic between component A and the compound melts first. As heating is continued, more and more of the compound melts (and more and more of component A) until the peritectic point is reached at which it should disappear completely. This incongruent melting point can easily be recognized provided the compound can be distinguished from component B, either on morphological grounds or by reason of differences in appearance between crossed polars. The most certain test of a peritectic reaction is to heat a crystal film of the mixture of composition corresponding to that of the compound to the incongruent melting point. At this temperature, the compound "melts" with precipitation of component B since it is the temperature at which we have equilibrium between compound, component B and liquid. The process is analogous to the peritectic reaction which arises in the Bakhuis-Roozeboom Type IV system (p. 452) under some conditions of limited miscibility in the solid state and which involves the equilibrium between two solid solutions and the liquid. For example, the fluorene, m.p. 114°C, —2,4,6-trinitroresorcinol (styphnic acid), m.p. 177°C, system has a eutectic temperature at 98°C and a peritectic temperature of 130°C.

(c) Partial miscibility of the liquid phases. This situation has already been referred to in connection with the failure of a foreign additive to cause a melting point depression (p. 375) and with the identification of liquids (p. 404).

Mixtures of polar and non-polar components are the most likely to exhibit some degree of immiscibility which is manifest as a mixing gap over part of the range of composition on the phase diagram [Fig. 45(a)]. The contact preparation should not be observed between crossed polars in the usual way but between uncrossed or parallel polars (if the construction of the microscope permits this) or even in unpolarized light. When the contact preparation is heated, the eutectic melts followed by component A. On further heating, the melt, which is a solution rich in A, will dissolve component B until it is saturated. When this happens, a second melt, a solution rich in B, forms alongside the remaining solid B. The boundary between the

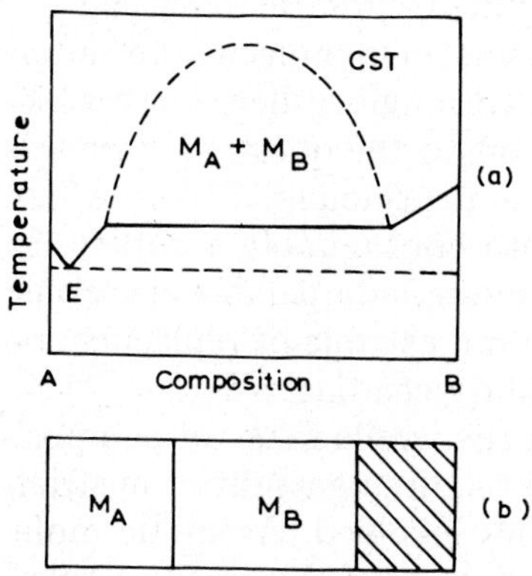

Fig. 45. Phase diagram (a) and contact preparation (b) of a system with a miscibility gap between the liquid phases.

two liquid phases can be seen as a sharp dark line [Fig. 45(b)]. Further heating leads to the melting of the remainder of component B and eventually to complete mixing of the liquids but the critical solution temperature itself is difficult to determine because diffusion across the boundary between the liquids greatly reduces its visibility. The system benzil, m.p. 95° C—methylurea, m.p. 102° C, is an example. The eutectic temperature is 92°C and immiscibility appears at 97° C.

(2) Systems with solid solutions

(a) Isomorphism, isodimorphism and isopolymorphism. The formation of solid solutions or mixed crystals is intimately associated with isomorphism [8,161,162]. The term is applied to substances which, because they have the same or very similar structures, can replace one another in the building of a crystal lattice and so are able to form mixed crystals. The requirements are that the molecules be of similar size and shape and van der Waals attractions to immediate and more distant neighbours. The ideal case of isomorphism arises when there is no limit to the degree of replacement of the molecules of one component by those of the other and the phase diagram shows an uninterrupted series of solid solutions over the whole range of composition. If the components are capable of taking up molecules of each other only to a limited extent, then the isomorphism is imperfect and mixed crystals can only be formed within certain ranges of composition separated by a mixing or immiscibility gap.

The question of which groups of atoms in a molecule are interchangeable has been studied by Grimm, although earlier workers like Bruni [163] had given some consideration to the question. Grimm's work resulted in the "hydride displacement principle" [164], which enables isomorphism to be predicted with considerable accuracy. In Table 9, atoms and atom groups which possess comparable electronic configurations and are, according to Grimm, capable of replacing one another in a crystal lattice are arranged in descending order.

For formation of solid solutions over the whole range of composition, the chemical and crystallographical requirements differ in strictness according to the kinds of molecules involved. Aromatic molecules (because of their elaborate structures) have the smallest tolerance and quasi-spherical molecules the most; linear molecules are intermediate. In the camphor—borneol system, for instance, the

TABLE 9

Hydride displacement principle according to Grimm

No. of H atoms						
0	C	N	O	F	Ne	Na^{+}
1		CH	NH	OH	FH	
2			CH_2	NH_2	OH_2	FH_2^{+}
3				CH_3	NH_3	OH_3
4					CH_4	NH_4^{+}

oxygen and the hydroxy groups are mutually replaceable in spite of the hydride displacement principle rule [165] while, in accordance with the rule, the same groups in such large molecules as those of the steroid hormones cannot be interchanged. For example, pregnenolone and progesterone are completely immiscible in the solid state and form an addition compound which melts incongruently [166]. Barbiturates, on the other hand, are surprisingly tolerant of admixture as the investigation of about three hundred pairs has demonstrated. In mixtures of the substances of this class, cyclic and acyclic groups can replace one another and also straight and branched groups, and so on [35,36]. Nevertheless, isodimorphism and isopolymorphism do influence the behaviour if these systems.

The very great number of substances exhibiting polymorphism has been mentioned in the section on that topic (p. 410). In the result, not only may the stable modifications of two related compounds be isomorphous and capable of forming mixed crystals, but unstable forms may also do so. Isodimorphism is said to occur if two dimorphic substances form two series of mixed crystals and isopolymorphism if several modifications of the two substances form several series of mixed crystals.

(b) Solid solutions over the whole range of composition. The contact preparation method is excellent for the recognition of continuous series of mixed crystals in the phase diagram. The property of isomorphism operates so that, in the mixing zone, one component is able to grow forward into the melt of the other component (or, rather, into the mixed melt) and retain its original crystallographic

form and orientation. If a contact preparation is completely melted on the hot stage and the heating switched off, then, after cooling sufficiently, crystallization can be induced below the melting point of the higher melting component by scratching the edge of the cover slip or by seeding. The crystal front thus initiated grows first through the melt of pure component B, then, after the temperature has fallen sufficiently, through the mixing zone and finally through the melt of component A. If component B can be supercooled sufficiently, crystallization can be induced in the lower melting component, A, and the crystal front grows through the preparation as before but in the opposite direction. The "isomorphic penetration" of the mixing zone is regarded as proof that mixed crystals form over the whole of the concentration range.

Solid solution binary phase diagrams may be divided into five types according to the form of the liquidus and solidus curves. Bakhuis-Roozeboom [167] enumerated three of these (types I, II, and III) and two others have been found by the Innsbruck school in the course of investigations into the behaviour of pairs of barbiturates [34—36]. Since Bakhuis-Roozeboom's time, the numbers IV and V have been allotted to systems not having continuous series of mixed crystals so the two new types have been designated IIa and IIb.

All five types, having continuous series of mixed crystals, may be recognized by heating the homogeneous solidified contact preparation.

(1) Type I. The equilibrium freezing and melting points all lie between the melting points of the pure components (Fig. 46). When the contact preparation is heated, pure component A melts first, followed by the mixed crystals and finally by the pure component B as the temperature is raised. The chloracetamide, m.p. 120° C, —bromoacetamide, m.p. 90° C, system provides an example.

(2) Type II. The solidus and liquidus curves pass through a common maximum (Fig. 47). When the homogeneous contact preparation is heated, pure component A melts first, then pure component B, while the mixed crystals whose composition corresponds to the maximum melting point remain as a strip of crystals between the melts. If the preparation is now cooled, isomorphic regrowth is observed from both sides of this strip. The system, (+)-kawain, m.p. 111° C, —(—)-kawain, m.p. 111° C, provides an example.

(3) Type III shows a minimum (Fig. 48). On heating, the homogeneous mixed crystals in the middle of the mixing zone melt first as

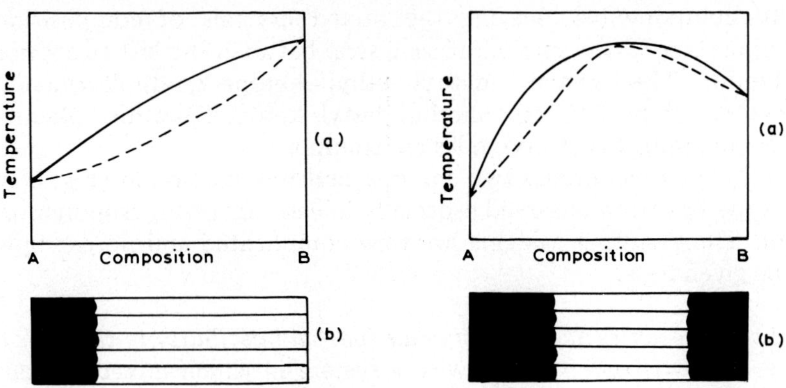

Fig. 46. Phase diagram (a) and contact preparation (b) of a type I system.

Fig. 47. Phase diagram (a) and contact preparation (b) of a type II system.

shown by the azobenzene, m.p. 68° C, —tolan, m.p. 60° C, system.

(4) Type IIa represents a combination of type II and Type III in that the solidus and liquidus curves pass through a common maximum and minimum. The melting sequence in the homogeneous contact preparation according to Fig. 49 on heating is: mixed crystals of composition corresponding to the minimum, component B and

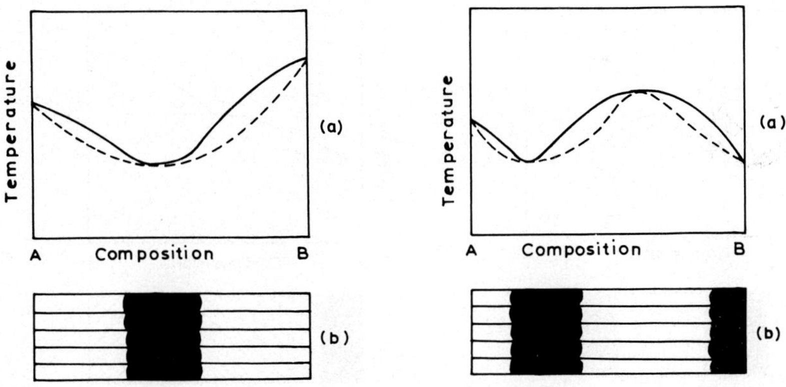

Fig. 48. Phase diagram (a) and contact preparation (b) of a type III system.

Fig. 49. Phase diagram (a) and contact preparation (b) of a type IIa system.

finally component A, leaving the mixed crystals of composition corresponding to the maximum as a strip between the last two strips of liquid. The system methyl ethyl ketone-2,4-dinitrophenylhydrazone, m.p. 117°C, —methyl butyl ketone-2,4-dinitrophenylhydrazone, m.p. 110°C, provides an example.

(5) Type IIb contains two minima and one maximum (Fig. 50). This type has been observed especially in cases involving isopolymorphism. The resulting systems are very complicated and no example will be given here.

(c) Limited series of mixed crystals (partial miscibility in the solid). There are two basic reasons why a system in which mixed crystals appear may show only partial miscibility in the solid. On the one hand, the molecules of the two components may not be sufficiently alike in size, shape and bonding characteristics so that the tolerance limits for uninterrupted mixed crystal formation are overstepped. On the other hand, there may be the complication of isodimorphism or -polymorphism. The classical example of the first case is the KI—KCl system [168] but isodi- or isopolymorphism is the main reason for limited miscibility in the case of pairs of organic substances [8].

Bakhuis-Roozeboom distinguished two types of limited miscibility system based on the form of the binary phase diagram curves.

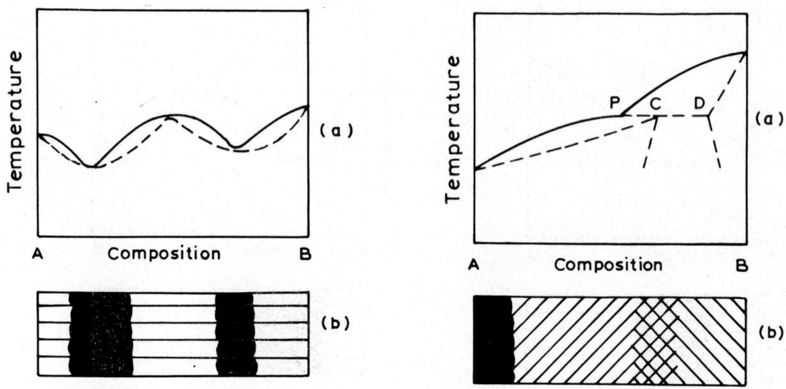

Fig. 50. Phase diagram (a) and contact preparation (b) of a type IIb system.

Fig. 51. Phase diagram (a) and contact preparation (b) of a type IV system.

(1) Type IV. The melting point of component A is raised by the addition of small amounts of B. At the temperature of the line PCD [Fig. 51(a)] the melt of composition corresponding to point P is in equilibrium with both mixed crystal phases represented by C and D. At this peritectic point, there are two solid phases represented by C and D in equilibrium with the melt and because of this there are discontinuities in the solidus and liquidus curves. There is a miscibility gap in the concentration range between C and D and the crystals in this region consist of a conglomerate of both solid A saturated with B and solid B saturated with A. In the contact preparation, the continuous growth of crystals of B in the melt of component A does not occur. Component A either crystallizes spontaneously or it must be induced by seeding or otherwise. Sometimes, the nucleation of mixed crystals rich in A takes place on the crystal front of B and this could possibly be confused with isomorphic continuation of growth through the mixing zone. However, in this case, the growth of crystals rich in B stops while crystals rich in A nucleate suddenly at a few places on the crystal front. This gives rise to outgrowths with an orientation different to that of the original crystals and the crystal front becomes irregular in the contact zone.

When a solid contact preparation is heated, melting does not begin in the generally clearly visible line of contact between the two different kinds of crystals but in the pure component A [Fig. 51(b)]. Subsequently, the mixed crystals rich in A begin to melt, the last of them going at the peritectic temperature. Occasionally, an actual peritectic reaction, the crystals rich in A melt while those rich in B grow, may be observed.

(2) Type V is characterized by a eutectic between mixed crystals rich in component A and component B, respectively [Fig. 52(a)]. The contact preparation [Fig. 52(b)] cannot be distinguished from that for a simple eutectic system [Fig. 41(b)] and the type V phase diagram cannot be recognized by this method unless crossed isodimorphism can be demonstrated. The diagram can only be elucidated by the examination of individual mixtures. Over a considerable range of concentration, these will begin to melt at the one low eutectic temperature, but mixtures very rich in the individual components will begin to melt at higher temperatures, the solidus rising to the melting points of the pure components.

It was stated above that mutual solubility in the solid may be due to the pure components being dimorphous or polymorphous. For

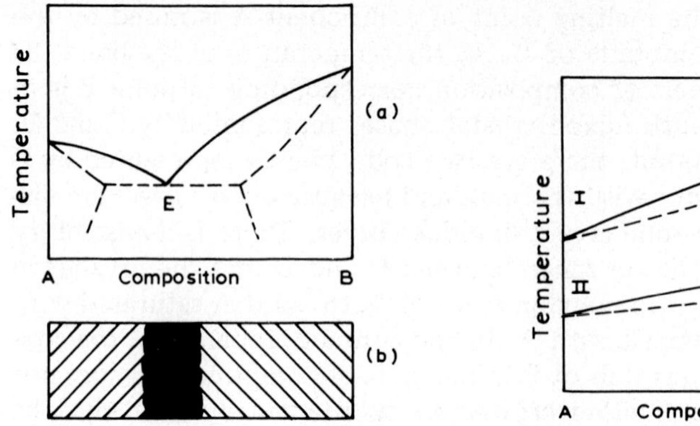

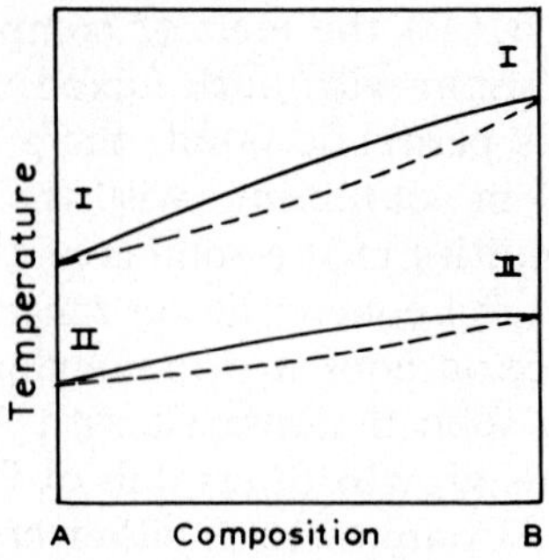

Fig. 52. Phase diagram (a) and contact preparation (b) of a type V system.

Fig. 53. Phase diagram of a system showing parallel isodimorphism.

simplicity, we shall consider first the effect of dimorphism. Now it may be that the stable modification of one component is isomorphous with the stable modification of the second and the lower melting modifications are also isomorphous. This synchronization leads to parallel isodimorphism and is illustrated in Fig. 53. The commoner case is that in which the stable modification of one component is isomorphous with an unstable modification of the other and vice versa. The result is crossed isodimorphism and, according to the position of the point of intersection of the liquidus curves, we get systems resembling type IV or type V.

In crossed isodimorphism, part of the series of mixed crystals is stable and the other part is unstable (Figs. 54 and 55). Sometimes, both series can be followed right through to the unstable modifications themselves; in other cases, only one of the two series can be followed completely or even only partly. Generally, it is easier to allow modification I of the higher melting component B to grow into the melt rich in A there producing modification II. Modification I of A often starts to form by spontaneous nucleation or can be induced to do so by scratching the cover slip and this transforms modification II of A and the mixed crystals of this lattice structure until the peritectic point is reached in the case of a type IV system (Fig. 54) or the eutectic composition in the case of a type V system (Fig. 55).

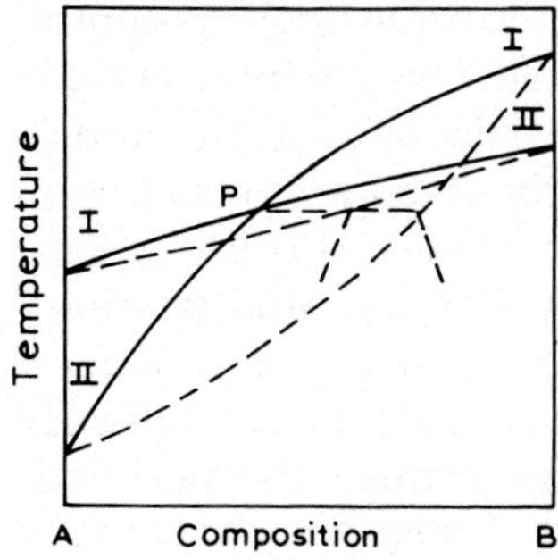

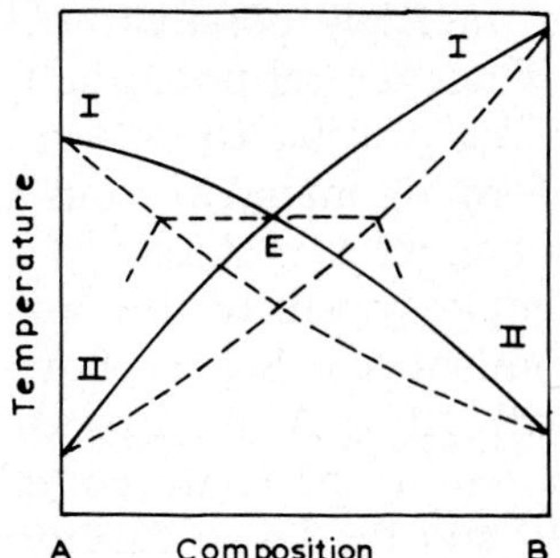

Fig. 54. Phase diagram of a system showing crossed isodimorphism with a peritectic point.

Fig. 55. Phase diagram of a system showing crossed isodimorphism with a eutectic point.

In our experience, there are many more cases of partial miscibility in the solid resulting from intersections of series of solid solutions than as a result of the limits of solubility being exceeded. In principle, every kind of complete mixed crystal series, including types II and III, could be involved in such systems.

In addition to these various possibilities due to isodimorphism,

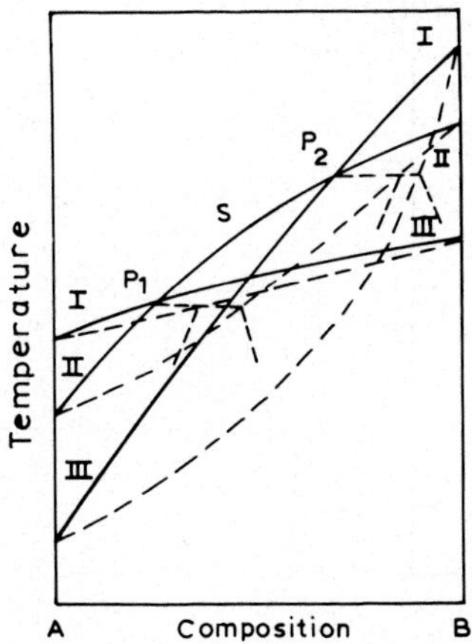

Fig. 56. Phase diagram of a system showing crossed isotrimorphism.

there are multiple possibilities which could be due to isopolymorphism. When several potential series of mixed crystals intersect one another, it is possible for one or more of the pairs of unstable modifications forming mixed crystals to be partially stabilized, resulting in what is known as "stabilized intermediate phases" [169]. A relatively simple system of this sort is shown in Fig. 56. Modification I of component A is isomorphous with III of B and I of B is isomorphous with III of A. As a result, the mixed crystals formed between II of A and II of B are completely stable between the peritectic points P_1 and P_2. To go further, four series of mixed crystals might intersect to form two stabilized intermediate phases thus giving rise to three peritectic points or combinations of eutectic and peritectic points. Schemes of the many possible combinations of isodi-, isotri- and isotetramorphic systems have been published elsewhere [35].

Complicated systems such as these cannot be elucidated by the contact preparation method alone; quantitative mixtures must also be used. Figure 57 shows the result of one such investigation into the phase relationship of dipropylbarbituric acid and phenobarbital. Two

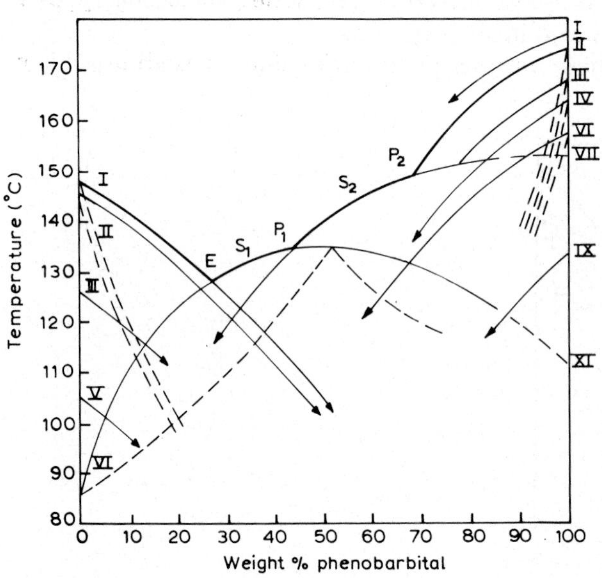

Fig. 57. Phase diagram of the dipropylbarbituric acid—phenobarbital system.

stabilized intermediate phases, S_1 and S_2, are formed. Isomorphic growth through contact zones due to isomorphism with other barbiturates has revealed six modifications of phenobarbital [35] in addition to the five already known from crystals films of the pure substance. Many new modifications of other barbiturates and of the sulphonamides [170] have been discovered through contact preparations.

(B) THE DETERMINATION OF TEMPERATURE—COMPOSITION RELATIONSHIPS

In the ordinary course of events, phase diagrams are plotted from the results of successive trials. The phase relationships between metals and, in particular, between inorganic substances have been investigated by cooling mixtures and plotting the arrest points on the cooling curves [171,172]. In the past, this method was also used for pairs of organic substances but, as supercooling can be very considerable and the thermal conductivity is poor, it was abandoned. Stock [173] began instead to determine the temperatures corresponding to the beginning and the end of melting by heating mixtures in capillaries as an alternative to the temperatures at which solidification set in on cooling. Stock's method, later refined by Rheinboldt [174], was used for many years but was eventually superseded by Kofler's introduction of the thermomicroscopic method [175].

In recent years, the differential scanning calorimeter has begun to compete with the hot-stage microscope, at least in the less complicated cases. The microscope, however, remains indispensable for the clarification of complicated two-component systems in which polymorphism plays a part, but it must be conceded that the best results are obtained by a combination of both methods. For instance, it is often easier to delimit the range of immiscibility in the solid by taking a DSC thermogram than by determining the beginning of melting in a crystal film. Nevertheless, if one had to opt for one particular method for investigating phase diagrams, then thermomicroscopy would be preferred to differential scanning calorimetry. Only the hot-stage method will be discussed here.

We begin, as in all methods, by making a series of trials, but the advantage of the thermomicroscopic method is that it has already given us, by way of the contact preparation, a semi-quantitative picture of the phase diagram. Thus, if the contact preparation indicates that the system is of the simple eutectic type, very few mixtures need be prepared to elucidate the whole diagram. Of course, in the investigation of complicated diagrams, more mixtures must be taken. These should be no more than 10% apart on the composition axis, often with additional mixtures no more than 5% apart in the vicinity of peritectic points and the like. Even when the basic outline of the phase diagram has been found in this way, it is sometimes necessary to examine mixtures only 1 or 2% apart in order to locate some of the crucial points with accuracy.

Although, in the determination of the eutectic temperature by itself (p. 376), only the beginning of melting has to be observed and it is enough to grind the two substances together in any proportions, when obtaining data for the binary phase diagram, components must be weighed exactly and mixed as well as possible by grinding together in a mortar. A total weight of mixture of about 0.1—0.2 g is generally sufficient. The addition of a drop of two of ether often assists in the grinding process. The observations are made on the crystal film (p. 423) because, when a powder preparation is heated, the eutectic melt flows to the edge of the cover slip and spoils the homogeneity of the mixture. Only in special circumstances, as for example when prior melting might cause decomposition, are powder preparations used directly and then at the expense of precision.

To prepare a crystal film, as much of the finely ground mixture is taken as will fill the whole space between slide and cover slip when melted. If the risk of loss by sublimation is considerable, the material is covered with a cover slip before melting; otherwise, it is covered after melting so as to achieve a more uniform solution and crystal film. The preparation should not be heated too slowly; this can be avoided by pre-heating the hot stage to the melting point of the higher melting component or by using the hot bench (p. 339). After the crystals have melted completely, the slide is transferred to a metal cooling block to solidify and the edge of the cover slip should be scratched if necessary to induce crystallization. Rapid crystallization results in a finer grain and more uniform distribution throughout the preparation than is the case if it is allowed to take place slowly. On no account should the preparation be allowed to cool very slowly on the hot stage because this will allow the component in excess to grow into very large crystals accompanied by the sequestration of the other component in the remaining melt with consequent lack of uniformity. Glassy solidified melts, if formed, can be devitrified by reheating and scratching. A field in the middle of the preparation is chosen for observation and the temperatures at which melting begins and ends are determined.

If the phase diagram contains a eutectic (e.g. Fig. 41), the beginning of melting at the eutectic temperature is marked by the sudden fusion of fragments of the crystallizate over the whole field of view. The component in excess remains in the solid particles which are distributed uniformly through the melt as separate grains or as a lattice-like aggregate. On further heating, the residual crystals dissolve gradually. The end of melting is determined, if possible, as the temperature at which the last crystal is in equilibrium with the melt, but otherwise as that at which the last crystals just disappear. The observations are made with the preparation between crossed polars so that the residual crystals can be seen easily as bright fragments against the dark melt as background. The form of the crystals which grow from the residual grains should be noted since their shape will often help to decide which of the two components is in excess and crystallizing out as the solid phase.

If the eutectic composition has not been clearly resolved by means of a few mixtures, then it can be found by comparing the closer mixtures with one or other pure substance by contact preparation. If the pure substance and the mixture lie on opposite sides of the eutectic, a molten strip of eutectic will appear between them on heating. If the pure substance and the mixture lie on the same branch of the liquidus curve, the molten strip of eutectic does not form.

It is, of course, important to have a thorough knowledge of the polymorphism of the individual components to avoid misinterpreting the values found. This is especially important as the quick cooling of the molten mixtures favours the formation of polymorphic modifications.

Systems containing several addition compounds may be very complicated as, for example, the benzocaine—picric acid system [61] shown in Fig. 58. It can be stated without reservation that the positions of the curves in this phase diagram could only have been determined by the thermomicroscopic method described. The same applies to the construction of the diagrams of mixed crystal systems, especially those involving crossed isopolymorphism with their intersections of several series as described above (p. 455). To be able to interpret the data in order to construct diagrams for systems like these, great experience and great patience is needed, especially when

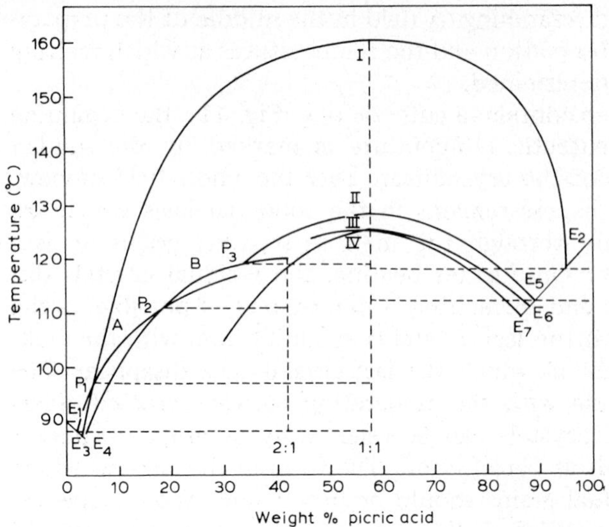

Fig. 58. Phase diagram of the benzocaine—picric acid system.

the crystallization velocity is very low. For the clarification of the phase relationships in these systems, contact preparations between the pure substance on the one hand and the mixture on the other are often needed. In this way, the crystallization of a substance in a structure which would never rise spontaneously can be induced by isomorphic growth through the mixing zone of the preparation.

Several examples can be found in the literature which show that, even in this age of differential thermal analysis, the contact preparation cannot be ignored. Quite recently, the phenanthrene--fluorene system has been investigated and reported on for at least the fifth time. The investigators, Kotula and Rabczuk [176], have come to the same conclusion as Liplawk [177] and Krawczenko [178] that the system involves a series of mixed crystals with a minimum (Bakhuis-Roozeboom type III), while Klochko and Jovnir [179] claimed to have found a simple eutectic system. The system had already been investigated in 1950 by the author [180] who discovered it to be a most interesting case involving a stabilized intermediate phase (S in Fig. 59). This is stable only above 90°C, however, and breaks up into the two pure components on cooling. The behaviour of the contact preparation was considered so striking that it was

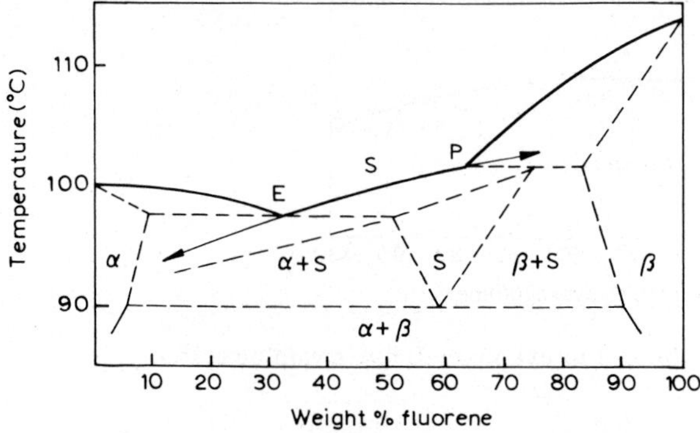

Fig. 59. Phase diagram of the phenanthrene—fluorene system.

included in a film made at that time [181]. It seems that some of the progress brought by thermomicroscopy decades ago might again be lost by over-enthusiasm for differential thermal analysis.

(C) THE PHASE DIAGRAMS OF ENANTIOMERS

It is very well known that the optically inactive form of a pair of enantiomers may consist of a mechanical mixture or of mixed crystals, although in practice most people associate the term "racemate" with a molecule addition compound having a 1 : 1 ratio. In general experience, even if this was not the case, at least the phase relationship between the pure enantiomers would be of a simple sort (for example, Bakhuis-Roozeboom type II). Investigations, however, have now shown that surprising and complicated systems can arise due to polymorphism. That this should be so with compounds so long known lends added interest to the matter.

Consider the phase diagram of L-hyoscyamine and atropine. This is not of great complexity, but nevertheless surprising [25]. L-Hyoscyamine and D-hyoscyamine form a series of mixed crystals following the Bakhuis-Roozeboom II type of diagram (Fig. 60) so atropine is not a "true" racemate (i.e. an addition compound) but a racemic mixed crystal. The (+)- and (—)-dihydrokawain [182] system also involves mixed crystals. In this case, the antipodes form a continuous series with a maximum between two minima, but on this is superimposed a stabilized intermediate phase which also consists of

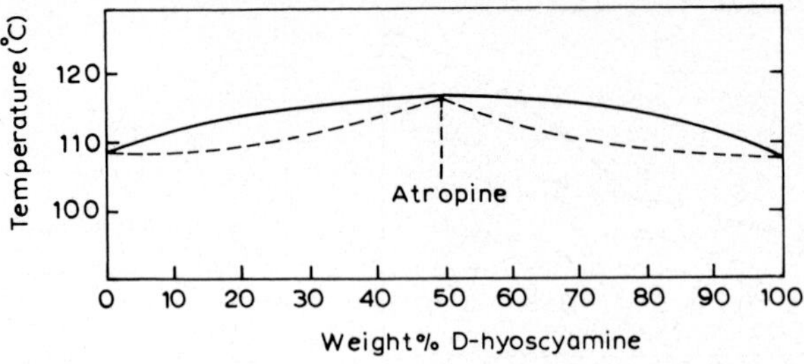

Fig. 60. Phase diagram of the D-hyoscyamine—L-hyoscyamine system.

mixed crystals and which derives presumably from two previously unrecognized unstable forms (Fig. 61).

The (+)-menthol—(-)—menthol system is even more complicated [183] as a result of isotrimorphism. Because of this, we have the intersection of three series of mixed crystals with stabilized intermediate phases (Fig. 62). Modification II of (±)-menthol, a racemic mixed crystal, corresponds to the minimum of a series of solid solutions (Bakhuis-Roozeboom type III) between the modifications II of the antipodes. This phase is completely stable from 18 to 32% and from 68 to 82% (+)-menthol. This means that mixtures, including that having the equimolar composition, crystallize spontaneously on cooling to the crystal phase of modification II. On the other hand, there is very little tendency to crystallize in the series of solid solutions formed by the inactive compound and the stable modifications of the antipodes.

Great difficulties are presented by the D- and L-malic acid system

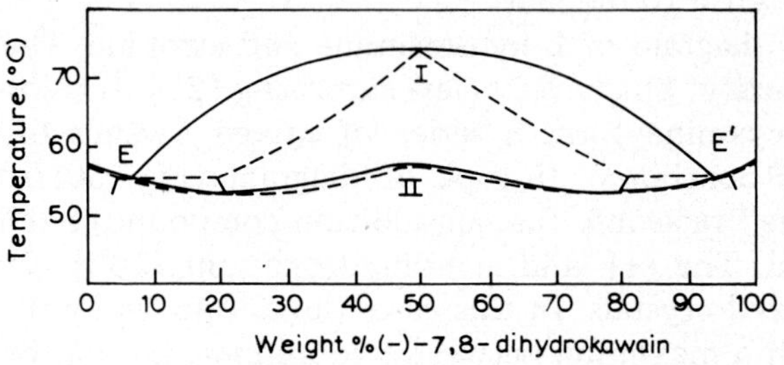

Fig. 61. Phase diagram of the (+)-dihydrokawain—(-)-dihydrokawain system.

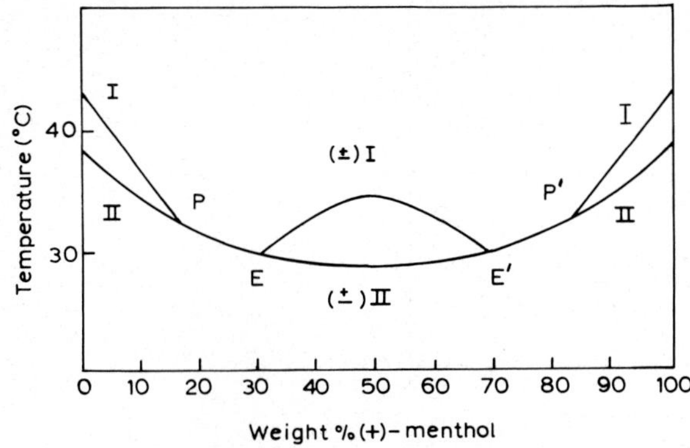

Fig. 62. Phase diagram of the (+)-menthol—(-)-menthol system.

which, if one were to believe the literature [184], involves anomalous racemates. The difficulty was that the dimorphism of DL-malic acid could not be detected microscopically. Only after it was found possible to crystallize the two modifications from different solvents and identify them by infrared spectroscopy was the problem solved [185]. Modification II of DL-malic acid forms a stabilized intermediate phase due to the fact that its structure differs from that of modification I and this was the reason for the assumption that anomalous addition compounds (having formulae in the proportions 1 : 3 or 3 : 1) existed and had incongruent melting points. The phase diagram for the system is given in Fig. 63.

Finally, the drug methyprylon deserves notice. This, like hyoscyamine in the racemic form atropine, is used as a racemate which

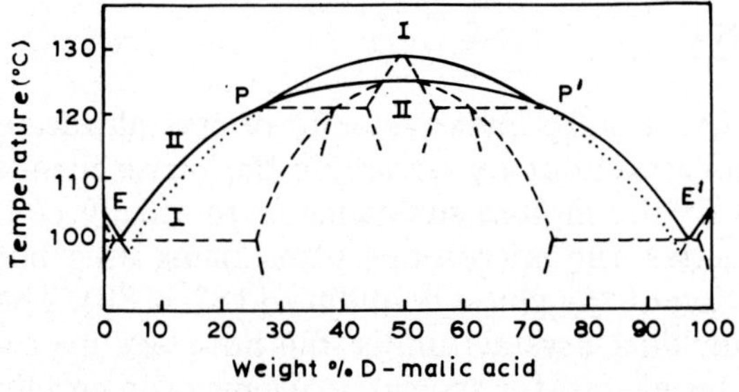

Fig. 63. Phase diagram of the D-malic acid—L-malic acid system.

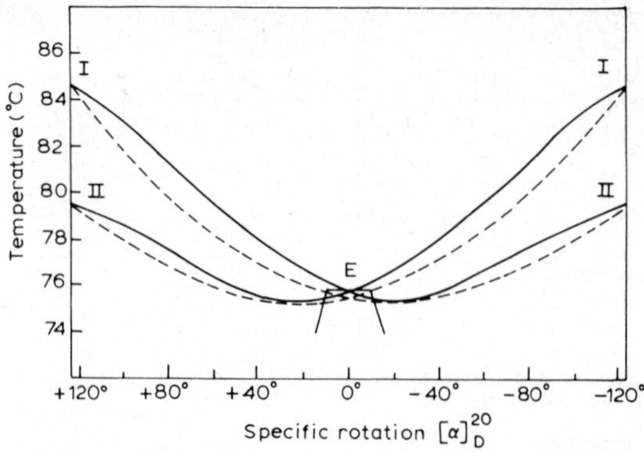

Fig. 64. Phase diagram of the (+)-methyprylon—(-)-methyprylon system.

arises because of an unusual binary system (Fig. 64). Both the antipodes are dimorphous and the stable modification of each antipode is isomorphous with the unstable form of the other. The two series of solid solutions, both of which are characterized by a minimum, intersect to produce a eutectic of zero optical rotation. The minima lie to the right and left of the eutectic in the unstable region. Consequently, the substance consists of a conglomerate of mixed crystals [186].

6. Some special applications of thermomicroscopy

(A) THE DETERMINATION OF CRYSTALLIZATION VELOCITY AND TRANSFORMATION VELOCITY

(1) Crystallization velocity

The determination of the average linear velocity of crystallization (C.V.) has generally been carried out by observing the phase change in macroscopic amounts of the molten substance in thermally controlled glass tubes or cuvettes, the microscope often being used but only to estimate the position of the phase boundary [187—189]. The direct observation of individual crystals under the hot-stage microscope has been less popular except for special problems or in certain temperature regions. Quite recently, the latter method has received more detailed attention and its advantages and limitations carefully examined [190,191].

Of the commercially available hot stages, Mettler's FP 52 (see p. 336) seems to be the best for the purpose since it is provided with electronic temperature control which allows the temperature of the object chamber to be maintained within ±0.1° in the low and ±0.2° in the high temperature region.

Since the diameter of the field of view of the microscope is only about 1 mm, only substances with a suitably low C.V. can be measured. This is a serious restriction and it is the reason why macroscopic methods have been favoured for the determination of C.V. versus temperature curves. In addition, extremely slow crystallizing substances are also unsuited to direct measurement because the rate of growth is too slow for the investigator to measure.

For the determination of C.V., films of the utmost possible uniformity are required. These are prepared in the same way as for investigations of polymorphism (see p. 423). The film of melt is prepared on the Kofler hot bench or other hot surface and crystallization is allowed to start at the edge of the cover slip by cooling or seeding. The slide is then inserted into the FP 52 which has been adjusted to the required temperature. Measurement is delayed until a continuous crystal front has formed and has moved in from the edge of the cover slip because measurements close to the edge must be avoided; the delay involved depends on the substance under investigation. If a continuous crystal front is not formed, care should be taken that measurements are made only on broad crystals of similar size. It should be noted that it is only in one favoured direction that growth can occur and be measured; growth at an angle to the favoured direction is prevented by that of parallel crystals.

Measurements are made with the aid of a micrometer eyepiece whose scale contains 100 divisions each usually equal about 10 μm and which can be read to half a scale division (giving a precision of ±5 μm). The movement of the crystal front is timed and the distance it moves per minute calculated. At least five measurements made at different parts of the growing front should be averaged to obtain the final C.V. value.

To determine a C.V.—T curve, it is sufficient to make measurements at 5° intervals between the temperature of first observable movement and the melting point. In particular cases, e.g. when a sharp maximum occurs, it may be necessary to measure at closer intervals.

Since previous investigations have shown that C.V. is influenced by the thickness of the layer of melt, investigations have been carried out in this direction in order to see if the usual method for the preparation of crystal films (see p. 423) is adequate or if exact weighing-out is necessary. Because about 8 mg are usually needed for a film of melt measuring 15 × 15 mm, this and an especially thin (2 mg) and an especially thick (16 mg) film were prepared and the C.V.s for the three different layer thicknesses were measured. It appeared that the C.V. in the thin (2 mg) film was as much as 4% lower than in the normal (8 mg) film and in the thick films up to 4% higher. Since the individual determinations by this method show greater deviations between themselves than the deviations of the average values between thicknesses, actual weighing of the amount of substance to be used seems not to be necessary. In any case, the practised microscopist, when preparing his film, will take an amount of substance so as to give him approximately the optimum thickness.

It must be pointed out that, with this method no less than with conventional macroscopic methods, absolute values of C.V. are not obtainable because the growth of crystals from the melt is influenced by too many factors. What is of interest is, rather, the possibility of comparison of the C.V.—T curves for different substances or polymorphic modifications. Such curves show the temperature at which a substance or particular modification has its maximum C.V. and the limits where it does not grow at measurable rates. In addition, information on purity can be obtained [148,192].

In Fig. 65, the C.V.—T curves for three modifications of testosterone cyclopentyl propionate are shown. The melting points of the three crystal forms are 100—102°C, 90—91°C and 74—75°C, the unstable forms having been revealed in the course of C.V. measurements on modification I [191].

In addition to the factors which disturb the determination of C.V. considered above, we must add the phenomenon of bubble formation. Air bubbles sometimes formed during crystallization cause great irregularity in the progress of the crystal front at the points where they lie. This phenomenon is explained by Neumann and Micus [187] as being due to the poor heat conductivity of the air bubbles so that the crystal front lying behind a bubble is warmer than elsewhere. Thus, bubbles cause a reduction in C.V. if the supercooling is slight, but an increase if the supercooling is great. The reduction

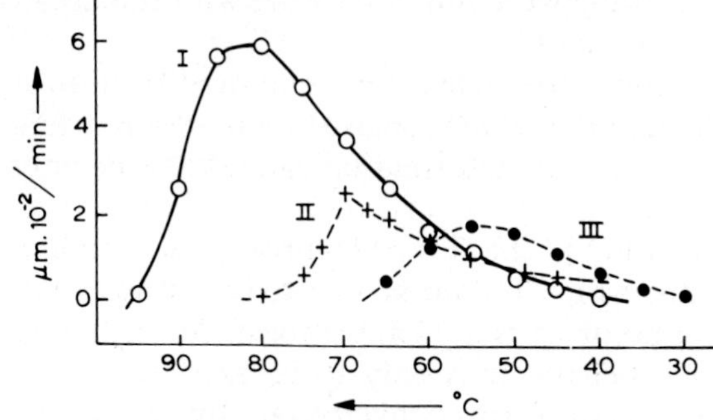

Fig. 65. Crystallization velocity versus temperature curves for three modifications of testosterone cyclopentylpropionate.

effect is not so obvious since the air bubble will be passed and enclosed by the faster growing front in neighbouring regions. If, however, the C.V. behind the bubble is increased, it is pushed forward by the crystal front forming behind and is further enlarged by continuing separation of gas. Thus, on the one hand, the bubble's heat protecting effect increases due to its enlargement and on the other, due to the increased C.V., the bubble acquires more heat of crystallization, so the local temperature may rise until the maximum C.V. is reached. By pressing on the cover slip, undesirable air or gas bubbles caused by decomposition can often be removed from the film of melt.

(2) Transformation velocity

The determination of transformation velocity (T.V.) between polymorphic modifications is undoubtedly of greater practical importance than the determination of C.V. since, in pharmacy, metastable crystal forms are often used because of their greater solubility and bioavailability. In this connection, their stability at storage temperature is of importance as is the T.V. at different temperatures.

In contrast to the determination of the C.V., the microscope was used from the very beginning (by Lehmann [1]) for the determination of the T.V. Many investigators of polymorphism experimented also with T.V. [192—195] but, in most cases, at particular temperatures of special interest and they were not concerned with the determination of complete T.V.—T curves.

Monotropic systems. In monotropic systems, one modification is (at a given pressure) stable over the whole range of temperature while the second is unstable. Thus, the transformation can take place only in one direction (see p. 412).

According to Tammann [171], the T.V.—T curve shows a maximum between the low rates which obtain at low temperatures and, again, close to the melting point of the unstable form. According to Kofler and Kofler [7], such behaviour is only to be expected if the vapour pressure curves of the two forms converge rapidly so that the virtual point of intersection (i.e. their notional transition point) is not far outside the range of observation. In this case, the situation is similar to the ordinary C.V.—T variation. On the other hand, if the vapour pressure curves are parallel or only slowly converging, a continual increase in T.V. is to be expected over the temperature range accessible to experimental observation. As an example of the first case, the T.V.—T curve for salicylic acid guaiacol ether (m.p. modification I, 71°C; modification II, 68°C) is shown in Fig. 66. An example of rising curve is shown by the transformation of modification II to modification I of N-methylurea (m.p. modification I, 102°C; modification II, 100.5°C) in Fig. 67.

The measurement of T.V. is similar in principle to that of C.V. so the criteria for ease of observation with respect to rate of movement of the crystal front are the same.

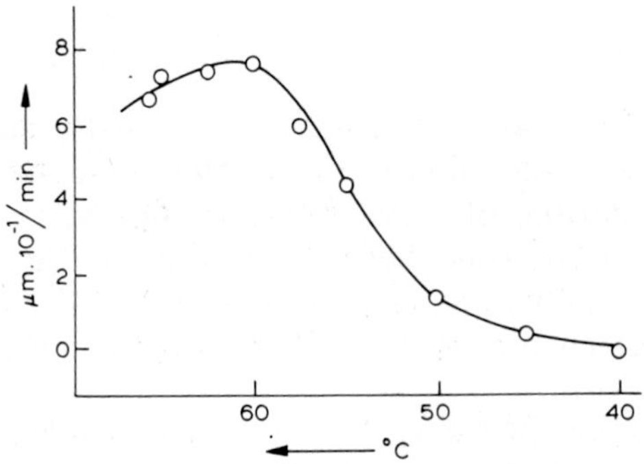

Fig. 66. Transformation velocity versus temperature curves for salicylic acid guaiacol ether (monotropic system).

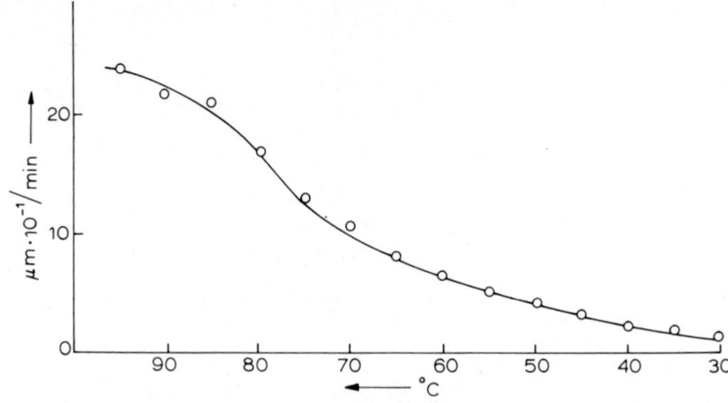

Fig. 67. Transformation velocity versus temperature curves for *N*-methylurea (monotropic system).

Enantiotropic systems. In an enantiotropic system, the transformation can proceed in either direction and there is a transition point (see p. 411). Below the transition point, the form stable at higher temperatures (modification I) is converted to the low temperature form (modification II) and vice versa.

In transformations between enantiotropes, there is only one kind

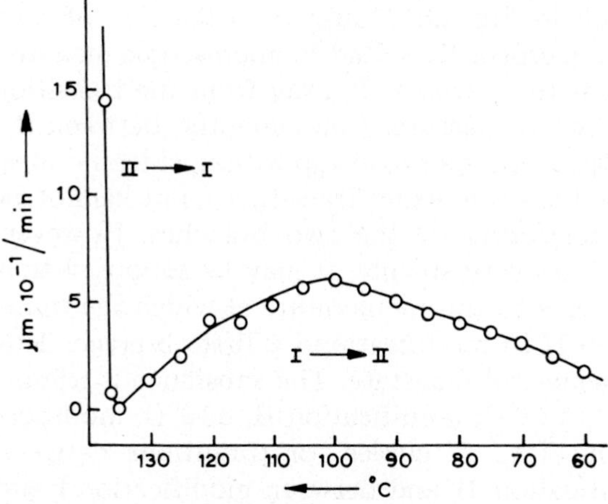

Fig. 68. Transformation velocity versus temperature curves for caramiphen hydrochloride (enantiotropic system).

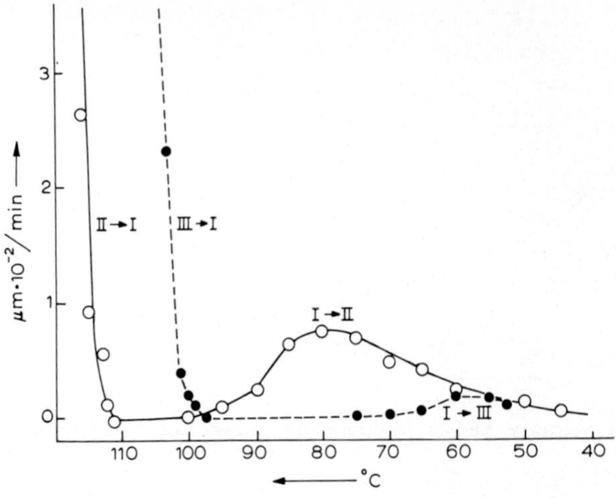

Fig. 69. Transformation velocity versus temperature curves for dienestrol diacetate (enantiotropic system).

of T.V.—*T* curve and it consists of two branches. The branch covering the transition of modification I to modification II passes through a maximum and the other for the transition from modification II to modification I rises continuously and steeply. Ideally, the two branches meet at the transition point as shown by those for caramiphen hydrochloride in Fig. 68. However, in the case of sluggish transitions (such as are especially suited to microscopic measurement) it is often the case that, even well away from the transition point, the T.V. is too low to measure. Consequently, between the two branches of the curve there is a broad gap within which no measured values can be found and the exact transition point cannot be determined from the intersection of the two branches. However, since the second branch rises very steeply, it may be supposed that the transition point lies close to the temperature at which the transformation of modification II to modification I is first observed. This behaviour is shown by dienestrol diacetate. The substance is trimorphic (m.p. modification I, 123°C; modification II, 119°C; modification III, 112°C) and the T.V.—*T* curves for transitions between modification I and modification II and between modification I and modification III have been determined (see Fig. 69).

Disturbing factors in the determination of the T.V. Besides those which influence the C.V. (layer thickness, position of measurement, gas bubbles, impurities, etc.), additional factors must be considered as possibly influencing the reproducibility of determinations of T.V., both by microscopic and macroscopic methods.

Contraction fissures which appear on cooling (often in a characteristic pattern, see p. 365), are particularly important. When, in the course of a transformation, the front reaches such a point, its progression is interrupted and some time elapses before it can pass beyond the fissure. It is absolutely necessary to avoid specimens with microscopically visible fissures as they greatly falsify the T.V. Similar effects can be observed affecting the T.V. inside and beyond the edges of spherulites [192]; the edge of the spherulite acts as a barrier and causes retardation.

Another factor is the orientation of secondary to primary crystals (see p. 425). According to Weygand [131], transformation takes place preferentially in some direction of least resistance, e.g. along the line of the fibres of spherulites, but is retarded across the direction of growth.

The preliminary treatment of the melt of the substance to be transformed often exerts a considerable influence on the T.V. So, for example, one observes a higher T.V. if, in a crystal film held at a higher temperature in order to produce a particular modification, measurements are made at that same temperature without cooling than if they had been made after cooling to room temperature and then reheating [190]. Probably the reason for this behaviour is that, on cooling, microscopically invisible fissures form and these impede the transformation. When a crystal film preparation, having been held at a higher temperature, is transferred to the hot stage, cooling should be avoided.

Crystal habit can exert an extraordinary influence on the T.V. For example, in the case of the monotropic transition of protheobromine from modification II (m.p. 130—132°C) to modification I (m.p. 140—142°C), we have two kinds of crystals of the same stalk-like appearance but with different face combinations at the ends, designated habit A and habit B. Habit A is characterized by rectangular pointed ends and habit B is pointed to one side. Several investigations have shown without doubt that the two habits are of one and the same modification [191]. When the T.V. in crystals of the two habits are measured under exactly the same conditions, the monotropic type T.V.—T curves in Fig. 70 are the result. They show that,

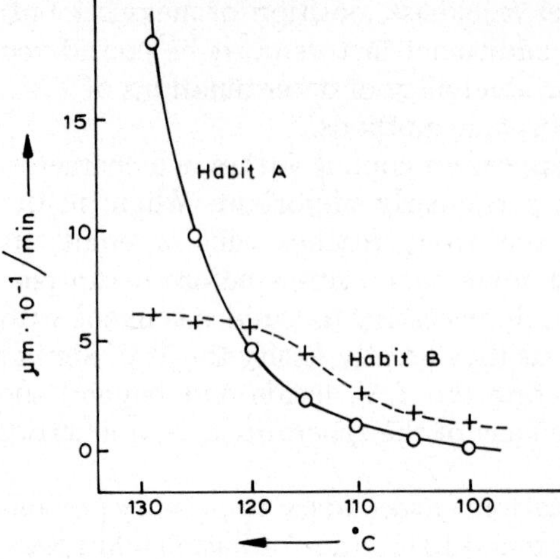

Fig. 70. Transformation velocity versus temperature curves for two different crystal habits of protheobromine.

a little below the melting point, the T.V. of habit A is about three times that of habit B.

It is clear from the foregoing that the T.V. is not a reproducible property and its value is subject to many influences [125]. However, by following carefully standardized preliminary treatment and taking into account the many possible disturbing factors, it is possible to obtain values which are useful for purposes of comparison.

(B) DEMONSTRATION OF THE GROWTH SPIRALS ON ORGANIC CRYSTAL SURFACES

Growth spirals were occasionally observed before 1949, some as early as the beginning of the century [196,197], but since they had no place in the accepted theories of crystal growth, the phenomenon seemed to be without importance.

It was only when the theory of the growth of imperfect crystals due to screw dislocations was put forward by Burton and his co-workers [198—200] that the growth spiral became of interest. The essence of the earlier ideas concerning the growth of perfect crystals were, according to Frank, as follows: in the theory which derives from Gibbs [201] and was further developed by others, especially Volmer [202], Kossel [203] and Stransky [204], additional growth on the surface of a crystal is by planes, each of which has to be freshly nucleated, and the rate of formation of the surface nuclei is an especially sensitive function of supersaturation. This is such that at low supersaturations (below some critical minimum) the rate of formation of nuclei is so reduced that the rate of crystal growth falls to zero. This is not in accord with the observation that crystals do actually grow, and at rates proportional to supersaturation, well below the critical supersaturation of the surface nucleation theory. Burton and co-workers [198—200] proceeded from the assumption that imperfect rather than perfect crystals are the rule and the former possess screw dislocations, these being defects of a special kind which lead to the formation of steps and are of importance not only for the understanding of crystal growth but also for the hardening process of metals, plastic flow of crystalline substances, the migration of grain boundaries and for chemical reactions in solids [205].

In Frank's [206] model, dislocations lead to the formation of steps (Fig. 71) which grow continuously in the form of spirals (Fig. 72). This can occur even at very low supersaturations because the need for nucleus formation is obviated.

Organic substances are very suitable for demonstrating this mode of crystal growth, especially when growing from the vapour phase, since many spirals are formed and these are visible in the light microscope [207,208]. The best subjects for this study are plate-like crystals formed by sublimation on a cover slip. Once such crystals grow

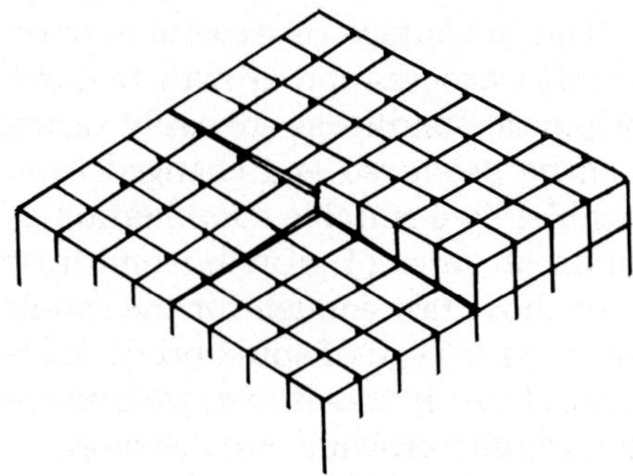

Fig. 71. The end of a screw dislocation (after Frank).

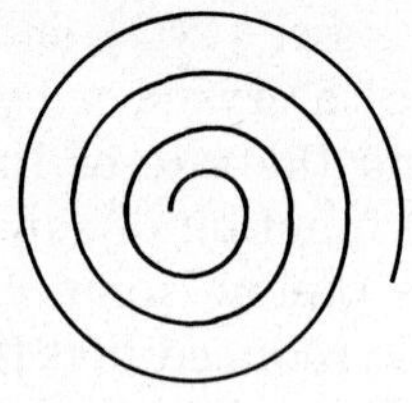

Fig. 72. Model of a simple growth spiral.

beyond a certain thickness, tensions develop which cause dislocations to form and the resulting spiral growth can be observed in nearly all organic substances [209,210].

In the case of very small crystals which show only one spiral, this may well follow the form of the crystal. On bigger crystal faces, dislocations are more numerous and they may be related to each other in different ways. For example, isolated dislocations form single spirals (Fig. 73), two neighbouring dislocations with the same sense of rotation may combine to form a double spiral (Fig. 74), or, if the sense of rotation is opposed, a closed region may result (Fig. 75). The growth of a spiral is such that the centre winds continuously inward due to uniform enlargement of the step while the spiral as a whole grows outwards and appears to rotate. Regions of neighbouring spiral growths knit together so that the crystal face as a whole advances.

In principle, the same types of spirals occur in the growth of organic substances as have been observed in inorganic materials, viz. circular and angular spirals of different forms (square, rectangular, rhombic, hexagonal, etc.) [211]. The advantage of experiments on organic compounds is that the growth process and growth features are easy to observe and the experimental conditions are easily varied so that various types of spirals can be produced and changed from one to another. Moreover, it is possible, by a suitable combination of substance and temperature, to obtain all rates of growth from those so low as to be barely perceptible to those fast enough for the spirals to appear to rotate rapidly. In general, the type of spiral produced is governed by the degree of supersaturation; if this is low, we observe slow growing angular spirals, if high, rapidly growing circular ones.

The procedure for the demonstration of spiral growth is very simple. Some crystals of the substance to be investigated are placed on

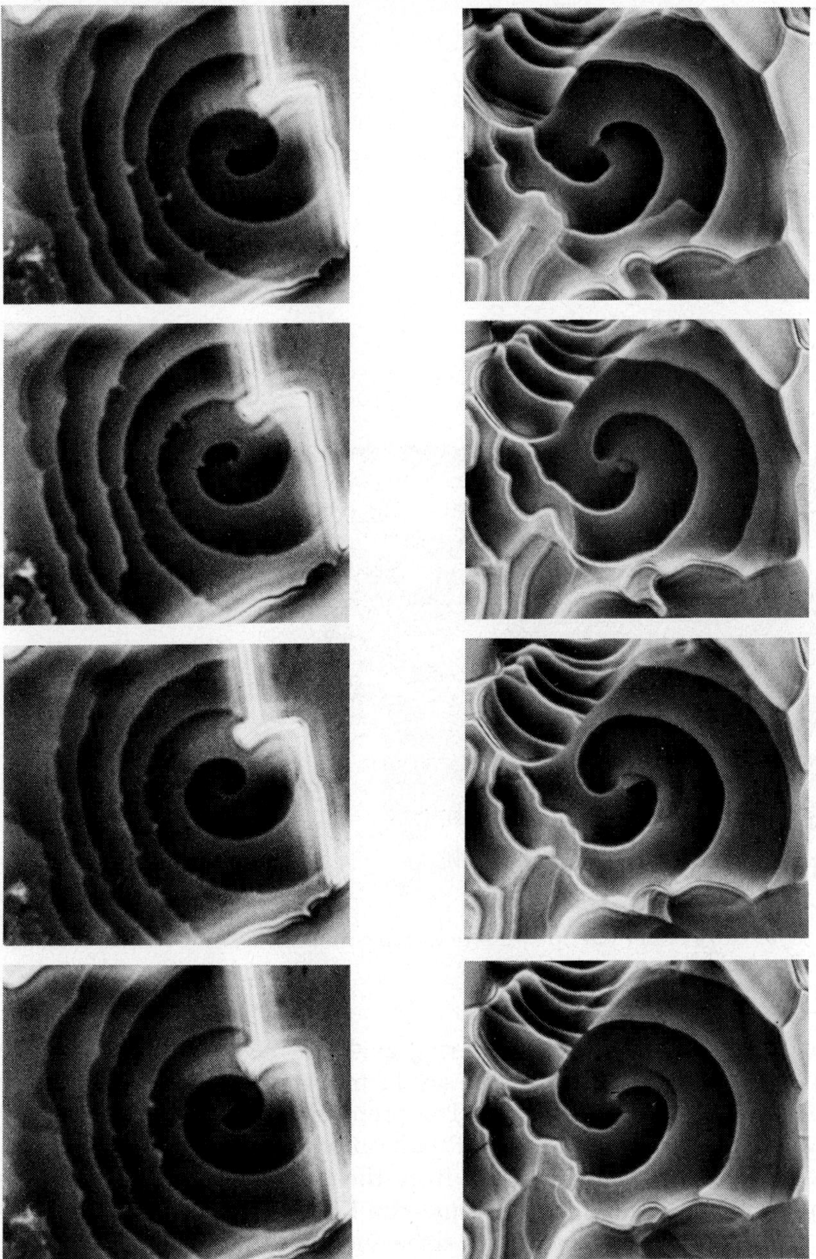

Fig. 73. Growth of a simple spiral: diphenylamine (phase contrast).

Fig. 74. Growth of a double spiral: phenanthrene (phase contrast).

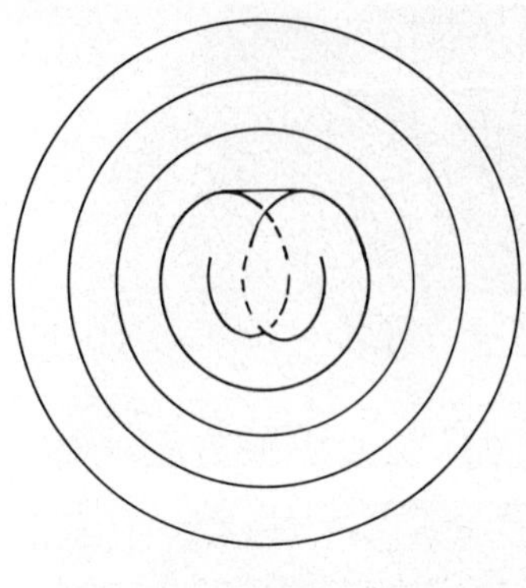

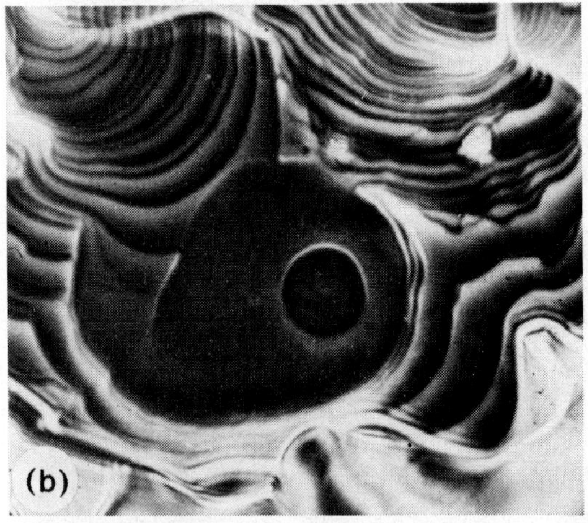

Fig. 75. Growth hill due to two screw dislocations of opposite sense: (a) model, (b) borneol (phase contrast).

the object slide, inside a glass ring 2—4 mm high and about 15 mm internal diameter. A cover slip, say 18 mm square, is placed on top of the ring to serve as receiver. The preparation is slowly heated on a Kofler hot stage (the Mettler FP 52 cannot be used for this purpose) over the range of temperature from the beginning of sublimation to the vicinity of the melting point. The Kofler stage's glass cover plate is not used in this operation since the resulting temperature drop between slide and cover slip assists the sublimation process, higher power objectives can be employed, and precise temperature measurements are not required anyway.

A more powerful objective than the usual ×10 low-power drive, for example a ×40, may be used to advantage in this work, especially since spirals and the growth process can be observed only at very low apertures with ordinary transmitted illumination. It is unfortunate that phase contrast illumination cannot be easily adapted for use with the hot stage, as this is a great advantage. However, the lack of phase contrast is not a serious disadvantage and low-aperture transmitted illumination is adequate in some cases, e.g. borneol. Spirals can be observed on the sublimate crystals of this substance from 90°C and they are particularly easy to observe between 110 and 120°C. At the higher temperature growth is extremely rapid and the spirals appear to rotate rapidly. Closed growth steps as described above may also be seen in this preparation.

Although, as in the case of borneol, it is often possible to follow the process without the aid of phase contrast, this depends on the depth of the spiral step. If the step is very thin, ordinary low aperture illumination is inadequate or completely ineffectual and phase contrast makes for a substantial improvement in visibility. If the growth process is to be followed in such a case, the cover slip receiver will need to be transferred from the hot stage to phase contrast microscope several times during the course of sublimation and again after sublimation is complete when the specimen can be examined at leisure.

In order to be able to follow the growth of especially delicate spirals directly in the phase contrast microscope, we have attached a long metal strip to its stage and heated the free end of this with a micro-burner. The arrangement works well with low-melting substances but, before starting the experiment, it is necessary to seed the cover slip to overcome the very sluggish nucleus formation which obtains in such cases. This is done by rubbing crystals on the cover slip and wiping off lightly (so that some seeds still stick) before putting the cover slip on the glass ring. Well developed crystals then form at low temperatures. Although, at the start of sublimation, crystals first form with a principal crystallographic face against the cover slip, later crystals grow at different angles and they are best blown off since they impede observation of the growth spirals.

It should be noted that the reproducibility of these experiments is influenced by differences in room temperature (since the hot stage is open), the height of the sublimation ring, the amount of substance taken, and by the rate of heating. Some substances reveal spirals readily under all conditions, and they may be demonstrated without great difficulty for the majority of easily sublimable organic substances, but others (e.g. benzoic acid, phenacetin) are very susceptible to small changes in experimental conditions. This is particularly so with substances whose sublimates adopt an acicular or bladed rather than a platy habit.

For more extensive studies of spiral growth, we recommend a special design of long working distance phase contrast condenser which may be used directly with the hot stage. We have used such a condenser for ciné photomicrography [208].

(C) THE STUDY OF EPITAXY

By epitaxy we mean the structure-dependent orientated overgrowth and intergrowth of the crystalline phases of two different chemical compounds, one being designated the host and the other the guest. The development of guest crystals on the host crystal may come about in different ways: from solution, from the melt, from the gas phase, or by electrolysis. With regard to the nature of host and guest, there are also four possibilities: both inorganic, inorganic host and organic guest, organic host and inorganic guest, or both organic [212].

For a long time, epitaxy was of interest mainly to mineralogists [213–217], but today research into the phenomenon is important in the investigation of many chemical, physical and technical problems.

The Kofler hot-stage microscope is especially suited to the study of the epitaxy of organic guest crystals forming on host crystals by sublimation since direct observation of the depositing material is possible. Formerly, cleavage faces or freshly formed surfaces of cultivated single crystals were generally used as the host surface and, because of this, the number of substances investigated was very limited [218]. By using crystal films instead of single crystals, it is possible to produce for epitaxy experiments clean host surfaces of any substance which melts without decomposition [219]. The substance selected as host is melted between slide and cover slip, allowed to crystallize, and the cover slip is then removed. The crystallized material will stick more to the slide or cover slip depending on whether the preparation was cooled rapidly (on a cooling block) or slowly in air. The procedure for sublimation is as described in the preceding section. Some crystals of the guest substance are placed on a slide within a glass ring, 2—4 mm high and about 15 mm diameter, and the cover glass or slide with the exposed host film is placed on top. The whole preparation is then placed on a Kofler hot stage pre-heated to a temperature suitable for sublimation. (Minimum sublimation temperatures are to be found in Kofler identification tables [7, 11]). During sublimation, the glass cover plate is left off the stage and the preparation is watched through the microscope. Figure 76 shows a typical epitaxial deposit on a crystal film surface. Most organic substances are sufficiently volatile for the procedure described and it is not necessary to resort to vacuum sublimation. In most cases, it is possible to interchange host and guest so long as both constituents can be sublimed. The experimental arrangements described are also suitable for investigations with inorganic host substances. For example, thin split gypsum plates can be used as host and receiver instead of the usual cover slip.

When sublimation is not possible, substances can be epitaxially deposited from solution. For this purpose, the guest substance is dissolved in a suitable solvent and the solution carefully dropped in tiny droplets on the crystal faces of the host substance supported on a slide. The orientated overgrowth is to be seen most easily at the edges of evaporating drops.

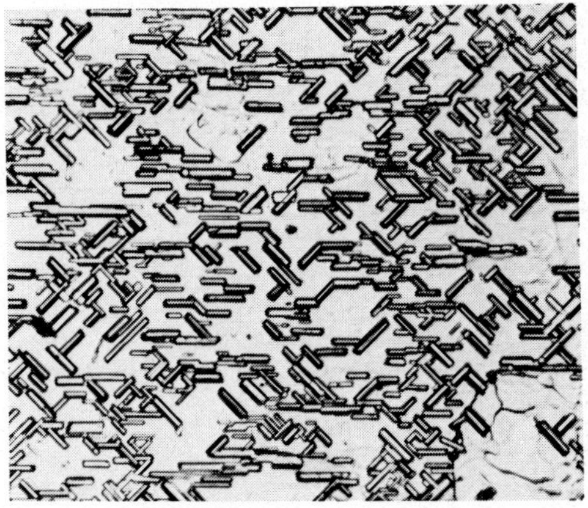

Fig. 76. Orientated overgrowth of hexaethylbenzene on a crystal film of anthracene (001 face).

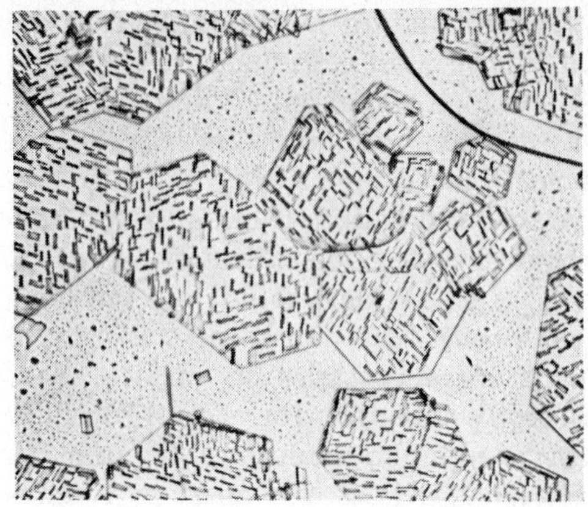

Fig. 77. Orientated overgrowth of benzamide on single crystals of anthracene (001 face).

Host crystal surfaces can also be prepared by sublimation instead of melting and recrystallizing. If the habit is suitable, e.g. in the case of anthracene, very large single platy crystals may be produced and the guest component again deposited by sublimation as shown in Fig. 77. In spite of the earlier view that the surface of the host crys-

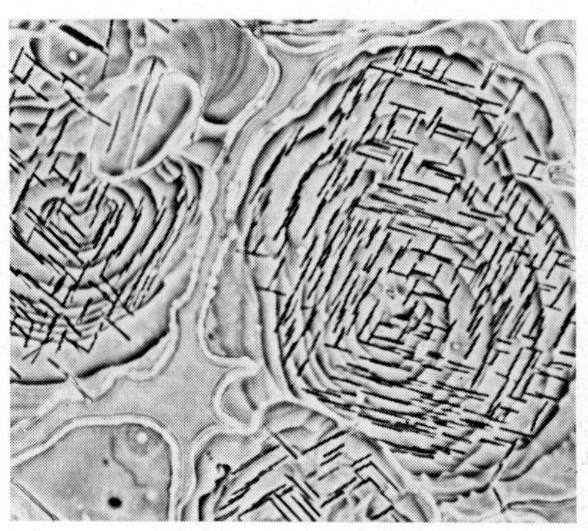

Fig. 78. Orientated overgrowth of bromisovalum on growth spirals of phenanthrene (001 face) (phase contrast).

tal must be strictly planar, tests have shown that epitaxial growth occurs readily on the growth spiral hills which characterize many sublimed surfaces [220]. Figure 78 shows spiral hills of a phenanthrene surface, grown by sublimation, on which there is an epitaxial deposit of needles of bromisovalum. It is obvious that the spiral step edges have encouraged deposition.

It can be stated as a general rule that, although temperature affects epitaxy, it influences the behaviour of the host and guest substances differently. If the host crystal surface temperature is increased, the orientation of the deposit is more irregular, unless guest and host form a molecular compound in which case higher host temperatures favour regularity. The temperature at which the guest component is sublimed has no effect in itself on the regularity of the deposit, nevertheless, for each pair of substances there is an optimum temperature at which the rate of nucleus formation and consequent size of crystals gives the best results [221].

Finally, epitaxy is a widespread phenomenon between organic compounds and the currently available data (for several hundred pairs [212]) could be increased many times over.

(D) MOLECULAR WEIGHT DETERMINATION

The cryoscopic method of molecular weight determination has today been largely replaced by other methods but there are still special circumstances in which it is relevant. A short description is given of it here.

The cryoscopic method depends on the accurate determination of the melting or freezing point of the solvent and of the temperature corresponding to the end of melting or beginning of freezing of a solution, the solute of which is the subject of the investigation. The calculation assumes that there is a simple eutectic phase relationship (p. 443) between solute and solvent and that the solution under observation is dilute enough to be regarded as ideal.

The well-known method of Rast [222] is best suited for adaption to the hot-stage microscope. Rast's method is based on camphor which has a very high molar freezing point depression constant. The solution is made up to contain about 10 wt.% of solute. Because of camphor's great volatility, the solution cannot be prepared as a crystal film but must be observed in a sealed glass melting-point tube [223], an internal diameter of about 1.8 mm being suitable. With the aid of a thin glass capillary, a small amount of sample is introduced followed by ten to twenty times as much camphor, the relative weights of the components being found as accurately as possible by difference. The tube is sealed with a microflame and its contents mixed by melting and rotating. It is then put into a windowed metal block which fits into the object guide (p. 332) of the hot stage (224, 225]. The heating of the tube on the hot stage might be expected, in the ordinary course of events, to lead to separation because the camphor would be expected to sublime to the cooler, uppermost part of the tube but this is prevented by the metal heating block completely surrounding it, except for the observation window [226—228].

Since camphor is not birefringent at the temperatures concerned (the rhombic form stable at room temperature is transformed to the isotropic cubic modification at 101.5°C [229]), care must be taken to adjust the substage iris diaphragm to its optimal setting because the camphor crystals differ little in refractive index from the melt and they will not be clearly visible if the diaphragm is opened too far. There is not likely to be any difficulty if the iris is correctly adjusted but phase contrast illumination (p. 370) and also the use of a blue filter have reported advantages [227].

Changes due to crystallization are easier seen when the solution is cooling because the camphor crystallizes in the form of stars which quickly grow to dendrites (Fig. 79). In Ekborn's [223] opinion, also, the temperature of spontaneous crytallization on cooling is more reproducible than that at which the last crystal disappears on heating but, in fact, if both determinations are carried out at the same rate of temperature change (about 1°C min^{-1}), the reproducibility is the same. The accuracy, however, differs. Pure camphor freezes spontaneously at 1—1.5°C below the true melting point, but the supercooling error in the case of 10% mixtures is about 2—2.5°. This shows that the addition of the solute lowers the ability of the liquid to crystallize spontaneously. To avoid error, it is recommended that the temperature corresponding to the end of melting on heating be taken together with the corresponding value for pure camphor. Since we are concerned only with the difference between the two readings, it is not important that both are too high because of the position of the metal heating block. The use of a special thermometer with 0.1°C divisions is not necessary [228].

The determination is generally repeated several times on the same

Fig. 79. Spontaneous crystallization of camphor.

tube and the average temperature used to calculate the molecular weight. The calculation is based on the equation

$$M = 1000\, K\, w_1/w_2\, \Delta t$$

where K is the molar freezing point depression constant of the solvent (in this case camphor for which $K = 40$), w_1 and w_2 are the weights (in the same units) of sample and solvent taken, and Δt is the observed difference between the melting points of solvent and solution.

7. The handling of thermochemical data

Thermomicroscopic data (melting points, eutectic and refractive index temperatures) can be used both to check the identity of supposedly known substances and to identify unknown organic compounds, but only in so far as the data are available and suitably tabulated. If the problem is the former, as it might be in a pharmacy with the checking of a labelled substance, then it is best to have an alphabetic arrangement of substances. Table 10 shows a section from the International Pharmacopoeia [230], which contains a table for checking the identity of pharmaceuticals. Of course, if an authentic sample were available, a mixed melting point determination might be carried out instead of or in addition to the refractive index and

TABLE 10

Section of a table for checking the identity of pharmaceuticals

Substance	Melting range (°C)	Eutectic temperature * (°C)	Refractive index	Temperature (°C)		Special remarks
				Red light	Sodium light	
Pethidini hydrochloridum	184—189	Benzan. 119 Sal. 134	1.5101	163—160	163—160	Above 110°, subl. of grains, leaflets and rhomboids. Later droplets interfering with vision. Equilib.: hexagons, grains, spindles. On longer heating gas-bubble formation. Melt glassy on solidification; at 140°C, crystalline unstable spherulites with grey interference colours m.p. II 162—166°, in addition to stable spherulites with higher interference colours. Refractive index must be determined with falling temperature, rises on repetition. On cooling, almost rectangular crystals precipitate from melt.
Phenacetinum	133—135	Acet. 92 Phen. 133	1.5101	133—134	138—139	Above 100°C, subl. of needles, leaflets, columns. Equilib.: rectangles, columns. Melt solidifies giving mod. II spherulites. On

Phenazonum	109—111	Benzil Acet.	70 51	1.5611	118—120	123—124	Above 85°C, subl. of needles, rods, leaflets. Equilib.: grains. Melt solidifies slowly giving stable and unstable spherulites.
Phenindamini tartras	160—167	Phen. Benzan.	125 141	Decomposition			The main mass of crystals melts at 160—163°C, a few residual crystals melting at 167°C. The melt becomes brown in colour and forms bubbles.
Pheno-barbitalum	174—177	Benzan. Sal.	140 154	1.5299	176—177	179—180	Above 135°C, abundant subl. of needles and rhomboids. Original substance occurs as mod. II, as well as part of subl. (m.p. 174°C). During melting, stable rods immediately precipitate out, m.p. 176°C. Equilib.: rhomboids, rods. Melt glassy on solidification, on heating, spherulites form (several mod.)
Phenolum	35— 40	Azob. Benzil	25 19	1.5403	36— 37	43— 44	Above 25°C, subl. of needles. Equilib.: needles. Melt radiate on solidification. Substance hygroscopic; completely anhydrous, m.p. 41—43°C.

* Benzan. = Benzanilide; sal. = salophen; acet. = acetanilide; phen. = phenacetin; azob. = azobenzene.

TABLE 11

Section of an identification table adapted from ref. 11

M.p. (°C)	Compound	Eutectic temp. (°C) with		Glass powder	Temp. (°C)		Special remarks
		Acetanilide	Phenacetin		Red light	Sodium light	
121	4,4′-Dimethyldiphenyl $C_{14}H_{14}$	100	112	1.5403	117—118	123—124	Subl. of grains, rhomboids, rectangles and prisms from 90°C. Equilib.: rectangles and prisms. Melt solidifies to plates.
121.5	Clofedanol $C_{17}H_{20}ClNO$	95	108	1.5299	128—129	130—131	Subl. of needles from 110°C. Equilib.: needles and rods with feeble birefringence. Melt solidifies to a glass and at 60°C stable spherulites form with unstable quadrangles and rosettes between them. Transformation on further heating.
117—122	17α-Hydroxyprogesterone caproate $C_{27}H_{40}O_4$	72	93	1.4936	132—134	134—136	Residual crystals grow sluggishly into grains, prisms, and beams. Melt solidifies to a glass; after prolonged standing at 80°C or on seeding aggregates of broad stems are formed.
120—122	Chloral formamide $C_3H_4Cl_3NO_2$	79	93	1.4842	115—117	117—118	Subl. of grains, pentagons and hexagons, and ribbons from 55°C. Melts sluggishly, soon decomposing. Melt solidifies

120—122	Estradiol methyl ether $C_{19}H_{26}O_2$	86	101	1.5405	114—116	117—118	Subl. of individual rods and needles from 110°C. Residual crystals grow into rectangular or hexagonal prisms and stems. Melt solidifies to a glass; at 90°C with seeding to broad-rayed aggregates.
122	*m*-Amino-phenol C_6H_7ON	82	94	1.5795	124—126	132—133	Subl. of grains from 65°C. Equilib.: prisms and stems. Melt solidifies to spherulites.
122	Butyraldehyde-2,4-dinitrophenyl-hydrazone $C_{10}H_{12}N_4O_4$	86	98	1.6353	112—113	140—141	Yellow substance. Subl. of grains from 100°C. Equilib.: stalks and broad stems. Melt solidifies to the tessellated mod. II, on heating transformation or m.p. 110°C.
122	2,6-Dichloro-benzoquinone $C_6H_2O_2Cl_2$	82	96	1.5403	121—122	129—130	Yellow substance. Subl. of needles, stalks, stems, and beams from 60°C. Transforms completely by sublimation. Equilib.: stalks and stems. Melt solidifies to a stalk-like aggregate. Colour of eutectic melt orange to brown.
122	Picric acid $C_6H_3N_3O_7$	71	89	1.6011	118—119	131—132	Yellow substance. Subl. of needles from 115°C. Equilib.: prisms and beams. Melt solidifies to lamellar or feathery aggregates.
122	Testosterone propionate $C_{22}H_{32}O_3$	74	93	1.5000	117—119	122—123	Equilib.: grains and prisms. Melt solidifies sluggishly to stable and unstable spherulites with subsequent transformation.

TABLE 11 (continued)

M.p. (°C)	Compound	Eutectic temp. (°C) with		Glass powder	Temp. (°C)		Special remarks
		Acetanilide	Phenacetin		Red light	Sodium light	
122.5	Benzoic acid $C_7H_6O_2$	76	89	1.5000	132—133	138—139	Subl. of needles, beams, and oblique-angled leaflets from 60°C. Equilib.: beams and prisms. Melt solidifies in rays.
122.5	β-Naphthol $C_{10}H_8O$	60	69	1.6011	134—135	143—144	Subl. of thin and lobed leaflets from 90°C. Equilib.: stems and spindles. Plates of mod. II crystallize spontaneously from the melt; these undergo tranformation into clusters of mod. I; m.p. of mod. II 122°C.
120—123	Pholedrine formate $C_{10}H_{15}NO$ CH_2O_2	86	102	1.5204	116—118	120—122	Subl. of needles, small spindles, and stars of needles from 100°C. Residual crystals grow sluggishly into spindles, prisms, and plates.
121—123	Trichloro-lactic acid $C_3H_3O_3Cl_3$	33	52	1.4683	126—127	126—127	Characteristic smell. Subl. of needles, stalks, and prisms from 80°C. Residual crystals grow into needles and stems. Melt solidifies to unstable spotted aggregates, on heating transformation.
123	Carbimazole $C_7H_{10}N_2O_2S$	85	102	1.5299 1.5403	134—135	117—118	Subl. of rods, rhombs, hexagons, and trapezoids from 80°C. Equilib.: hexagons, prisms, and plates. Melt solidifies in spheru-

123 (119, 111–112)	Diacetyl-dienestrol $C_{22}H_{22}O_4$	95	108	1.5204	121–122	126–127	Occurs as mod. III or mod. II. Mod. III partial melting at 111–112°C, partial transformation into mod. I. Equilib.: grains and pointed prisms. If mod. II occurs: Equilib.: grains and prisms. Just before or during melting sometimes partly transformation into mod. I. Melt solidifies to spherulites of mods. I and III.
123	Diethyl-stilbestrol dimethyl ether $C_{20}H_{24}O_2$	103	115	1.5299	113–114	118–119	Subl. of small hexagons, rhombs, and rods from 110°C. Equilib.: rhombs, hexagons and prisms. Melt solidifies to two kinds of spherulites, transformation on heating.
123	Picryl bromide $C_6H_2BrN_3O_6$	83	100	1.5795	133–135	142–143	Light yellow substance. Subl. of grains and hexagons from 90°C. Equilib.: hexagons and prisms. Melt solidifies to unstable spherulites; transformation on heating; m.p. of mod. II 121°C.
123	β-Progesterone $C_{21}H_{30}O_2$	73	91	1.5204	117–118	121–122	β-Progesterone is mod. II of progesterone. Equilib.: rectangles, rhombs, hexagons, and prisms. Melt solidifies to a glass; on heating to 60°C, several mods. crystallize.

eutectic temperature determinations. If the data are to serve for the identification of unknown substances, then they are best arranged in order of increasing melting point. Table 11 shows part of an identification table adapted from ref. 11, the use of which will now be described.

The melting point, whose determination is the most essential step in the procedure, is first determined approximately under conditions of rapidly rising temperature (or on the hot bench) in order to save time. Having obtained a rough value, an accurate determination is made, preferably by the equilibrium method if this has been mastered or, in the case of unstable or sluggishly melting substances, the melting range is found. The sample is observed closely during heating and note taken of the phenomena described in the final column of the table. After it has melted completely, the sample is placed on the cooling block to observe whether or not the melt solidifies and, if so, in what form.

On the basis of the melting point, now determined, the appropriate reference substances for the eutectic temperature determinations are chosen. In order not to cool the stage unnecessarily far but at the same time not to start reheating at too high a temperature, it is best to find the lowest eutectic temperature for the substances melting within one or two degrees of the determined melting point and to reheat from about five degrees below this value. With the melting point and two eutectic temperatures in hand and the knowledge of any peculiarities of behaviour during the heating or cooling processes, closer study of the table will narrow the possibilities down to a very few or even a single substance. Final identification is confirmed by determination of the temperature at which the refractive index of the melt matches that of the glass reference powder given in the table for the substance (p. 367). In the case of high melting substances, the temperature of refractive index match cannot be found because of decomposition and this confirmatory test is replaced by the preparation of a derivative (p. 388) or the determination of the UV absorbance or by the determination of some other eutectic temperatures, e.g. with phenolphthalein as given for some substances in the Kofler tables [7].

As an example, suppose an unknown substance is found to melt at 122°C and the eutectic temperatures with the appropriate reference substances, acetanilide and phenacetin are 75 and 88°C, respectively. Of all the substances for which data are given and which are known to melt between 121 and 123°C, there is only one whose eutectic temperatures with the reference substances are within one degree of those found and this is benzoic acid. The reference glass powder for confirmation of identification is that having refractive index 1.5000. The melt would be found to match this in red light at 132—133°C. Other confirmatory tests and supporting characteristics are described in the final column of Table 11.

It is, of course, presupposed that the substance to be checked or identified is included in the tables. The original Kofler tables [7] contain data for about 1200 compounds and these data are also given (but without the "Special Remarks") by McCrone [148]. An extensive tabulation of more recent date, more especially for pharmaceuticals, is to be found in ref. 11. This contains data for about 1000 compounds. The International Pharmacopoeia referred to above [230] contains 200 entries. Both refs. 11 and 230 give separate tables for identifications using the hot bench (p. 339). Numerous supplements containing data for a further 600 or more compounds have been published by Kuhnert-Brandstätter and Kofler with their co-workers [231,232].

References

1 O. Lehmann, Molekularphysik, Engelmann, Leipzig, 1888.
2 O. Lehmann, Das Kristallisationsmikroskop, Viehweg, Brunswick, 1910.
3 H.H. Emons, H. Keune and H.H. Seyfarth, Chemische Mikroskopie, VEB Deutscher Verlag für Grundstoffindustrie, Leipzig, 1973; English version, this volume pp. 1—328.
4 E.M. Chamot and C.W. Mason, Handbook of Chemical Microscopy, Vol. I, Wiley, New York, 3rd ed., 1958.
5 N.H. Hartshorne, The Microscopy of Liquid Crystals, Microscope Publications, London, 1974.
6 W. Kofler, A. Kofler and L. Kofler, Mikrochemie, 38 (1951) 218.
7 L. Kofler and A. Kofler, Thermomikromethoden, Wagner, Innsbruck, 1954.
8 L. Kofler and A. Kofler, in F. Hecht and M.K. Zacherl (Eds.), Handbuch der Mikrochemischen Methoden, Springer, Vienna, 1954.
9 D. Harangozo and H. Suter, GIT Fachz. Lab., 14 (1970) 1255.
10 L. Kofler and W. Kofler, Mikrochemie, 34 (1949) 374.
11 M. Kuhnert-Brandstätter, Thermomicroscopy in the Analysis of Pharmaceuticals, Pergamon Press, Oxford, 1971.
12 M. Kuhnert-Brandstätter, in E. Graf and R. Preuss (Eds.), Gadamers Lehrbuch der Chemischen Toxikologie und Anleitung zur Ausmittelung der Gifte, Vandenhoeck und Ruprecht, Göttingen, 1966.
13 A. Kofler, Mikroskopie, 13 (1958) 82.

14 N.H. Hartshorne and A. Stuart, Crystals and the Polarising Microscope, Arnold, London, 1970.
15 N.H. Hartshorne and A. Stuart, Practical Optical Crystallography, Arnold, London, 1964.
16 C.C. Fulton, Modern Microcrystal Tests for Drugs, Interscience, New York, 1969.
17 M. Kuhnert-Brandstätter and L. Müller, Microchem. J., 13 (1968) 20.
18 R. Linder, Dissertation, Innsbruck, 1976.
19 A. Kofler and J. Kolsek, Mikrochim. Acta (Vienna), (1969) 1038.
20 F. Pröll, Dissertation, Innsbruck, 1978.
21 M. Kuhnert-Brandstätter, A. Kofler, A. Vlachopoulos and A. Lobenwein, Sci. Pharm., 38 (1970) 154.
22 M. Brandstätter-Kuhnert and W. Schöniger, Mikrochim. Acta (Vienna), (1962) 1075.
23 R. Fischer, Mikrochemie, 31 (1944) 296.
24 W. Biltz, Ber. Dtsch. Chem. Ges., 40 (1907) 2182.
25 M. Kuhnert-Brandstätter and R. Linder, Mikrochim. Acta (Vienna), (1976 I) 513.
26 M. Kuhnert-Brandstätter and W. Heindl, Sci. Pharm., 44 (1976) 18.
27 E. Lindpaintner, Mikrochemie, 27 (1939) 21.
28 F. Bernauer, "Gedrillte" Kristalle, Borntraeger, Berlin, 1929.
29 L. Kofler, Mikrochemie, 22 (1937) 241.
30 G. Linck and E. Köhler, Chem. Erde, 4 (1930) 458.
31 M. Brandstätter-Kuhnert, Mikroskopie, 17 (1962) 17.
32 L. Kofler, Ber. Dtsch. Chem. Ges., 76 (1943) 1096.
33 J.D.M. Ross and I.C. Somerville, J. Chem. Soc. (London) (1926) 2770.
34 M. Brandstätter, Z. Phys. Chem. Abt. A, 191 (1942) 227.
35 M. Brandstätter-Kuhnert and M. Aepkers, Mikrochim. Acta (Vienna), (1962) 1041; (1962) 1055; (1963) 360.
36 M. Kuhnert-Brandstätter and A. Vlachopoulos, Mikrochim. Acta (Vienna), (1967) 201.
37 L. Kofler and M. Brandstätter, Ber. Dtsch. Chem. Ges., 75 (1942) 496.
38 M. Brandstätter-Kuhnert and E. Junger, Mikrochim. Acta (Vienna), (1964) 238.
39 M. Brandstätter-Kuhnert, Oesterr. Apoth. Ztg., 14 (1960) 252.
40 L. Kofler and A. Kofler, Angew. Chem., 53 (1940) 434.
41 M. Kuhnert-Brandstätter and G. Friedrich-Sander, Sci. Pharm., 41 (1973) 117.
42 M. Kuhnert-Brandstätter and R. Kronbichler, Sci. Pharm., 46 (1978) 122.
43 Österreichisches Arzneibuch, 9 Ausgabe, Österreiche Staatsdruckerei, Vienna, 1960.
44 A. Kofler and L. Kofler, Ber. Dtsch. Chem. Ges., 74 (1941) 1394.
45 T.Y. Huang, Acta Pharm. Int., 2 (1951) 43.
46 L. Kofler and R. Wannenmacher, Ber. Dtsch. Chem. Ges., 73 (1940) 1388.
47 J. von Braun, W. Gmelin and A. Schultheiss, Ber. Dtsch. Chem. Ges., 56 (1923) 1338.
48 J. Lindner, Monatsh. Chem., 44 (1923) 337; 46 (1925) 231; 72 (1939) 330, 354, 361; Ber. Dtsch. Chem. Ges., 74 (1941) 231; Mikrochemie, 34 (1949) 382; 35 (1950) 205.

49 R. Fischer, Mikrochemie, 38 (1951) 342.
50 A. Martinek, Mikrochim. Acta (Vienna), (1971) 877.
51 H. Lieb and W. Schöniger, in F. Hecht and M.K. Zacherl (Eds.), Handbuch der Mikrochemischen Methoden, Springer, Vienna, 1954.
52 R. Fischer, Mikrochemie, 15 (1934) 247.
53 H. Molisch, Mikrochemie der Pflanze, Fischer, Jena, 1923.
54 M. Brandstätter-Kuhnert and E. Junger, Mikrochim. Acta (Vienna), (1963) 506.
55 R. Fischer and A. Kofler, Arch. Pharm. Ber. Dtsch. Pharm. Ges., 270 (1932) 207.
56 L. Rosenthaler, Toxikologische Mikroanalyse, Borntraeger, Berlin, 1935.
57 H. Meyer, Analyse und Konstitutionsermittlung Organischer Verbindungen, Springer Vienna, 1938.
58 R. Fischer, Arch. Pharm. Ber. Dtsch. Pharm. Ges., 271 (1933) 466.
59 L. Kofler and F.A. Müller, Mikrochemie, 22 (1937) 43.
60 D.T. Meredith and C.O. Lee, J. Am. Pharm. Assoc., 28 (1939) 369.
61 L. Borka and M. Kuhnert-Brandstätter, Arch. Pharm., 307 (1974) 377.
62 A. Mayrhofer, Mikrochemie der Arzneimittel und Gifte, I and II, Urban and Schwarzenberg, Vienna, 1923 and 1928.
63 L. Kofler and M. Brandstätter, Angew. Chem., 54 (1941) 322.
64 H. Behrens and P.D.C. Kley, Organische Mikrochemische Analyse, Voss, Leipzig, 1922.
65 A. Kofler, Z. Phys. Chem. Abt. A, 187 (1941) 363.
66 E. Lindpaintner, Arch. Pharm., 277 (1939) 398.
67 F. Reimers, Dan. Tidsskr. Farm., 14 (1940) 219; Z. Anal. Chem., 122 (1941) 404.
68 L. Kofler and M. Baumeister, Z. Anal. Chem., 124 (1942) 386.
69 L. Kofler and D. Prause, Pharm. Zentralhalle, 88 (1949) 129.
70 M. Brandstätter-Kuhnert and J. Obkircher, Mikrochim. Acta (Vienna), (1960) 836.
71 H.J. Lennartz, Pharm. Zentralhalle, 87 (1948) 225.
72 H.J. Lennartz, Z. Anal. Chem., 127 (1944) 5; 128 (1948) 271.
73 M. Lehmann, Chem. Ztg., 37 (1914) 402.
74 D. Holde and W. Bleyberg, Z. Angew. Chem., 43 (1930) 897.
75 L. Kofler, Chem. Ztg., 68 (1944) 43.
76 F. Emich, Lehrbuch der Mikrochemie, Bergmann, München, 1926.
77 A. Siwoloboff, Ber. Dtsch. Chem. Ges., 19 (1886) 795.
78 R. Fischer, Chemie, 55 (1942) 244.
79 M. Brandstätter and H. Thaler, Mikrochemie, 38 (1951) 358.
80 J.S. Wiberley, R.K. Siegfriedt and A.A. Benedetti-Pichler, Mikrochemie, 38 (1951) 471.
81 R. Fischer, Mikrochemie, 28 (1940) 173.
82 R. Fischer and T. Reichel, Mikrochemie, 31 (1943) 102.
83 R. Fischer and G. Karasek, Mikrochemie, 33 (1948) 316.
84 R. Fischer and E. Neupauer, Mikrochemie, 34 (1949) 319.
85 J. Harend, Monatsh. Chem., 65 (1935) 153.
86 R. Fischer and Th. Kartnig, Arzneim. Forsch., 7 (1957) 365.
87 R. Fischer and J. Horner, Mikrochim. Acta (Vienna), (1953) 386.

88 R. Fischer, Fette, Seifen, Anstrichm., 67 (1965) 748.
89 R. Fischer, E. Pinter and H. Auer, Pharmaz. Zentralhalle, 99 (1960) 299.
90 R. Fischer, in N.D. Cheronis (Ed.), Microchemical Techniques, Interscience, New York, 1962, p. 977.
91 F. Gölles, Sci. Pharm., 31 (1963) 105.
92 H. Behrens, Chem. Ztg., 26 (1902) 1126.
93 R. Fischer, Mikrochemie, 13 (1933) 123.
94 C. Griebel and F. Weiss, Mikrochemie, 5 (1927) 146.
95 C. Griebel, Z. Unters. Nahr. Genussm., 47 (1924) 438.
96 R. Fischer and W. Paulus, Mikrochemie, 17 (1935) 356.
97 R. Opfer-Schaum, Angew. Chem., 62 (1950) 144.
98 R. Opfer-Schaum, Mikroanalyse Organischer Arzneimittel und Gifte, Editio Cantor, Aulendorf i. Württ., 1953.
99 L. Müller, Dissertation, Innsbruck, 1968.
100 J. Haleblian and W.C. McCrone, J. Pharm. Sci., 58 (1969) 911.
101 E. Mitscherlich, Ann. Chim. Phys., 19 (1821) 350.
102 E. Mitscherlich, Berl. Akad. Abh., 43 (1822).
103 A. Findlay, The Phase Rule and its Applications, Dover, New York, 1951.
104 O. Lehmann, Z. Kristallogr., 1 (1877) 97.
105 G. Tammann, Aggregatzustände, Voss, Leipzig, 1923.
106 P.W. Bridgman, Proc. Am. Acad. Arts Sci., 52 (1916) 57, 96.
107 L. Deffet, Bull. Soc. Chim. Belg., 44 (1935) 87.
108 G. Tammann, Z. Phys. Chem., 84 (1913) 257.
109 G. Tammann, Z. Phys. Chem., 88 (1914) 57.
110 M.J. Buerger and M.C. Bloom, Z. Kristallogr. A, 96 (1937) 182.
111 M.J. Buerger, Trans. Am. Crystallogr. Assoc., 7 (1971) 1.
112 M.J. Buerger, in R. Smoluchowski, J.E. Mayer and W.A. Weyl (Eds.), Phase Transformations in Solids, Wiley, New York, 1951, Chap. 6.
113 H.G.F. Winkler, Struktur und Eigenschaften der Kristalle, Springer, Vienna, 1955.
114 E.F. Westrum Jr. and J.P. McCullough, in D. Fox, M.M. Labes and A. Weissberger (Eds.), Physics and Chemistry of the Organic Solid State, Vol. 1, Interscience, New York, 1963.
115 W. Nowacki, Tech. Ind. Schweiz. Chem. Ztg., 26 (1943) 33.
116 W.C. McCrone, Polymorphism, in D. Fox, M.M. Labes and A. Weissberger (Eds.), Physics and Chemistry of the Organic Solid State, Vol. 2, Interscience, New York, 1965.
117 K. Schaum, Chem. Ztg., 38 (1914) 185; Ann. Chem., 411 (1916) 161.
118 C.N.R. Rao and K.J. Rao, Phase Transition in Solids, McGraw-Hill, New York, 1978.
119 J. Bernstein and A.T. Hagler, J. Am. Chem. Soc., 100 (1978) 673.
120 J. Bernstein and A.T. Hagler, Mol. Cryst. Liq. Cryst., 50 (1979) 223.
121 J. Bernstein, Acta Crystallogr. Sect. B, 35 (1979) 360.
122 Yu. V. Mnyukh and N.N. Petropavlov, J. Phys. Chem. Solids, 33 (1972) 2079.
123 Yu. V. Mnyukh and N.A. Panfilova, J. Phys. Chem. Solids, 34 (1973) 159.
124 Yu. V. Mnyukh, N.A. Panfilova, N.N. Petropavlov and N.S. Uchvatova, J. Phys. Chem. Solids, 36 (1975) 127.

125 A.R. Verma and P. Krishna, Polymorphism and Polytypism in Crystals, Wiley, New York, 1966.
126 St. R. Byrn, J. Pharm. Sci., 65 (1976) 1.
127 J. Reisch, Dtsch. Apoth. Ztg., 119 (1979) 1.
128 A.J. Aguiar and J.E. Zelmer, J. Pharm. Sci., 58 (1969) 983.
129 H. Andersgaard, P. Fischolt, R. Gjermundsen and T. Hoyland, Acta Pharm. Suec., 11 (1974) 239.
130 A. Burger, Sci. Pharm., 45 (1977) 269.
131 C. Weygand, Chemische Morphologie der Flüssigkeiten und Kristalle, Akademie Verlagsgesellschaft, Leipzig, 1941.
132 K. Schaum, Z. Anorg. Chem., 120 (1922) 241.
133 A.E. van Arkel and M.G. van Bruggen, Z. Phys., 42 (1927) 795.
134 E. Cohen and A. van Lieshout, Z. Phys. Chem. Abt. A, 173 (1935) 1.
135 L. Deffet, Répertoire des Composés Organiques Polymorphs, Desoer, Liége, 1942.
136 M. Kuhnert-Brandstätter and E. Junger, Spectrochim. Acta Part A, 23 (1967) 1453.
137 M. Kuhnert-Brandstätter and F. Bachleitner-Hofmann, Spectrochim. Acta Part A, 27 (1971) 191.
138 M. Kuhnert-Brandstätter and P. Gasser, Microchem. J., 16 (1971) 419, 577, 590.
139 M. Kuhnert-Brandstätter and H. Winkler, Sci. Pharm., 44 (1976) 177, 191, 288.
140 A. Burger and R. Ramberger, Mikrochim. Acta (Vienna), (1979 II) 259, 273.
141 F. Machatschki, Grundlagen der Allgemeinen Mineralogie und Kristallchemie, Springer, Vienna, 1946.
142 H. Ott, Ann. Phys., 85 (1928) 81.
143 F. Halla, Kristallchemie und Kristallphysik Metallischer Werkstoffe, Barth, Leipzig, 1939.
144 A. Kofler, Ber. Dtsch. Chem. Ges., 76 (1943) 871.
145 M. Brandstätter, Mikrochemie, 32 (1944) 33.
146 M. Kuhnert-Brandstätter and I. Moser, Mikrochim. Acta (Vienna), (1979) 125.
147 A. Kofler and L. Kofler, Monatsh. Chem., 78 (1948) 13.
148 W.C. McCrone, Fusion Methods in Chemical Microscopy, Interscience, New York, 1957.
149 F. Reinitzer, Monatsh. Chem., 9 (1888) 421.
150 O. Lehmann, Ann. Phys., 56 (1895) 786.
151 O. Lehmann, Die Neue Welt der Flüssigen Kristalle, Akademie Verlagsgesellschaft, Leipzig, 1911.
152 G. Friedel, Ann. Phys., 18 (1922) 273.
153 N.H. Hartshorne, The Microscopy of Liquid Crystals, Microscope Publications, London, 1974.
154 G.W. Gray, Molecular Structure and the Properties of Liquid Crystals, Academic Press, London, 1962.
155 H. Sackmann and D. Demus, Mol. Cryst., 2 (1966) 81; Fortschr. Chem. Forsch., 12 (2) (1969) 349.

156 G. Pelzl and H. Sackmann, Mol. Cryst. Liq. Cryst., 15 (1971) 75.
157 CH. Wiegand, Z. Naturforsch. Teil B, 4 (1949) 249.
158 G.W. Gray, J. Chem. Soc., (1956) 3733.
159 G. Meier, E. Sackmann and J.G. Grabmaier, Applications of Liquid Crystals, Springer, Berlin, 1975.
160 A. Kofler, Z. Phys. Chem. Abt. A, 188 (1941) 201.
161 J.W. Retgers, Z. Phys. Chem. Abt. A, 3 (1889) 552.
162 W. Nernst, Theoretische Chemie, Enke, Stuttgart, 1900.
163 G. Bruni, Feste Lösungen, Enke, Stuttgart, 1901.
164 H.G. Grimm, Naturwissenschaften, 17 (1929) 535.
165 M. Brandstätter and H. Wachter, Monatsh. Chem., 87 (1956) 595.
166 L. Friedl, Dissertation, Innsbruck, 1979.
167 H.W. Bakhuis-Roozeboom, Z. Phys. Chem., 10 (1892) 145.
168 H.G.F. Winkler, Struktur und Eigenschaften der Kristalle, Springer, Berlin, 1955.
169 A. Kofler, Z. Elektrochem., 50 (1944) 104.
170 M. Kuhnert-Brandstätter and S. Wunsch, Mikrochim. Acta (Vienna), (1969) 1297, 1308.
171 G. Tammann, Lehrbuch der Heterogenen Gleichgewichte, Vieweg, Brunswick, 1924.
172 R. Vogel, Die Heterogenen Gleichgewichte, Akademie Verlagsgesellschaft, Leipzig, 1959.
173 A. Stock, Ber. Dtsch. Chem. Ges., 42 (1909) 2059.
174 H. Rheinboldt, J. Prakt. Chem., 111 (1925) 262.
175 A. Kofler, Z. Physik. Chem. Abt. A, 187 (1940) 201.
176 J. Kotula and A. Rabczuk, J. Therm. Anal., 15 (1979) 343.
177 I.L. Liplawk, Fiz. Chim. Swoj. Chim. Prod. Koksow, Kamiennych Ugl., Gos. Naucz.-Tech. Izol. Lit. Moskwa, 1954 (quoted from ref. 176).
178 B.M. Krawczenko, Zh. Prikl. Chem., 25 (1952) 662.
179 Quoted from J. Kotula and A. Rabczuk, J. Therm. Anal., 15 (1979) 343.
180 M. Brandstätter, Monatsh. Chem., 81 (1950) 806.
181 A. Kofler and M. Brandstätter, Isomorphie und Isodimorphie, Mikrofilm H. Pacher and Co., Innsbruck, 1951.
182 M. Kuhnert-Brandstätter and L. Langhammer, Arch. Pharm., 301 (1968) 351.
183 M. Kuhnert-Brandstätter, R. Ulmer and L. Langhammer, Arch. Pharm., 307 (1974) 497.
184 M. Andersson and A. Fredga, Acta Chem. Scand., 20 (1966) 1060.
185 M. Kuhnert-Brandstätter and L. Friedl, Mikrochim. Acta (Vienna), (1977 II) 507.
186 M. Kuhnert-Brandstätter, K. Schleich and K. Vogler, Monatsh. Chem., 101 (1970) 1817.
187 K. Neumann and G. Micus, Z. Phys. Chem., 2 (1954) 25.
188 I. Gutzow, Krist. Tech., 7 (1972) 769.
189 L.O. Melesko, Krist. Tech., 7 (1972) 1341.
190 M. Kuhnert-Brandstätter and O. Dietmaier, Mikrochim. Acta (Vienna), (1979) 207.
191 O. Dietmaier, Dissertation, Innsbruck, 1978.

192 H. Lautz, Z. Phys. Chem., 84 (1913) 611.
193 K. Schaum, Liebigs Ann. Chem., 462 (1928) 194.
194 N.H. Hartshorne, G.S. Walters and W.O.M. Williams, J. Chem. Soc., (1935) 1860.
195 D.G. Eade and N.H. Hartshorne, J. Chem. Soc., (1938) 1636.
196 H.A. Miers, Proc. R. Soc. London, 71 (1903) 439.
197 A.W.C. Menzies and C.A. Sloat, Nature (London), 123 (1929) 348.
198 N.Cabrera and W.K. Burton, Discuss. Faraday Soc., 5 (1949) 40.
199 W.K. Burton, N. Cabrera and F.C. Frank, Nature (London) 163 (1949) 398.
200 W.K. Burton, N. Cabrera and F.C. Frank, Philos. Trans. R. Soc. London Ser. A, 243 (1951) 299.
201 J.W. Gibbs, Collected Works, Vol. I, Longmans, London, 1928.
202 M. Volmer, Kinetik der Phasenbildung, Steinkopf, Dresden and Leipzig, 1939.
203 W. Kossel, Nachr. Ges. Wiss. Goettingen, Math. Phys. Kl., (1927) 135.
204 J.N. Stranski, Z. Phys. Chem., 136 (1928) 259.
205 N.F. Mott, Nature (London), 165 (1950) 295.
206 F.C. Frank, Z. Elektrochem., Ber. Bunsenges. Phys. Chem., 56 (1952) 429.
207 M. Brandstätter, Z. Elektrochem., Ber. Bunsenges. Phys. Chem., 56 (1952) 968.
208 M. Brandstätter, Spiralwachstum der Kristalle. Wissenschaftlicher Film C.681/1954 aus dem Institut für den Wissenschaftlichen Film, Göttingen.
209 M. Brandstätter, Z. Elektrochem. Ber. Bunsenges. Phys. Chem., 57 (1953) 438.
210 M. Brandstätter, Naturwissenschaften, 40 (1953) 272.
211 A.J. Forty, Philos. Mag. Suppl., 3 (1954) 1.
212 M. Gebhardt and A. Neuhaus, in Landolt-Börnstein, Physikalisch-chemische Tabellen, Neue Serie 1972, III/8.
213 M.L. Royer, Bull. Soc. Fr. Mineral., 51 (1928) 7.
214 I.W. Gruner, Amer. Mineral., 14 (1929) 227.
215 K. Spangenberg and A. Neuhaus, Chem. Erde, 5 (1930) 437.
216 H. Seifert, Fortschr. Mineral., 20 (1936) 349.
217 A. Nauhaus, Fortschr. Mineral., 29/30 (1950/51) 136.
218 J. Willems, Naturwissenschaften, 32 (1944) 324.
219 M. Brandstätter, Mikrochemie, 33 (1948) 184.
220 M. Brandstätter, Naturwissenschaften, 42 (1955) 643.
221 M. Brandstätter-Kuhnert, Z. Kristallogr., 110 (1958) 1.
222 K. Rast, Ber. Dtsch. Chem. Ges., 55 (1922) 1051, 3727.
223 G. Ekborn, Collect. Pharm. Suecica I, (1946).
224 N.D. Cheronis, Micro and Semimicro Methods, Interscience, New York, 1954.
225 F. Pregl, Quantitative Organische Mikroanalyse, Springer, Vienna, 1958.
226 L. Kofler and M. Brandstätter, Mikrochemie, 33 (1948) 20.
227 M. Brandstätter and L. Kofler, Mikrochemie, 34 (1949) 364.
228 M. Brandstätter, Mikrochemie, 36/37 (1951) 291.
229 M. Brandstätter and H. Frischmann, Sci. Pharm., 21 (1953) 264.

230 Specifications for the Quality Control of Pharmaceutical Preparations. Second Edition of the International Pharmacopoeia. World Health Organization, Geneva, 1967.
231 M. Kuhnert-Brandstätter et al., Sci. Pharm., 38 (1970) 154; 42 (1974) 150, 234; 46 (1978) 54, 180; 48 (1980) 250; Microchem. J., 17 (1972) 719, 739; Pharm. Acta Helv., 50 (1975) 360.
232 A. Kofler and J. Kolsek, Mikrochim. Acta (Vienna), (1969) 408, 1038; (1970) 367, 1063; (1971) 848; (1974) 85.

Symbols and abbreviations

Acet.	acetanilide
Azob.	azobenzene
Benzan.	benzanilide
CST	critical solution temperature
C.V.	crystallization velocity
C.V.—T curve	crystallization velocity—temperature curve
decomp.	decomposition
E	eutectic point
ET	eutectic temperature
equilib.	equilibrium
FP 52	Mettler hot stage
M	melt
mod.	modification
m.p.	melting point
n_D	refractive index (sodium light)
P	peritectic point
Phen.	phenacetin
S	stabilized intermediate phase
Sal.	salophen
Sil Gel	silicone gel
subl.	sublimation
temp.	temperature
t.p.	transition point
T.V.	transformation velocity
T.V.—T curve	transformation velocity—temperature curve

Index